WITHDRAWN
UTSA Libraries

River Basin Trajectories:
Societies, Environments and Development

Comprehensive Assessment of Water Management in Agriculture Series

Titles Available

Volume 1. Water Productivity in Agriculture: Limits and Opportunities for Improvement
Edited by Jacob W. Kijne, Randolph Barker and David Molden

Volume 2. Environment and Livelihoods in Tropical Coastal Zones: Managing Agriculture–Fishery–Aquaculture Conflicts
Edited by Chu Thai Hoanh, To Phuc Tuong, John W. Gowing and Bill Hardy

Volume 3. The Agriculture Groundwater Revolution: Opportunities and Threats to Development
Edited by Mark Giordano and Karen G. Villholth

Volume 4. Irrigation Water Pricing: the Gap Between Theory and Practice
Edited by François Molle and Jeremy Berkoff

Volume 5. Community-based Water Law and Water Resource Management Reform in Developing Countries
Edited by Barbara van Koppen, Mark Giordano and John Butterworth

Volume 6. Conserving Land, Protecting Water
Edited by Deborah Bossio and Kim Geheb

Volume 7. Rainfed Agriculture: Unlocking the Potential
Edited by Suhas P. Wani, Johan Rockström and Theib Oweis

Volume 8. River Basin Trajectories: Societies, Environments and Development
Edited by François Molle and Philippus Wester

River Basin Trajectories:
Societies, Environments and Development

Edited by

François Molle
Institut de Recherche pour le Développment, France

and

Philippus Wester
Irrigation and Water Engineering Group, Wageningen University, The Netherlands

CABI is a trading name of CAB International

CABI Head Office
Nosworthy Way
Wallingford
Oxfordshire OX10 8DE
UK

Tel: +44 (0)1491 832111
Fax: +44 (0)1491 833508
E-mail: cabi@cabi.org
Website: www.cabi.org

CABI North American Office
875 Massachusetts Avenue
7th Floor
Cambridge, MA 02139
USA

Tel: +1 617 395 4056
Fax: +1 617 354 6875
E-mail: cabi-nao@cabi.org

© CAB International 2009. All rights reserved. No part of this publication may be reproduced in any form or by any means, electronically, mechanically, by photocopying, recording or otherwise, without the prior permission of the copyright owners.

A catalogue record for this book is available from the British Library, London, UK.

Library of Congress Cataloging-in-Publication Data
River basin trajectories : societies, environments and development / edited by François Molle and Philippus Wester.
    p. cm. -- (Comprehensive assessment of water management in agriculture ; no. 8)
  Includes bibliographical references and index.
  ISBN 978-1-84593-538-2 (alk. paper)
  1.Water resources development--Case studies. 2. Water-supply--Political aspects--Case studies. I. Molle, François. II. Wester, Philippus. III. C.A.B. International. IV. Series: Comprehensive assessment of water management in agriculture series ; 8.

HD1691.R58 2009
333.73--dc22
                              2009000346

ISBN-13: 978 1 84593 538 2

Typeset by Columns Design Ltd, Reading, UK.
Printed and bound in the UK by the MPG Books Group, Bodmin.

The paper used for the text pages in this book is FSC certified. The FSC (Forest Stewardship Council) is an international network to promote responsible management of the world's forests.

Contents

Contributors		vii
Series Foreword		xii
Acknowledgements		xiv
1	**River Basin Trajectories: an Inquiry into Changing Waterscapes** *François Molle and Philippus Wester*	1
2	**Squeezed Dry: the Historical Trajectory of the Lower Jordan River Basin** *Mauro Van Aken, François Molle and Jean-Philippe Venot*	20
3	**Are Good Intentions Leading to Good Outcomes? Continuities in Social, Economic and Hydro-political Trajectories in the Olifants River Basin, South Africa** *Douglas J. Merrey, Hervé Lévite and Barbara van Koppen*	47
4	**From Half-full to Half-empty: the Hydraulic Mission and Water Overexploitation in the Lerma–Chapala Basin, Mexico** *Philippus Wester, Eric Mollard, Paula Silva-Ochoa and Sergio Vargas-Velázquez*	75
5	**Managing the Yellow River: Continuity and Change** *David Pietz and Mark Giordano*	99
6	**The Colorado River: What Prospect for 'a River No More'?** *Douglas J. Kenney*	123
7	**Sharing Scarce Resources in a Mediterranean River Basin: Wadi Merguellil in Central Tunisia** *Patrick Le Goulven, Christian Leduc, Mohamed Salah Bachta and Jean-Christophe Poussin*	147

8	**Water Competition, Variability and River Basin Governance: a Critical Analysis of the Great Ruaha River, Tanzania** *Bruce A. Lankford, Siza Tumbo and Kossa Rajabu*	171
9	**Buying Respite: Esfahan and the Zayandeh Rud River Basin, Iran** *François Molle, Iran Ghazi and Hammond Murray-Rust*	196
10	**Rural Dynamics and New Challenges in the Indian Water Sector: the Trajectory of the Krishna Basin, South India** *Jean-Philippe Venot*	214
11	**Pumped Out: Basin Closure and Farmer Adaptations in the Bhavani Basin in Southern India** *Mats Lannerstad and David Molden*	238
12	**Much Ado about the Murray: the Drama of Restraining Water Use** *Hugh Turral, Daniel Connell and Jennifer McKay*	263
	Index	293

Contributors

Mohamed Salah Bachta has a PhD in agronomic science from the University of Louvain, Belgium. He is Professor of Rural Economy at the Institut National Agronomique de Tunis. He specialises in agricultural economics and policies. *Institut National Agronomique de Tunisie, Avenue Charles Nicolle 43, Tunis Mahrajène 1082, Tunisia. E-mail: bachta.medsalah@inat.agrinet.tn*

Daniel Connell has a joint appointment at the Australian National University, with the Crawford School of Economics and Government and the ANU Water Initiative. Previously he worked as a journalist in the Australian Broadcasting Corporation, the Department of Foreign Affairs and Trade and the Murray–Darling Basin Commission. He is now researching water governance in large cross-border hydrological systems and public participation in environmental policy. This includes a comparative study of water management institutions in federal political systems in Australia, South Africa, the United States, Europe, India and China. His recent book, *Water politics in the Murray–Darling Basin*, examined the National Water initiative and the debate about institutional change in Australia's Murray–Darling Basin. *Room 214, Crawford Building # 13, Ellery Crescent, The Australian National University, Canberra ACT 0200, Australia. E-mail: daniel.connell@anu.edu.au*

Iran Ghazi was Senior Geographer at the University of Esfahan, Iran. Dr Ghazi is a scholar specializing in river-basin development planning. She is now retired. *1242 Saeb Avenue, Esfahan, Iran, 81848. E-mail: iranghazi@yahoo.com*

Mark Giordano is a geographer and economist with broad experience in water management and agriculture. He is currently Principal Researcher and Leader of the Water and Society Theme at the International Water Management Institute, headquartered in Colombo, Sri Lanka. Mark has been a frequent visitor to, and sometimes resident of, China for the past 2 decades. *International Water Management Institute, PO Box 2075, Colombo, Sri Lanka. E-mail: d.molden@cgiar.org*

Doug Kenney is a senior research associate of the Natural Resources Law Center, University of Colorado, where he also directs the Western Water Policy Program. His research is primarily focused on legal and policy arrangements for the allocation, management and use of western USA water resources, given stressors associated with population growth and climatic change. *University of Colorado, Natural Resources Law Center, UCB 401, Boulder, CO 80309-0401, USA. E-mail: douglas.kenney@colorado.edu*

Bruce A. Lankford is Senior Lecturer in Natural Resources at the School of Development Studies, University of East Anglia, UK. He has 25 years' experience in irrigation and water management, with an emphasis on sub-Saharan Africa. He is particularly interested in the linkages between irrigation systems and the behaviour and management of river basins. *School of Development Studies, University of East Anglia, Norwich NR4 7TJ, UK. E-mail: b.lankford@uea.ac.uk*

Mats Lannerstad is a PhD scholar at the Department of Water and Environmental Studies, Linköping University, Sweden. His educational background includes an MSc in Environmental Engineering and an MSc in biology, with specialization in limnology. His PhD studies include a basin closure case study analysing the water use dynamics in the Bhavani basin, India, and an analysis of present and future water demand for food production in agriculture. *Welanders väg 7, SE – 112 50 Stockholm, Sweden. E-mail: mats@lannerstad.com*

Christian Leduc is Research Director at IRD (Institute of Research for Development) and Deputy Director of the G-EAU research unit in Montpellier (France). His research focuses on hydro(geo)logy in semi-arid and Mediterranean areas, where he has spent many years. He has a special interest in the impacts of climatic and environmental changes on water resources, considered at multiple scales of space and time (from the season to the millennium). *Hydrogéologue, Directeur de Recherche IRD, BP 64501, 34394 Montpellier, Cedex 5, France. E-mail: christian.leduc@ird.fr*

Patrick Le Goulven is a hydrologist and Senior Researcher at the IRD (Institut de Recherche pour le Développement), with 35 years of experience in hydrology, irrigation and water resources management in Africa, South America, Maghreb and France. Co-founder of laboratory G-EAU (water management, actors and uses), he currently directs a research project on the effects of climatic changes on the water management in the Ecuadorian Andes and on the adaptations necessary. *Hydrologue, Directeur de Recherche IRD, UMR G-EAU (Gestion de l'Eau, Acteurs, Usages), IRD Apartado 17 12 857, Quito, Ecuador. E-mail: patrick.legoulven@ird.fr*

Hervé Lévite is a water engineer. He was previously a researcher at the South African Office of the International Water Management Institute. He is now Senior Officer at the International Programme for Technology and Research in Irrigation and Drainage, FAO, Rome. He is interested in the implementation of innovations in the water sector, mainly in Africa. *Senior Officer (Water Management), Land and Water Division, Natural Resources Departement, FAO, Viale delle Terme di Caracalla 0153 Rome, Italy. E-mail: herve.levite@fao.org*

Jennifer McKay is Professor of Business Law and Director of the Centre for Comparative Water Policies and Laws at the University of South Australia. She researches water law and policy in Australia and India, and has over 80 publications on these topics; she has received awards from the Australian Water Association, the Government of South Australia and a senior Fulbright fellowship. She studies the impact of water law on rural and urban communities and the processes to achieve law reforms, and provides advice to law reform commissions. She has supervised the work of nine PhD students on these topics to date and has established (with the Chief Librarian and others) one of the best collections of water law and policy books at City West Library in Adelaide. *Professor of Business Law and Director Centre for Comparative Water Policies and Laws, University of South Australia, Way Lee Building, North Terrace, Adelaide, South Australia 5000, Australia. E-mail: jennifer.mckay@unisa.edu.au*

Douglas J. Merrey is a social scientist who has worked on water and irrigation institutional issues since the mid-1970s. Now an independent consultant based in South Africa, he was Director for Research in a southern African policy research and dialogue network (FANRPAN) from 2006 to 2008. Employed by the International Water Management Institute from 1985 to 2006, he was Deputy Director General for Programs from 1998 to the end of 2000, and then became the first Director of IWMI's new Africa regional office. His research focuses on institutional arrangements and policies at local, river basin, national and regional levels for management of water resources, especially for irrigation. He has lived and worked in Pakistan, India, Sri Lanka, Indonesia and Egypt, before moving to South Africa, and has done shorter-term assignments in many African and Asian developing countries. *Natural Resources Policy and Institutions Specialist, 170 Dorado Street, Waterkloof Ridge 0181 Pretoria, South Africa. E-mail: dougmerrey@gmail.com*

David Molden is Deputy Director General for Research at the International Water Management Institute (IWMI). He has lived and worked closely with local communities and governments in Lesotho, Egypt, Botswana, India and Nepal. Now in Sri Lanka, he enjoys interdisciplinary and cross-cultural teamwork, covering a range of water issues. Recently, David coordinated a global programme to produce a Comprehensive Assessment of Water Management in Agriculture (www.iwmi.org/assessment), documented in this book series and other reports. *International Water Management Institute, PO Box 2075, Colombo, Sri Lanka. E-mail: d.molden@cgiar.org*

Eric Mollard is Senior Researcher in Environmental Sociology at the French Research Institute for Development (IRD) based in Montpellier. He first studied rural development in Latin America and West Africa. He then focused his research on associative water management, conflicts from local to transboundary level, and negotiations in Thailand and Mexico. His current research is on democracy as the way to make negotiations and environmental protection work. *Centre IRD, 911 Avenue Agropolis BP 64501, 34394 Montpellier, Cedex 5, France. E-mail: eric.mollard@ird.fr*

François Molle is a senior researcher at the Institut de Recherche pour le Développement, France, and holds a joint appointment with the International Water Management Institute. He has 25 years' experience in irrigation and water management in South America, Africa, Asia and the Middle East. He currently focuses his research on institutional and political aspects of river basin management and on water policy; he is co-editor of *Water Alternatives*. *Institut de Recherche pour le Développement, 911 Avenue Agropolis BP 64501, 34394 Montpellier, Cedex 5, France. E-mail: francois.molle@ird.fr*

Hammond Murray-Rust is a water specialist with long-standing experience in irrigation and river basin management. He has been a research leader with the International Water Management Institute and specializes in integrated modelling for water management, determinants of water productivity, irrigation performance assessment and use of remote sensing for water management. He is now Senior Associate at ARD Inc. in Burlington, Vermont. *ARD Inc., 159 Bank Street, Suite 300, Burlington, VT 05401, USA. E-mail: hmurray-rust@ardinc.com*

David Pietz is Associate Professor of Modern Chinese History and Director of the Asia Program at Washington State University. His publications include *Engineering the State: the Huai River and Reconstruction in China, 1937–1945* (Routledge), and *State and Economy in Republican China* (co-editor, Harvard). His current research focuses on the social context of Yellow River hydraulic engineering after 1949. *Department of History, Box 4030, Washington State University, Pullman, WA 99163, USA. E-mail: pietz@wsu.edu*

Jean-Christophe Poussin is a researcher in Agronomy at the French Institute for Research and Development (IRD). His current research focuses on the agricultural systems in irrigated areas (in France and in Sahelian and North African countries) and on the modelling of farmers' practices. *Hydrogéologue, Directeur de Recherche IRD, BP 64501, 34394, Montpellier, Cedex 5, France. E-mail: jean-christophe.poussin@ird.fr*

Kossa Rajabu was a water resources management officer working with the World Wide Fund for Nature – Tanzania Programme Office (WWF–TPO). He had over 18 years' experience in irrigation and water management in Tanzania. In 2007, he obtained his PhD in surface water hydrology and water resources management in river basins from the Sokoine University of Agriculture, Morogoro, Tanzania. His other areas of competency were rainwater harvesting and policy analysis. *WWF–Tanzania, Rujewa, Tanzania. (Deceased).*

Paula Silva-Ochoa is a hydraulic engineer and consultant with CH2M HILL, USA. She has worked with small-scale and large-scale irrigation systems in Mexico, as a researcher for the International Water Management Institute, as a rural development specialist with the Jalisco state government, and as external consultant for the World Bank and the Food and Agriculture Organization. She is now focusing her work on water resources management and modelling in the southern California region. *CH2MHILL, Water Business Group, 402 West Broadway Suite 1450, San Diego, CA 92101, USA. E-mail: paula.silva@ch2m.com*

Siza Tumbo is Senior Lecturer in Agricultural Technology at Soil–Water Management Research Group at the Sokoine University of Agriculture, Morogoro, Tanzania. He has over 20 years' experience in precision methods for agriculture and resource computer modelling, plus research interests and project management experience in rainfed agriculture and rainwater harvesting. *Soil–Water Management Research Group, Sokoine University of Agriculture, Morogoro, Tanzania. E-mail: sdt116@suanet.ac.tz*

Hugh Turral is an irrigation and water resources engineer, with 27 years' experience in research, consulting and long-term technical cooperation, mostly in Asia and Australia. He was theme leader for basin water management at the International Water Management Institute from 2003 to 2007, and prior to that worked at the University of Melbourne, Australia, and at the Overseas Development Institute in London. He is now trying to establish a new career as a film-maker. *28 Newry Street, North Carlton, Melbourne, Victoria 3054, Australia. E-mail: hugh.turral@gmail.com*

Mauro Van Aken is a researcher and lecturer in Cultural Anthropology and Anthropology of Development at the University of Milan-Bicocca (Italy) and Associate Researcher at IFPO (Institut Français du Proche Orient, Amman). His main interests have been displacement and socio-cultural change, and cultural dynamics in resource management in the Middle East; zat present he is conducting research on cultural and political dynamics in irrigation management. *Fraz. Valdamonte 17, S. Maria della Versa (PV) 27047, Italy. E-mail: freedabkeh@yahoo.it*

Barbara van Koppen is Principal Researcher, Poverty, Gender, and Water with the International Water Management Institute, based in the Southern Africa Regional Program, Pretoria, South Africa. She has specialized in the institutional and legal aspects of water development for multiple uses in Africa and Asia. Her research includes policy dialogue and capacity building. She is editor, author, and co-author of three books and over 70 other international publications. *Principal Scientist, International Water Management Institute, Southern Africa Regional Office, 141 Cresswell Street, Weavind Park 0184, Pretoria, South Africa. E-mail: b.vankoppen@cgiar.org*

Sergio Vargas-Velázquez is a researcher at the Mexican Institute of Water Technology, Mexico. He has worked for 19 years on water conflicts, the organization of water users and the evaluation of socio-economic aspects of water management in Mexico. He was involved in the assessment of the irrigation management transfer, and subsequently in the constitution and organization of river basin and aquifer councils, commissions and committees. *Barrio del Sumidero Lote 6 Mz 22, Fracc. Las Fincas, CP 62565 Jiutepec, Morelos, México. E-mail: kuirunhari@yahoo.com.mx*

Jean-Philippe Venot is a post-doctoral scientist in political geography/ecology with the International Water Management Institute in Accra, Ghana. He was formerly with IWMI in Hyderabad (India), where he carried out his PhD research in human geography (University of Paris Ouest-Nanterre) on basin water management and allocation in South India (Krishna basin). His current research involves assessing the opportunities and constraints for collective action and small-scale irrigation development in West Africa. He is particularly interested in assessing the human dimensions of natural resources management and in understanding how societies adapt across scales to broader environmental change. *International Water Management Institute, PMB, CT 112 cantonments, Accra, Ghana. E-mail: j.venot@cgiar.org*

Philippus Wester is Assistant Professor Water Reforms at Wageningen University, the Netherlands. Trained as an interdisciplinary, socio-technical water management researcher, he has 15 years' experience in studying water politics and governance in Senegal, Pakistan, the Netherlands, Bangladesh and Mexico. His current research focuses on water politics, collaborative processes in river basin governance, and environmental and institutional change processes. *Irrigation and Water Engineering Group, Center for Water and Climate, Wageningen University, PO Box 47, 6700 AA Wageningen, The Netherlands. E-mail: flip.wester@wur.nl*

Series Foreword: Comprehensive Assessment of Water Management in Agriculture

There is broad consensus on the need to improve water management and to invest in water for food, as these are critical to meeting the Millennium Development Goals (MDGs). The role of water in food and livelihood security is a major issue of concern in the context of persistent poverty and continued environmental degradation. Although there is considerable knowledge on the issue of water management, an overarching picture on the water–food–livelihoods–environment nexus is missing, leaving uncertainties about management and investment decisions that will meet both food and environmental security objectives.

The Comprehensive Assessment of Water Management in Agriculture (CA) is an innovative, multi-institute process aimed at identifying existing knowledge and stimulating thought on ways to manage water resources to continue meeting the needs of both humans and ecosystems. The CA critically evaluates the benefits, costs and impacts of the past 50 years of water development and challenges to water management currently facing communities. It assesses innovative solutions and explores consequences of potential investment and management decisions. The CA is designed as a learning process, engaging networks of stakeholders to produce knowledge synthesis and methodologies. The main output of the CA is an assessment report that aims to guide investment and management decisions in the near future, considering their impact over the next 50 years in order to enhance food and environmental security to support the achievement of the MDGs. This assessment report is backed by CA research and knowledge-sharing activities.

The primary assessment research findings are presented in a series of books that will form the scientific basis for the Comprehensive Assessment of Water Management in Agriculture. The books will cover a range of vital topics in the areas of water, agriculture, food security and ecosystems – the entire spectrum of developing and managing water in agriculture, from fully irrigated to fully rainfed lands. They are about people and society, why they decide to adopt certain practices and not others and, in particular, how water management can help poor people. They are about ecosystems – how agriculture affects ecosystems, the goods and services ecosystems provide for food security and how water can be managed to meet both food and environmental security objectives. This is the eighth book in the series.

Effectively managing water to meet food and environmental objectives will require the concerted action of individuals from across several professions and disciplines – farmers, fishers, water managers, economists, hydrologists, irrigation specialists, agronomists and social scientists. The material presented in this book represents an effort to bring a diverse group of people together to present a unique assessment of river basin management throughout the world. The complete set of books should be invaluable for resource managers, researchers and field implementers. These books will provide source material from which policy statements, practical manuals, and educational and training material can be prepared.

The CA is done by a coalition of partners that includes 11 Future Harvest agricultural research centres, supported by the Consultative Group on International Agricultural Research (CGIAR), the Food and Agriculture Organization of the United Nations (FAO) and partners from some 80 research and development institutes globally. Co-sponsors of the assessment, institutes that are interested in the results and help frame the assessment, are the Ramsar Convention, the Convention on Biological Diversity, FAO and the CGIAR.

For production of this book, financial support from the governments of the Netherlands and Switzerland for the Comprehensive Assessment is appreciated.

David Molden
Series Editor
International Water Management Institute
Sri Lanka

Acknowledgements

The editors would like to extend their warm thanks to various friends and colleagues who have helped in the reviewing of chapters or parts of this volume: Nicolas Faysse, Akissa Bahri, Abdelkader Hamdane, Rudolf Orthofer, Habib Ayeb, Dominique Rollin, Barbara Schreiner, Mike Muller, Jacqui Goldin, Jon Barnett, Matthew McCartney, Wu Baosheng, Michael Cohen, Mel Neave, Robert Ward, Scott Rozelle, Line Crase, Richard Kingsford, Rudolph Orthofer, Geoff King and Gerardo van Halsema.

In addition, we would like to thank Kingsley Kurukulasuriya for his valuable editorial assistance and Mala Ranawake for her secretarial support.

1 River Basin Trajectories: an Inquiry into Changing Waterscapes

François Molle[1]* and Philippus Wester[2]**

[1]*Institut de Recherche pour le Développement, Montpellier, France;*
[2]*Wageningen University, Wageningen, The Netherlands;*
e-mails: *francois.molle@ird.fr; **flip.wester@wur.nl

Introduction

This book is concerned with 'river basin trajectories', loosely defined as the long-term interactions between societies and their environments, with a focus on the development and management of water and associated land resources (Molle, 2003). A basin trajectory encompasses human efforts to assess, capture, convey, store, share and use available water resources, thereby changing waterscapes and turning parts of the hydrological cycle into a hydro-social cycle (Wester, 2008). It also includes human efforts to deal with the threats posed by particular 'shock events', such as droughts, floods and contamination incidents, and to achieve a degree of environmental sustainability. Last, a basin trajectory includes institutional change and the shifting relations of power that govern access to, and control over, water resources. While this book focuses on human-induced environmental and hydrological transformations, its chapters also show how environmental change impacts on society and influences policy making. This includes the generation and particular social distribution of costs and risks, and shifts in the very conception of, and values attached to, nature.

The idea that the river basin is the 'natural' and most appropriate unit for water resources development and management has strongly influenced water–society relationships in the past 150 years (Molle, 2006; Warner *et al.*, 2008). Late in the 19th century it nurtured utopias and political struggles concerning the relationships between central and local power in countries such as Spain, France and the USA (Molle, 2006). Based on colonial experiences with water resources development in the Indus (van Halsema, 2002) and the Nile (Willcocks, 1901) basins in the early 20th century and the establishment of the Tennessee Valley Authority (TVA) in the USA during the 1930s (Lilienthal, 1944), the river basin became the unit where 'unified' or 'comprehensive' water resources development was to take place. This approach focused on the full utilization of rivers, multi-purpose dams, and wider regional development planning (White, 1957).

With time, and partly in reaction to significant modifications of river systems by hydraulic infrastructure and human water use, the river basin became the pivotal geographical unit for integrated water resources management (IWRM). The aim of this approach is to take into account, and reconcile conflicts arising from, the interactions between surface water and groundwater, water quantity and quality, human use and environmental functions, and scales and sectors of management (GWP, 2000; Grigg, 2008). More particularly, questions of river basin governance, with the vexing issue of cross-scale interaction and integration, came to the fore, as water problems were increasingly recognized as managerial, societal and political (Molle et al., 2007). Watershed movements and river basin organizations (RBOs) of various stripes have emerged to address these concerns.

The choice of the river basin as the management or governance unit is not undisputed. While there is an obvious (physical) logic for working with hydrological units in which the generation and use of water resources are largely coterminous, it is also well recognized that river-basin-based approaches suffer from 'tunnel vision' (Molle et al., 2007). Many drivers and consequences of river basin dynamics can be observed outside the basin, where solutions to local problems may also lie. In addition, even on a physical plane, river basin boundaries may not be relevant, for example in the case of small islands, deltas, flood plains or coastal areas. The occurrence of aquifer systems that are non-coterminous with river basins, or of interbasin transfers, is also frequent and demands consideration of linkages with adjacent basins. Yet all these particular situations can be treated as extensions of the river basin concept, and the influence of external factors can be considered through specific examination of the interactions of a river basin with its physical, economic and political 'environment'.

Water challenges, in the form of scarcity, excess or pollution, can be responded to in many different ways. Although droughts seem to call for dams, floods for dikes, and water pollution for treatment plants, response options are often much broader. Flood damage can be controlled locally by infrastructure (upstream dams, dikes, pumping stations) and also by more careful land-use planning (avoiding settlement in flood-prone areas), efficient flood warning, changes in upstream land cover, restoration of buffer areas, etc. Situations of water scarcity can be responded to in three different ways: supply augmentation (more water mobilized through dams, canals or pumps); demand management (including reducing absolute demand or saving water to expand uses); and (re)allocation (redefining access to a given amount of water) (Molle, 2003). Although the term 'river basin trajectory' may suggest there is a simple linearity in the development of river basins from supply augmentation, through demand management to water (re)allocation (Molden et al., 2005), the chapters in this book show that these three responses occur simultaneously and at different scales.

Technical and economic rationality have long inspired ways to select among available options by proposing various types of sophisticated cost–benefit analyses and other impact assessments. The history of water resources development (and that of public investment in general), however, abundantly shows that 'good intentions are not enough' (Green, 1996) and that these techniques are value laden, prone to distortion, and often justifications of projects that have (already) been decided upon, on political or other grounds (Berkoff, 2002). It also shows that options are never equivalent and that they entail flows of benefits and costs (financial, political or otherwise), and risks that accrue to particular sectors or groups of society. The identification of risks and costs is made more complex by the fact that interventions in the hydrological cycle tend – and increasingly so when pressure on water resources rises – to generate externalities in terms of modifications of the hydrological regime that affect users or residents elsewhere in the basin (Molle, 2007).

The question of political power and decision making – what are the options and who decides – is at the core of the 'shape' of a particular basin trajectory. The distribution of decision-making power and the political clout of different groups of stakeholders in society – in other words a particular power configuration or governance regime – are key to defining allocation or dam management rules, the decision to build another dam, or the establishment of particular water-related institutions. A defining characteristic of river basin trajectories is the political struggles surrounding the ways water is owned, allocated and managed, and 'over the right to define what a water right entails' (Boelens and Zwarteveen, 2005).

One particular and generic aspect of a basin trajectory is the closure of a basin. *Basin closure* occurs when the quantity of water abstracted is too high to ensure regular supply to downstream users or sufficient outflow to dilute pollution, control salinity intrusion, flush sediments and sustain healthy ecosystems at the mouth of the river (Seckler, 1996; Molle, 2003; Molden et al., 2005; Molle et al., 2007). This phenomenon (illustrated in Fig. 1.1) can be transient when it occurs only in a few dry months, and the basin is said to be closing, or almost permanent, when the basin is said to be closed. Basin closure occurs due to the 'overbuilding' of water infrastructure in river basins for the extraction of surface water and groundwater, to the point that more water is consumed by agriculture, industry and humans than is renewably available (Molle et al., 2007). Rivers no longer reaching the sea or contracting lakes are the most visible signs of basin closure, as exemplified by the Colorado River and the Aral and the Dead Seas.

The process of river basin closure induces increased competition between water use(r)s, and water scarcity reaches such a level that the exploitation limits become evident. However, using the term 'water scarcity' to describe situations of water overexploitation is dangerous, as it obscures issues concerning unequal access to, and control over, water (Bakker, 1999; Mehta, 2001). For most people, water scarcity is caused by competition between water uses and by political, technological and economic barriers that limit their access to water, rather than by physical water scarcity. Water scarcity is caused not only by variability in supply (supply-induced scarcity) or increases in population (demand-induced scarcity) but also by the overdevelopment of water resources, the selective entitlement of water rights and resource capture by better-off people, which Homer-Dixon (1999) terms structural scarcity. The design and social control over water technologies such as dams, pipelines and irrigation canals lead to what Vincent (2004) terms designed water scarcity, which influences who gets access to water.

Basin closure and water overexploitation tend to spur water quality decline, intersectoral water transfers, inequitable water allocation and reduced access to water (Molle et al., 2007). The inequality in access to water and the conflicts between the different users of water call for new approaches to water management (Mehta, 2001). The construction of large dams, irrigation schemes, interbasin transfer schemes and groundwater pumps create path dependency and lock-in situations (Sexton, 1990). The socio-ecologies that become dependent on these technologies and the water resource base are formidable and very difficult to reverse (Shah et al., 2003). While the overbuilding of river basins results in a situation that constrains the scope for reducing water use, it also radically alters the role that hydrocracies need to play, from centralized water resource developers to regulators and facilitators of decentralized water governance.

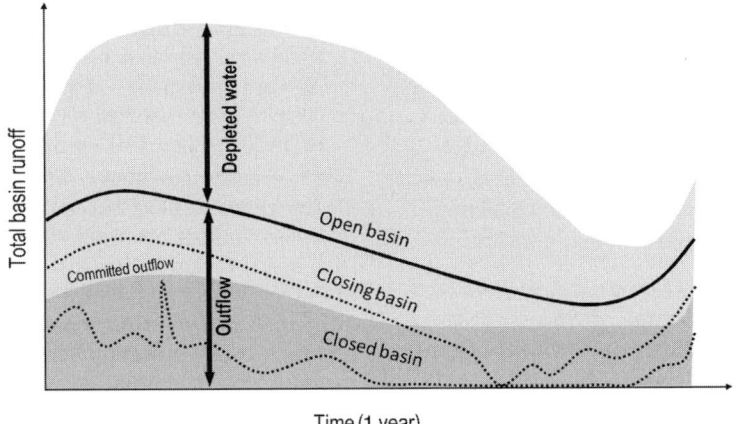

Fig. 1.1. The process of basin closure.

This book presents a rich analysis of 11 river basin trajectories. Each chapter provides a historical perspective on river basin development, highlighting the particular set of physical and human features that have shaped basin trajectories. All the authors have faced the double challenge of providing historical depth to their account while, at the same time, combining analyses of both environmental and institutional transformations. Because of the scale chosen, that of medium river basins, it was not possible to include the details of more local processes, such as changes in the management or governance of irrigation systems.

The 11 river basins investigated are mostly located in one country (the Zayandeh Rud in Iran, the Krishna and the Bhavani in India, the Merguellil in Tunisia, the Lerma–Chapala in Mexico, the Yellow in China, the Ruaha in Tanzania, and the Murray–Darling in Australia); other basins include the Olifants (South Africa) and the Colorado (USA) basins, which have their lower tips located in Mozambique and Mexico, respectively, and the Jordan basin, whose study is limited to Jordan. Five basins are located in federal countries (USA, Mexico, Australia, India), where relationships between the federal and state governments appear to be a crucial dimension of basin management and governance. The 11 river basins all face conditions of water scarcity, with a few particularly acute cases (Jordan, Zayandeh Rud, Lerma–Chapala).

This chapter presents general findings and reflections drawn from the river basin trajectories analysed in this book, occasionally enriched by evidence drawn from other basins in the world. It attempts to both identify commonalties and emphasize the specificity of each basin. It starts with a discussion on ideologies and models of river basin management and then describes four widely observed processes related to river basin trajectories. The responses of society to the issues raised by basin trajectories are then discussed. Last, conclusions are drawn.

Drivers of Change and Competing Paradigms

River basin development has long been predicated on an ideology of domination of nature, where 'conquering', 'harnessing' or 'taming' the wilderness were touted as a civilizing mission made possible by science and advances in technology. The development of irrigation was central in wider state settlement policies, whether it was to settle a nomadic population, as in Jordan (Chapter 2) or in Tunisia (Chapter 7), provide jobs after the two World Wars to returning servicemen in Australia (Chapter 12) and South Africa (Chapter 3), break up haciendas and colonize them with a new type of industrious farmer devoted to 'revolutionary irrigation' in Mexico (Chapter 4; Aboites, 1998), or strategically occupy land (as in the USA, Chapter 6; or Israel, Lipchin, 2003). In the post-World War II period, irrigation held the promise of feeding the masses, raising rural income and – in the particular context of the Cold War – enlisting 'development' and food self-sufficiency in the struggle against communism. Projects were churned out based on the expectation of large increases in yields, optimistic cropping intensities, and adoption of cash crops.

The transition from local water control to large-scale water resources development by the state, based on river basins, was intimately linked to the 'hydraulic mission' of the hydraulic bureaucracies (hydrocracies) created in the 19th and 20th centuries. Wester (2008) defines the hydraulic mission as:

> the strong conviction that every drop of water flowing to the ocean is a waste and that the state should develop hydraulic infrastructure to capture as much water as possible for human uses. The carrier of this mission is the hydrocracy who, based on a high-modernist world-view, sets out to control nature and 'conquer the desert' by 'developing' water resources for the sake of progress and development.

The hydraulic mission era, which ended in the 1970s in most affluent countries, was marked by the growth of powerful state hydrocracies, such as in Mexico (Chapter 4), where the logo of the Ministry of Hydraulic Resources was *Por la Grandeza de México* (For the Greatness of Mexico). Many of the senior hydrocrats manning the hydrocracies were educated in the West, notably in the USA, where the Bureau of Reclamation trained 'a new generation of Mexican hydraulic engineers' (Chapter

4) as well as engineers of many other countries, where the export of the TVA model was attempted (see Ekbladh, 2002; Molle, 2006).

The hydraulic mission era was characterized by a massive injection of public money in all countries and 'blatant subsidies and political favours' in the USA (Chapter 6; Worster, 1985; Reisner, 1993). These subsidies were a result of the recognition of the failure of private irrigation initiatives at the end of the 19th century, such as in Australia (Chapter 12), India (Chapter 10) or the western USA (Chapter 6), and also of the overriding political goals attached to irrigation development. In the USA this phase was associated with 'a "private commodity" paradigm, featuring an emphasis on water development and the rights of individual rights-holders' (Chapter 6).

This first phase of agricultural growth and modernization clearly marked the period from 1960 to 1990 in the Ruaha basin in Tanzania (Chapter 8). It was later substituted by a narrative of efficiency, environmentalism and water reallocation during the period 1995–2005. While in the former period, water and land were seen to be abundant, the latter drew from a growing perception of water as a finite supply and concerns over power cuts. A similar shift emerged in most basins, albeit at slightly different times. In the USA, a 'public value' paradigm, emphasizing resource protection, value pluralism, and democratic (i.e. collective and participatory) decision making, took root (Chapter 6). In the Murray–Darling basin (Chapter 12), the water reforms beginning in the late 1980s were also the product of changing ideas about how public institutions should be organized and operated. There was a widespread feeling that decision making could no longer be left to small groups of engineers who had spent their careers dealing mainly with water resources infrastructure. Under the new arrangements, the basin's river system was to be managed to conserve biodiversity and improve sustainability as well as for production. In the Olifants basin (Chapter 3), environmental and social considerations were incorporated into the 1998 Water Law, which triggered attempts at broadening participation of stakeholders and quantifying environmental flows. In China (Chapter 5), the Ministry of Water Resources brought forward ideas for the conceptual transformation of water resource development and management from engineering-dominated approaches to approaches based on demand management and the value of water resources (a shift from emphasis on *gongchengshuili*, engineering water benefits, to *ziranshuli*, broader water resources benefits).

These changes were the result of a change in societal values linked to growing affluence and awareness of environmental degradation. In the Colorado basin (Chapter 6), the national goal of western settlement based on water resources development also created something heretofore missing from the region: an urban constituency drawn to the aesthetic and environmental amenities of the region, supportive of public lands and other collective resources, and emphasizing quality of life over return on investment. As Kenney notes (Chapter 6), the inherent incompatibility of the two paradigms suggests that they have evolved sequentially and incrementally rather than simultaneously. In China, however, the two attitudes are linked to competing philosophies and seem to have always coexisted (Chapter 5): Confucianism and the Naturalist school of thought sought to explain nature on the basis of the complementary cosmic principles of yin and yang and saw man as a natural master of nature. Taoism, on the other hand, saw water as 'the supreme moral example of the stricture to find harmony with "the way" (tao), (...) as an object of contemplation intending to reveal moral truths ... something to be admired rather than controlled, ... with gardens as a place of contemplation where it was possible to connect with the ultimate realities of nature, and to escape worldly concerns.'

With the growing recognition of the associated social and environmental costs, and also with the decreasing availability of suitable dam sites, the hydraulic mission ran out of steam in most affluent countries in the 1970s (Barrow, 1998). Priority shifted towards water quality and environmental sustainability, setting the stage for a resurgence of the river basin concept in the 1990s. This resurgence was strongly inspired by the ecosystem approach, in which a river basin is seen as an ecosystems continuum and water as an integral part of ecosystems (Marchand and Toornstra, 1986). In many ways, this is a reaction to the construction bias of the hydraulic mission era, but proponents of the ecosystem approach are adamant that 'water resources should be managed on the basis of river or drainage basins in an integrated fashion, with a continued and deliberate effort to maintain and restore ecosystem functioning within both catchments and the coastal and marine ecosystems they are connected with' (IUCN, 2000). In the early 1990s, the centrality of river basins for environmental governance was reflected in the Dublin Principles (ACC/ISGWR, 1992) and the formulation of IWRM approaches, and was later formalized by the European Union in its Water Framework Directive (EU, 2000).

Major Processes at Work in River Basin Trajectories

River basins are very different from one another. However, the 11 story-lines that follow, as well as the wider bibliography on river basin development and management, allow us to identify generic processes that are at work in most river basin trajectories. These are: (i) the overbuilding of river basins; (ii) the overallocation of entitlements; (iii) the overdraft of reservoirs and aquifers; and (iv) the double squeeze of agricultural water use, due to declining water availability and quality on the one hand and rising urban and environmental needs on the other.

Overbuilding of river basins

The overbuilding of river basins is a socially constructed process that generates basin closure through the overextension of the water abstraction capacity, in general for irrigation. Decision makers are faced with powerful incentives for continued public investments in irrigation infrastructure. Politicians, whether at the local or government level, have long identified iconic, large-scale projects as the best way to build up constituencies and state legitimacy with public funds. Hydrocracies vie to maintain and expand their bureaucratic power (sustained budgets and fringe benefits, upholding of professional legitimacy, etc.). Private consulting and construction firms, often linked to particular politicians/parties, look for business opportunities. Last, development banks and cooperation agencies also have vested interests in maximizing the disbursement of funds (Chambers, 1997).

The overdevelopment of water-use infrastructure, principally irrigation schemes, generates water scarcity 'mechanically'. When most available resources are committed and little 'slack' remains in the hydrological regime of a particular river basin, any substantial drop in available resources below average values is likely to result in shortages for some users. With a growing hydrological variability due to climate change and a tendency to mismanage carry-over stocks in reservoirs (managers being under pressure to generate electricity or to release water at the cost of mid-term reserves and security of supply), the frequency and intensity of such shortages are increasing. Crises result in public outcry, media coverage of farmers with withering crops, newspapers stamped with pictures of cracked soils, and tales of looming disasters. Politicians are prompt to seize such crises to promise more populist projects aimed at tapping more water. New irrigated areas are often necessary to make dam or diversion projects economically more attractive and also to achieve the 'buy in' of provinces or populations that will be affected by new reservoirs or projects. The vicious circle of overdevelopment thus becomes self-sustaining (Molle, 2008).

Augmenting supply maximizes benefits to what has been termed the 'iron triangle' in the western USA (Reisner, 1993; McCool, 1994) and often minimizes short-term political stress, compared with options where supply to existing users must be reduced or reorganized. Logrolling (Chapter 6) is a political behaviour that fuels overbuilding, whereby 'legislators from various jurisdictions all agree to support each other's proposed projects in their home districts. In this way, a project with only local appeal can gain the support of a broad base of legislators.'

The process of basin overbuilding is well illustrated by the case of the Zayandeh Rud (Chapter 9), where each new import of water into the basin is justified by water shortages and accompanied by an expansion of irrigation and out-of-basin transfers. Instead of stabilizing water use in the basin, providing more 'slack' and security to users, whatever additional water is made available is committed to expanding irrigation areas. This process is also illustrated by the Lerma–Chapala basin (Chapter 4) and other case studies from central and north-east Thailand, and from the Bhavani basin (Chapter 11).

Other critical drivers of basin overbuilding appear in our case studies. In the Colorado basin (Chapter 6), the upper states, and later Arizona, partly pursued development as a means of securing their entitlements and claims by effectively diverting water. In the Krishna basin (Chapter 10), as the award (basin-sharing agreement) of 1976 was to be revised in 2000, the states sharing the Krishna water 'engaged in massive development of their hydraulic infrastructure (with serious economic and fiscal damage) to lay claim on water resources and ensure they would be holding a prevailing position when the award would be renegotiated' (Gulati et al., 2005). Politically motivated concerns for regional equity also fuel basin overbuilding. Preventing regional tensions and threats of state implosion under the pressure of independence claims from all three regions of Andhra Pradesh state have been major drivers of infrastructural development in the lower Krishna basin (Chapter 10; Venot et al., 2007). Although irrigation is first expanded in favourable areas, it leads to later claims from other (poorer) regions that they have not only been discriminated against but also need such investments for their development. This often leads to the expansion of costly infrastructure in marginal areas.

Politicians are used to resorting to overriding justifications that close or 'securitize' the debate (Warner, 2008): new projects are indispensable and cannot be delayed because 'poverty demands that we do something', development is needed and requires 'sacrifice', national or food security is at stake, or growing energy needs make the development of hydropower 'unavoidable'. These concerns are legitimate and often truly pressing. But by closing the debate, decision makers also make it impossible to discuss alternatives, to examine in detail the social and environmental costs of projects, and to reveal the frequent absurdity of supply augmentation projects when seen through the lens of investment costs (soon to become cost overruns).

Overallocation of water entitlements

Basin overbuilding is also made possible by the fuzziness or absence of water rights, which means that many projects are, in fact, partly predicated upon water that is already committed to other (generally downstream) areas. Such a problem may occur not only because of uncontrolled expansion of private irrigation, as in the Ruaha (Chapter 8), Lerma–Chapala (Chapter 4), Zayandeh Rud (Chapter 9) and Krishna basins (Chapter 10), but also because of state-initiated anti-erosion works, as in the Merguellil (Chapter 7) and Yellow River (Chapter 5) basins, or even public irrigation schemes, as in the Zayandeh Rud and Chao Phraya (Thailand) basins.

River basins with stricter control of hydrological conditions and definition of water rights and entitlements should theoretically avoid this trap. Experience shows that this is not the case. Overbuilding through private investments is paralleled by an overallocation of water entitlements that creates similar patterns of scarcity. In the Colorado basin, apportionment of water among riparian states has been based on optimistic average hydrological data, without considering either evaporation losses in reservoirs to be built years later (now totalling 2 billion m^3) or native Indian rights. In the Murray–Darling basin, notably the state of New South Wales, licences have been granted despite recognition of the ticking time bomb represented by large contingents of 'dozers and sleepers' who only use their rights occasionally or pay their fees without using water. This has led to a water allocation that amounts to 65% of all entitlements, on average, and to a reduction in security and predictability. In the Olifants basin (Chapter 3), all water was allocated, making it virtually impossible to grant new rights to black communities. In the Lerma–Chapala basin, the 1991 treaty on surface water allocation was based on an optimistic assessment of annual water availability (with two dry periods excluded from the hydrological model underlying the treaty) and no attempt was made to reduce the volumes of water concessioned to water users.

The overallocation of water entitlements is an obvious political expedient to reduce tension, avoid denying access to resources, and satisfy a maximum of existing (or would-be) users in particular constituencies (Allan, 2006). This, of course, occurs at the cost of supply security to all. More recently, overallocation was made more critical because of prolonged droughts (Murray–Darling, Colorado, Lerma–Chapala), dwindling runoff (Yellow River), and painful expectations of climate change (Murray–Darling). On top of these concerns, preoccupation with aquatic ecosystem health put environment flows on top of the agenda. Attempts to reallocate water to the environment from existing users have been largely frustrated, and this remains an unresolved issue. In the Olifants basin, environmental flows (eflows) have been much discussed but have so far remained on paper. In the Colorado basin, federal laws generally defer to the tradition in state water law of allowing water users to consume rivers in their entirety. Western states now provide some mechanisms for granting water rights to instream flows, but these tend to be very limited in scope, often relying on water rights that are junior to traditional consumptive users. In the Murray–Darling basin, attempts to reduce entitlements to enhance environmental flows have also not been popular, and states have been forced to resort to a (still limited) buy-back of water rights. In the Zayandeh Rud and Jordan basins, the environmental objective of maintaining terminal sinks (the Gavkhuni lake and the Dead Sea) has been simply written off. The Lerma–Chapala (Chapter 4) offers an example of reallocation away from irrigation with the aim of sustaining the level of the Chapala lake, but this objective was mainly dictated by urban supply objectives downstream of the lake.

Overdraft of reservoirs and aquifers

As a consequence of basin overbuilding and/or the overallocation of entitlements, the case studies confirm a widely observed tendency for managers and users to 'overtap' reservoirs and aquifers. Reservoirs generally have several purposes but are pivotal in providing interannual regulation and carry-over storage. Storing water allows managers to ensure supply in dry years. Water security, measured as the capacity to withstand a number of successive dry years, is largely dependent upon storage capacity. The Murray–Darling and Colorado basins are famous for storage capacities that are much higher than the average annual runoff: dams can store 2.8 and 3.5 times annual runoff, respectively. Conversely, the lack of storage in basins such as the Ruaha and the Jordan means that users have to face greater irregularity and risk.

Under pressure from users and politicians, managers frequently release more water in a given year than would be expected if carry-over storage were managed prudently. This increases risk and does indeed generate or magnify crises. The case of the Zayandeh Rud basin (Chapter 9) shows how careless releases in 1999 and 2000 contributed to an exceptional crisis in 2001. Likewise, in 2000, the managers of the Nagarjuna Sagar dam in the lower Krishna basin took a gamble and released all the available water, paving the way for the ensuing crisis (Chapter 10). In the Ruaha basin (Chapter 8), pressure to generate hydroelectricity at the national level also led to lowering of dam water levels beyond what risk management dictated, and to subsequent major power cuts in the capital. In the Lerma–Chapala basin (Chapter 4), the 1991 surface water allocation treaty was based on the assumption that the carry-over storage in reservoirs would increase with time if the treaty was adhered to. Instead, carry-over storage was largely depleted to comply with annual water allocations as river runoff was less than predicted by the hydrological model underlying the treaty.

Overdraft of aquifers is a better-documented and more familiar problem. Almost all basins show a long-term drawdown of water tables. This is particularly worrying in basins where groundwater provides a 'buffer' in case of insufficient supply of surface water, such as in the Zayandeh Rud, Lerma–Chapala and lower Yellow River basins. Indeed, as surface deliveries become more uncertain, users develop conjunctive use and turn to groundwater in compensation. In the Lerma–Chapala basin, groundwater-based irrigation also developed as a market response to opportunities for producing vegetables for the USA market. Ten years ago, water tables were dropping at rates that would bring aquifers to exhaustion, but these have been partly replenished by exceptional rainfall. The Merguellil, Jordan and Zayandeh Rud basins are typical cases where aquifers are declining and where authorities have found no way of reversing this process. The Jordan highlands suggest that price-based regulation is illusory and that where enforcement of quotas is not realistic the only solution is buying back wells and controlling further drilling. The Merguellil case illustrates the contradiction between long-term sustainability concerns and the short-term needs of food and income generation, which explains why authorities often turn a blind eye to private drilling and aquifer overdraft (a decline of between 0.25 and 1 m a year since the 1980s).

Reallocation from agriculture to cities (and the environment)

Another lesson drawn from many river basin trajectories is that agriculture – often after a phase of overexpansion due to basin overbuilding – ends up constrained by a double squeeze (see Fig. 1.2). On the supply side, water availability is sometimes reduced by long-term trends due to climate change or otherwise. Predictions for the Colorado basin by 2100 point to reductions anywhere between 11 and 45%, while the Murray–Darling basin expects reductions in mean annual flow in the order of 20–30%. Degradation of water quality is also a trend that contributes to reducing freshwater availability, with some river or drainage water unfit for use in domestic supply and even in agriculture.

On the demand side, the large historical share of agricultural use now collides with urbanization and environmentalism. All water-short basins, although sometimes buying respite by continued supply augmentation, end up facing the issue of water reallocation. It is always politically very sensitive to take water away from existing users to serve expanding urban constituencies; it is even more challenging – in a closed basin – to set water apart for 'environmental use', i.e. to sustain or restore ecosystem health. Figure 1.2 shows how irrigation gets squeezed by these trends in supply and demand and how the variability of freshwater supply induces increasingly severe shortages, which tend to primarily affect environmental and agricultural uses.

The case of the Lerma–Chapala basin (Chapter 4) illustrates how the hydro-social networks constituted around, and by, the hydraulic infrastructure in the basin make it difficult to reduce consumptive water use, even if a range of water reforms are attempted and serious efforts are made to arrive at negotiated agreements on surface water allocation

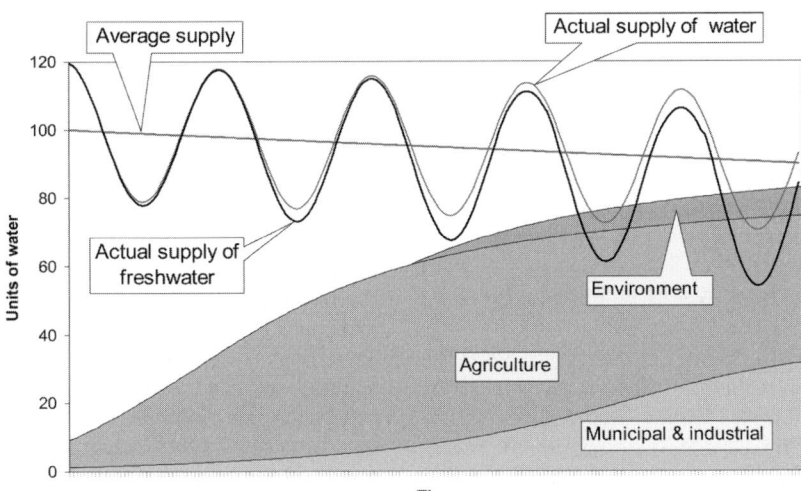

Fig. 1.2. River basin 'double squeeze'.

mechanisms. In the Colorado basin, the recipe of 'drawing on surplus flows in wet years, transferring water from agricultural to urban users in normal years, and tapping reservoir storage in dry years' has reached its limits, as storage reached critical lows and transfers faced a series of difficulties. Market mechanisms allow a degree of reallocation to cities, and several direct agreements between urban and irrigation areas can also be noted: San Diego buying water from the Imperial Valley irrigation district (supplied from the lower Colorado), Melbourne acquiring rights to 75 Mm^3 of the lower Murray–Darling in exchange for investments, and Chinese cities in the Yellow River basin transacting with irrigation districts. In other basins (diversions to Amman in the Jordan basin, to Hyderabad in the Krishna basin, to Tirrupur and Coimbatore in the Bhavani basin, to coastal cities in the Merguellil basin), transfers have been decided by administrative fiat. This was also the case in the Lerma–Chapala basin, where, in 1999, because of critically low levels in Lake Chapala and to secure Guadalajara's water supply, the CNA (National Water Commission) transferred 200 Mm^3 from the Solis dam, the main water source of the largest irrigation district in the basin, to Lake Chapala. A second transfer of 270 Mm^3 followed in November 2001, as lake levels continued to decline.

Keeping water in lakes and rivers is even more challenging. In the Olifants basin, the establishment of environmental flows has remained largely theoretical, with different approaches tested to determine environmental requirements. The gridlock as to how to reduce agricultural use is likely to be eventually eased by constructing a new dam and therefore developing more resources. Such a way out is also visible in the Mexican case (with a new dam on the upper Santiago River to serve Leon city and a new dam on the Santiago River near Guadalajara to supply its urban water) and the Zayandeh Rud case (interbasin transfer). Whenever possible, and often regardless of costs, supply augmentation is still a favoured option, which minimizes political stress but, of course, only buys time and eventually compounds basin closure.

Expectations of reduced supply are taken very seriously in the Murray–Darling basin. The main challenges for the future concern the best way to reduce overall allocation in the basin and, more importantly, to make sure that each state will take its share of the burden. It is no longer merely a question of complying with the 1994 cap on abstraction but of adjusting to significantly reduced allocations for the irrigation sector. The pressure to do this is mostly driven by current environmental allocation concerns, plus the expectation of reductions in mean annual flow in the order of 20–30% by 2100 under a range of climate change scenarios.

Major Societal Responses and Issues

Several major issues, associated with the four processes highlighted above, can be singled out and illustrated by our case studies. One issue concerns the 'politics of blame,' which is the way crises are explained, handled and used to justify specific policies and further particular agendas. Other issues concern the actual responses to basin closure, the impact of water scarcity on water-use efficiency and equity, and basin governance.

The politics of blame

Water-related problems (floods, shortages, contamination, etc.) are often accompanied by efforts by stakeholders, managers and politicians to find explanations and apportion blame. The way blame is apportioned to different causes is important because it not only reflects the distribution of power (and the capacity of particular stakeholders to get their message across in the media) but also paves the way for what will be done next, the money that will be spent, and the options that will be favoured. As such, it is an exercise of power.

Predictably, climatic vagaries or El Niño are convenient scapegoats, which, indeed, often bear part of the 'responsibility', but irrigation, its large share of water diversion, highlanders (responsible for deforestation) and pastoralists (associated with overgrazing) are also primary targets. During the second Lake Chapala crisis (Chapter 4), water authorities blamed the desiccation of the lake on the drought and the high levels of evaporation from the lake, although the extractions from the lake by Guadalajara city of at least 240 Mm^3 a year contributed strongly to the decline of the lake. In the Ruaha basin, water shortages experienced in the Mtera–Kidatu hydropower complex (which resulted in power cuts in the capital and other cities) were blamed on upstream irrigators and pastoralists. A series of analyses demonstrates that, despite claims by power-generation authorities, the power cuts experienced from 1992 onwards were largely due to improper dam operation rather than to upstream depletion of water. In 2004, for example, the situation was so critical that the Mtera reservoir was operated by utilizing the dead storage, despite advice to the contrary from the Rufiji Basin Water Office and the ministry responsible for water. This advice was not heeded, resulting in higher risks and showing the economic and political importance of maintaining power generation at any risk and cost.

In the Mekong basin, the floods in the summer of 2008 were used to critique the dams built by the Chinese in the upper basin and the lack of transparency concerning dam releases, although evidence of their responsibility is dubious. Floods in central Thailand or the Ganges basin have also been associated with land management practices by highlanders, although scientific evidence of a correlation is at best weak (Forsyth and Walker, 2008). In the Thai case, accusations have been blended with ethnic stereotypes and conveniently justified expansion of state enclosures (in the guise of national parks, reserves, etc.), afforestation by private companies and, in some cases, expulsion of hill tribes (Walker, 2003).

Whether justified or not, such accusations are active elements of negotiation processes (if any) and/or state decision making. In the Lerma–Chapala basin, the *Grupo de Trabajo Especializado en Planeación Agrícola Integral* (GTEPAI, Specialized Working Group on Integral Agricultural Planning) attempted to strengthen the negotiating position of irrigators in the river basin council. Its strategy was to show that the irrigated agriculture sector was serious about saving water and hence a credible negotiating partner. However, the stigma of irrigation being a wasteful use of water was too strong, and the farmers continued to be blamed for the desiccation of Lake Chapala by urban dwellers and environmentalists.

Conversely, proponents of particular solutions must paint them in a positive mode. The Red–Dead project in Jordan, which proposes to bring water from the Red Sea into the Dead Sea, generate hydropower and desalinate water, and pump part of it to Amman and other cities (Chapter 2), is alternatively painted with environmental (save the Dead Sea), religious (the cradle of three religions) or political (the peace conduit) arguments. Other mega-projects, such as the diversion of the São Francisco in Brazil (Alves, 2008) or the Water Grid in Thailand (Molle and Floch, 2008), also emphasize 'eradication of poverty', enhanced rural incomes and abundant water, while typically disregarding costs and investment alternatives.

Responses to basin closure

Basin closure and associated water scarcity, decline of water quality and environmental degradation – as mentioned earlier – give way to three types of responses: supply augmentation, demand management and (re)allocation. It has been hypothesized that these three types of responses occur sequentially along the basin closure trajectory (Molden et al., 2005). While it is true that early phases of basin development are almost exclusively typified by supply augmentation, case studies of closing or closed basins show that – under pressure and in the face of recurring crises – the three options are pursued concurrently.

The blend of options selected depends on the physical, financial and political features of each option. Physical constraints refer to the accessibility of water resources and clearly set a limit to what is possible. Yet such constraints are typically qualified by financial and political considerations, as shown by the interbasin transfers through tunnels in the Zayandeh Rud basin and by the Red–Dead project in Jordan. If the costs of such works are shifted to the country as a whole and/or, partly, to the international community, then they may be eventually realized. Likewise, the acceptance of federal policies in the Murray–Darling and the Colorado basins was strongly linked to billions of dollars of federal subsidies in various guises (e.g. for Land Care groups in Australia, or water diversions and dams in the USA). Interbasin transfers may be opposed by 'donor basins', and imposition by the central government may involve lots of political manoeuvring and arm-twisting, as seen in the current project to divert the water of the São Francisco River in Brazil (Alves, 2008). While in some cases project costs are an impediment, in other cases higher costs may be seen as desirable by unchecked private interests.

While most infrastructural projects are costly, other measures are financially more attractive. Technical improvements or conservation policies, whether physical (e.g. canal lining or retrofitting of home appliances) or not (e.g. awareness campaigns), may be cost-effective. Fine tuning of management may also result in savings. In the Colorado basin, the reservoir operations and shortage-sharing rules were the most debated elements in the recent audit process (Chapter 6). The water level in the dams governs not only the head (hydropower generation) and the flood-control capacity but also the size of the water body and thus its evaporation losses. New rules may better account for hydrological changes and desired levels of security, and better balance priorities (e.g. environment versus human use).

Political constraints refer to the political benefits and costs associated with particular options. Options impacting key supportive or strong constituencies are likely to be discarded. This is clearly demonstrated in the case of the Jordan basin (Chapter 2), where regulation of groundwater use in the highlands and charging for water in the valley (notably in citrus and banana farms) are poised to damage the support of certain tribes and entrepreneurs to the King and the government. In the Olifants basin (Chapter 3), redistributive and participatory policies are adverse to white economic interests and have made little progress. Other types of policies meet with little popular support but they seem to go ahead out of bureaucratic inertia or ideology, as the intriguing case of water-harvesting structures in the Merguellil basin suggests (Chapter 7).

As a result of such complex sets of constraints, responses are often diverse and shifting but more or less efficient. The Colorado basin has seen the emergence of an unusually rich suite of strategies for increasing yields and avoiding (overcoming) limits, highlighted by efforts to eliminate reservoir spills (and associated 'over-deliveries' to Mexico), marketing of water salvaged through conservation programmes, the eradication of water-loving tamarisk and Russian olive trees, weather modification (i.e. cloud seeding), desalination, the proposed importation of water from neighbouring basins, and compensated fallowing of agricultural land. In Jordan (Chapter 2), policies have also mixed all kinds of conservation incentives with supply augmentation (dams, import of groundwater from distant aquifers) and forced reallocation of water (from agriculture in the valley to cities in the highlands).

In the past, the key to positive-sum bargaining in river basins was to expand the available benefits (i.e. water and power) at public cost, with little consideration of environmental and other public values. Today, opportunities for new storage or diversions are limited, civil society at large has gained political space and clout, and decisions are increasingly debated in wider and more contested arenas. Yet this clearly varies from one basin to another, and unilateral state decision making still prevails in many countries.

Hydrological pathologies

The hydrology of closing basins is problematic. Because most flows, including return flows from existing uses, are tapped, there is little 'slack' in the basin hydrological system to dampen or buffer natural hydrological variability, and perturbations thus strongly reverberate on the whole system. The pathology of closed river basins has been the subject of many works, which have emphasized the concept of river basin efficiency, as opposed to local user or system efficiency (Seckler, 1996; Molle and Turral, 2004; Perry, 2007). They have shown how local 'inefficiencies' associated with leaky canals, reservoir spills, inefficient irrigation practices and other system losses are often the primary source of water for other users or for ecosystems.

More generally, interventions in the hydrological cycle generate externalities in terms of water quantity, water quality, sediment load or timing that travel across the basin. These externalities are heightened by the process of closure but are also sometimes difficult to seize or appreciate as they involve time lags and two-way interactions between surface water and groundwater resources. Deforestation in the Murray–Darling basin has altered runoff and groundwater recharge, resulting in the phenomenon of dry-salinity. Afforestation in the upper Olifants basin has reduced natural runoff to the point that forest areas are considered as a water user and forestry companies have to pay fees accordingly. Development of diffuse water-harvesting structures and shallow wells in the Krishna and Merguellil basins has critically curtailed runoff and benefits to downstream water users. In the Zayandeh Rud basin, several hydrological interactions have also been evidenced, including reverted net flows between the river-bed and adjacent aquifers. In the Yellow and Lerma–Chapala basins, reduced river base flows due to groundwater over-exploitation have also been observed.

Unless they save water that goes to sinks, such as saline aquifers or the sea (all ecosystem functions of river outflows being considered), conservation efforts tend to amount to disguised reallocation. This is a zero-sum game, with reallocation from public environmental interests to water users, or from one user to another, merely robbing Peter to pay Paul. The deal between San Diego and the Imperial Valley Irrigation district, supplied from the lower Colorado, is a textbook example of a zero-sum game branded as a 'win–win agreement'. The 100 Mm^3 of water 'saved' by lining the All-American canal and reallocated to San Diego have merely been subtracted from the flows reaching the Salton Sea and replenishing the Mexicali aquifer, on which Mexican farmers on the other side of the border depend (Cortez-Lara and García-Acevedo, 2000; Cortez-Lara, 2004).

Kendy et al. (2003) have also highlighted the hydrological nature of closed basins in the North China Plain, where virtually all annually renewable water is used (depleted) and groundwater tables are falling with agricultural and urban expansion. While water might be used and reused more wisely or reallocated within the basin, little water reaching the sea means that all resources are depleted and that reducing demand can only come from reduced use (i.e. mostly reduced evapotranspiration). With almost no water reaching the sea, it could be argued that the same holds true for the Yellow River in general.

The lesson drawn from all these examples is that the management of river basins becomes increasingly difficult with closure. Arid basins are somewhat easier to manage, in that most of the resource mobilized is stored in a few reservoirs or aquifers, which are potentially amenable to quantification. In basins such as the Yellow or Krishna, where rainfall is more frequent and better distributed throughout the year, supply and demand vary a lot and the spatial and temporal distribution of flows is harder to grasp and control. In all cases, supply augmentation, conservation and reallocation appear to be clearly scale dependent. What is stored or conserved at one point is often a reallocation when seen at a larger scale. Managing such externalities and interconnectedness is challenging in both technical and governance terms.

Adding further complexity to the hydrology of closed river basins is the variability of rainfall. There is no such thing as an 'average' hydrological year, although many treaties on surface water are based on calculations of long-term averages. However, the periods for which rainfall data are available have proven to be too short to calculate robust averages; assuming this is still meaningful in a context of climate change, where the future will not look like the past. In both the Colorado and Lerma–Chapala basins, treaties on surface water were based on calculations of average runoff that later proved to be too high. With climate change it appears that variability in rainfall will increase, further weakening the reliability of estimates of average runoff.

Family/subsistence farming versus entrepreneurial capitalism

As competition increases, water tends to be gradually reallocated towards uses with higher economic value. This is achieved through administrative decisions, negotiations between users, or market mechanisms. An important and ubiquitous question is the allocation of water within the agriculture sector and the fate of irrigated agriculture as water becomes more valuable. Following the Dublin principle on water as an economic good, maximizing aggregate welfare has become a commonplace recommendation, but it is apparent that this principle also tends to conflict with that of ensuring equity or livelihoods for the poorest.

Most basins present a contrast between two broad types of agriculture: the first type is family based, sometimes partly devoted to subsistence agriculture, with limited links to markets and a lack of capital or knowledge, which prevents farmers from intensifying or embarking on more market-oriented and risky ventures. The second type is entrepreneurial, market oriented or export oriented, and owners – frequently absentee owners – often manage their farms through hired managers and labourers. This dichotomy is a simplification and does not do justice to hybrid types of farms: smallholders fully integrated to the market (e.g. peri-urban vegetable farming in the Merguellil plain) or absentee owners keeping low-value prestige olive tree plantations in Jordan. Yet it is useful in highlighting governments' dilemmas in allocating water and other resources.

Many state policies, indeed, are predicated on transforming the former type into the latter, often with little understanding of the constraints faced by farmers and with optimistic assumptions on how they will respond to 'incentives'. In particular, it is often inferred that higher water prices would trigger a shift towards higher-value crops, an assumption that runs into contradictions since these higher-value crops are already available to farmers; they have not opted for them for good reasons, which are often poorly understood.

The contrast between smallholder and agribusiness agriculture is particularly apparent in the Olifants basin, where discourses on economic efficiency and policies to redress inequalities of the past are at loggerheads. In the Colorado basin, agribusinesses that produce vegetables exported to distant states are indirectly pitted against extensive rearing of dairy cows in Wyoming. In Brazil's São Francisco basin, public irrigation schemes designed to settle poor farmers have been abandoned in favour of wealthy and corporate investors coming from the south and abroad.

In the Lerma–Chapala basin, the boom in export agriculture (primarily vegetables) has been fed by expensive groundwater, while support for land reform communities was discontinued in the early 1990s.

In the Krishna basin, two sets of policies have translated into two different modes of access to, and use of, water in different parts of the basin (Chapter 10). Broadly, the first group of policies aims at 'efficiency in development' and concentrates financial and institutional investments on those social groups and areas that offer the highest potential for development. They are the technologies of the Green Revolution, adopted in medium and large irrigation projects, and more recently they have attempted integrating agriculture into agribusiness chains. The second group aims at 'equity in development' and advocates rural development programmes through strong state planning and public investments in remote areas. They are watershed and tank rehabilitation programmes, and minor irrigation projects in upper secondary catchments (Landy, 2008). This need to balance economic efficiency and equity in rural development has been a major driver of the spatial distribution of water use in the Krishna basin over the last 50 years.

Although vegetable and fruit production typically provides higher farm revenues, it tends to be capital intensive and a risky venture that is unfit for smallholders. In any case, this production only makes up 9% of the world's total cropping area and it cannot be expected to displace other grain, oil or fibre crops. Modernization of more extensive farms devoted to such crops is a problem experienced in many countries (including European countries such as Spain and Italy). It is clear that productivity gains cannot be satisfactorily achieved through negative incentives such as pricing but must come through subsidies to help farmers invest and intensify. Adoption of micro-irrigation, for example, is almost invariably made possible by generous public subsidies.

Basin governance

All the hydrological and socio-political complexities of river basin development and management discussed above must be addressed by relevant decision-making and governance structures. Although the establishment of RBOs has become a standard prescription, the diversity of physical and historical contexts militates for a less normative approach (Molle et al., 2007; Warner et al., 2008). However, the belief that a river basin agency should deal with all the water problems in a river basin is deeply rooted in the water sector. This reflects the modernist conviction that strong government agencies staffed by scientifically trained experts should be delegated responsibilities for policy design and implementation in natural resources management (Norgaard, 1994). For hydrocracies, the river basin forms an ideal territorial unit over which they can rule, based on the argument that nature has determined this to be the scale at which water should be managed.

Thus, a central element of river basin trajectories is the process of turning river basins into domains of water governance, a 'scale-making project' (Tsing, 2000) frequently pursued by hydrocracies. However, this process is hidden from view, as recourse is made to the 'naturalizing metaphor' of the river basin (Bakker, 1999). This leads to a neglect or denial of the political dimensions of river basin management, through the reification of 'natural' boundaries, the emphasis on 'neutral' planning and the search for optimal management strategies (Molle, 2006). Frequently, the situation before the creation of new river basin institutions is treated like a *tabula rasa*, while, in effect, many organizations and institutions and the technologies for controlling water are already in place (Warner et al., 2008). The chapters in this book show that the delineation of river basin boundaries, the structuring of stakeholder representation and the creation of institutional arrangements for river basin management are political processes revolving around matters of choice. An explicit recognition of the political dimension of river basin management is necessary so that institutions and procedures may be designed in a more democratic and inclusive manner.

International basins, multi-state basins in federal countries and national basins clearly appear as distinct cases. We focus here on the latter two. Federal countries exhibit a tension between the states overlapping within the basin and the central federal government. States tend to have a large autonomy in managing their water resources, but it is clear that the sum of uncoordinated state-centred interests is unlikely to lead to sustainable river basin management. The case of India shows that states pursue antagonistic expansion strategies that are poorly checked by the existing sharing agreement. Interstate regulation in the Krishna, Colorado and Lerma–Chapala basins is largely achieved through water-sharing agreements and through the management of the main infrastructures by federal agencies.

In Australia, salinity, and, more recently, environmental and drought-related problems, have triggered federal interventions. The institutional challenge is whether a more active and dominant role by central government will deliver arrangements that are better than existing ones. Although the Murray River Basin Commission has been credited with a successful mediation role, negotiated and voluntary water sharing and custodianship of the basin have been slow to react in front of pressing needs and environmental degradation. 'The belief of Federal government is that it has the intellectual horsepower, political muscle and financial resources to succeed where it (and others) believes that the Murray River Basin Commission has failed. This is probably a belief that is common to many central government elites, and their immediate technocracies, and often leads to impatience with detail and the preservation of considerable secrecy and minimal transparency' (Chapter 12).

In the Olifants basin, attempts at establishing a catchment management agency (CMA) have been stalled. Officials initially had high hopes for CMAs as 'the key vehicles to implement the new water management paradigm' (Schreiner et al., 2002), but underestimated the requirements to make the initial consultation process genuinely inclusive, given the highly unlevel playing field, with the large public and private water users well organized to defend their interests (Wester et al., 2003). Similar difficulties had been faced by the Olifants River Forum, established in 1993 to promote cooperation for conservation and sustainable use of the river. The forum was founded by white representatives of large mining firms, Kruger National Park and the Department of Water Affairs and Forestry in order to influence the formation of the planned CMA, with local communities not well represented, and signalled a continuation of the 'white water economy' (van Koppen, 2007).

In the Lerma–Chapala basin, a river basin council was formed in the 1990s, initially only with government representatives, and later also with water-user representatives. However, this council had very few decision-making powers, and was not delegated the authority to approve the budgets of the federal water agency's river basin office. Although proposals to move to a bimodal form of river basin management have been debated since 1992, they have been successfully resisted by the federal water agency during the various revisions of the national water law. While more space has been created for the participation of water users and state governments in river basin management, the federal government remains in control.

In many cases, participatory policies are initiated by government agencies with the implicit intent to keep control of river basin management. The Lerma–Chapala case, however, shows that such processes also create a political space that stakeholders can use to challenge the dominant power of the state. This has not yet happened to a significant degree in the Olifants and Ruaha basins, but could change with time.

The Yellow River Conservancy Commission is another type of RBO where central power seems to be overriding. The Esfahan Water Agency is also an example of centralized water administration that concentrates decisional power. Likewise, little direct representation of users in decision making is observed in the Jordan, Krishna or Merguellil basins. The resilience of civil-engineering-dominated water bureaucracies is clearly one of the main obstacles to change in these water sectors. Their water resources governance structure and policies remain characterized by centralization, hierarchy, specialization in infrastructural planning and secretive, top-down decision making.

As mentioned earlier, with regard to shifting paradigms, ideologies and societal values, water management is – or should be – in a constant flux to accommodate these changes. The Murray–Darling basin provides a good example of where water management is constantly evolving and adapting to changing needs, biophysical influence and public expectation.

Conclusions

The chapters in this book illustrate the diversity of both the water challenges that societies face and their responses to these challenges in varied physical and historical contexts. Although crucial water issues include flood management, urban water supply and sanitation, and pollution control, the dominant process is that of basin closure, whereby available water resources are invariably gradually tapped and depleted beyond the level required to ensure the sustainability of aquatic ecosystems and minimize the conflicts caused by supply variability. With river basin closure the interdependencies among stakeholders, the water cycle, aquatic ecosystems and institutional arrangements increase. These interdependencies manifest themselves in alterations of the water cycle that create positive and negative externalities to different categories of users and the environment. These externalities are not always easy to foresee or quantify and often result in amplified turbulence and greater complexity in terms of water governance mechanisms.

Despite the diversity of contexts presented by the case studies, four generic processes can be singled out. First, the process of overbuilding, which directly fuels the closure of basins, reveals a number of societal and political mechanisms by which the development of water-use capacity and infrastructure tends to outstrip resources and thus to generate 'scarcity'. Second, this overcommitment of resources also affects systems of allocation, whether formal – through a system of rights – or otherwise, which signals that it is politically always easier to downplay hydrological realities by overallocating one 'pie' than by excluding some constituencies (or nature) from accessing it. Third, pressure over resources translates into the 'overtapping' of both superficial (lakes and dams) and underground (aquifers) reservoirs. Fourth, basin closure makes the issue of water allocation critical, and a 'double squeeze' of agriculture is widely observed: the share of agriculture is under pressure from both growing non-agricultural needs and a widening awareness of, and call for, a need to increase environmental flows, since nature, the residual user, bears the brunt of variability in supply.

Indeed, the lack of possibilities to develop new water supplies, and the perception that agriculture is a 'low-value' use of water, lead to increasing intersectoral water transfers: one-way (frequently extra-legal) transfers from agriculture to industry and domestic use, as well as intrasectoral transfers in agriculture to economically higher-value crops and from small farmers to large commercial farmers. Most governments face the need to reconcile the antagonistic objectives of privileging economic efficiency and supporting the livelihoods of the poorest. Plans to transform subsistence farmers into market-oriented producers make light of issues of risk, marketing, and access to capital, labour and information.

The overexploitation of water sources leads to environmental degradation through the destruction of aquatic ecosystems, the depletion of aquifers and the generation of polluted wastewater flows (both industrial/urban effluents and agricultural drainage effluents). In closed river basins, these trends can principally be reversed by consuming less water and making judicious use of wastewater; but creating new 'hydraulic property' (Coward, 1986), even where only marginal and costly solutions remain available (distant dams, interbasin transfers, desalination), is often preferred and, in many cases, pursued in parallel with demand-management options.

Response options are diverse and always in competition. This book clearly shows how politically contested decision making is, both with regard to the selection of these options in general, and to water allocation in particular. The era of water resources development was characterized by a consensus on the desirability of the hydraulic mission, by the need to 'make the desert bloom', and the problems it dealt with could be classified as 'tame', i.e. amenable to solution by construction of hydraulic infrastructure and injection of technology and expertise (Lach et al., 2005). Many problems can now be characterized as 'wicked', with a multiplicity of viewpoints, interests and uses that demand new governance mechanisms. Conventional water bureaucracies or RBOs, which were instrumental in (over)building river basins, need to change their operating paradigms to be able to deal with basin closure.

The chapters in this book show that the cognitive, social and political complexities in closed basins are such that no easy-to-implement blueprints are available to resolve wicked water resources management problems. They take us through very rich and instructive stories that make explicit the deeply political and contentious nature of river basin management, and the need to start from this recognition as a necessary first step for working towards a socially and environmentally just governance of water resources.

References

Aboites, L. (1998) *El Agua de la Nación. Una Historia Política de México (1888–1946)*. Centro de Investigaciones y Estudios Superiores en Antropologia Social, Mexico City.

ACC/ISGWR (United Nations Administrative Coordination Council Inter-Secretariat Group on Water Resources) (1992) The Dublin Statement and Report of the Conference. Prepared for the International Conference on Water and the Environment: Development Issues for the 21st Century, 26–31 January 1992, Dublin.

Allan, J.A. (2006) IWRM: the new sanctioned discourse? In: Mollinga, P.P., Dixit, A. and Athukorala, K. (eds) *Integrated Water Resources Management: Global Theory, Emerging Practice and Local Needs*. SAGE Publications, New Delhi, pp. 38–63.

Alves, J. (ed.) (2008) *Toda a Verdade Sobre a Transposição do Rio São Francisco*. Mauad X, São Paulo, Brazil.

Bakker, K. (1999) The politics of hydropower: developing the Mekong. *Political Geography* 18(2), 209–232.

Barrow, C.J. (1998) River basin development planning and management: a critical review. *World Development* 26(1), 171–186.

Berkoff, J. (2002) Economic valuation: why is it often unsatisfactory? And does it matter? With reference to the irrigation sector. Paper presented at the ICEA meeting, 19 June 2002.

Boelens, R. and Zwarteveen, M. (2005) Prices and politics in Andean water reforms. *Development and Change* 36(4), 735–758.

Chambers, R. (1997) *Whose Reality Counts? Putting the First Last*. Intermediate Technology Publications, London.

Cortez-Lara, A.A. (2004) Enfoques encontrados en la gestión de recursos hidráulicos compartidos. El revestimiento del Canal Todo Americano y el Valle de Mexicali: ¿equilibrio estatico de mercado ó equilibrio de Nash? In: Sánchez Munguía, V. (ed.) *El Revestimiento del Canal Todo Americano: Competencia o Cooperación por el Agua en la Frontera México-Estados Unidos*. Playa y Valdez, Mexico City, pp. 273–293.

Cortez-Lara, A.A. and García-Acevedo, M.R. (2000) The lining of the All-American Canal: the forgotten voices. *Natural Resources Journal* 40(2), 261–279.

Coward, W.E. Jr (1986) State and locality in Asian irrigation development: the property factor. In: Nobe, K.C. and Sampath, R.K. (eds) *Irrigation Management in Developing Countries: Current Issues and Approaches*. Westview Press, Boulder, Colorado and London, pp. 491–508.

Ekbladh, D. (2002) 'Mr. TVA': grass-roots development, David Lilienthal, and the rise and fall of the Tennessee Valley Authority as a symbol for U.S. overseas development, 1933–1973. *Diplomatic History* 26(3), 335–374.

EU (European Union) (2000) Directive 2000/60/ec of the European Parliament and of the Council of 23 October 2000. Establishing a framework for community action in the field of water policy. *Official Journal of the European Communities* 43(L327), 1–72.

Forsyth, T. and Walker, A. (2008) *Forest Guardians, Forest Destroyers: the Politics of Environmental Knowledge in Northern Thailand*. University of Washington Press, Seattle.

Green, C.H. (1996) Water and economics: what does experience teach us so far? In: Howsam, P. and Carter, R. (eds) *Water Policy: Allocation and Management in Practice*. E & FN Spon, London, pp. 213–220.

Grigg, N.S. (2008) Integrated water resources management: balancing views and improving practice. *Water International* 33(3), 279–292.

Gulati, A., Meinzen-Dick, R. and Raju, K.V. (2005) *Institutional Reforms in Indian Irrigation*. Sage, New Delhi.

GWP (Global Water Partnership) (2000) *Integrated Water Resources Management*. TAC Background Papers No. 4. GWP, Stockholm.

Homer-Dixon, T.F. (1999) *Environment, Scarcity and Violence*. Princeton University Press, Princeton and Oxford.

IUCN (International Union for Conservation of Nature) (2000) *Vision for Water and Nature: a World Strategy for Conservation and Sustainable Management of Water Resources in the 21st Century*. International Union for Conservation of Nature, Gland, Switzerland.

Kendy, E., Molden, D.J., Steenhuis, T.S., Liu, C.M. and Wang, J. (2003) *Policies Drain the North China Plain: Agricultural Policy and Groundwater Depletion in Luancheng County, 1949–2000*. IWMI Research Report 71. International Water Management Institute, Colombo, Sri Lanka.

Lach, D., Rayner, S. and Ingram, H. (2005) Taming the waters: strategies to domesticate the wicked problems of water resource management. *International Journal of Water* 3(1), 1–17.

Landy, F. (2008) *Feeding India. The Spatial Parameters of Food Grain Policies*. Manohar, New Delhi.

Lilienthal, D.E. (1944) *TVA: Democracy on the March*. Harper and Brothers Publishers, New York and London.

Lipchin, C. (2003) Water, agriculture and Zionism: exploring the interface between policy and ideology. Paper presented at the 3rd conference of the International Water History Association, December 2003, Alexandria, Egypt.

Marchand, M. and Toornstra, F.H. (1986) *Ecological Guidelines for River Basin Development*. Centrum voor Milieukunde Rijksuniversiteit Leiden, Leiden, The Netherlands.

McCool, D. (1994) *Command of the Waters: Iron Triangles, Federal Water Development and Indian Water*. The University of Arizona Press, Tuscon.

Mehta, L. (2001) The manufacture of popular perceptions of scarcity: dams and water-related narratives in Gujarat, India. *World Development* 29(12), 2025–2041.

Molden, D., Sakthivadivel, R., Samad, M. and Burton, M. (2005) Phases of river basin development: the need for adaptive institutions. In: Svendsen, M. (ed.) *Irrigation and River Basin Management: Options for Governance and Institutions*. CAB International, Wallingford, UK, pp. 19–29.

Molle, F. (2003) *Development Trajectories of River Basins: a Conceptual Framework*. IWMI Research Report 72. International Water Management Institute, Colombo, Sri Lanka.

Molle, F. (2006) *Planning and Managing Water Resources at the River-basin Level: Emergence and Evolution of a Concept.* Comprehensive Assessment of Water Management in Agriculture Research Report 16. International Water Management Institute, Colombo, Sri Lanka.

Molle, F. (2007) Scales and power in river basin management: the Chao Phraya River in Thailand. *Geographical Journal* 173(4), 358–373.

Molle, F. (2008) Why enough is never enough: the social determinants of river basin closure. *International Journal of Water Resources Development* 24(2), 217–226.

Molle, F. and Floch, P. (2008) Megaprojects and social and environmental changes: the case of the Thai 'Water Grid'. *Ambio* 37(3), 199–204.

Molle, F. and Turral, H. (2004) Demand management in a basin perspective: is the potential for water saving overestimated? Paper prepared for the International Water Demand Management Conference, June 2004, Dead Sea, Jordan. Minister of Water and Irrigation, Jordan/United States Agency for International Development, Washington, DC.

Molle, F., Wester, P. and Hirsch, P. (2007) River basin development and management. In: Molden, D. (ed.) *Water for Food, Water for Life: a Comprehensive Assessment of Water Management in Agriculture.* Earthscan, London; International Water Management Institute, Colombo, Sri Lanka, pp. 585–625.

Norgaard, R.B. (1994) *Development Betrayed: the End of Progress and a Coevolutionary Revisioning of the Future.* Routledge, London.

Perry, C. (2007) Efficient irrigation; inefficient communication; flawed recommendations. *Irrigation and Drainage* 56(4), 367–378.

Reisner, M. (1993) *Cadillac Desert: the American West and its Disappearing Water.* (Revised and updated). Penguin Books, New York.

Schreiner, B., van Koppen, B. and Khumbane, T. (2002) From bucket to basin: a new water management paradigm for poverty eradication and gender equity. In: Turton, A.R. and Henwood, R. (eds) *Hydropolitics in the Developing World: a Southern African Perspective.* Africa Water Issues Research Unit, Centre for International Political Studies, University of Pretoria, Pretoria.

Seckler, D. (1996) *The New Era of Water Resources Management: from 'Dry' to 'Wet' Water Savings.* International Irrigation Management Institute (IIMI) Research Report 1. International Water Management Institute, Colombo, Sri Lanka.

Sexton, R. (1990) *Perspectives on the Middle East Water Crisis: Analysing Water Scarcity Problems in Jordan and Israel.* ODI/IIMI Irrigation Management Network Paper 90/3f. Overseas Development Institute, London/International Irrigation Management Institute (now International Water Management Institute), Colombo, Sri Lanka.

Shah, T., Roy, A.D., Qureshi, A.S. and Wang, J. (2003) Sustaining Asia's groundwater boom: an overview of issues and evidence. *Natural Resources Forum* 27(2), 130–141.

Tsing, A. (2000) The global situation. *Cultural Anthropology* 15(3), 327–360.

van Halsema, G. (2002) Trial and re-trial: the evolution of irrigation modernisation in NWFP, Pakistan. PhD dissertation, Wageningen University, Wageningen, The Netherlands.

van Koppen, B. (2007) Institutional and legal lessons for redressing inequities from the past: the case of the Olifants Water Management Area, South Africa. Paper presented at the HELP Southern Symposium. CD HELP Southern Symposium. Help in Action. Local solutions to global water problems. Johannesburg, 4–9 November 2007.

Venot, J.P., Turral, H., Samad, M. and Molle, F. (2007) *Shifting Waterscapes: Explaining Basin Closure in the Lower Krishna Basin, South India.* IWMI Research Report 121. International Water Management Institute, Colombo, Sri Lanka.

Vincent, L. (2004) Science, technology and agency in the development of droughtprone areas: a cognitive history of drought and scarcity. PhD dissertation, The Open University, Milton Keynes, UK.

Walker, A. (2003) Agricultural transformation and the politics of hydrology in northern Thailand. *Development and Change* 34(5), 941–964.

Warner, J. (2008) The politics of flood insecurity: framing contested river management projects. PhD dissertation, Wageningen University, Wageningen, The Netherlands.

Warner, J., Wester, P. and Bolding, A. (2008) Going with the flow: river basins as the natural units for water management? *Water Policy* 10(S2), 121–138.

Wester, P. (2008) Shedding the waters: institutional change and water control in the Lerma–Chapala basin, Mexico. PhD dissertation, Wageningen University, Wageningen, The Netherlands.

Wester, P., Merrey, D.J. and de Lange, M. (2003) Boundaries of consent: stakeholder representation in river basin management in Mexico and South Africa. *World Development* 31(5), 797–812.

White, G.F. (1957) A perspective of river basin development. *Law and Contemporary Problems* 22(2), 156–187.

Willcocks, W. (1901) *The Nile Reservoir Dam at Assuân and After.* E&FN Spon, London.

Worster, D. (1985) *Rivers of Empire: Water, Aridity, and the Growth of the American West.* Oxford University Press, Oxford.

2 Squeezed Dry: the Historical Trajectory of the Lower Jordan River Basin

Mauro Van Aken,[1]* François Molle[2]** and Jean-Philippe Venot[3]***

[1]*University of Milan-Bicocca, Italy;* [2]*Centre IRD, Montpellier, France;*
[3]*International Water Management Institute, Accra, Ghana;*
e-mails: *mauro.vanaken@unimib.it; **francois.molle@ird.fr; ***j.venot@ggiar.org

Introduction

The lower Jordan River basin (LJRB) provides a fascinating tale of coupled social and environmental transformations of a waterscape. In this semi-arid to desert area, water is an essential determinant of life, cultural values, social structures, economic activities, power and politics. The trajectory of this basin from a nomadic agro-pastoral Bedouin culture to an urbanized region where water circulation is highly artificial, illustrates how a particular resource endowment is valued, mobilized, shared, used and fought for.

This chapter first recounts past water resource development in the LJRB – defined as the Jordanian part of the Jordan River basin, downstream of Lake Tiberius – and dwells on the specific relationships between water, local culture and national/regional politics. The historical evolution of supply and demand is then expressed in terms of water balances that quantify the degree of closure of the basin.[1] Water challenges and response options are then addressed through the lens of the distribution of the benefits and costs they entail, and of their linkages with the current distribution of decision making and political power. Basin closure induces increased interconnectedness between water users and ecosystems through an increasingly manipulated water cycle: response options are interdependent and reveal the political and contested nature of resource sharing and water management (Molle *et al.*, 2007). This chapter describes how these processes, constrained by the drastic natural conditions of the basin, have unfolded since the late 1950s and explores possible futures.

Features of the Lower Jordan River Basin

The Jordan River is an international river which drains a total area of about 18,000 km². Its three headwater tributaries originate in Lebanon and Syria and flow into Lake Tiberius, a freshwater reservoir now used almost exclusively by Israel (Fig. 2.1). The Jordan River then flows southward before discharging into the Dead Sea.

Ten kilometres downstream of Lake Tiberius, the lower Jordan River receives water from its main tributary, the Yarmouk River, which originates in Syria. The Zarqa River and several temporary streams of lesser importance, named side-wadis, come from the two mountainous banks and feed the lower Jordan River (Fig. 2.1). Prior to water development projects, the original flow of the Jordan River into the Dead Sea varied between 1100 and 1400 Mm³/year (El-Nasser, 1998; Klein, 1998; Al-Weshah, 2000).

© CAB International 2009. *River Basin Trajectories: Societies, Environments and Development*
(eds F. Molle and P. Wester)

Fig. 2.1. The lower Jordan River basin in Jordan.

This chapter focuses on the LJRB and does not dwell on the geopolitical issues related to water sharing between the riparian states of the Jordan River (Lebanon, Syria, Israel, Jordan). The Yarmouk River and the upper Jordan are thus considered as contributing inflow to this basin. Moreover, the other streams draining to the Dead Sea from the south and from Israel are also not analysed.

The LJRB represents 40% of the entire Jordan River basin but only 7.8% of the Jordanian territory (cf. Fig. 2.1). The basin so defined is nevertheless the wettest area in Jordan, is home to 83% of the population, supplies 80% of the national water resources, and encompasses most irrigated areas. The basin, like the country, is divided into two main areas (see Fig. 2.2):

- The Jordan valley is a 110 km stretch between the Yarmouk River in the north and the Dead Sea in the south. Its altitude varies from 200 m (in the north) to 400 m below sea level (in the south). The valley can be considered as a natural greenhouse, with moderate temperatures during winter and high records during summer, commonly exceeding 45 °C. Rainfall ranges from 350 mm/year in the north to 50 mm/year near the Dead Sea (Fig. 2.3). The Jordan River flows in a 30–60 m deep gorge through a 0.2–2 km wide fertile alluvial plain, locally called *Al Zhor* (Fig. 2.2). The rest of the valley, *Al Ghor*, is a 4–20 km wide area with deep and fertile colluviums.
- The highlands comprise a mountain range running alongside the Jordan valley (named

Uplands hereafter) and a *badia* (desert plateau) extending eastwards to Syria and Iraq (Fig. 2.2). About 30 km wide, with an altitude reaching 1000 m above sea level, these mountains receive around 400–600 mm of rain per year, while snowfall can also be observed during winter (Fig. 2.3). Historically, they were covered with forests (essentially composed of Mediterranean conifers), but are now mostly composed of rangelands with olive trees and stone-fruit trees.

The eastern plateau has an average altitude of 600 m, and rainfed cereals are grown near the mountains, in the area where rainfall is still sufficient and where main urban agglomerations (Amman, Irbid, Al-Baq'ah, Jerash, Ajloun) are concentrated. Eastward, precipitation becomes scarcer (between 200 and 300 mm/year), and only nomadic livestock farming and some groundwater-irrigated farms can be found.

Total precipitation in the LJRB is estimated at 2235 Mm^3. In crude terms, 88% of this precipitation is directly evaporated (40% of this evaporation being beneficial, i.e. consumed by irrigated and rainfed crops or domestic and industrial uses), 5% flows into the rivers, and the remaining 7% infiltrates to recharge the aquifers (and is then pumped to meet human demands).

The flow at Lake Tiberius, which averaged 605 Mm^3/year before the 1950s (Klein,

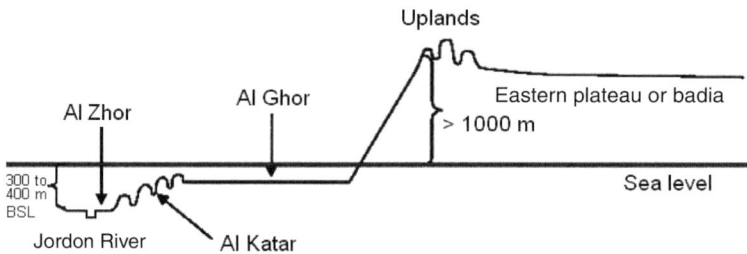

Fig. 2.2. Topography of the lower Jordan River basin in Jordan.

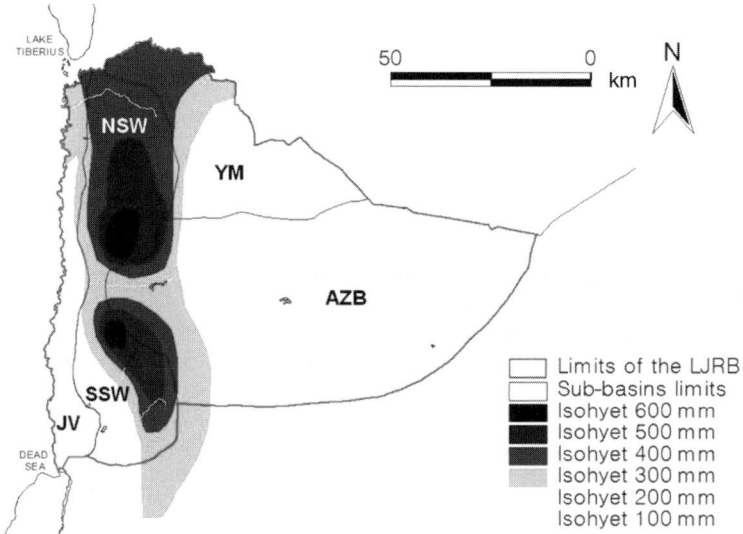

Fig. 2.3. Average rainfall distribution in the lower Jordan River basin in Jordan.[2]

1998), is now diverted by Israel to its National Water Carrier. The Yarmouk River is thus the main source of surface water: its flow averaged 470 Mm^3/year in the 1950s (Salameh and Bannayan, 1993), while the side-wadis and the Zarqa River originally contributed 120 and 90 Mm^3/year, respectively (Baker and Harza, 1955).

The main aquifers closely dovetail with the five main sub-basins (Fig. 2.3): the Yarmouk basin (YM), which drains northward to the Yarmouk River; the Zarqa basin (AZB), which drains most of the *badia* towards the valley; the northern and southern side-wadi basins (NSW and SSW, respectively), which pool lateral wadis north and south of the Zarqa River; and the Jordan valley (JV) itself. Annual recharge of the aquifers is estimated at 155–160 Mm^3/year (THKJ, 2004).

A Chronology of Water Resources Development

Ancient settlements and early land development

The lower Jordan River basin is at the heart of historical transformations in the Middle East, due to its central position 'as a land bridge for animals and humans between Africa and Eurasia; a Levantine corridor, a transit route for large and small migrant groups but also an area pinned between powerful states: Egypt to one side, Northern Syria/Mesopotamia to the other' (van der Koij and Ibrahim, 1990: 14).

Large settlements like Ain Ghazal (near today's Amman) are associated with the Neolithic period (c.8000–6000 BC). In this period, plants and animals (sheep, goats, cattle, pigs) were domesticated. Rainfed farming of wheat, barley and legumes expanded later, in the fourth millennium BC, to lentils, bitter vetch, sesame, olives, flax, dates and grapes. Rock basins and pools collecting natural water were utilized for storage for domestic and agricultural uses (Lancaster, 1999).

Later, 900–300 BC was a flourishing period for the Arabic kingdoms and a peaceful time in the LJRB. The first urban settlements were established in this era. The Nabataeans moved from the Arabian Peninsula into southern Jordan, where they established themselves in the eastern steppe and, with the help of ingenious hydraulic infrastructures, were able to farm the land at Petra while maintaining important trading activities. After the conquest of the region by the Romans, the economy came to rely on a flourishing irrigated agriculture, trade and Christian pilgrimages (Lancaster, 1999).

Through ups and downs, the region witnessed the Islamic conquest, the Ummayads, the Abbasids, the Crusades, the Ayyubid–Mamluk era (1187–1516) and the Ottoman conquest in 1516. The Jordan valley reached the peak of its agricultural development during the first period of the Mamluks (14–15th century). Irrigation developed wherever possible, and sugar mills, powered by water, were built in many spots in the valley. The Ottoman administration period, in contrast, was characterized by instability and depopulation in both the valley and the highlands. In 1956, the population of the east bank of the Jordan River (Transjordan) was estimated at 52,000 (Abujaber, 1988).

In Western travel accounts of the 19th century, the Jordan valley appears as a wild and dangerous place (with the threat of malaria and the fear of attack and robbery by Bedouin tribes) but, at the same time, as a biblical region with impressive, exotic scenery. The valley was a large grazing ground and an important region intersecting the tribal land of several Bedouin tribes. Up to World War II, surface irrigation was practised along the wadi valleys and, most prominently, at the point where wadis formed alluvial fans in the Jordan valley (Lancaster, 1999; Suleiman, 2004). Management was communal, under the authority of the tribes' sheikh, but coexisted with forms of private ownership of land and even of collective ownership of spring water and well water (Shryock, 1997).

The first planning interventions: 1921–1973

Transjordan was placed under temporary British administration (*Mandate*) in 1921 and became fully independent in 1946, as the Hashemite Kingdom of Jordan (THKJ). The British initiated cadastral registration of land titles and fiscal surveys from 1929 onward, demarcating village boundaries, state domains and forests in agricultural lands. The mandate period allowed Zionist projects, also based on irrigation schemes, to expand on the east bank of the Jordan River (the East Bank), making this region a security area (Goichon, 1967). Moneylenders and merchant families increased their investments in agriculture and their ownership of land, forming the basis for later capital investments in agriculture, in parallel with the decreasing power of the Bedouin tribes.

As one of the regions with the highest potential for agricultural expansion, the Jordan valley has been the object of numerous hydraulic and agricultural feasibility studies since the end of the 19th century. For foreign experts the valley was a symbol of high productivity wasted for lack of attention, which thus required urgent external intervention, prompting Merril (1881: 139) to declare that 'The American farmer would look with envious eyes upon the fertile portions of this valley.' Projects were fuelled by technical optimism and by a new ideology of irrigation as a transfer of resources and expertise from outside that would solve the problems of a local population depicted as 'conservative, ignorant, wretchedly poor, unable to contend with the forces of nature' (Gottman, 1937: 556).

In 1948, following the creation of Israel, 774,000 Palestinians were displaced (UN, 1949), of whom 70,000–110,000 escaped directly to the East Bank, which at the time had an indigenous population of about 440,000 (Brand, 1995). Refugee displacement in 1948 added to the urgency of developing irrigation: the resettlement programme in the Jordan valley was highly influenced by USAID and the International Bank for Reconstruction and Development (later the World Bank) and inspired by the 'integrated development' scheme of the Tennessee Valley Authority (TVA) in the USA, the icon of large-scale hydraulic planning projects (Molle, 2006). This model of development clashed with previous British foreign policies in the Middle East focused on the development of agricultural cooperation. British interventions favoured small-scale projects built on local indigenous expertise that could bypass the regional political gridlock regarding the use of water resources in the Jordan basin (Kingston, 1996). By the mid-1950s the Jordanian Division of Irrigation had completed dams on several of the eastern wadis draining to the Jordan River (except the southernmost wadi Shu'ayb and the larger Zarqa River), boosting irrigation in the Jordan valley (Kingston, 1996).

Already, in the 1930s, the first wells were dug in the highlands and water was pumped from Azraq (an oasis in the desert located about 150 km east of Amman, outside the LJRB) (Lancaster, 1999). Significant exploitation of groundwater started in the 1950s and 1960s, with the introduction of diesel motor pumps. Several international organizations (UNESCO, FAO, ILO) launched or promoted sedentarization programmes for Bedouin tribes, which included plans for developing irrigated agriculture and settlements, viewed as an essential step to economic integration, 'modernization, stability in the region and control of rangeland (the *badia*)' (Bocco, 2006). The area irrigated with groundwater gradually increased and was multiplied fourfold between 1965 and 1980. Government licences and soft loans for drilling private wells led to a frontier 'moving ever eastward into an increasingly ecologically and economically marginal environment' (Millington *et al.*, 1999).

The introduction of tractors, water tanks and water pumps during the 1950s induced crucial changes in water management. The Jordan valley was home to various systems of irrigation (wadis, diversions, canals, reservoirs, springs, pumps) and multiple actors were involved (such as the United Nations Relief and Works Agency for Palestinian refugees (UNRWA), the Jordanian state, British consultants, the World Bank, USAID). The construction of the East Ghor canal, which was to distribute water diverted from the Yarmouk all along the East Bank, started in 1957, but was halted several times due to warfare. The first 69 km were completed in 1966. Between June and September 1967, 395,000 Palestinians crossed the Jordan River, due to the occupation of their land by Israel. Israel occupied the West Bank, and the Jordan valley became for some years a battleground between Palestinian fighters and Israel and, in 1970, between Palestinian fighters and the Jordanian army. The extension of the canal to the south resumed after 1971. Irrigated agriculture developed on a large scale (13,500 ha) through the East Ghor concrete canal (later renamed King Abdullah canal, or KAC), in parallel with a land reform (1962), and several projects of urbanization and settlements (Courcier et al., 2005).

The development phase: 1973–1995

In 1977, the Ministry of Water and Irrigation (MWI) published a global assessment of water resources in Jordan (THKJ, 1977). A first harsh reality was the dramatic loss of the upper Jordan water to Israel: the inflow to the LJRB had decreased from 605 to 70 Mm^3/year (Klein, 1998). Because of the combined water uses in Israel, Syria and Jordan, only 40% (505 Mm^3/year) of the historical flow of the Jordan River still reached the Dead Sea in 1975 (Courcier et al., 2005).

The exploitation of water resources further increased between 1975 and 1995. In the Jordan valley, irrigated agriculture was expanded through the construction of several hydraulic facilities: extension of the KAC (with 3400 ha of land newly irrigated), installation of pressurized water distribution networks, storage dams on the Zarqa River and other side-wadis. In the early 2000s, past investments in the water sector in Jordan, mainly financed by international aid, were estimated to total US$1500 million (Nachbaur, 2004; Suleiman, 2004). With new techniques of production (greenhouses, drip irrigation, plastic mulch, fertilizer, new varieties, etc.), the availability of Egyptian force and market opportunities (at least until the first Gulf War), irrigated agriculture in the Jordan valley enjoyed a boom in production and economic profitability, described by Elmusa (1994) as the 'Super Green Revolution'. The particular climate of the Jordan valley allows many small entrepreneurial farmers to produce vegetables almost all year round (and especially during winter), as well as some fruits that can withstand heat in summer (citrus and bananas).

In the highlands, private wells provided unlimited access to good-quality groundwater. Wealthy and dynamic entrepreneurs (of both Transjordanian and Palestinian origin), emulating or replacing past Bedouin or peasant (*fellahin*) settlements, made massive investments and developed an irrigated agriculture which supplied Jordan and the Gulf countries with fruits and vegetables during summer.

During the same 1975–2000 period, the urban population within the basin was multiplied by roughly 2.5 (DoS, 1978–2003), with urban groundwater use consequently growing fivefold to reach 150 Mm^3/year (records of the MWI–Water Resources Department). This demand was met by both increasing the number of wells in the surroundings of the cities and transferring more groundwater from distant areas and surface water from the KAC to urban areas in the highlands (Darmane, 2004). This latter transfer, initiated at the end of the 1980s, was expanded after the massive inflow of Jordanian-Palestinians returning to Jordan after the first Gulf War (1991) and is now the main source of water for Amman (almost 100 Mm^3/year by 2008). This transfer was made possible because of the concomitant treatment of wastewater from Amman: effluents are collected in the King Talal reservoir (built between 1971 and 1977 on the Zarqa River) and mixed with freshwater, and this blended water is then used to irrigate the middle and the south of the Jordan valley.

Further reduction in the water coming from the Yarmouk and reaching the LJRB was observed after the late 1970s. During the 1980s, water use doubled in Syria, with 35 middle-size dams built in the upper Yarmouk basin and direct pumping from rivers and wells (El-Nasser, 1998). In the early 2000s, the Yarmouk contributed 270 Mm^3/year to the Jordan River (THKJ, 2004), of which about 110 Mm^3/year flowed uncontrolled to the lower Jordan River until the recent completion of the Wehdah dam (2007). The peace treaty signed between Jordan and Israel in 1994 also specified that the 25 Mm^3 pumped each winter by Israel from the Yarmouk would be returned to the KAC during the year, an agreement loosely implemented so far and which does not consider issues of water quality. With all these changes, the inflow to the Dead Sea was reduced to less than 20% of the historical flow of the Jordan River, resulting in a drop of its water level by 20 m since the late 1950s, showing a dramatic degradation of the environment of the entire Jordan River system (Orthofer et al., 2007) and threatening the local tourist industry.

1995 onward: the rise of the water challenge in Jordan

In the 1990s, water rose to the top of the nation's political agenda. Concerns shifted from refugees in the 1950s, towards land management in the 1970s, and finally water in the 1990s. In 1997, the Jordanian government adopted a new Water Strategy Policy (THKJ, MWI, 1997), setting allocation priorities to the urban sector, then to the industrial and tourist sector, and finally to the agriculture sector: policy reforms aimed at meeting the challenges faced by the country.

Physical scarcity of water resources is an obvious challenge, compounded by rapid population growth. The rapid increase in water needs is due to an improvement in living standards and to a high demographic growth of 2.9%, notably in urban areas (nearly 80% of the population is concentrated in cities) (DoS, 2003). Migration, in particular the sudden waves of Palestinian refugees in 1948 and 1967, has had a major impact on water use in the country. So did the wave of around 300,000 people of Palestinian origin who had to return to Jordan from Kuwait after the Gulf War of 1990–1991, 95% of whom resettled in the LJRB area (de Bel-Air, 2002). The recent migration of Iraqis escaping from the embargo and the war – estimated at 1,300,000 – is now a major challenge for the country.

Groundwater overuse causes degradation of the groundwater resources, both in the short term (direct pollution due to infiltration of pesticides and fertilizers: see JICA, 2004) and in the long term (salinization of groundwater due to a drop of water tables: ARD and USAID, 2001; Chebaane et al., 2004). Overabstraction has also led to the drying of springs and, in particular, to the disappearance of the Azraq oasis, a Ramsar wetland. The measures taken to abate groundwater use for agriculture from private wells in the highlands have been unsuccessful. Abstraction limits have never been respected and too many licences have been issued. The Groundwater Control Bylaw No. 85, passed in 2002 and further amended in 2004, was designed to regulate groundwater abstraction through the establishment of a quota of 150,000 m^3 per year per well and a block tariff system for any use beyond that quota. However, this quota is much higher than the limits mentioned in the original well licences. It was reported that farmer interest groups obtained the cancelling of the former lower limits against the acceptance of the principle of taxing volumes abstracted above a higher limit (Pitman, 2004). Upper (optimistic) estimates of the reduction in gross water abstraction due to the bylaw point to a potential decrease of 4%, i.e. 5.5 Mm^3/year, a drop in an ocean of overabstraction and quite short of the 40–50 Mm^3 hoped for (Venot and Molle, 2008).

Water management challenges in the basin are linked not only to the expansion of Amman but also to the process of suburbanization of the countryside around the capital, near Irbid and in the *badia* (Lavergne, 1996). Farms have become secondary residences, new villas have increased land fragmentation, and the habitat in the highlands countryside gradually resembles that of Amman. A similar dynamic can be observed in the Jordan valley, where fenced fruit orchards often hide a villa – and sometimes a swimming pool – used at weekends and where the value of prestige and status is higher than the economic productivity of the farm itself.

Urban development and the lack of untapped resources have led to a policy of transferring increasing volumes of freshwater from irrigated agriculture to urban uses, thus affecting the stability of the agriculture sector. During dry years, 2000–2002 for example, the Jordanian government froze the quantity of water reserved for cities, while drastically reducing the amount allocated to agriculture in the Jordan valley. This reallocation from the Jordan valley to the highlands has been partly compensated for by an ever-increasing supply of treated wastewater (TWW) to the south of the valley (McCornick et al., 2002; THKJ, MWI, WAJ, 2004). The hazards associated with a generalized use of TWW in agriculture remain poorly known and include workers' and consumers' contamination, soil degradation, clogging up of irrigation system emitters, disappearance of certain sensitive crops (strawberries, beans, citrus, etc.), consumers' lack of confidence in the quality of the products, drop in prices and loss of some export markets (Grattan, 2001; McCornick et al., 2002). In the Jordan valley, there is also growing evidence of water pollution by nitrates and soil degradation (Orthofer, 2001).

Water conservation and the quest for greater end-use efficiency have also spurred several policies and measures in the urban and irrigation sectors. These include, for example, modernization and physical improvement of urban distribution networks, reduction of the volumes of water unaccounted for, and transfer of Amman's water supply and wastewater collection to a private company (Darmane, 2004).[3] In the Jordan valley, measures include completing the conversion from the earlier gravity network to pressurized systems, incentives to adopt micro-irrigation at the plot level, reduction of per hectare quotas, and increases in the cost of water to farmers. At a collective level, a German cooperation programme (GTZ) supported efforts at building up the first water-user associations in the Jordan valley.

Most of these policies have met with limited success. In the Jordan valley, quotas are low and farmers use their full allowance in all conditions: technical interventions improve irrigation efficiency not because water use is reduced but because better uniformity and timing of water application enhance crop evapotranspiration and yields (Molle et al., 2008). Agricultural water prices in the Jordan valley have been raised several times but with negligible impact on water demand (World Bank, 2003; Molle et al., 2008), especially for high-value fruits and vegetables. If prices were further raised they would substantially dent the net revenue of citrus and banana farmers and encourage/force them to reconsider the benefits, risks and constraints of adopting new crops and technologies. The poorest vegetable farmers would be bankrupt, at the risk of high social and political consequences. Quotas appear to be the only straightforward measure for reducing diversions. The 1997–1999 period was marked by a severe drought, which forced reductions in allocation, which were extended from 1999 to 2003, although adjusted each year, and made permanent in 2004. At a regional scale, this generated total freshwater savings of about 20.2 Mm^3/year, reallocated to domestic use in Amman.

Another important and sensitive issue is the operation and maintenance (O&M) costs of infrastructure. Until now, for both urban and irrigation supply, emphasis has been placed on obtaining international funding for implementing modern systems rather than on O&M recurring costs. Degradation and fiscal austerity call for better coverage of these costs. The increase in water tariffs in the Jordan valley has allowed two-thirds of O&M costs to be recovered, and studies show that full O&M cost recovery is achievable and commensurate with farmers' income (Molle et al., 2008).

The future of irrigated agriculture raises a complex set of social, economic and political questions that largely lie outside of the water sector itself. The two major issues are the treatment of prestige agriculture and the question of economic sustainability. Irrigated agriculture in the highlands has mainly developed during the last three decades through large private investments: the investors concerned belong to high society (MPs, senators, entrepreneurs, sheikhs, etc.). Their social importance and their influence on government decisions suggest that all the measures aiming at reducing their water use will be conflict prone and will take a long time to implement. While part of this agriculture is highly capital intensive and profitable, around 30% of irrigation water is used in low-value olive-tree farms. These orchards are a legacy of a time when the drilling of wells was subsidized, and are held for reasons of prestige, as a means of keeping ownership and control of land. Likewise, many citrus plantations in the Jordan valley are held by absentee owners (often urbanites) who are not interested in complex farm management, prefer low-return, extensive agriculture, and partly transform their farms into leisure places. Banana farms in the north of the valley are also linked to politically powerful tribes and partly thrive on higher water quotas and import barriers.

More generally, agriculture is facing declining profitability. Marketing constitutes the main problem that agricultural producers face (ASAL, 1994; World Bank, 1999). Jordanian irrigated agriculture mainly developed during a period (1975–1990) of strong regional demand for fresh products. Products could be sold at a high price because of the payment capacity of the Gulf countries and limited competition in the region. At that time, investments in agriculture (greenhouses, irrigation systems, wells, equipment, etc.) provided a handsome return within a few years and attracted many investors. After 1985, the quick development of production in Jordan and in the region (Syria, Lebanon, Gulf countries) led to a drop in prices and in the profitability of investments (Nachbaur, 2004). Moreover, the first Gulf war of 1991 worsened this situation, since the Gulf markets, which constituted a major outlet for Jordanian products, were lost (Jabarin, 2001) as a result of the Jordanian state's support of the invasion of Kuwait by Iraq. In addition, Jordan has favoured the development of new economic sectors (tourism, services, industry) and signed several agreements[4] which could undermine the profitability of certain agricultural products still protected in Jordan (e.g. bananas and apples). This situation could also reverberate on the country's trade balance. Fruits and vegetables and their export represent, on average, 12% of the value of Jordanian exports (THKJ, MoA, 2001), and any reduction in the production would raise macro-economic concerns.

Negative impacts would be passed on to the more vulnerable rural groups, notably low-income Jordanian categories (refugees, Jordanian tribes of low status, female labourers) and male migrants (two-thirds of whom come from Egypt).

Water, People and Politics

These socio-economic and technical aspects of water use in the country are closely linked to cultural and regional politics and to the changing relationships between Jordanian society and water.

The social fabric and changing perceptions of water

In the past, land and water were controlled by the *ashira* (tribe) represented by the *sheikh* (tribal leader) and were linked to the notion of *dirah*, which played a central role in resource management. The term dirah derives from *dar*, which literally means 'house', which may be a cement construction as much as a tent, and refers to the tribal territory, together with a system of exchange organized around the *khuwa* (the payment to tribes to obtain their protection). Access to resources was allowed to other tribes depending on demographic pressure, climatic conditions, resource scarcity and existing alliances. Thus, the border and the geographical extension of a dirah were often flexible but within perceptions of land that still persist nowadays (Bocco, 1987). This notion of territory is thus interlinked with indigenous ideas of resource property. As Lancaster (1999) showed, 'ownership comes through access, use, action and is validated by defence and reputation.' In fact, the notion of *ihya almawat* (vivification of land) through ameliorations and work, and not the ownership of land by itself, granted the rights and the control of land. This also applied to water resources since the ownership of water was a 'function of claims and access to resources, rather than a system of control and absolute right of disposal' (Lancaster, 1999).

The disruption of this tribal resource management in the last century and the displacement of Palestinians have transformed the units of belonging. Development institutions took charge of functions and responsibilities that were previously exercised by the tribe, such as management of land and water. Local agricultural and irrigation knowledge was displaced from the extended family and tribe to the experts and the administration. New international borders have severed pastoral routes, and the cement blockhouse has replaced the goat-wool tent. Bedouins have shifted from pastoralism to army employment, irrigated agriculture, outmigration, commercial and transport activities, or development administration.

Tribal political organization, forced displacement of Palestinians, labour migration, warfare in the Jordan valley, and intensive planning have shaped today's social structure and water projects. In this dynamic context, the developmental concept of a 'farmer community' has worked to unify, in one common category, communities who perceive themselves as diverse and are characterized by social heterogeneity and a diversified economy. The external representation of the heterogeneous population in the Jordan valley as a homogenous group of 'Jordanian farmers' within irrigation projects has definitely depoliticized a tense region and has been part of a wider process of incorporation into the nation. Through water development, new 'farmer settlements' have been set up. *Muzar'e* (farmer) is a category which can be understood only within the context of the new irrigated agribusiness developed in the LJRB: it refers to an occupational category within the new economic segmentation and differs from the traditional *fellahin* (peasant), who is understood as a wider moral and political belonging.

Modernization did not by itself mean the disappearance of tribal solidarity – on the contrary, it has readapted to new political and ecological environments. Bedouin values and tribal belonging have been mobilized and reproduced as part of a process of nation building within a demographic context marked by a large population of Palestinian origin (which fuelled a separate sense of belonging – inherent in the Palestinian national struggle), a contested border in the Jordan valley, and the Hashemite Kingdom's need for legitimacy in a tense environment. Tribal solidarity has often overlapped with the national administrative structure, shaping the new bureaucratic apparatus and national identity, and playing a central role as a form of patronage and a basis for affiliation (Brand, 1995). Therefore, 'tribal identity has become politicized as it continues to be the basic channel for allocation of resources by the central government' (Shami, 1982).

Other actors have been less visible. Since 1970, Egyptian migrants, together with minor groups of Syrian men and Pakistani families, have provided cheap labour to labour-intensive agricultural systems. Jordanian male wage labour in agriculture has nearly disappeared in recent decades, since men seek employment with better wages and less drudgery outside agriculture. This is counterbalanced by a feminization of agricultural labour, although notwithstanding the large number of women working as labourers today, the responsibility for irrigation has remained in the hands of men since it symbolizes the control over the wider production process (Shami and Taminiam, 1990).

The building of the state and regional geopolitics

In the past, periods of development and stagnation have often been related to the presence or absence of a strong authority that could offer security and protection to the Jordan valley (Khouri, 1981) and could allow a growing population to thrive. Periods of intense settlement have often been followed by periods with a sparse population, abandonment of agricultural settlements and insecurity.

In the Ottoman period (1516–1921) the state tried to establish firm control by introducing new concepts of land and water, supporting the immigration of agricultural settlers from the Caucasus, the setting up of a bureaucratic apparatus, the emergence of merchant elites in the agriculture sector (many from Syria or from Circassian, Chechen and Turcoman communities) and an increase in the export of agricultural products. Yet the communal patterns of resource management remained effective.

In 1933, the Land Settlement Law promulgated by the British administration opened the way to cadastral registration of land titles and to fiscal surveys. Agricultural development through irrigation projects for Bedouins was viewed by the British as the first step of a wider detribalization process that would help stabilize the country and settle 'new farmers to a neglected land' (Lowdermilk, 1944), as this area was erroneously portrayed, an essential step to economic integration, social emancipation and stability within nation construction.

At first instrumental in settling pastoral groups, water development would soon (after 1948) be aimed at resettling refugees. The development of water on the east bank of the Jordan valley in the 1950s led to the establishment of a new power structure and engendered a water bureaucracy, the Jordan Valley Authority (JVA). The tribal hierarchical system of distribution gave way to centralized planning of water, and high subsidies for irrigated water became a political tool, allowing state penetration in a crucial and unsettled rural area. Project implementation was characterized by a lack of participation and involvement of the local population, and the JVA introduced a new system of loyalty, through a centralized administration.

As de Bel-Air (2002) has shown, the state–citizen relationship has been intimately linked to the rentier nature of the Jordanian economy, based on an indirect rent (external aid from the Gulf States or the USA, remittances from migrants, etc.) and reproducing a clientelistic pattern of redistribution of resources, in which water is embedded. The structure of power has thus been linked to this redistribution of rents and to patronage: in this context, the economic value of agriculture, often criticized nowadays due to its large use of precious water and limited profitability, cannot be detached from the political and social meanings that agricultural development has acquired in Jordan, in terms of political stability, tribal and refugee settlement, and national incorporation of rural and arid areas.

Waterscape transformations and the new social environment

Today, irrigation in the LJRB has become an arena of struggle among different interest groups. In the Jordan valley, for example, the struggle between the extensive bureaucracy and its computerized distribution system, on the one hand, and the illegal methods reinvented daily by farmers to get access to water, on the other, express different and contrasting 'projects' with regard to water management. Centralized water management and the shift from surface irrigation to micro-irrigation and pressurized pipelines date back to the 1980s and have, in a very short time, radically changed the ways of thinking about and using water. This context of change is at the core of today's struggles and conflicts that arise around water at the local level. We face a situation of legal pluralism, with an overlapping of formal and informal water property rights systems and claims, in a context characterized by a lack of local participation and increasing water scarcity.

A first major change has been the transition from a water allocation based on the household head and tribal representatives to an allocation based on land use, controlled by a water administration vested with a new authority. This technical and bureaucratic presence has led to a wider process of secularization and materialization of, and disenchantment with, water (Hamlin, 2000): detached from the tribal community, water has become a technical affair, often artificially separated from its attendant social and cultural dimensions.

A second main consequence is the changed idea of water in relation to time. The water share was traditionally connected to an idea of a personalized and socialized time related to specific local ecological contexts and connected to the lineage system, where every part of the larger community received its time-share, which could be exchanged and adapted along social and neighbourhood relationships. With the establishment of a central bureaucracy, a new characterization of water in terms of quantity (cubic metres) and pressure of supply has been introduced.

Third, with state irrigation, a new idea of territory in relation to local communities has taken place. With drip irrigation and the introduction of pressurized collective networks during the 1990s, water has also gone underground and is not directly available or physically visible. Besides, water users in the LJRB are more and more hydraulically interconnected within pressurized systems covering larger areas than was the case when water flowed through traditional earthen canals.

Last, expert knowledge has become more important today than before in managing water since it has introduced a specific culture of organization and is linked to a resource management regime whereby, as Waller (1994) put it, 'water managers use their expertise to portray them [i.e. the changes] as technical rather than political decisions.' This is a major issue in the Jordan valley, where water management is described as an issue for experts, too complex for farmers to handle. This technical emphasis has both depoliticized the actual decisions made in relation to water and neutralized wider public debate on water issues in Jordan.

The Lower Jordan River Basin at the Crossroads

Embedded in a society in the making, with Bedouins, peasants, Palestinian refugees and foreign powers fused in a chaotic regional setting, water is a guiding element of historical transformations. Its importance in the future may even increase, as Jordan's scarce water resources cannot keep up with needs and immigration. This section first dwells on historical changes in land use and in the components of the water balance, giving a detailed account of the water flows in 2000, before expanding on the different solutions at hand.

Basin closure and the water squeeze

Changes in land use

The first notable evolution since the late 1950s is that of land use. Rainfed cropping areas significantly increased in the 1950–1975 period, with cereals providing work and food to a growing population (Fig. 2.4). This extensive type of agriculture later declined, with a shift in the economy towards non-agricultural activities. Irrigated olive-tree orchards in the highlands dramatically increased, from 420 ha in 1950 to 11,000 ha in 2000, i.e. making up close to half of the irrigated areas in the highlands (the other half consisting of vegetables and stone-fruit trees) (Courcier *et al.*, 2005). Figure 2.4 highlights the structural differences between the Jordan valley and the highlands: cultivated areas are much larger in the highlands (a total area of 143,900 ha in 1950) than in the valley (32,300 ha), which reflects the large areas of rainfed cereals and olive trees planted in the former. Irrigated areas increased from around 10,200 ha in 1950 to 45,800 ha in 2000.

Water accounting in 2000

The description of the transformation of the LJRB given in the preceding sections can be paralleled by a more quantitative accounting of the resulting (im)balance between water supply and demand. The net inflow to the LJRB includes rainfall, interbasin transfers, and possible net overdraft of the aquifers and reservoirs. This total inflow is partly transformed through evapotranspiration of crops (irrigated, rainfed and also natural vegetation) and evaporation from water bodies, and through municipal and industrial (M&I) processes. The balance flows to the Dead Sea, considered as a sink since maintaining its level is not considered as a management objective.

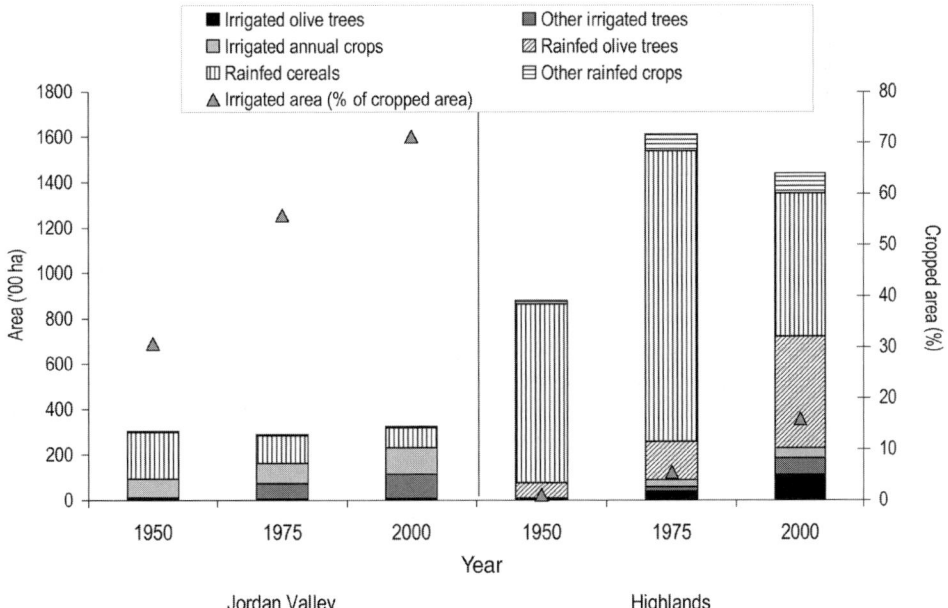

Fig. 2.4. Crop- and region-wise evolution of cropped areas in the lower Jordan River basin in Jordan since 1950.

Water is depleted, or consumed, by four generic processes: evaporation, flows to sinks (e.g. a saline aquifer), pollution and incorporation into a product (e.g. plant tissues) (Molden, 1997). In 2000, the LJRB consumed 86% of its net inflow through the above processes. About a third of the remaining outflow to the Dead Sea was coming from uncontrolled Yarmouk water (Courcier et al., 2005), which is now partly stored in the Wehdah dam. *Beneficial depletion* (evapotranspiration from irrigation, rainfed agriculture, and M&I uses) accounted for 33.5% of the net inflow, *low beneficial depletion* (evapotranspiration from natural vegetation and forest) for 14.5% and *non-beneficial depletion* (evaporation from bare land, deserts and water bodies) for the remaining 38% (2% of the net inflow was exported to other basins and 12% as runoff). In the LJRB, irrigation accounts for 18% of the total depleted fraction. Data also indicate that despite all the allocation conflicts between the cities and agriculture, the share of M&I depletion is negligible, representing only 3% of the total depleted fraction in the LJRB. This share, however, rises to 16% when compared with the amount of water depleted by irrigation, and to 10% when expressed in terms of withdrawals.

These basin-level figures prompt some remarks on the question of efficiency in water use. Groundwater-based irrigation efficiency in the highlands has increased in the last two decades, with an almost complete shift from surface water irrigation to micro-irrigation (Elmusa, 1994; THKJ, 2004). In many cases, farmers continued pumping the same amount of water and have expanded their irrigated area (Venot and Molle, 2008). This not only increased farmers' incomes but also resulted in higher evapotranspiration and lower return flow to the aquifer, thus compounding the net overdraft. There is evidence that percolation losses from irrigation in the highlands return to the aquifer (JICA, 2001) and therefore do not affect the net water balance significantly (although pumping costs and the low quality of return flows are issues). Areas irrigated by diversion of wadis along the main valleys also have high efficiencies because return flows are quickly reintegrated to the main stream. In the valley, the shift to micro-irrigation owes more to the intensification of agriculture than to water scarcity per se, since it started 15 years before talks of a water crisis emerged. Cultivation of vegetables under plastic mulch that controls weeds makes micro-irrigation necessary and also allows better application of water and nutrients (*fertigation*). Other more extensive crops (notably citrus) as well as part of the banana crop are still irrigated by gravity, but the defined JVA quotas keep application losses to a minimum since quotas are less than full crop requirements in months when the overall demand exceeds supply.

Water balances can also be expressed vis-à-vis the *controlled renewable blue water* (CRBW), i.e. the sum of surface water, aquifer recharge and imports from both distant aquifers and surface water, from which have been deducted the (few) resources which cannot be controlled and are of 'no use': a few flash floods exceeding the capacity of the dams[5] as well as brackish flows from Israel. The overall water use in the basin, considered as a system, has continuously increased, with depleted withdrawals accounting for 11% of the CRBW in 1950, for 37% in 1975, and for 87% in 2000 (Courcier et al., 2005).[6] The LJRB is a closed river basin, where little water is left to be mobilized and used. This sets a drastic limit to what can be achieved through conservation means (see next section).

Evolution of the terms of the water balance in the lower Jordan River basin

From the situation in the 1950s, when few of the surface water and groundwater resources were used, to the current situation of overexploitation, the terms of the water balance have obviously varied from one extreme to the other. The net inflow into the basin moved from over 3300 Mm^3/year in 1950 to around 2600 Mm^3/year in the following periods, because of upstream diversions by Israel and Syria. Deducting rainfall water directly evaporated from crops and bare soil, renewable blue water shows a similar drop by 50%, with a slump at 671 Mm^3 in 2000 (Fig. 2.5). The CRBW is significantly lower, since uncontrolled and/or brackish flows from the Yarmouk (now controlled by the Wehdah dam) or Israel are discounted. Strikingly, withdrawals (gross diversions of surface water plus abstracted groundwater minus interbasin transfers) now amount to 660 Mm³/year, or 121% of CRBW, because of groundwater overdraft and multiple diversions (return flows from wadi irrigation or from Amman are reused downstream). Annual withdrawals have continuously and dramatically increased in the last 50 years, from 101 Mm³ in 1950 (20% of the CRBW) to 316 Mm³ in 1975 (58% of the CRBW), and 660 Mm³ in 2000. In 2000, only 130 Mm³/year of controllable blue water made it to the Dead Sea (Courcier et al., 2005). Figure 2.5 also shows the evolution of the intended beneficial depletion (irrigation and M&I), which almost equated to CRBW in 2000. Overdraft of aquifers now reaches 32 Mm³/year (Courcier et al., 2005). Figure 2.6 shows the terms of water balances.

This water accounting pools together four different kinds of water sources – groundwater, surface water (controlled by dams), stream water (uncontrolled flows that are diverted) and efficient rainfall (used by irrigated and rainfed crops) – in a single category: water use (or withdrawals). These four categories of water sources are, however, not equivalent because the degree of control managers/users have over them varies highly (in decreasing order in the above list) (Molle, 2003). It is therefore instructive to disaggregate water use into these four categories and to plot these fractions against time. By so doing, and including projections for 2025,[7] we obtain a view of both their relative importance and time dynamics: Figure 2.7 first shows that (effective) rainfall on rainfed crops constitutes the major category of beneficial water, even in such arid conditions. It is also striking that groundwater abstraction in the LJRB now appears as a source of greater magnitude than (controlled) gross diversions of surface water (275 Mm³/year against 120 Mm³/year in 2000), although this will be reversed when the Red–Dead project is in operation (see later). Surface water follows the construction of the dams, while stream water includes side-wadis and Yarmouk diversions: stream water increases with the construction of the KAC (supplied by water diverted from the Yarmouk) but decreases as dam construction shifts water from the stream water category to the surface water category.

Sectoral water use

Sectoral water use has changed widely over the last 50 years (together with the projection for 2025). While total agricultural withdrawals have levelled off since the mid-1970s, M&I withdrawals reached 31% of total withdrawals in 2000 (M&I depletion amounts to 10% of total withdrawals) and are expected to hit 52%

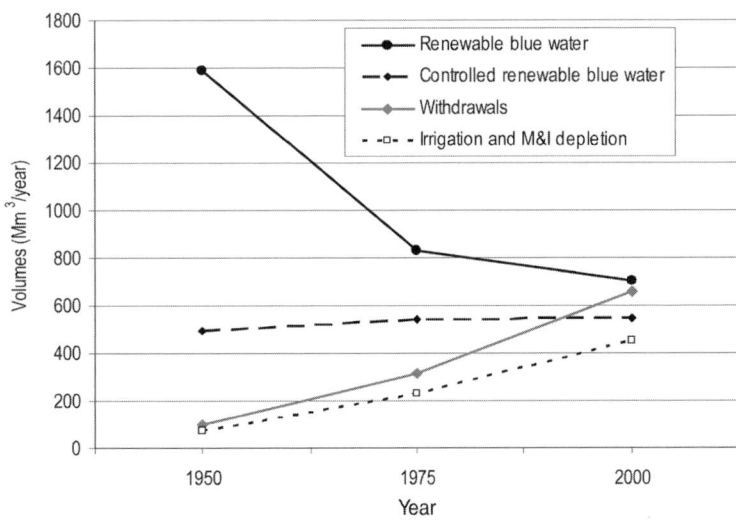

Fig. 2.5. Evolution of net inflow and available water in the Jordanian part of the lower Jordan River basin in Jordan. (M&I = municipal and industrial.)

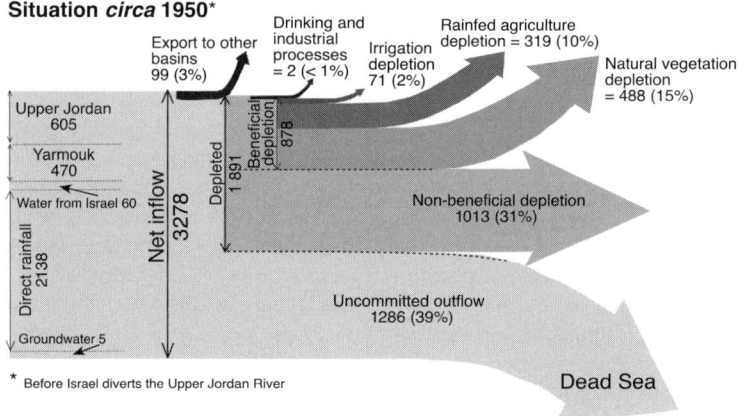

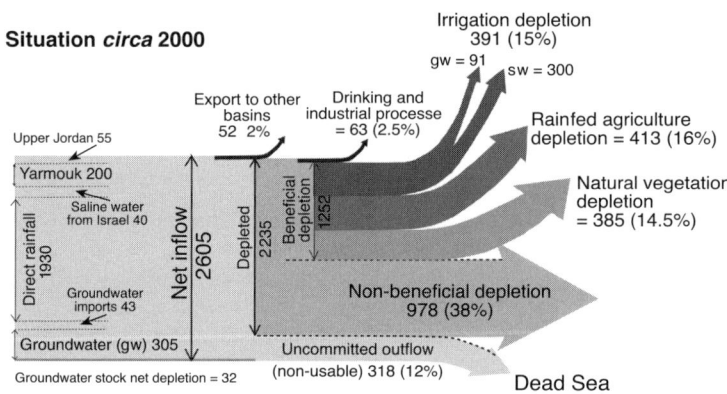

Fig. 2.6. Finger diagrams of water balances in the lower Jordan River basin in Jordan in 1950 and 2000.

in 2025 (Courcier et al., 2005). This evolution will reproduce that observed in Israel, where agricultural water use remains, by and large, stable, but increasingly relies on treated wastewater, while M&I uses benefit from increases in supply and eventually supersede agriculture. The share of groundwater in M&I is dominant but this situation will also be inverted with the supply of the Red–Dead project.

Water options and the distribution of benefits and costs

Faced with conditions of water scarcity, societies have three broad types of options at their disposal (Molle, 2003). First, they may increase the amount of water that is controllable for human use; this is the conventional supply augmentation option. Second, they may try to conserve water, either by reducing demand or by serving more users with the same amount of water abstracted. Third, they may keep the current level of withdrawals but reallocate water among uses and users. These three options, water resources development, conservation and allocation, can be resorted to at different scales, typically those of country (national policies), basin or local levels. When considering nested scales it becomes clear that these categories are not 'waterproof' (for example, mobilizing water locally by, say, small dams may be seen as reallocation at the basin level) and that the three options are resorted to concomitantly.

Which of the three options is selected depends on the respective costs and benefits

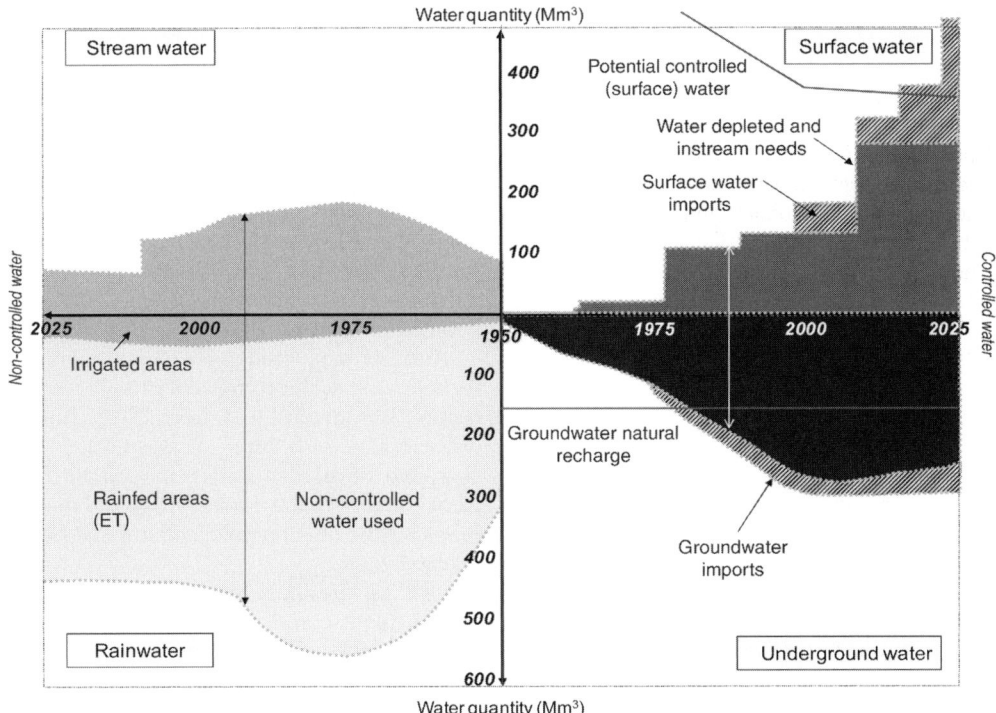

Fig. 2.7. Water withdrawal trends in the lower Jordan River basin in Jordan from 1950 to 2025 according to four water 'categories'.

attached to these options and on the social distribution of these costs and benefits among concerned parties (politicians, private companies, marginal groups, irrigators, cities, development banks, etc.) (see Molle et al., 2007 and Chapter 1, this volume). Costs are not only financial but also political, environmental, or expressed in terms of risk, health impact and benefits foregone. Likewise, benefits are not only monetary but often political, or expressed in terms of amenity and prestige, for example. It is the relationship between the distribution of decision-making power and the potential social distribution of the costs/benefits attached to each option that largely determines which actions are taken. This section examines the main options offered to the Jordanian society through this lens.

Supply augmentation

Most water sources have now been tapped and the costs of mobilizing additional water resources are ever increasing. These costs, which have until now been supported by the government and international aid (expensive dams, long-distance transfers, elevation costs, desalination, etc.; see GTZ, 1998; Nachbaur, 2004), may have to be increasingly borne by the population in the next decades.

The last reservoirs, which are likely to be built on side-wadis, are generally far from consumption centres, smaller and expensive. One of these, located on the wadi Mujib, which flows directly to the Dead Sea, has recently added a capacity of 35 Mm^3/year. The last main reservoir to have been built (after being delayed for several decades) was the Wehdah (unity) dam on the Yarmouk. It has a storage capacity of 110 Mm^3, for an annual inflow of 85 Mm^3/year (THKJ, 2004).[8] The consequence is a nearly complete disappearance of the lower Jordan River flow as well as that of lateral flows (Mujib dam) reaching the Dead Sea. The resources made available will be diverted mostly to cities.

In the situation of extreme scarcity characteristic of the country, large transfers have long been envisaged (transfer of fresh water from Lebanon, Iraq, Syria and even Turkey; transfer of sea water from the Mediterranean Sea to the Dead Sea (GTZ, 1998)); but these transfers have never been implemented because of the regional political instability and their very high costs in terms of investment and O&M. A large transfer from the fossil aquifer of Disi, located about 325 km south of Amman, has now been funded (at a purported cost of US$990 million), and finally entrusted to a Turkish company. The project is to provide 100 Mm^3/year, extracted by 65 wells at a depth of 500 m. According to the Water Minister, Raed Abu Soud, 'The capital [Amman] will get water from the aquifer for the coming 100 years' (Terra Daily, 2008b), while experts speak of 50 years (Terra Daily, 2008a). Uncertainty about the yield of the aquifer is paralleled by doubt about the fate of the existing irrigation based on the same aquifer (which consumes 80 Mm^3/year), as experts explain that the two uses are incompatible (IRIN, 2007). The issue is considered as 'sensitive' because 'there are tens of farms owned by former high-ranking officials with thousands of employees, the majority of them expatriates' and the government is still 'considering revoking licences for many farms in the area' (IRIN, 2007).

The Disi project is dwarfed by a US$5 billion plan to transfer seawater from the Red Sea to the Dead Sea over a distance of 180 km. A large transfer (1500 Mm^3/year in total) is planned in order to supply the main cities of Jordan, Palestine and Israel, but a first phase of 800–1000 Mm^3/year should bring to the Dead Sea a volume close to that historically contributed by the Jordan River. Seawater would be desalinated on the shores of the Dead Sea using the electricity generated by the natural difference in altitude (400 m) (Harza, 1998). The project has been alternately justified as the mother of all solutions, a means to restore the Dead Sea and its value to the three main monotheist religions, a means to counter environmental degradation and salvage the tourism industry, a solution to urban water shortages, and a way of fostering regional collaboration and contributing to the peace process in the region (the project is also known as the 'peace conduit'). The World Bank and several country donors are supporting a feasibility study of this multi-billion dollar project, launched in December 2006. This project bears all the characteristics of mega-projects: a relatively secretive planning and design, and an array of justifications that borrow from discourses on state building, national security and peace building, and is likely to face massive cost overruns (Flyvbjerg et al., 2003). Other local desalination projects of smaller scale are also being planned or implemented.

Another supply augmentation option is to reclaim wastewater to make it reusable in agriculture. It is forecast that, from 2025, Amman will produce 100 Mm^3 of wastewater each year. We have seen earlier that more TWW would be sent to the Jordan valley and that this raises a host of economic, cultural and health-related issues.

Last, a marginal increase in freshwater supply might come from the implementation of the 1994 Peace Treaty between Jordan and Israel: Israel is bound by the treaty to desalinate the 20 Mm^3 of saline water it now dumps each year into the Jordan valley, below Lake Tiberius, and will transfer half of this volume to Jordan.

Conservation

Water crises serve to increase scrutiny of the 'losses' occurring in man-made conveyance and distribution networks. Urban supply networks, notably those of Amman, have been targeted by several recent projects. Unaccounted-for water, which includes losses by leakage (and non-payment), was around 50% and is supposed to have been reduced to around 30% in Amman, after network rehabilitation and better management. A new US$250 million plan to rehabilitate old water networks is also underway (Terra Daily, 2008b). Additional measures have included public campaigns to raise awareness and encourage rationing in times of drought, and relative increases in prices. Recent announcements (April 2008) of further hikes have, however, caused havoc, showing the social sensitivity of price increases in a context of rising commodity prices.

Agriculture is also often designated as a wasteful user, and alleged low efficiencies of irrigation networks have also been targeted. In the highlands, as mentioned earlier, the various policies implemented have had limited impact, and improving irrigation technology, at best, reduces return flows to the aquifer (and therefore creates no net savings) or, at worst, leads farmers to capitalize on lower per hectare water requirements to expand cultivation (since land is not a constraint), thus increasing total water depletion and worsening the status of the aquifer. Raising prices to disqualify low-value olive trees will only lead to wells being sold to farmers engaged in a capital-intensive agriculture that is associated with higher depletion rates. Consequently, the only way to effectively curb abstraction is to buy wells back from farmers, and offer compensation for discontinuing licences, a measure considered positively by many farmers in the Amman–Zarqa basin (Chebaane et al., 2004). Reaching a sustainable level of groundwater exploitation would require discontinuing all agricultural groundwater abstractions. It seems unlikely, however, that from a role of producer–exporter of fresh products Jordan will become a net importer of these products. Moreover, as seen earlier, policies to reduce irrigated agriculture in the highlands are likely to face fierce opposition and to be delayed and only partially implemented, if at all.

The scope for conservation in the valley is somewhat larger, although equally limited. Part of the inefficiency comes from dysfunctional distribution at the level of the collective pump stations, and pilot projects have shown that redefinition of water turns could improve reliability, while other efforts have been devoted to building up water user associations (GTZ, 2001, 2002; Van Aken, 2004; MREA and JVA, 2006; San Filippo, 2006). Quotas for vegetables are already so low that it is hard to imagine any substantial gains in efficiency; some improvements can still be achieved in citrus and banana plantations, which enjoy larger quotas, and retrofitting of on-farm distribution networks has been found to be profitable because of the improved application of water and resulting gains in yield and quality of products. All these gains, however, are achieved with a full consumption of quotas. Decreasing demand through a hike in prices is not feasible: with two-thirds of O&M costs recovered in the valley it is hard to imagine price increases much beyond the O&M level, at which elasticity of demand is negligible. Here too, as shown by the rationing implemented in the past decade, effective cuts in water diversions were obtained by reducing quotas, not by price incentives (Molle et al., 2008).

(Re)allocation

Reallocation is the most delicate option and arguably the most politically stressful. Irrigated agriculture consumes two-thirds of the national water resources (THKJ, 2004) and competes with domestic and industrial uses, which have been officially declared a priority (THKJ, MWI, 1997). The competition with agriculture in the highlands is indirect, because water from most distant wells can hardly be transferred to cities, and in the long term, because the actual overdraft of the aquifers decreases the resources potentially available for future urban use, as well as their quality, and implies that more costly alternative resources will have to be tapped.

Water has been, and will be, reallocated out of agriculture in the valley, although the impact has been smoothed by the supply of TWW. An important aspect of this sectoral competition is the growing vulnerability of agriculture to climatic vagaries. As the overall residual water user, agriculture in the valley bears the brunt of the variability in supply (see Chapter 1). Compensation measures for fallowing land or in case of reduced supply (as in 2001) need to be considered in order to avoid financial and livelihood breakdowns. As seen earlier, further reallocation can be effected through reduction of quotas. In 2004, however, in contradiction to its policy to reduce demand, the JVA legalized citrus orchards planted illegally between 1991 and 2001, granting them the higher citrus allocation instead of the vegetable quota they received earlier. This illustrates the political clout of the Ghzawi tribe, well established in the northern part of the valley, and the way tribal solidarity and national policies may overlap or conflict with each other.[9]

Reallocation of water among farmers in the Jordan valley can be envisioned if, following the completion of the Wehdah dam, annual quotas can replace the actual monthly quotas;

the possibility of trading water would then enhance both irrigation and economic efficiency, but this would require a quite elaborate system of monitoring and computing of individual water diversions, with effective valley-wide mechanisms to move allocations from one user to another. Fine tuning of irrigation supply would do away with overirrigation in (the rare) times of excess supply but would substantially reduce leaching of salt. There are serious reservations and worries about whether this might also have an impact on soil salinization (McCornick et al., 2001).

Reallocation from low-value to high-value agriculture can also be obtained by incentives to farmers to either change their cropping patterns or lease/sell their farms to entrepreneurs. Higher water prices, or removal of tariff barriers, would, for example, decrease the profitability of marginal, subsidized and/or thirsty crops, including 'luxurious' olive trees in the highlands, and citrus or banana in the Jordan valley.[10] Two types of farmers are concerned, with two corresponding obstacles to policy implementation. Some of these crops are grown by (sometimes wealthy) absentee owners who are interested in prestige or leisure and not in agricultural returns, and who are therefore insensitive to price incentives. In addition, as mentioned earlier, these landowners are linked to influential tribes or to political elites and are likely to oppose such policies. A second group is composed of vulnerable farmers with little capital, labour or willingness to face the risk of intensifying their practices. High-value, profitable crops are already an option for them and there are good reasons/constraints why they have not opted for them earlier. Positive incentives that reduce capital and risk constraints, offering subsidies for improving irrigation technology, attractive cropping alternatives, and exit options with compensation should be implemented if prices are to be raised (Venot et al., 2007).

A political ecology of responses to water scarcity

While all these options are on hand and have been floated for a number of years, most have not been implemented or have met with limited success. The overall decision-making process is highly political and is based on a constant reassessment of the costs and benefits incurred by the different categories of actors and by the environment, in both the short and the long term. Financial and economic costs that form the heart of conventional cost–benefit analyses only capture part of the story. Political arbitration remains central to decision making.

This is not the place to make a detailed analysis of Jordanian society, but a few groups have already appeared in the preceding discussions: the royal entourage, the different Bedouin tribes, Bedouin and Palestinian entrepreneurs in the valley and the highlands, impoverished farmers, urban-based landowners, migrant labourers, the aid industry and national/transnational expert systems. These categories are interlinked by relationships of economic power, patronage and social stratification. The constant confrontations of interests, ideologies and power at the interface of these groups define which actions are taken or not. Complexity is added by the fact that these confrontations are not restricted to water decisions, but include other hot issues (land, economic liberalization, shrinkage of the state sector, agreement with WTO, the Palestinian question, etc.), which signal the embeddedness of water policy within the wider political arena.

Table 2.1 illustrates how the different options reviewed earlier translate into specific costs and benefits to particular actors.[11] Supply augmentation options tend to be the favoured solution of most quarters. They are attractive to development banks, politicians (works are visible political landmarks), water bureaucracies (professional legitimacy and sustained budgets) and the private sector (business opportunities) (Molle, 2008). Because they are capital intensive they also frequently open the way to corruption and private benefits. Costs tend to be shifted to weak or silent constituencies (typically the environment and the next generations) and shifted to the country as a whole (public investments), although the involvement of private investments means that consumers will share an unknown part of the burden. Treated wastewater is a compensation to farmers deprived of fresh water in the valley but entails hidden costs in terms of health hazards for producers and consumers.

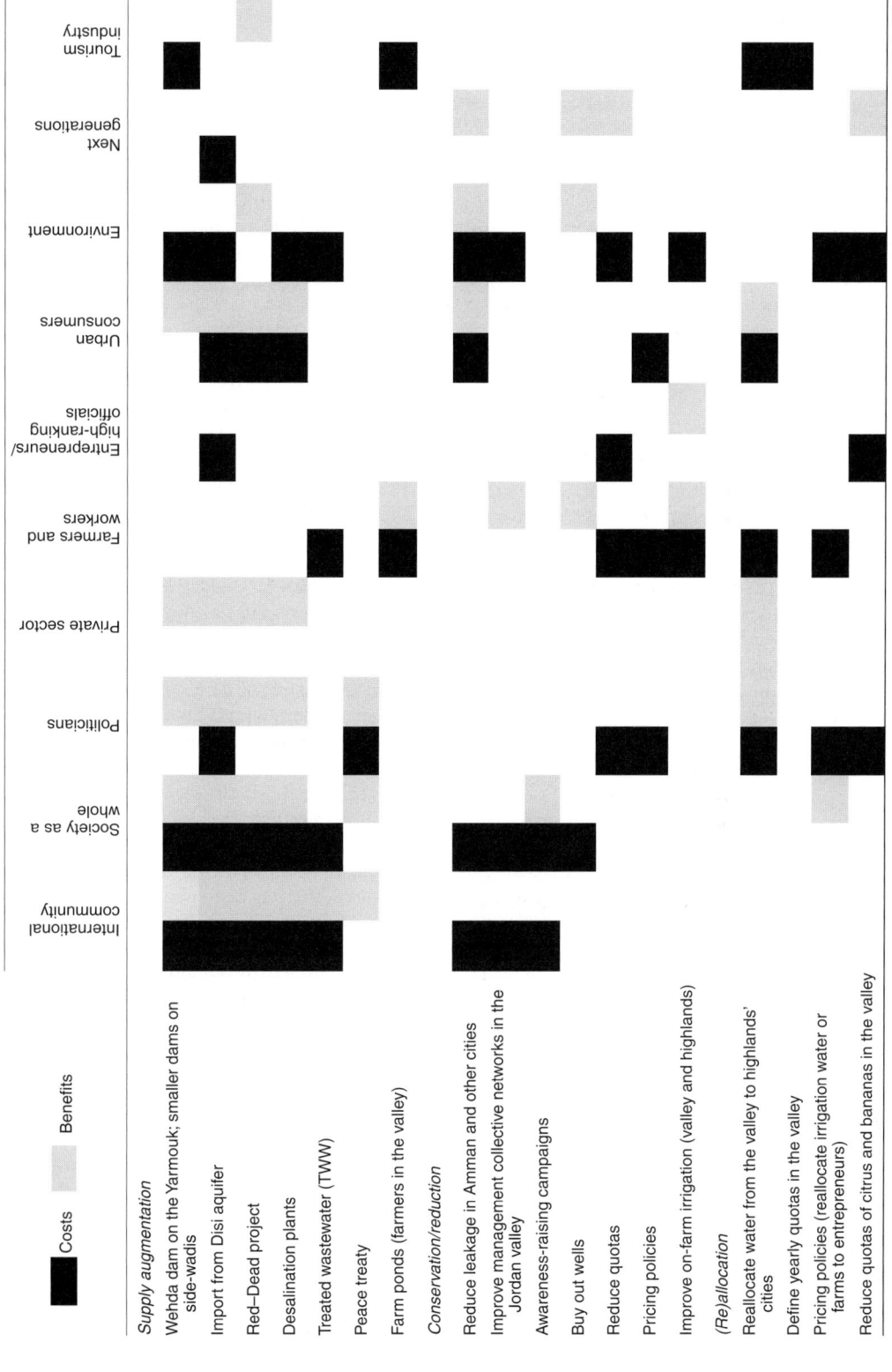

Table 2.1. Water sector reforms and interventions in Jordan: a multi-actor perspective on the distribution of costs and benefits.

The cost of conservation through technological improvement (in Amman's networks, collective or on-farm irrigation) largely depends on who shoulders the capital costs. For high-value crops, entrepreneurs pay for technology because its effect on produce quality, yield and labour makes it profitable. For less risky and capital-intensive crops, adoption of technology is not attractive and depends on government financial support and promotion by extension services. Negative incentives alone, through price increases or reduced quotas, are likely to spark opposition or social unrest and to give way to negotiations and weakening of the measures (as for the by-law on groundwater use).

In order to understand the contemporary political framework in relation to water it is useful to be reminded here of some events. In 1988, King Hussein declared Jordanian disengagement from the West Bank, an important political act towards the Jordanian population of Palestinian origin. In April 1989, riots exploded in the town of Ma'an, in southern Jordan, when subsidy reductions on certain basic items were announced in accordance with a debt-rescheduling agreement with the IMF. Riots and opposition also shook southern Jordan in the 1996 'bread riots', and later in 2003. It is feared that recent increases in the price of commodities will lead to further demonstrations (Al-Jazeera, 2008; LA Times, 2008). It is important to note that these demonstrations developed in areas dominated by tribes once highly loyal to the Hashemite regime but who felt marginalized in the redistribution of resources.

As Richards (1993) put it:

> In Jordan, all government decisions must be viewed through the lens of His Majesty, who must balance contentious internal and external forces. It is the calculus of the 'balancing act', not economic logic that determines all economic (and other) policies. The costs of offending important political actors, whether domestic or foreign, must be offset by tangible benefits.

Agriculture is viewed mainly as a source of patronage for key constituencies, whose support is essential to achieve domestic stability/foreign policy goals. Some landowners in the highlands and many farmers in the valley belong to influential Bedouin tribes that provide important support for the King, such as the Adwani tribe in the valley, whose members are well represented in the army and government bureaucracy. Maintaining their support, in particular against young urban Islamist radicals, is very important for the King, and sectoral or economic policies must therefore offer 'packages' in which compensations are extended to adversely affected constituencies (Richards, 1993). This provides hints on why many agricultural or pricing policies are watered down, circumvented or delayed. Relationships between the King and Palestinians are also important 'given the Palestinian private sector/Transjordanian public sector divide, and given the fact that economic liberalization targets a shrinkage in the state sector and an encouragement in the private sector, it is not surprising that Transjordanians felt threatened by the economic restructuring' (Brand, 1995). Past conflicts with Palestine – and the lack of a solution to the problem of refugees – still loom large.

Regional politics (not addressed in this chapter) also appear in several issues: the peace agreement between Israel and Jordan and the transfers of water attached to it, and also the lurking competition with Israel on whether the Red–Dead project will prevail, as opposed to alternatives to transfer water from the Mediterranean Sea, over which Israel would have full control. Because of the regional political situation and financial needs, the Red–Dead project will have to be facilitated by international aid or funding agencies, and bringing up environmental (save the Dead Sea), religious (the cradle of three religions) or political (the peace conduit) arguments may allow Jordan and Israel to shift parts of the costs to the 'international community.

Conclusion

This chapter illustrates the gradual anthropogenization and complexification of the lower Jordan River basin over a time-span of 60 years. It describes a striking transformation from the situation around 1950, when only 10,000 ha were irrigated, groundwater was untapped and abundant water flowed to the Dead Sea, to the current situation, when nearly all surface resources are diverted and committed and groundwater is being severely overexploited. This trajectory has revealed a drastic concomitant change of societies and waterscapes in an arid region subject to dramatic political tensions and socio-technical change.

Mobility of social groups has had, and continues to have, a major impact in framing the trajectory of the basin: the tradition of transhumance and nomadism of tribal pastoral groups, interconnected with agricultural settlements and fluxes of rural labour, the past slave trade, the two main shocks due to the forced migration of Palestinian refugees, the migration of workers from Pakistan and Egypt and the hundreds of thousands of Iraqis who recently found shelter in Jordan have been major drivers of the changes of the waterscape of the LJRB. In addition to migrations and displacement, mobility also refers to contacts with the 'outside', including fluxes of exogenous ideologies and institutional actors in the management of resources and flux of capital to a rentier economy (from remittances and the aid industry and also, more recently, from economies at war and from Gulf countries).

Water projects have constituted a main tool in the search for stability, both with regard to regional competition for this scarce resource and in terms of internal stability: an instrument to settle and 'root' nomadic populations and to depoliticize a tense context; a vehicle for building up bureaucracies, which would be pivotal in the distribution of resources and the development of patronage; a form of consensus building and modernization of the nation; a way to 'solidify' the border in a disputed frontier area; and a means to cement regional peace and obviate wars. All these have been determining elements in shaping patterns of water resources development and management and in defining new relationships between the state and citizens, between tribes and the state, and between farmers and engineers.

The waterscape of the LJRB, first occupied by Bedouins and small rural settlements, witnessed the emergence, or the occasional presence, of actors as diverse as Palestinian or Iraqi refugees, Pakistani or Egyptian workers, sheikhs from the Gulf region, peace negotiators, greenhouse entrepreneurs, irrigation bureaucrats, foreign development experts, researchers, international bankers, tourists, Islamic fundamentalists, urban absentee owners with swimming pools in their orchards, Bedouin farmers using desalination plants, and prestige olive-tree gardens watered in the middle of the desert. These actors have contributed to the peculiar trajectory of the LJRB.

Bedouin tribes who controlled natural resources, interlinked with peasant settlements, were the first to be targeted by irrigated settlement schemes (in both the highlands and the Jordan valley), construed as a basis for nation building: their incorporation into the state apparatus has been the counterbalance to the disruption of their pastoral economy and also the basis of the adaptation of tribal solidarity within the new political system.

Rural livelihoods have shifted from livestock, rainfed cereals and olive trees, with spots of seasonal irrigated farming, to an artificial, 'plastic' and intensified agriculture, partly linked to export markets and also to rentier strategies (irrigated olive trees in the highlands and some citrus orchards in the valley). Palestinian technical knowledge, foreign aid and immigration of foreign labourers (the often 'invisible' water users in agriculture, under the dependence of their patrons and managers) have been pivotal in agricultural development. Technological change, in particular micro-irrigation and pressurized networks, has made water users interdependent in a social context characterized by social/ethnic heterogeneity and by the fragmentation of previous social networks and forms of cooperation. Waterscapes have been reshaped from small springs and streams diverted to family gardens and communal patterns of distribution of land and water in the integrated agro-pastoral management to a centralized bureaucratic system with water pumps and pipes lifting water 1000 m up from the valley bottom and from distant aquifers to cities. Water is thus largely de-territorialized, since it has lost its ancient linkages with land and local communities. Both the valley and the highlands, on the one hand, and agricultural areas and cities, on the other, are thus interconnected and interdependent. This interdependence manifests itself in terms of competition (water quantity) and also more and more in terms of water quality.

Around 2000, 64% of surface runoff and groundwater annual recharge was depleted through irrigation and M&I depletion, and this percentage springs up to 83% if we disregard the uncontrolled flow of the Yarmouk to the Jordan valley. At present, the basin is closed, as most of the water is mobilized and depleted. Because of the reuse of water and of current groundwater overdraft, withdrawals amount to 121% of controllable blue water. Resulting environmental change has included depletion of aquifer systems, springs drying up in oases and salinization of groundwater, as well as the lowering of the level of Dead Sea by over 20 m. It is also important to note that the high percentages of controlled and depleted volumes are obtained even though we have considered the Dead Sea as a sink with no 'needs'. Environmental considerations have de facto been written off as a result of the diversion of the upper Jordan by Israel but are back on the agenda, as illustrated by the debate around the Red–Dead project.

A new cycle of pressure over water resources, continued concentration of power and water use in urban centres, and capital investments is being triggered by the recent inflow of Iraqi refugees. Among competing solutions, conservation offers a limited prospect: the Wehdah dam has brought controlled blue water resources to the level of 93% of the total renewable blue water, irrigation efficiency has been drastically improved through micro-irrigation, and percolation losses in highland agriculture largely return to the aquifer. Consequently, the scope for water savings at the local and basin levels is much reduced. Control of leakage in Amman and further efficiency gains in the valley are desirable, but they will not radically alter the facts that a ceiling has been reached and that demand-management options may only alleviate the actual situation without providing long-term solutions. Typical capital- and technology-intensive supply augmentation projects, namely large-scale interbasin transfers (Disi, Red–Dead) and desalination, may therefore be the sign not only of a lasting dominance of the engineering approach but also of the exhaustion of resources in the face of a new boom in population.

Yet the permanence of the use of scarce resources in low-value agriculture (olive trees and citrus), in subsidized thirsty crops (bananas) or by rich private entrepreneurs (Disi's fossil water) constitutes an economic 'anomaly' that makes water import or desalination projects look suspicious, since the cost of water will be much higher than its opportunity cost in these agricultural activities. Political objectives and constraints, as is often the case, override economic considerations, and agriculture keeps a role in buying loyalty from some Bedouin tribes and rewarding high-level officials. Threats to vested interests inherent in demand-management measures raise the political costs of these policy options. The gradual intensification of agriculture towards a capital- and knowledge-intensive activity also has implications for weaker segments of the population in terms of social stratification, access to land and distribution of benefits, and stresses the importance of compensations and of the availability of alternative activities for those who are pushed to give up agriculture.

As a closing basin, the LJRB is characterized by an increasing interconnectedness of uses and users through a hydrological cycle reshaped by human technology. But technology allows a reversal of gravity and water to be pumped from the valley to the highlands, in a manifestation of the sectoral competition over water, and of the economic and political power of urban users. In agriculture, entrepreneurs, family farms and rentiers also compete for water, in both the valley and the highlands, with their respective strategies and assets – financial, political or otherwise. The weaker and the most downstream 'user', the Dead Sea, is the ultimate loser. These intricate influences of social, economic and political factors in the shaping of the LJRB's future trajectory illustrate the inherent and strong coupling of the evolution of both societies and waterscapes.

Acknowledgments

The authors would like to thank the French Regional Mission for Water and Agriculture (MREA) for its support over the last few years. Thanks are also extended to our partners at the Jordan Valley Authority and in the Ministry of Water and Irrigation.

Notes

1 The water-accounting exercise presented here draws on the categories of water balance proposed by Molden (1997). For more details on data sources refer to Courcier et al., 2005 and Van Aken et al., 2007.
2 Average rainfall distribution is adapted from EXACT (1998).
3 Recently, the process of privatization has been slowed down. For example, after 2006 a public company (Mihayuna) replaced the private one in charge of the management of Amman's water utilities. At the same time, several projected privatizations have been delayed (notably one concerning the privatization of the Jordan Valley Authority, the public agency in charge of water management in the Jordan valley).
4 The WTO, the Jordan–EU Agreement, the Great Arab Free Trade agreement establishing a free trade area between the Arab states, and several bilateral agreements, notably with the USA and Israel.
5 At the time of our accounting, flash floods included 110 Mm^3 from the Yarmouk River, which could not be stored and flowed to the Dead Sea. Those are now (2008) captured by the recent Wehdah dam, constructed upstream of the intake of the KAC.
6 The CRBW amounted to 493, 543 and 545 Mm^3/year in 1950, 1975 and 2000, respectively. With the completion of the Wehdah dam CRBW is 655 Mm^3/year.
7 These include: the recent Wehdah dam on the Yarmouk River, increasing water imports from outer basins, the construction of several desalination plants, the extension of irrigation (with treated wastewater) in the south of the Jordan valley, the reduction of agricultural groundwater abstraction in the highlands, and the first transfers of desalinated water through the Red Sea–Dead Sea project (100 Mm^3/year; against a provision of 570 Mm^3/year for Jordan at completion of the project (Harza, 1998)).
8 In 2007 and 2008, inflows in the Wehdah dam were, however, much lower than expected (this could be due to increased water use in the upper Yarmouk basin).
9 As illustrated by recurring conflicts between members of the Adwani tribe in the southern Ghor and high-ranking officials of the JVA.
10 Alternatives include different types of vegetables and fruits, such as date palm in the valley, which are low water consuming, relatively salt resistant and highly profitable but are capital and management intensive, risky and require good control of marketing.
11 These categories of actors are, of course, simplifications. In reality none of them is homogeneous but consideration of inner diversity is beyond the scope of this work.

References

Abujaber, R. (1988) *Pioneers over Jordan: the frontier of settlement in Transjordan, 1850–1914*. Tauris, London.
Al-Jazeera (2008) Jordan 'set to face more riots' 15 April 2008. http://english.aljazeera.net/news/middleeast/2008/04/200861423382668811 6.html
Al-Weshah, R.A. (2000) Optimal use of irrigation water in the Jordan Vvalley: a case study. *Water Resources Management* 14, 327–338.
ARD/USAID (United States Agency for International Development) (2001) *Hydrogeological Impacts of Over Pumping and Assessment of Groundwater Management Options in the Amman–Zarqa Highlands.* Amman, Jordan.
ASAL (Agricultural Sector Adjustment Loan) (1994) Jordan: issues in water pricing. Draft Working Paper.
Baker, M. Inc. and Harza Engineering Company (1955) The Hashemite Kingdom of Jordan–Yarmouk–Jordan Valley Project – Master Plan Report. Amman, Jordan.
Bocco, R. (1987) La notion de dirah chez les tribus bédouines en Jordanie. Le cas de Bani Sakhr. In Cannon, B. (dir.) *Terroires et Sociétés au Maghreb et au Moyen-orient*. Série Etudes sur le Monde Arabe, No. 2, pp.195–215. Maison de l'Orient, Lyon.
Bocco, R. (2006) International organizations and settlements of nomads in the Arab Middle East, 1950–1990, In: Chatty, F. (ed.) *Nomadic Societies in the Middle East and North Africa: Entering the 21st Century*. Brill, Leiden, The Netherlands.
Brand, L. (1995) In the beginning was the state ...: the quest for civil society in Jordan. In: Norton, A.R. (ed.) *Civil Society in the Middle East, Vol. 1*. Brill, New York.

Chebaane, M., El-Naser, H., Fitch, J., Hijazi, A. and Jabbarin, A. (2004) Participatory groundwater management in Jordan: development and analysis of options. Groundwater: from development to management. *Hydrogeology Journal* 12(1), 14–32.

Courcier, R., Venot, J.P. and Molle, F. (2005) *Historical Transformations of the Lower Jordan River Basin (in Jordan): Changes in Water Use and Projections (1950–2025)*. Comprehensive Assessment Research Report 9. Comprehensive Assessment Secretariat, Colombo, Sri Lanka.

Darmane, K. (2004) *Gestion de la Rareté: le Service d'Eau Potable d'Amman entre la Gestion Publique et Privée*. IWMI–IFPO–MREA Working Paper. French Regional Mission for Water and Agriculture, Amman, Jordan.

de Bel-Air, F. (2002) Population, politique et politiques de population en Jordanie 1948–1998. Thèse de doctorat. Ecole des Hautes Études en Sciences Sociales, Paris.

DoS (Department of Statistics, Jordan) (1978–2003) *Statistical Yearbook* – different volumes: 1978, 1983, 1984, 1985, 1986, 1987, 1992, 1993, 1994, 1995, 1996, 1997, 1998, 1999, 2000, 2001, 2002, 2003. Department of Statistics, Amman, Jordan.

Elmusa, S.S. (1994) *A Harvest of Technology, the Super Green Revolution in the Jordan Valley*. Center for Contemporary Arab Studies, Georgetown University, Washington, DC.

El-Nasser, H. (1998) The partition of water resources in the Jordan River basin: history and current development. Paper presented at the Conference on Water in the Mediterranean Countries: Management Problems of a Scarce Resource, Naples, 4–5 December 1997.

EXACT (Executive Action Team) (1998) *Temporal Trends for Water-resources Data in Areas of Israeli, Jordanian, and Palestinian Interest*. Middle East Water Data Banks Project.

Flyvbjerg, B., Bruzelius, N. and Rothengatter, W. (2003) *Megaprojects and Risk: an Anatomy of Ambition*. Cambridge University Press, Cambridge, UK.

Goichon, A.A. (1967) *Jordanie Réelle*. Maisonneuve, Paris.

Gottman, J. (1937) The pioneer fringe in Palestine. *Geographical Review* 27, 550–565.

Grattan, S.R. (2001) *Impact of Increasing Supplies of Recycled Water on Crops, Soils and Irrigation Management in the Jordan Valley*. Technical Report. ARD United States Agency for International Development, Amman, Jordan.

GTZ (Deutsche Gesselshaft für Technische Zusammenarbeit) (1998) *Middle East Regional Study on Water Supply and Demand Development*. GTZ Evaluation Report. Amman, Jordan.

GTZ (2001) *Water Resources Management for Irrigated Agriculture. Annual Progress Report, June 2000–May 2001*. Amman, Jordan.

GTZ (2002) *Water Resources Management for Irrigated Agriculture. Annual Progress Report, June 2001–May 2002*. Amman, Jordan.

Hamlin, C. (2000) Waters or water? Master narratives in water history and their implications for contemporary water policy. *Water Policy* 2, 313–325.

Harza JRV Group (1998) *Jordan Rift Valley Integrated Development Study. Red Sea–Dead Sea Canal Project. Prefeasibility Report. Volume 1 – Main Report*. Harza JRV Group, Amman, Jordan.

IRIN (Humanitarian News and Analysis) (2007) Jordan: US$600 million project to end water shortage. www.irinnews.org (accessed 15 May 2008).

Jabarin, A. (2001) *Curtailment of Groundwater Use for Irrigated Agriculture in the Amman–Zarqa Basin Uplands: a Socio-economic Analysis*. Ministry of Water and Irrigation/United States Agency for International Development – Water Resource Policy Support, Amman, Jordan

JICA (Japan International Cooperation Agency) Study Team (2001) Understanding of Present Conditions of Water Resources Management in Jordan. JICA 2nd Seminar, April 2001. Amman, Jordan.

JICA (2004) The study on water resources management in the Hashemite Kingdom of Jordan. Draft Report Vols I and II plus annexes. Prepared for the Ministry of Water and Irrigation, Amman, Jordan. Yachiyo Engineering Co. Ltd, Tokyo, Japan.

Khouri, R.G. (1981) *The Jordan Valley: Life and Society Below Sea Level*. Longman, London and New York.

Kingston, W.T. (1996) *Britain and the Politics of Modernization in the Middle East*. Cambridge University Press, Cambridge, UK.

Klein, M. (1998) Water balance of the upper Jordan River basin. *Water International* 23(4), 244–248.

Lancaster, W. (1999) *People, Land, and Water in the Arab Middle East*. Harwood Academic Publishers, Amsterdam.

LA Times (2008) Food crisis creates an opening for Muslim fundamentalists. 18 May 2008. www.latimes.com/news/printedition/front/la-fg-food18-2008may18,0,6184648.story

Lavergne, M. (1996) *La Jordanie*. Karthala, Paris.

Lowdermilk, W.C. (1944) *Palestine. Land of Promise*. Victor Gollanez Ltd, London.

McCornick, P.G., Haddadin, M. and Sabella, R. (2001) *Water Reuse Options in the Jordan Valley*. Water Reuse Component, Water Policy Support Activity, United States Agency for International Development, Amman, Jordan.

McCornick, P.G., Taha, S.S.E. and El Nasser, H. (2002) Planning for Reclaimed Water in the Amman–Zarqa Basin and Jordan Valley. American Society of Civil Engineering – Environmental & Water Resources Institute, Conference & Symposium on Droughts & Floods, Roanoke, Virginia.

Merril, S. (1881) *East of the Jordan*. Richard Bentley and Son, London.

Millington, A., Al-Hussein, S. and Dutton, R. (1999) Population dynamics, socioeconomic change and land colonization in northern Jordan. *Applied Geography* 19, 363–384.

Molden, D. (1997) *Accounting for Water Use and Productivity*. SWIM Paper 1. International Irrigation Management Institute, Colombo, Sri Lanka.

Molle, F. (2003) *Development Trajectories of River Basin: a Conceptual Framework*. Research Report No. 72. Comprehensive Assessment of Water Management in Agriculture. International Water Management Institute, Colombo, Sri Lanka.

Molle, F. (2006) *Planning and Managing Water Resources at the River-basin Level: Emergence and Evolution of a Concept*. Comprehensive Assessment Research Report No. 16. Comprehensive Assessment Secretariat, Colombo, Sri Lanka.

Molle, F. (2008) Why enough is never enough: the societal determinants of river basin closure. *International Journal of Water Resource Development* 24(2), 247–256.

Molle, F., Wester, P. and Hirsh, P. (2007) River basin development and management. In: Molden, D. (ed.) *Water for Food, Water for Life: a Comprehensive Assessment of Water Management in Agriculture*. Earthscan, London and International Water Management Institute, Colombo, Sri Lanka.

Molle, F., Venot, J.P. and Hassan, Y. (2008) Irrigation in the Jordan valley: are water pricing policies overly optimistic? *Agricultural Water Management* 95(4), 427–438.

MREA (Mission Régionale Eau et Agriculture, French Embassy) and JVA (Jordan Valley Authority) (2006) *Irrigation Optimization in the Jordan Valley (IOJoV Project). Feasibility Study of Phase II: Extension to All North Conversion Project*. Mission Régionale Eau et Agriculture/Jordan Valley Authority, Amman, Jordan.

Nachbaur, J.W. (2004) *The Jordan River Basin in Jordan: Impacts of Support for Irrigation and Rural Development*. MREA Working Paper. French Regional Mission for Water and Agriculture, Amman.

Orthofer, R. (2001) Options for a More Sustainable Water Management in the Lower Jordan Valley. Workshop: Rewriting the Water History in Palestine, Oxford Environment Week, Environmental Change Institute, Oxford University, Oxford, October 24, 2001.

Orthofer, R., Gebetsroither, E. and Lehrer, D. (2007). Scenarios for a more sustainable water management in the Dead Sea basin. In: Lipchin, C. et al. (eds) *Integrated Water Resources Management and Security in the Middle East*. Springer, Dordrecht, pp. 297–321.

Pitman, G.T.K. (2004). *Jordan: an Evaluation of Bank Assistance for Water Development and Management. A country assistance evaluation*. World Bank, Washington, DC.

Richards, A. (1993) *Bananas and Bedouins: Political Economy Issues in Agricultural Sector Reform in Jordan*. Democratic Institutions Support DIS Project, United States Agency for International Development, Washington, DC. Memorandum Paper.

Salameh, E. and Bannayan, H. (1993) *Water Resources of Jordan: Present Status and Future Potentials*. Royal Society for the Conservation of Nature/Friedrich Ebert, Stiftung, Amman, Jordan.

San Filippo, F. (2006) *Evaluation of the Possibilities of Participation of the WUAs of the Jordan Valley to the Operation, Management and Maintenance Procedures Recommended by IOJoV. Mission Report*. Société du Canal de Provence–French Regional Mission for Water and Agriculture, Amman, Jordan.

Shami, S. (1982) Ethnicity and leadership: the Circassians in Jordan. Unpublished PhD dissertation, University of Berkeley, California.

Shami, S. and Taminian, L. (1990) Women's participation in the Jordanian labour force: a comparison of urban and rural patterns. In: Shami, S., Taminian, L., Morsley, S.A. and El Bakri, Z.B. (eds) *Women in Arab Society. Work Patterns and Gender Relations in Egypt, Jordan and Sudan*. Berg Publishers, Oxford.

Shryock, A. (1997) *Nationalism and Genealogical Imagination. Oral History and Textual Authority in Tribal Jordan*. University of California, Berkeley.

Suleiman, R. (2004) *The Historical Evolution of the Water Resources Development in the Jordan River Basin in Jordan*. MREA–IWMI Working Paper. French Regional Mission for Water and Agriculture, Amman, Jordan.

Terra Daily (2008a) Thirsty Jordan scrambles to find new water resources. March 20, 2008. www.terradaily.com/reports/Thirsty_Jordan_scrambles_to_find_new_water_resources_999.html

Terra Daily (2008b) Multi-million dollar plan to pump water to Jordan's capital. April 20, 2008.

THKJ (The Hashemite Kingdom of Jordan) (1977) *National Water Master Plan*. German Agency for Technical Cooperation, Amman, Jordan.

THKJ (2004) *National Water Master Plan*. German Agency for Technical Cooperation, Jordan.

THKJ, MoA (Ministry of Agriculture) (2001) *Agricultural Sector Development Program 2001–2010*. Ministry of Agriculture, Amman, Jordan.

THKJ, MWI (Ministry of Water and Irrigation) (1997) *Jordan's Water Strategy*. Ministry of Water and Irrigation, Amman, Jordan.

THKJ, MWI, WAJ (Water Authority of Jordan) (2004) Feasibility study for the re-use of treated wastewater in irrigated agriculture in the Jordan valley. Project financed by the KFW, in cooperation with GITEC Consult GmbH, AHT International GmbH and Consulting Engineering Centre. Ministry of Water and Irrigation/Water Authority of Jordan, Amman, Jordan.

UN (United Nations) (1949) *First Interim Report – Survey Mission for Middle East*. UN Document A/1106, 17 November 1949. United Nations, New York.

Van Aken, M. (2004) *Social and Cultural Aspects of Current and Future Governance for the Management of Water Resources in the Jordan River Valley*. Project Report, French Regional Mission for Water and Agriculture, Amman, Jordan.

Van Aken, M., Courcier, R., Venot, J.P. and Molle, F. (2007) *Historical Trajectory of a River Basin in the Middle East: the Lower Jordan River Basin (in Jordan)*. International Water Management Institute/French Regional Mission for Water and Agriculture, Amman, Jordan.

van der Koij, G. and Ibrahim, M.M. (1990) *Picking up the Threads. A Continuing Review of Excavations at Deir Alla*. University of Leiden, Leiden, The Netherlands.

Venot, J.P. and Molle, F. (2008) Groundwater depletion in Jordan highlands: can pricing policies regulate irrigation water use? *Water Resource Management* 22, 11.

Venot, J.P., Molle, F. and Hassan, Y. (2007) *Irrigated Agriculture, Water Pricing and Water Savings in the Lower Jordan River Basin*. Comprehensive Assessment Research Report 18. Comprehensive Assessment Secretariat, Colombo, Sri Lanka.

Waller, T. (1994) Expertise, elites and resources management reform. Resisting agricultural water conservation in California's Imperial Valley. *Journal of Political Ecology* 1, 13–42.

World Bank (1999) *Jordan Agricultural Development Project. Final Report*. Study by Agridev-Agricultural Development Company, Amman, Jordan. World Bank, Washington, DC.

World Bank (2003) *World Bank Water Resources Sector Strategy: Strategic Directions for World Bank Engagement*. Washington, DC.

3 Are Good Intentions Leading to Good Outcomes? Continuities in Social, Economic and Hydro-political Trajectories in the Olifants River Basin, South Africa

Douglas J. Merrey,[1]* Hervé Lévite[2] and Barbara van Koppen[3]*****

[1]*Waterkloof Ridge 0181, Pretoria, South Africa;* [2]*FAO, Rome, Italy;*
[3]*International Water Management Institute, Pretoria, South Africa;*
e-mails: *dougmerrey@gmail.com; **herve.levite@fao.org; ***b.vankoppen@cgiar.org

Overview of the Argument

Beginning in the early 19th century, land and water resources in South Africa's Olifants basin were systematically mobilized to benefit commercial agriculture, mines and industries owned by a tiny minority of the population. During the 20th century, the majority African population was increasingly confined to small areas of the basin having little agricultural potential or access to water. This resulted in dramatic contrasts between the wealthy minority and the extremely poor majority. Since the early 1990s, under the new democratic regime, South Africa's constitution, with its basic rights guarantees, including access to water, and its world-famous Water Act, intended both to reverse the wrongs of the past and to conserve scarce water resources for future generations, have raised high expectations. The Water Act is being implemented by politicians and professionals whose good intentions cannot be questioned. However, to date, access to water remains highly inequitable in the Olifants basin, and socio-economic well-being is improving very slowly.

Setting the Physical Scene

The Olifants water management area

The Olifants River is the largest tributary to the Limpopo, one of several transboundary rivers in Southern Africa. Shared by Botswana, Zimbabwe, South Africa and Mozambique, the Limpopo basin has an area exceeding 400,000 km^2 (45% in South Africa). Of a basin population of 14 million, 10.7 million are in South Africa (a quarter of the total population). Turton (2003) emphasizes the critical strategic importance of the Limpopo basin for all four riparian countries and the considerable ethnic diversity overlapping national boundaries.

The total area of the Olifants basin (including Mozambique and South Africa and two large northern tributaries, the Letaba and Luvuvhu) is 73,534 km^2, nearly 17% of the Limpopo basin (ARC and IWMI, 2003). 'Olifants' is the Afrikaans name for elephant. In Northern Sotho, the main language of the basin, it is 'Lepelle', 'the river that meanders along' (Bulpin, 1956). About 770 km long, the Olifants originates east of Johannesburg and flows north before curving gently to the east. Its upper reaches are in the 'highveld', over 1200 masl. Further east, the lower reaches are below a steep escarpment in the 'lowveld', at altitudes of less than 800 m. The Olifants crosses three provinces (Gauteng, Mpumalanga and Limpopo) into Kruger National Park, then flows into Mozambique, where it meets the Limpopo (Fig. 3.1).

In Mozambique, the Massingir dam, with 2840 Mm^3 of storage, is important for hydropower, irrigation (30,000 ha), flood control, and urban and rural water supply, as well as maintenance of low flows to prevent salt water intrusion at the mouth of the Limpopo (Carmo Vaz, 2000). There have been several devastating floods in recent years.

From the perspective of Mozambique, upstream South African water use is a vitally important issue, fraught with the potential for conflict. Low flows result in salt water intrusion and water shortages (FAO, 2004:87–88). In 2005, the Olifants stopped flowing into Mozambique for 78 days, causing considerable hardship. The implications for Mozambique of South African use of the Olifants have not been addressed by researchers and there is no specific international agreement on water flows. The South African Department of Water Affairs and Forestry (DWAF) is aware of this issue, although its assumptions about the amount that should flow to Mozambique may not be consistent with those of Mozambique officials.

The official Olifants water management area[1] in South Africa drains an area of 54,308 km^2. In 2005, the population of 3.2 million represented 7% of the national population. Of this population, 67% is rural, higher than the national average. Blacks are the majority (94%), with an illiteracy rate of 50%. Distribution of wealth and access to services are highly skewed between urban and rural areas, and between whites and blacks (Magagula et al., 2006). Population growth is slow, although shifting from rural to urban over time. There are seven major tributaries to the Olifants (Fig. 3.1). Based on DWAF's demarcation, the Olifants water management area is a 'primary drainage area' (McCartney et al., 2004), and includes seven secondary, 13 tertiary and 114 quaternary sub-basins. But the basin is normally divided into five distinct water management regions (McCartney et al., 2004; de Lange et al., 2005).

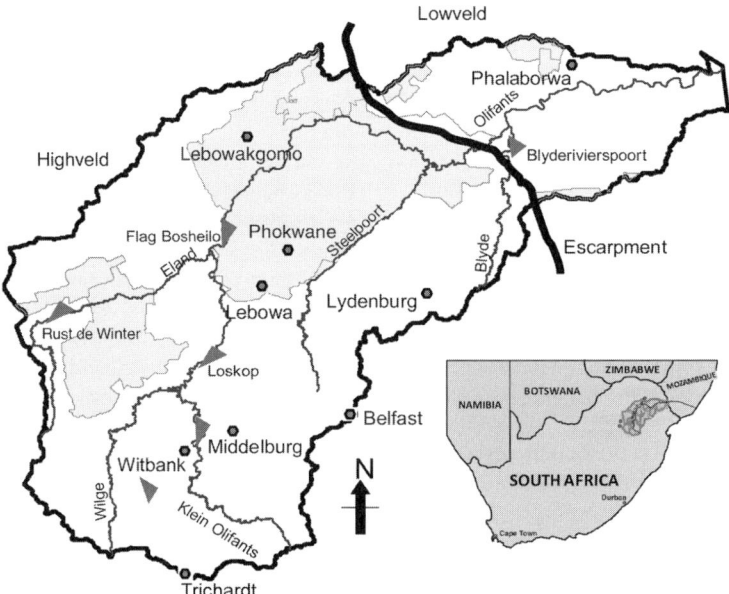

Fig. 3.1. Map of Olifants River, major dams (triangles), tributaries, towns (hexagons) and demarcation of former homeland areas (shaded areas). From McCartney and Arranz (2007).

Physical features

The geology of the basin is complex, and dominated by igneous and metamorphosed rocks. There is a relatively low-relief, gently undulating plateau and a steep escarpment roughly separating the lower Olifants region from the middle and upper regions. Land use consists primarily of cultivation (dry and irrigated), grazing, mining, industry, forestry, and rural and urban settlements. There are many tourist attractions in the basin, including the Kruger National Park, private game reserves, Blyde River Canyon Nature Reserve and several wildlife management areas. There are important fish hatcheries and trout farms, while some reservoirs are also used for recreation.

Climate, rainfall and hydrology

The basin is characterized by warm summers and mild winters, with temperatures influenced by altitude. In summer, maximum temperatures are 30–34°C and with a minimum of 18–22°C; in winter they are 22–26°C and 5–10°C, respectively. Frost occurs only in the southern and western portions of the basin (FAO, 2004).

The mean annual precipitation is 630 mm, with a range of 500–800 mm and coefficients of variation greater than 0.25 in all sub-basins. In the mountains to the east and on the escarpment, annual rainfall can exceed 1000 mm (McCartney and Arranz, 2007). The rainy season is from October to April, with heavy rainfall in December and January producing occasional floods. There are no months when rainfall exceeds potential evapotranspiration, and typically it exceeds 50% of potential evapotranspiration only in November–February (McCartney et al., 2004). Midsummer dry spells are common, making rainfed agriculture risky.

McCartney et al. (2004) studied the hydrology, complementing DWAF's work (Basson and Rossouw, 2003; van Vuuren et al., 2003; DWAF, 2004a). The naturalized mean annual flow (MAF) of the whole basin is 2040 Mm^3, only about 6% of the average annual rainfall (McCartney et al., 2004). However, this value masks considerable annual variability. Actual measured runoff, as influenced by human activities and exotic (i.e. alien) vegetation, reaches 1235 Mm^3 (de Lange et al., 2005). All studies agree that with total South African consumption at around 44% of the naturalized MAF and increasing, the basin is already stressed.

DWAF estimates that the total groundwater recharge is 3–6% of mean annual precipitation, which is about 1800 Mm^3. Others suggest that the average recharge is only half this amount, so values are not precise (McCartney et al., 2004). DWAF variously estimates total groundwater abstractions at 75–99 Mm^3, principally from mining, urbanization, stock-water and irrigation.

Estimates of average annual transfers into the basin as of 1990 (the official values have not changed in 18 years) vary slightly at around 196 Mm^3 (McCartney et al., 2004). Most of this (188.8 Mm^3) is used for cooling power stations operated by Eskom (Electricity Supply Commission). It leaves the basin as evaporation and has little impact on basin hydrology. Nearly all these interbasin transfers support large-scale commercial water users (van Vuuren et al., 2003:4, 2ff.). Transfers out of the basin are very small.

Agriculture, livestock and forestry in the basin

Commercial agriculture officially contributes only 7% of Gross Geographical Product (GGP) to the basin economy, but this is nearly twice the national level. Subsistence and small-scale agriculture, whose value is not measured, play a critical role in human survival, child nutrition and potential poverty alleviation.

South Africa generally classifies three farming types: (i) subsistence/semi-commercial farming (typically dryland); (ii) commercial dryland farming (large scale and highly mechanized); and (iii) commercial irrigated farming (export oriented, intensive) (Magagula and Sally, 2005). All three occur in the basin, with commercial dryland on more than 70% of the cultivated area of 1.17 million ha, and commercial irrigated covering around 11% (128,000 ha). Today, the average size of commercial farms in Limpopo Province is 972 ha (van Koppen, 2007). An estimated 70% of water withdrawals goes to irrigation (30% of which is groundwater) (Magagula and Sally, 2005). Estimated water requirements using the SAPWAT model range from 436.8 Mm^3 (DWAF data) to 569.5 Mm^3 (van Heerden, 2004). Of the estimated R 5.3 billion (approximately US$828 million) gross value of agricultural production in 2004, 60% was generated by commercial dryland and 37% by commercial irrigation (Magagula and Sally, 2005). High-value crops for export, such as citrus, are more common here than elsewhere in South Africa. Maize remains the dominant crop by area and is grown in summer under rainfed conditions.

In addition, there is a small-scale irrigation sector, mostly in the former homeland areas. The basin has around 72 small-scale irrigation schemes with a total command area of 9534 ha, 5564 farmers and an average plot size of 1.6 ha. However, many of these are either defunct or underutilized. More than half of the farmers are women and often elderly (Mpahlele et al., 2000; Kamara et al., 2002; van Koppen et al., 2006).

Large parts of the Olifants basin are used for livestock and game farming. Van Vuuren et al. (2003) estimate 337,006 livestock units, but there are no data from the former homelands. Cattle are the most common, but there are also sheep. Game (impala, kudu, waterbuck, gemsbok and rhino) is farmed for hunting and meat production, and is becoming popular. Nationally, the 'hunting industry' creates many jobs and a substantial income (www.phasa.co.za/index.php?pid=3).

Commercial forestry (mainly pine and eucalyptus) is an important water consumer; it is estimated to cover 400 km^2 (Le Roy, 2005:10). Non-indigenous trees were originally grown for mining needs, but today commercial forestry is mainly linked to paper production (Lévite et al., 2003), and is dominated by large national and international corporations. These plantations account for 28% of national commercial forestry. Non-indigenous species are seen as depleting far more water through evapotranspiration than indigenous forests. Therefore, DWAF charges companies for the additional 'stream flow reduction' at a rate of R 10 per ha (DWAF, 2004b). There are also about 1399 km^2 of indigenous forests in the Blyde River and lower Olifants regions. An assessment of actual evapotranspiration (ETa) in part of the middle Olifants during one day in January 2002, using a remote-sensing technique (SEBAL), found that agriculture accounted for only 24% of actual basin ETa, compared with over 58% through commercial forests (Ahmad et al., 2005).

Expansion of mining in the basin

Mining, a significant user and polluter of water, is the largest economic sector in the basin (22.1% of GGP versus 7% GDP nationally). Employment in mining is growing slowly in the Olifants: declines in gold mining balance growth in platinum mining. Manufacturing is largely a function of the relatively cheap supply of coal and electricity, much of it based on processing minerals. There are eight major coal-fired electricity power stations, generating more than 50% of South Africa's electric supply (van Vuuren et al., 2003). The downstream impact of coal mining from both decommissioned and functioning mines is a major problem, with the release of acidic leachate into both surface water and groundwater (Klarenberg, 2004).

Monopolizing Water and Creating Water Scarcity

From the early 19th century, the history of the Olifants River basin has been a story of resource capture by the powerful. By the late 20th century, a small race-based minority controlled nearly all the land, water and mineral resources and the wealth they produced, while the African majority was becoming increasingly poor and marginalized (van Koppen, 2007).

Warfare and competition on the eve of the Afrikaner Boers' arrival

In the early 19th century, the Olifants basin was inhabited by African ethnic groups, largely agropastoralists also engaged in trade with the Indian Ocean. Demand for ivory had led to a quantum leap in its export from Delagoa Bay (today

Maputo, Mozambique) (Reader, 1998:469–470). Rainfall patterns were critical for grazing and sorghum cultivation. The highveld and middleveld areas were more suitable for cultivation and summer grazing; the malaria- and tsetse-infested lowveld was only suitable for dry-season winter grazing and as a major source of ivory. Settlement was largely along streams. People spoke languages that are part of the Bantu language family, divided mainly into Sotho and Nguni languages (Earle et al., 2006:9–16). They were agro-pastoralists, highly mobile groups with loose political affiliations that easily assimilated other groups (Delius, 1983).

Even before the Afrikaner Boers arrived, there was rising competition for water and land, cattle raiding and more serious warfare. Fearing slave-traders (for Europe's colonies and the Cape Colony), waves of the population fled into the Limpopo and Olifants basins, seeking protection from the 1780s to the 1840s (Reader, 1998:464–478). The closing of the land frontier in the narrow coastal areas inhabited by Nguni-speaking people (i.e. Zulus), combined with serious periodic droughts, led to new and bloodier warfare. Conquering tribes came into the Olifants basin, raiding cattle, destroying assets and either subjugating inhabitants or driving them out. As this process (called *mfecane*) was occurring, the Boers began moving in from the south, and with their superior technology (guns, horses) defeated many dominant African groups. They too needed slaves for labour (euphemistically called 'apprentices') to farm (Reader, 1998:472–473). They grew the same crops using the same technologies as the Africans and were often dependent on the Africans' willingness to help them (Delius, 1983; Reader, 1998:480).

As African chiefs became more powerful, social differentiation grew. The Pedi chiefdom, with its centre in the Tubatse (Steelpoort) valley, ultimately controlled tens of thousands of Africans. By the 1840s, it controlled the main trade routes, buying cloth and guns from the coast in return for iron, copper beads, meat, ivory, horns and slaves. In 1876, near the present-day Flag Boshielo dam, Sekhukhune I defeated the Boers. However, 3 years later, his army was crushed when the British joined the Boers and Swazis against him (Delius, 1983).

Opening salvos: white expropriation of land, water and mineral resources, 1832–1913

Migration, alliances and conquests in the early 19th century

During the eight decades from 1832 to the early 1900s, three groups of whites, initially mutually hostile, encroached into the basin: the Boers, a small group of missionaries and the British. Both the Africans and the whites were seriously subdivided, but the whites exploited the cleavages among the Africans more effectively (Thompson, 2001). The early Boers competed directly with the Africans for water, land and trade routes. Although the black population exploded (becoming 20 times more numerous than the whites) over the century, this did not translate into political or economic power. Conflict over land grew, leading to clashes. The Pedi defeat of the Boers in 1876 and the annexation of the *Zuid–Afrikaanse Republiek* (ZAR) by the British in 1877 led to the reorganization of the Republic's administration, enabling it to defeat the Pedi in 1879. The Pedi heartland was put under classic British colonial 'indirect rule', as a 'location' in which black chiefs ruled, supervised by white magistrates (Delius, 1983).

In 1886, gold was discovered in Witwatersrand near the Olifants basin, as well as smaller deposits of gold and minerals within the basin. By 1895, the first coal mine in the upper Olifants region opened. Then the British and foreign-owned corporations wished to control all of Southern Africa. The ZAR, now led by Paul Kruger, vehemently resisted and sought to tax the mines, leading to the Second Anglo-Boer War (1899–1902).

Boers and British: white conquest and expropriation

Understanding the developments in the Witwatersrand (now the largest industrial and urban complex in sub-Saharan Africa) is critical to understanding the Olifants basin development trajectory (Turton and Meissner, 2002). The discovery of gold led to Johannesburg's rapid growth and placed enormous strains on a water supply previously perceived as plentiful. By 1900, African political power and control over water, land and mineral resources were nearly destroyed. The Boers controlled the most fertile lands and the best water supplies. British legislation backed by the British Army declared registered water and land to be white private property. A tiny proportion of the land was set aside for African occupation. Boer society was changing rapidly, becoming more inequitable and elitist. A group of new Afrikaner 'notables' became large landowners. Speculators, absentee landlords and companies from outside the basin owned 20% of the land by 1900. Well-watered land, often occupied by Africans, was the first to be controlled. Africans were forced to provide labour to these farms.

With rising market demand for maize and other food crops for miners, large-scale cropping, sometimes irrigated, was initiated. For decades, absentee white owners extracted rents from African tenants and sharecroppers; but as the market grew and railway facilities were constructed, there was a shift to capitalist wage labour arrangements for farm management (Bundy, 1988; Terreblanche, 2002). The Afrikaner notables and British mining interests now had a shared interest in a docile, low-wage labour force, leading to the 'alliance of maize and gold'. Many Boers who could not compete with large farms were also pushed into landlessness, forcing them to compete with cheap African labour.

Nevertheless, African farmers, often sharecroppers on white-owned land, responded effectively to the new food markets, adopting new strains of maize and irrigation. Some of these farmers used communal land and kinship relations as a base; some purchased land using legal loopholes; but most were tenants on white-owned land. Unfortunately, most of these 'peasant capitalists' were soon deprived of their access to land and markets (Bundy, 1988).

The process of creating an ideological and de facto basis for territorial and institutional segregation was consolidated by the South African Native Affairs Commission in 1905. Its purpose was to forge a black male migrant labour force with a black female subsistence base in the 'native reserves'; this labour was allocated proportionally to the mines and to Afrikaner farms. This segregation policy was further consolidated with the Native Land Act of 1913 (Thompson, 2001; Terreblanche, 2002).

The 1913 Act separated the Union into white areas (91% of the land), where Africans, coloureds and Indians were disenfranchised, and black reserves ruled by 'chiefs' as black administrators. The Development Trust and Land Act of 1936 consolidated this exclusionary process. These Land Acts also implicitly deprived Africans of any formal water rights, because riparian rights were tied to land ownership (van Koppen, 2007).

In 1910, with the establishment of the Union of South Africa, a Native Affairs Department was created, and later the Native Administration Act of 1927 formalized 'chiefs' as arms of the government. In 1936, the reserves were placed under the South African Native Trust (later the South African Development Trust), and legitimized the racially and gender-segregated labour market with extremely low wages for men. The apartheid government's homeland policies after 1948 entrenched these patterns more rigidly. Through the Homeland Constitution Act of 1971, existing reserves were reorganized and new ones established, based on nine officially recognized African ethnic groups. In the Olifants basin, the supposed 'Northern Sotho', including the Pedi, were included in Lebowa, created in 1973. Similarly, on the eastern highveld, KwaNdebele was created for the Ndbele, and Gazankulu for the Shangaan to the north-west border of the Olifants basin (see Fig. 3.2).

By the early 1900s, all of the ingredients for state-supported, race-based wealth accumulation were in place, and these greatly determined the Olifants basin development trajectory. These ingredients included:

- A Land Act excluding Africans from claims to most of the land, water and minerals.
- Native reserves as a reservoir of cheap labour.
- Repressive labour laws, enhancing employers' control over the black labour force.
- Discriminatory arrangements favouring white workers.

Henceforth, until late in the apartheid era, water development was used to further deepen the divide between privileged whites and the black majority, what Lévite et al. (2003:4) call

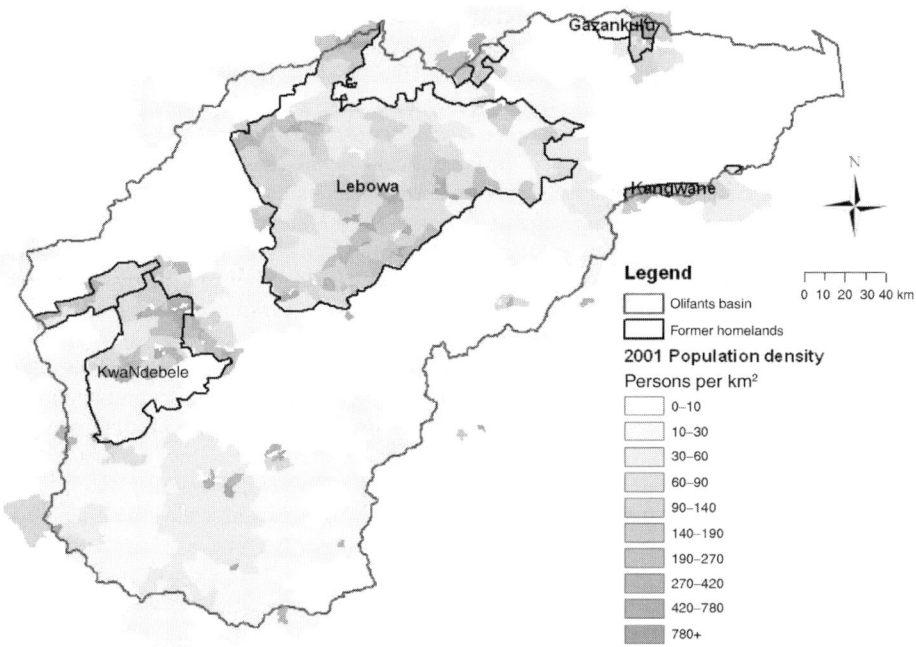

Fig. 3.2. Population densities and former homelands in the Olifants basin. From McCartney et al. (2004).

'race-based differentiation in basin development'. The state played a critical role in this hydraulic mission in the Olifants basin, initially mainly catalysing irrigation development, but from the 1970s onwards promoting centrally controlled, large-scale bulk water supplies, in particular to the Witwatersrand and the adjacent Olifants highveld. The era of engineers able to overcome all obstacles to increase the water supply to meet growing demand had arrived (Turton and Meissner, 2002:41; van Koppen, 2007).

State-supported water development in the 20th century

Irrigation development and the role of the state

There were three waves of investment in irrigation in South Africa: around the 1920s (with a peak in 1922), in the 1930s (with a peak of 5% of total state expenditure) and in the 1970s. Until the 1950s, the government exclusively supported irrigation development; support for other users started in the 1950s, and around 1970 priority shifted from agriculture to other uses (Department of Water Affairs, 1986).

The ZAR adopted its first irrigation law in 1884, revised it in 1908 and established an Irrigation Department in 1903 (van Koppen, 2006). By the late 19th century, the Transvaal had adopted the Roman–Dutch permit system (van Koppen, 2007). In 1912, the union government created a national Irrigation Department and promulgated the Union Irrigation and Conservation of Waters Act. This Act adopted the British riparian rights system, which tied water rights to land ownership. This continued until major revisions were made through the Water Act of 1956, when the Irrigation Department became the Department of Water Affairs (DWA). This Act further strengthened government control over water and broadened its scope to ensure industrial and mining interests, the new priority.

From the 1920s, another motivation was to employ poor unemployed whites and to settle potential farmers such as white war veterans. Smallholders were seen as more intensive and committed cultivators, and labour intensiveness was seen as a way of absorbing landless and unemployed whites. The policy also helped to secure white domination of productive land. Two such schemes were in the Olifants: the Loskop dam and the Rust de Winter scheme (Turton et al., 2004; van Koppen, 2007; see Fig. 3.1). The government encouraged both irrigation boards, i.e. schemes managed by white farmers but heavily subsidized, and government water schemes for white farmers.

The Loskop dam was built by and for poor white men during the depression era. Today, the area below the dam is intensively irrigated, growing, in particular, high-value crops (citrus and table grapes) for export. Most farms are large, modern and capital intensive, employing thousands of workers.

Seventeen irrigation boards were established in the basin (van Koppen, 2007). Public irrigation has been especially important in the middle Olifants, under the Loskop dam. As settlement of white farmers proceeded, Africans were forced to move. But there were a few cases where the South African Development Trust purchased white farms to 'rationalize' boundaries between white areas and homelands, including farms below the Flag Boshielo dam (Stimie et al., 2001:57–58; van Koppen, 2006).

The trajectory of dam construction

McCartney et al. (2004) estimate the basin has 37 major and another 300 'minor' dams, plus 3000–4000 small dams, with a total cumulative storage of about 1480 Mm^3 (85% in the major dams). The total storage capacity is 72% of the average annual naturalized flow. McCartney et al. (2004) also note that more than half are multi-purpose dams (often including irrigation), while 28% (38% of the storage) are solely for irrigation. Figure 3.3 is a timeline of storage development in the 20th century, distinguishing former homelands from former white areas (Republic of South Africa). There is a clear discrepancy, with nearly all dams aimed at benefitting white users until the 1980s, when two dams were built that also provided some benefits to former homeland areas (see also McCartney et al., 2004:27–31).

Water for mining, industry, energy, and rural and urban sectors

Until the 1940s, water development in the Olifants for urban uses, mining and industry was largely a private affair of municipalities and firms. These schemes were scattered physically, and generally their costs were low. The Water Act of 1956 changed the prioritization of water use and, for the first time, made some subsidies available to non-agricultural local bodies. Coal mining in the upper Olifants basin played a major role in this shift. Eskom (a parastatal created in 1919) constructed coal-fired electricity-generating plants in the upper Olifants highveld, and coal-based industries developed around iron and steel, using ore available locally. For these industries, which require large and highly secure quantities of water, dams were constructed in the upper Olifants from 1950, but demand quickly exceeded supply (van Koppen, 2007).

Mineral deposits had stimulated land speculation, prospecting and railway development. Phalaborwa and Steelpoort became two major mining areas. In Phalaborwa (in the lower Olifants: see Fig. 3.1), first copper and, later, phosphate were the most important minerals, but this has now diversified. Initially, small dams were built to supply water to the mines, white urban areas and black townships. The Phalaborwa Water Board was established in 1963, and after 1994 it was expanded and renamed the Lepelle Water Board. By the 1970s, the assurance of water supply during the dry months to most of these downstream areas had become risky.

The Steelpoort area is even richer in minerals (platinum, magnetite, chrome). Mining was also done within the Pedi native reserve, but under the firm legal control of the union government. Mines created jobs for men, although recruitment was from outside the region. By the 1970s, the appetite of the mining houses was whetted to further exploit the underground wealth in the Olifants basin, and the need to quench their thirst for water increased, a trend that has recently intensified.

Water policies on the eve of democracy: creating the 'white water economy'[2]

After 1970, water for the mining, industrial and white urban sectors became priorities – although support for irrigation continued. This entailed not only large-scale water works, including interbasin transfers, especially to the

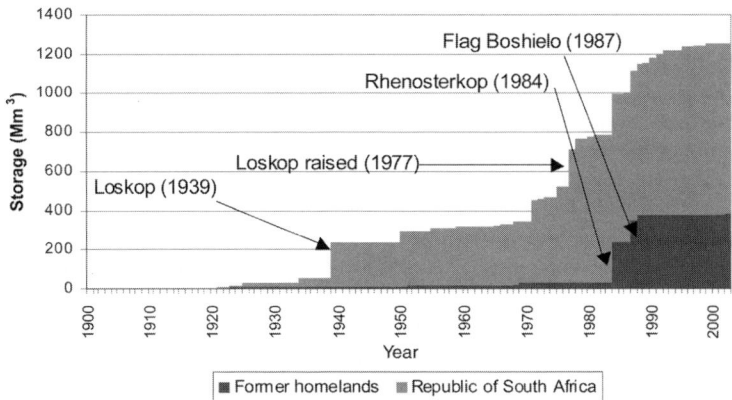

Fig. 3.3. Development of large dam storage in the Olifants basin. From McCartney et al. (2004).

upper Olifants for electricity generation, but also providing further assurance of supply to the Witwatersrand (started in the 1980s) through the Lesotho Highlands Project. Based on this, industrial development was promoted outside the white towns but near the homelands (Fig. 3.2) for their cheap labour. As a result, most of the total basin GGP is produced in the urban areas of the upper Olifants.

The same policy led to prioritizing water supplies to mining in Phalaborwa, justifying the construction of the multi-purpose Blydepoort (or Blyderivierspoort) dam in 1975. The third focus was supplying mines in the Steelpoort sub-basin. Stimie et al. (2001:38) estimate that the number of mines (around 100) was the primary driver for constructing the Flag Boshielo dam in 1987, although the dam also supports small-scale irrigation and water supply to Polokwane (then called Pietersburg). Agriculture was not neglected: in 1977, the Loskop dam was raised to increase its storage capacity, in tandem with new upstream dams in Witbank and Middleburg (see Figs 3.1 and 3.3).

Water for subsistence: irrigation in the former homelands

The creation of the 'homelands', combined with forced removals and rapid population growth, led to rising tensions and frustration.

Lebowa's population grew from 291,000 in 1970 to 629,000 in 1985. The tensions engendered by congestion and poverty further undermined the remaining community-based water management institutions.

From the 1930s, the government tried to minimize poverty by imposing urban-based models, for example by regulating grazing. The 1956 Tomlinson Commission recommended 'Betterment Schemes' as measures to 'develop' the homelands by concentrating access to land only on large-scale male farmers and moving the landless closer to settlements ('homeland towns'). Some domestic water schemes were developed, but in a top-down manner, ignoring the needs of black rural households (e.g. for livestock, gardening).

Black farmers had themselves initiated many small-scale irrigation schemes (around 36), especially along the middle Olifants River. Most of these were developed on lands formerly irrigated by whites, and, in most cases, the South African Native Trust had bought them to consolidate white–black segregation. Most homeland irrigable land was owned by the Trust and sometimes 'improved' with new water management infrastructure before plots were allocated. The plot size was usually 1.28 ha, considered by white definitions sufficient for a nuclear African family to farm full-time and earn a 'reasonable livelihood'. Plot holders were supposed to be males farming full-time, but by 1994 most irrigators on what was called the 'Olifants River Scheme' under the Flag Boshielo dam were, and remain, women. This was partly due to male migration for work and also reflected women's traditional role (van Koppen et al., 2006).

After 1969, plot holders needed 'Permission to Occupy' (PTO) certificates. In 1993, ownership of all but four 'farms' in the scheme was transferred from the South African Development Trust to the government of Lebowa and the infrastructure was improved by the Lebowa Agricultural Corporation; the Flag Boshielo (then called 'Arabie') dam was built by 1987. The irrigable area was over 2000 ha, controlled by 'white management and leadership', assumed to be the key condition for success. Management dictated crops (alternating wheat and maize), dates of ploughing, fertilizer and chemicals to be used, irrigation and harvesting schedules; provided ploughing services and inputs; and purchased the outputs. Service costs were deducted from the sale price before paying the cultivators. Shah et al. (2002:6) observe that farmers were hardly more than labourers on their own plots. These centrally managed schemes collapsed on the withdrawal of government support after 1994.

The Olifants on the eve of democracy: population, poverty and concentrated wealth

The stark differentiation between the poor and well off, blacks and whites, and rural and urban people is worse in the Olifants than at the national level. Sixty per cent of the population reside in the former homeland areas, constituting 26% of the basin area (Fig. 3.2). Two-thirds are in rural areas, mostly in scattered informal villages with limited commerce and services. There are few major urban centres within the basin, but important interactions exist with Pretoria and Johannesburg. Ninety-four per cent are black Africans. Most future population growth will be urban; the rural population is expected to stabilize because of HIV/AIDS (van Vuuren et al., 2003).

According to the 2001 census, 47% of the Olifants labour force is unemployed, with most available jobs outside the former homelands (Magagula et al., 2006). Nearly 50% of formal jobs are in government, 21% in mining and 19% in agriculture. Distribution of wealth is highly skewed between urban and rural areas (van Vuuren et al., 2003). Some 70% of the population live in poverty; 75% of them report they have no monthly income (Magagula et al., 2006).

Much of the area below Loskop dam (a region where International Water Management Institute (IWMI) researchers have worked intensively) is now in the Greater Sekhukhune District Municipality, which today combines prosperous as well as poor, formerly white areas with poor, predominantly black areas. This region contains some of the highest concentrations of heavy metals in the world (chromium, platinum, titanium, vanadium) (Ziervogel et al., 2006). Growth in mining in this area and in the Steelpoort region is enormous but has not yet reduced the municipality's 69% unemployment rate. The 2005 census recorded a population of 1.12 million living in the district, mostly in the former homeland areas. Commercial agriculture is the main employer there (Ziervogel et al., 2006:9–10). Only 30% of households have access to agricultural land.

Post Uiterweer et al. (2006) provide a poignant description of the problems characterizing Sekhukhune. In the 19th century, Sekhukhuneland had been a powerful kingdom; today, it is one of the poorest areas in the country and no longer well known. Over 40% of the villages did not have even a basic water supply in 2004.

The Post-1994 Dispensation: Trying to Achieve Equity without Reducing Large-scale Users' Access

The new dispensation in South Africa: constitutional guarantees and idealism

Remarkably, there was a peaceful, negotiated transition from the apartheid regime to a representative, constitutional government based on one person, one vote. The first fully democratic election was held in 1994, and elections have been held regularly since then. The new constitution, created through a wide-ranging public consultation process, has explicit provisions regarding citizens' rights to a healthy, sustainable environment and access to health care and 'sufficient food and water', and requires the government to take reasonable measures to progressively achieve these and other rights (de Lange, 2004).

A widespread, although white-dominated, consultative process during the mid-1990s led to the National Water Services Act (1997) and the National Water Act (NWA) of 1998 being adopted. This process is described in detail by de Lange (2004) and others (De Coning and Sherwill, 2004; Backeberg, 2005; Garduño and Hinsch, 2005; De Coning, 2006). Despite strong differences of opinion, the final bill was broadly supported by all major parties. This is remarkable considering the radical nature of some reforms: for example, the riparian rights system and private groundwater ownership were abolished, as well as the connection between land and water rights. Water is now a national resource, with the Minister of Water Affairs as its custodian on behalf of the government, and a system of licensing for specified periods has replaced water rights in perpetuity.

The NWA has been perceived by senior DWAF officials as an instrument to achieve the broader goals of the new South Africa, captured in the slogan 'a better life for all' (Muller, 2001; Schreiner et al., 2002). It is intended to provide a framework for achieving broad, constitutionally mandated goals, such as equity, productivity and environmental sustainability, as well as specific objectives, such as cost recovery, decentralized management, effective service delivery and flexibility to adapt to changes.

DWAF has been simultaneously carrying out numerous complex activities to implement the NWA while transforming itself structurally and in terms of gender and ethnic balance, and recruiting new expertise. It has carried out studies, prepared policy statements and implementation guidelines, and held many consultations with stakeholders, which have become increasingly race and gender balanced. It has also been pilot testing reforms.

DWAF has also given the highest priority to providing basic water and sanitation services as rapidly as possible to the estimated population of 12 million lacking these in 1994, and is making good progress: as of July 2008, 2.48 million still do not have water supply infrastructure and 13.38 million lack basic sanitation infrastructure (http://www.dwaf.gov.za/dir_ws/wsnis/, accessed 4 July 2008), but this situation is far better than it was a decade ago. Since 2006, this function has been a municipal responsibility. To implement the right to sufficient water, DWAF adopted a 'free basic water' policy, giving every household a right to 6000 litres per month without charge. Where good infrastructure is in place, this works well, but for most poor rural municipalities, implementation is difficult (Post Uiterweer et al., 2006; Muller, 2007). With the handover to the newly created local municipalities, domestic water service has become increasingly problematic without the temporary 'cushion' previously provided by DWAF's technical staff (van Koppen, 2007).

Implementation of the water act in the Olifants basin: institutional transformation?

The Olifants catchment management agency: a stalled process

The NWA provides for establishing catchment management agencies (CMAs) in each water management area, to decentralize and integrate river basin management and to provide stakeholder forums. A CMA is not expected to be fully democratic; its board should be broadly representative of basin interests but is appointed by the minister (Ligthelm, 2001). DWAF officials initially had high hopes for CMAs as 'the key vehicles to implement the new water management paradigm' (Schreiner et al., 2002:127): 'Catchment Management Agencies for poverty eradication in South Africa' is the title of a paper by a senior DWAF official (Schreiner and van Koppen, 2001).

The process of establishing an Olifants CMA was initiated in 1998 by a major consulting firm. IWMI was appointed as 'peer reviewer'. The process itself, pitfalls and proposed solutions are described from DWAF's perspective by Ligthelm (2001), who was the DWAF task manager.

Wester et al. (2003) assessed the process and compared it with a much different approach in Mexico. The draft CMA proposal (van Veelen et al., 2002) was submitted to DWAF, but not taken to the minister, although CMAs are being established in a few other (smaller) basins.

With hindsight, DWAF policy makers were probably overoptimistic about the efforts required to render the consultation process genuinely inclusive, given the highly unlevel playing field. The large public and private water users are well organized to defend their interests. However, the rural poor are not organized, and most were not even aware of the process (Stimie et al., 2001; Wester et al., 2003). There were serious cultural barriers: most of the consultants were white engineers who did not speak the local languages. Only summary translations were provided. Poor communities tended to raise issues such as lack of drinking water, only to be told these problems would be addressed by others. In short, as Wester et al. (2003:808) note, 'the effectiveness of the process in the poor rural areas is doubtful'.

Clearly, DWAF and its consultants did not address the core issues. The consultants focused on the organizational structure of the CMA, not on the critical issue of equitable voice and power capture by minority interests in setting the agenda of the CMA. The consultations were not designed to 'balance' political inequalities, for example by investing special efforts in dialogues with poor communities. Therefore, the CMA could never have achieved the government's equity objectives. There were similar experiences in other basins (Wester et al., 2003; Waalewijn et al., 2005; Simpungwe, 2006). In recent years DWAF has reached out to the new, upcoming local and provincial governments through Provincial Water Summits in 2005 and 2006; in the long run, municipalities are expected to fill the local void, while large-scale users will also cooperate with local and provincial governments. These developments, under the conceptual umbrella of 'Water for Growth and Development', have also served to begin closing the administrative gap between domestic and productive water services (van Koppen, 2007).

Catchment management forums (CMFs)

DWAF senior officials realized the dangers of replicating existing inequities and monitored the consultation processes carefully. A major challenge is involving poor communities, and especially women, in these processes (Schreiner et al., 2004). One solution was to pursue more bottom-up participation (Schreiner and van Koppen, 2001; Schreiner et al., 2002; Simpungwe, 2006). In three other water management areas, DWAF tried to enhance the skills of the poor, especially of women, by getting them involved in this participation (Schreiner et al., 2004). Some resources were also allocated in the Olifants to enable a grassroots organizer to demonstrate how this would work (Schreiner and van Koppen, 2001). She organized workshops in the local language, which addressed domestic and productive water issues. A suggestion emerged to organize multi-tiered, small-scale water users' forums as a way to ensure effective local representation in the future CMA governing board. Smallholder water user forums (SWUFs) were thus suggested in the draft Olifants CMA proposal, but this was never followed up.

These proposed SWUFs are not to be confused with the Olifants River Forum (ORF), established in 1993 to promote cooperation for conservation and sustainable use of the river (www.orf.co.za; see Schreiner and van Koppen, 2001; Klarenberg, 2004:89–91). The founders were mostly white representatives of large mining firms, the Kruger National Park and DWAF. Membership today is more varied, but local communities are not well represented. It is clear that this forum was intended, in part, to lobby DWAF and influence the formation of the planned CMA and water allocation processes, and in this sense it is a continuation of the 'white water economy' (van Koppen, 2007). Simpungwe (2006:15) claims that more than 200 CMFs have emerged in other South African catchments, and DWAF has formally endorsed their importance, even in the absence of supporting legislation (DWAF, 2004b:97–98). Like the Olifants River Forum, many of these recent CMFs are de facto dominated by government departments, other formal organizations and white economic interests, minimizing the potential to empower poor water users (Simpungwe, 2006).

Schreiner and van Koppen (2001) reflect on DWAF's high hopes that an inclusive CMA process could lead to institutions able to service the poor better. Unfortunately, there is little evidence that CMAs, or CMFs for that matter, have achieved this. In the Eastern Cape, Simpungwe (2006) found that CMFs have not been effective in achieving equity; while he remains optimistic, his cases suggest that they have not created a level playing field – differential political and economic power distort the outcomes. In the Olifants, DWAF halted the CMA process in favour of attempting to establish CMAs in other, usually smaller, basins, and is using its own authority to manage the basin. Institutional transformation through CMAs is stalled, although there is now greater attention to the role of local and provincial governments.

Water users' associations and transformation of irrigation boards

The NWA provides for establishing local cooperative associations to undertake water-related activities for their members' mutual benefit called water users' associations (WUAs). There are several approaches: transforming existing irrigation boards into more inclusive WUAs; establishing new WUAs on small-scale government schemes; or other water users, farmers or not, forming a WUA. In practice, most are organized around irrigation schemes.

Unlike irrigation boards, WUAs should include all water users, for example farm workers and informal water users. Therefore, in transforming the irrigation boards, whose members are nearly all white men, the board members must reach out to farm workers, neighbouring communities and local government, and give them a voice. The commercial farmers have invested substantially in what they consider as 'their' irrigation scheme; for them, the new rule is problematic as people who have made no investment can participate in decisions that affect the scheme's future (Faysse, 2004; Schreiner et al., 2004).

Comparing seven irrigation boards (two in the Olifants), Faysse (2004:14ff.) identifies two factors explaining the level and outcome of involving 'Historically Disadvantaged Individuals' (HDIs). First, commercial farmers' initiatives to open the management to HDIs occur only where upstream HDIs can affect downstream commercial farmers' water availability or where they are paying fees. Although DWAF policy states that all water users can participate in WUA management whether they pay or not, commercial farmers oppose this and discount non-paying members.

Second, there is a lack of clarity about WUA responsibilities and there are competing definitions of 'equity'. Irrigation boards were invariably set up with access to water, fees and votes based on the proportionality rule; therefore, commercial farmers feel emerging farmers' roles should be on an 'equal footing' under this rule. Emerging farmers, often supported by government departments, feel special treatment is 'equitable', given their inherent historical disadvantages.

Faysse (2004:18ff.) suggests preconditions for the effective inclusion of HDIs: representation based on organizing the HDI community, access to information, and stronger capacity to voice problems and influence decisions. To achieve this, Faysse (2004:23) emphasizes that DWAF must monitor progress and use its enforcement capacity where needed. Only a few irrigation boards have been transformed into WUAs to date. The underlying conceptual framework for WUAs is the same as for CMFs – using 'multi-stakeholder platforms' to level the playing field among stakeholders. Clearly, the assumptions behind this approach need to be questioned.

Transferring management of small-scale irrigation schemes to WUAs

Nearly all small-scale irrigation schemes are in former homeland areas. They were designed with entirely different objectives than commercial irrigation, and the problems they face reflect this history. Although some have older roots, many were built by the government in the 1950s, and farmers were basically contract labourers. Most schemes were highly subsidized and stopped operating when the management parastatals collapsed in the mid-1990s (Mpahlele et al., 2000; Shah et al., 2002; Machethe et al., 2004; Veldwisch, 2006).

In the late 1990s, the Limpopo (then 'Northern') Province tried to 'revitalize' some schemes. IWMI, the University of Pretoria and the University of Limpopo (then called the University of the North) became associated with this programme, concentrating on the small schemes below the Flag Boshielo dam. The problems of these schemes include low yields, small plot sizes, high operational costs and centralized management. With low and variable farm incomes, most plot-holders depend largely on other sources of income. Irrigated plots are a source of some security, but people do not invest in them. It is only on some vegetable schemes where (mostly) women have very small holdings that productivity and net income per ha are high, but the holdings are too small to provide sufficient household income (see also Mpahlele et al., 2000). In 2003, a much larger revitalization of smallholder irrigation schemes (RESIS) programme was launched throughout the Limpopo province (see the conclusions, below).

Water as an instrument of social reform: water allocation reform (WAR)

The context of glaring inequities between the poor, largely black, majority and the wealthy, largely white, minority is well recognized by government. A basic premise of reform has been that reversing inequities needs democratic institutions that give a real voice to the poor. However, the democracy-as-solution premise itself needs critical re-examination: can water reform really be the driver to reduce poverty and achieve equity, while preserving the economy, i.e. avoiding rapid radical changes in current ownership patterns?

For senior DWAF officials, 'water is seen as a tool in the transformation of society towards social and environmental justice' (Schreiner et al., 2002:129). They acknowledge the challenges and obstacles, but generally offer solutions within this 'new water management paradigm for poverty eradication and gender equity' (the subtitle of the paper by Schreiner et al., 2002). The new legislation did introduce the paradigm, and DWAF officials are seriously committed to meeting equity goals. But paradigms, whether new or old, carry their own implicit, often hidden, assumptions, which may not always be realistic.

While emphasizing the importance of radical water reform, DWAF also perceives a need to 'balance' equity with productivity and profitability. It is cautious about reallocating too quickly lest 'the country suffer economic or environmental damage as emerging users struggle to establish productive and beneficial use of water' (DWAF, 2005:3–4; see also Garduño and Hinsch, 2005:xi; Seetal and Quibell, 2005). Indeed, this caution is expressed in the minister's National Water Act speech to the National Assembly in 1998: 'Our water policy says that our aim in managing water is not just to ensure equitable access to the resource, not a crude dividing up of so many buckets per person. Our aim is to extract and exact the maximum benefit to society from its use.'[3]

However, Minister Kader Asmal goes on to say that 'The mischief we have to right in the economic use of water is to ensure that the benefits from the use of our common water are equitably shared.' Shortly thereafter he states, '... all South Africans have *equal* (emphasis added) rights of access to water resources.' A subsequent minister, Ms Buyelwa Sonjica, similarly emphasizes 'the need to introduce equity in water distribution', and water as 'one obvious tool for the eradication of poverty' (DWAF, 2004b:1–2). Elsewhere, the minister discusses the need for equity, efficiency and sustainability but does not address the underlying potential trade-offs and contradictions of these three policy 'principles'.

Over time, DWAF appears to have lost faith in using CMAs as a means to achieve equity; in the Olifants, DWAF chose not to forward the CMA proposal to the minister and to carry out the CMA functions itself for the indefinite future. To operationalize these intentions in other domains of its competence, DWAF started implementing a 'water allocation reform' (WAR). The NWA replaces the water rights system that previously combined rights tied to land and, in government water control areas, rights based on prior appropriation, with a fixed-period, tradable licensing system. Moreover, water allocation aims at redressing inequities of the past and allows for transferring water from the 'haves' to the 'have-nots'. In a technical and legal sense, WAR involves implementing this potentially radical transformation.

However, superimposing a licensing system is not necessarily appropriate with huge numbers of poor informal users, and alternative tools such as general authorizations are proposed instead (DWAF 2006; van Koppen, 2007).

DWAF (2005:8) notes that the WAR programme is being implemented because of the 'slow progress with, and little evidence of, redress as we enter the second decade of South Africa's democracy'. But the process proposed is careful, measured, 'balanced', and focused on water and not on land or support services. A major objective of WAR is to 'meet the water needs of HDIs and the poor'. The actions to achieve this include financial support to resource-poor farmers and compulsory licensing to support 'equitable (re)allocation of water' (www.dwaf.gov.za/war/).

The WAR position paper (DWAF, 2005) was discussed in all provinces. In the absence of effective forums, poor rural people will have little voice, placing the entire responsibility on DWAF. Investing in creating effective forums facilitated by DWAF to prevent elite capture might have been a way to achieve broad agreement around the programme. Current state-of-the-art views on promoting institutional reforms suggest the state must be the main driver of reform, but the process itself must be structured and designed to facilitate negotiations and create coalitions of stakeholders (Merrey et al., 2007).

Attempts by DWAF to achieve equity without radical reallocation are seeking to 'balance' factors that may really be incompatible or at least not amenable to water allocation reform alone. This is compounded by the government's lack of an integrated approach to agrarian and rural reform. Land reform and support to new emerging farmers are done with little coordination by the national Department of Land Affairs, the provincial agricultural departments and, to a lesser extent, local governments. Indeed, past water-sector reforms have often been attempted internationally without recognizing that they must be part of a larger inter-sector reform programme (Merrey et al., 2007). In sum, the evidence suggests that water reform alone is not enough. Land reform accompanied by water reform might have a greater impact on equity.

Household rainwater harvesting: reducing malnutrition while avoiding reform

DWAF is initiating a subsidized, household-level rainwater-harvesting programme based on the experience of the Water for Food Movement and systematic pilot testing. Growing fruit and vegetables has substantial benefits (Schreiner et al., 2004; de Lange, 2006:46–48). Grants are provided to build tanks and train women in nutrition and vegetable production and use of water for household purposes, livestock, etc. (DWAF, 2007).

This programme is clearly useful in assisting poor households to improve nutrition, child performance at school and possibly incomes. However, despite substantial short-term benefits for the poor, it does not address the fundamental equity problems or the need for more radical agrarian transformation, and may even divert attention from this.

Trade-offs' paralysis: environment, Mozambique, big business or the poor?

The NWA requires environmental protection. The reserve is the only water 'right' specified in the Act; it has priority over all other uses and must be strictly met before allocating water to other uses. The reserve comprises: (i) the basic human needs reserve, i.e. water for drinking and other domestic uses, consisting of less than 1% of mean annual rainfall (MAR); and (ii) the ecological reserve (i.e. water to protect aquatic ecosystems, requiring an estimated 23% of Olifants MAR) (McCartney et al., 2004; van Koppen, 2007).

The ecological reserve determination for the Olifants was based on the building block method (Tharme and King, 1998; DWAF, 1999; King et al., 2000; Louw and Palmer, 2001), and does not include basic human needs (Schreiner et al., 2002). Standards are set for different reaches of the river – heavily used sections have a lower standard than more pristine sections, which are seen as worthy of preservation.

Currently, average environmental flow requirements are met in most months, except in some locations during the dry season. Water resources do not match demand; therefore,

DWAF is not fully implementing the reserve to avoid damage to existing economic users. Instead, it plans to phase in full implementation over time. Meeting the reserve requirements while providing more water to mining and commercial agriculture is among the main motivations for infrastructural development (i.e. construction of the controversial de Hoop dam on the Steelpoort River and raising the Flag Boshielo dam; DWAF, 2004a). Implementation of the reserve could significantly improve dry-season flows through the Kruger National Park into Mozambique. We are not aware of any detailed assessment of the costs and benefits – and of losers and beneficiaries – of meeting the ecological reserve.[4]

Projections of water demand and supply: discourse of water scarcity trumps all

McCartney and Arranz (2007:1) assess three scenarios of 'future' water demand, based on plausible and internally consistent projections of water use in 2025. They use the water evaluation and planning (WEAP) model, based on water balance accounting, to build scenarios to answer 'what if' questions on changes in allocation, demand and efficiencies (see www.sei.se; SEI, 2001). After developing a 'historic' water demand (1920–1989) and a 'baseline' demand (1995) for each scenario, McCartney and Arranz (2007) assess the implications of constructing new infrastructure and implementing water conservation and demand management practices, and calculate levels of supply assurance; by combining water productivity data with estimated unmet demand, the authors estimate the economic cost of failing to supply water to each scenario.

The annual net demand in 1995 ranges from 577 Mm^3 to 995 Mm^3, depending on rainfall ('average' 744 Mm^3) (McCartney and Arranz, 2007:21). The basin experiences shortfalls annually, mostly for irrigation (approximately 26 Mm^3), and also smaller shortfalls for mining (in this scenario rural and urban supplies are assured at the 99.5% level, i.e. failure would occur less than once in 200 years). The annual cost of this unmet demand, based on figures from Prasad et al. (2006:24) varies from approximately US$6 to 50 million (0.2–1.5% of current GGP), mostly in agriculture. In this scenario, environmental flows are simulated as they are. Full implementation of the reserve would lead to shortfalls in both urban and rural sectors, and would reduce the assurance of supply to mining and irrigation, bringing the total costs to US$13 to 78 million (McCartney and Arranz, 2007:25). The analysis does not assess the benefits of meeting the reserve (there is no market basis for doing so) or the presumed benefits for the livelihoods of poor people.

The three future scenarios project low, medium and high water demand levels, depending on population growth, changes in per capita demand, mine openings and closings, commercial forestry practices and assumptions on implementation of the reserve. They assume no change in commercial irrigation, land use and livestock. Within each scenario, demand fluctuates annually, based on rainfall and hence irrigation requirements, from 625 to 1325 Mm^3 (McCartney and Arranz, 2007: 25).

For all scenarios in 2025, seasonal supply shortfalls occur every year, and since irrigation is given the lowest priority, it suffers the most. In the high-demand scenario, shortfalls occur annually in every sector. The estimated costs range from US$23–404 million (low demand), to US$92–1334 million (high demand), i.e. a range of 12 to 41% of GGP (McCartney and Arranz, 2007:30). The authors also assess the likely impacts of infrastructural development and measures of water conservation and demand management. New infrastructure and water demand management combined result in better levels of supply, although shortfalls are not eliminated; annual costs are reduced to between US$0.6 million (good rainfall in low-demand scenario) to US$191 million (poor rainfall in high-demand scenario) (McCartney and Arranz, 2007:35–36, Table 30).

These scenarios are indicative, offering a useful platform for discussion, and suggest further research, including an assessment of social consequences, the impact of groundwater development and full cost–benefit analyses (McCartney and Arranz, 2007: 33–34). Another gap is linking water productivity and equity with environmental sustainability and international flows to understand the exact nature of their relationship. Current implementation policies (such as water allocation reform) at least implicitly assume a zero-sum game: achieving greater equity will reduce overall productivity (DWAF, 2005). But there is no evidence to support this perspective for agriculture: smallholders can certainly achieve high levels of water productivity, and more equitable allocation of basic water supplies will undoubtedly have large impacts on local productivity and well-being. In other sectors, there may well be water productivity economies of scale; in this case, benefit sharing becomes crucial, as discussed below.

A more systematic socio-economic and political analysis is needed as a basis for integrated reform policies (e.g. land and water), and researchers could use tools such as WEAP to identify alternatives. Surprisingly, no investigations have assessed more radical alternatives. In future, demand will increase. Plausible scenarios indicate that even with low to medium growth (i.e. net water demand increasing to between 818 and 1073 Mm^3 by 2025), currently planned infrastructure will be insufficient to meet demands, including those of the reserve; shortfalls will occur every year, with irrigation suffering most (McCartney and Arranz, 2007:26–27, Table 20). Water conservation and demand management interventions must be implemented.

Outcomes to Date: Old and New Winners and Losers

We have discussed the extreme inequity in the Olifants basin, its history and drivers. In the mid-1990s, the former homeland areas, with 64% of the population, accounted for less than 3% of the total agricultural GGP, 2.35% of total mining GGP and 3.4% of manufacturing GGP (Lévite, 2003). This inequity continues and may not be improving. Researchers have applied three methodologies for measuring equity of both access to and benefits from water: the water poverty index, equity coefficient and Gini coefficient. All of these measures have limitations, but taken together they reinforce the observation of continuing high levels of inequity. Molle and Mollinga (2003) and Shah and van Koppen (2006) warn that such indicators must be used cautiously and complemented with local in-depth studies, but the findings do provide important insights.

Magagula et al. (2006) assess the impact of water scarcity and lack of water access using the 'water poverty index' (WPI), which is based on five component indices: resources, access, capacity, use and environment, each with various sub-indices and using a scale from 0 to 100.[5] A low score indicates high poverty. The WPI of the Olifants basin was 27.1 for 2001, half the national estimated WPI (52.2). The WPI is worst in and near the former homelands, as displayed in Fig. 3.4. Although WPI improved in many quaternaries between 1994 and 2005, Magagula et al. (2006) point out that many quaternaries changed very little, despite interventions by DWAF.

Prasad et al. (2006) use data from DWAF's Water-use Authorization and Management System and other sources to assess equity – 'who uses how much water, where, and for what purpose' (Prasad et al., 2006:67). They examine 13 tertiary sub-basins and four sectors – agriculture, industry, mining and water supply services – and calculate a measure of 'skewness', the degree of diversion from total equity (which they refer to as 'equity coefficient'), in terms of 'water use per capita' and 'water use per unit area'. The equity coefficient ranges from 0 to 1, zero being the least equitable.

They note the huge variation among sub-basins within all sectors. The equity coefficients for per capita water use are highly skewed and low. In agriculture, a few farmers receive most of the water. More striking is that the least equitable sector was basic water services, even in 2003. The water services and agriculture sectors are intended to serve individuals and numerous farms and therefore should be the most meaningful; industry and mining are in the hands of a few large firms, making the measure less useful. Figure 3.5 combines two measures for each sector, i.e. water use per capita and water use per unit area, to provide a composite score. By this measure, the basin-level average equity coefficient is a low 0.161. Agriculture is again the least inequitable and water supply the most inequitable.

Cullis and van Koppen (2007) use the Gini coefficient to assess inequality of access to water in the basin, to our knowledge the first attempt to do so in the world. In a perfectly

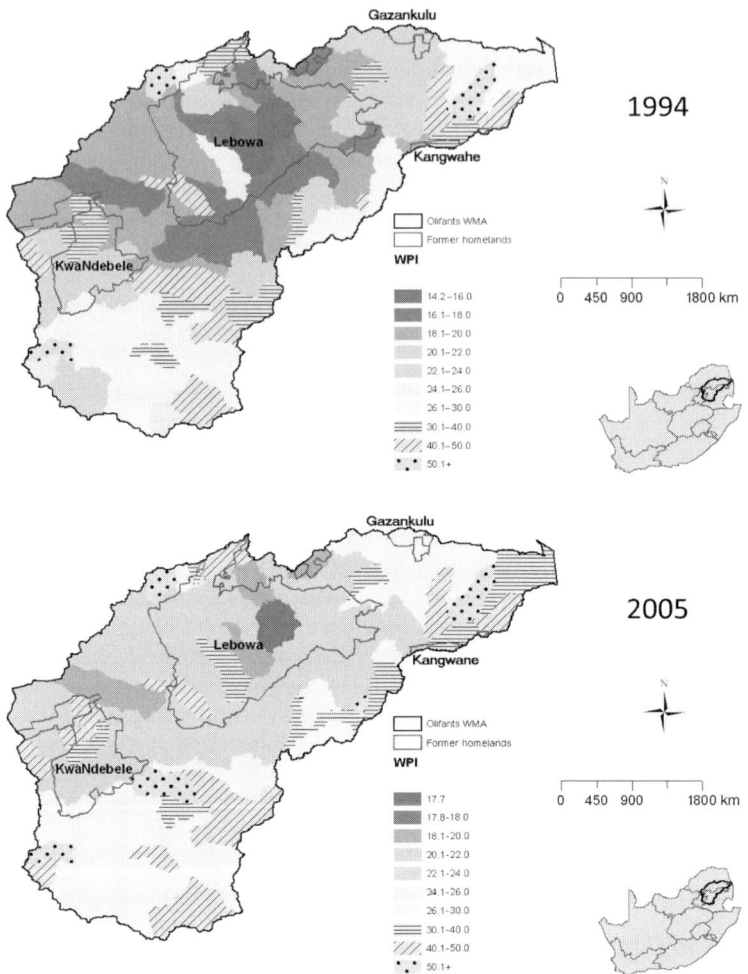

Fig. 3.4. Changes in the water poverty index (WPI), in the Olifants basin (Olifants Water Management Area (WMA)), 1994–2005. From Magagula et al. (2006).

equal situation, the Lorenz curve would be a straight line, termed the line of equality, and the Gini coefficient 0.0. In most cases, it diverges below the line of equality, showing the inequality of distribution of income, land or water, with the Gini coefficient moving to 1.0 for total inequality.

The Gini coefficient for South Africa's national income is the second highest among middle-income countries after Brazil, and has been increasing during the past decade, from 0.60 in 1995 to 0.64 in 2001 (Cullis and van Koppen, 2007). This distribution obviously reflects the historical legacy. Inequality of access to land is even worse than inequality of income, and is intimately related to the inequality of access to water and its benefits. Cullis and van Koppen (2007) measure the distribution of direct access to water by rural households and the distribution of indirect benefits of water use in the form of direct employment.

Using DWAF estimates, the Gini coefficient of direct rural water use is a shocking 0.96. The 1782 registered users claim to use 1550 Mm^3 per year, while the 290,000 rural households use an estimated (not 'claimed') 74 Mm^3 per year. Therefore, 99.5% of rural households use just 5% of the total water used, demonstrating an extremely inequitable distribution (Fig. 3.6). These findings may exaggerate the

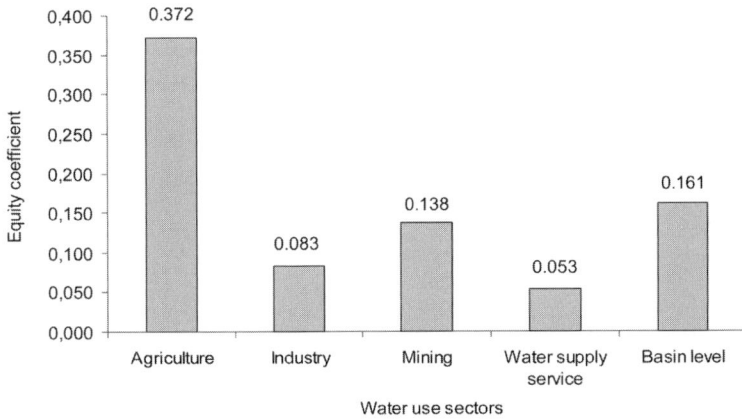

Fig. 3.5. Combined equity coefficients in the Olifants basin, 2003. From Prasad et al. (2006).

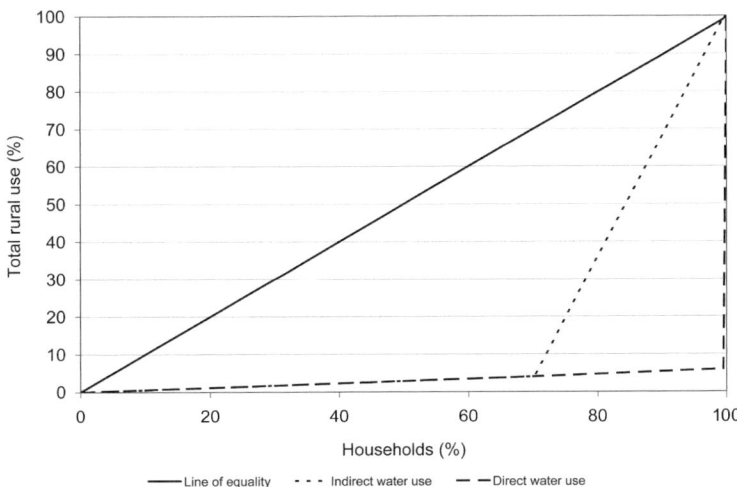

Fig. 3.6. Distribution of estimated direct and indirect rural water use in the Olifants basin. From Cullis and van Koppen (2007).

inequity. 'Claimed' water use is likely to be significantly higher than actual water use, as large-scale users attempt to maximize the amount they can obtain through registration.

Further, as alluded to in the minister's speech quoted above, extracting maximum benefits and sharing these equitably are more important than 'dividing up so many buckets per person'. Using official employment figures and assuming that all industries have equal levels of efficiency and all employed persons benefit equally (ensuring a 'best possible' but highly unrealistic case), Cullis and van Koppen (2007) plot the distribution in terms of employment. The Gini coefficient for the benefits of water use in rural areas is 0.64, better than the 0.96 for direct use but equal to the national Gini coefficient and still highly unequal.

Cullis and van Koppen (2007) also test two policy scenarios: (i) the impact on equality of revitalizing small-scale irrigation; and (ii) increasing the allocations to all rural households. Because it affects relatively few people, revitalizing small-scale irrigation has a marginal impact. This finding is confirmed in an adjacent basin by Hope et al. (2008). Increasing the direct allocation of water to unemployed households from the current approximately

255 m³ per household per year to 610 m³ per household per year would improve the amount of water available for domestic use and permit irrigation of a garden of 1000 m². Existing registered users would have to reduce their irrigation demand by just 6%. The water-use Gini coefficient would improve slightly for both direct water use (0.94 to 0.90) and distribution of benefits (0.65 to 0.58).

The Gini coefficient is potentially a useful tool to assess policy scenarios and measure outcomes, but as shown above mere 'tinkering' to improve equity in a 'balanced' manner will contribute only marginally to achieving the country's equity goals.

The current WAR process is intended to 'promote equity, address poverty, generate economic growth, and create jobs' (DWAF, 2005:1). A recent paper whose first two authors were senior DWAF officials has the intriguing title, 'Washing away poverty: water, democracy and gendered poverty eradication in South Africa' (Schreiner et al., 2004). However, the evidence to date does not support using water reforms as an entry point for wider socio-economic reforms. Reforms in other sectors, especially land, combined with strengthening the political voice of relatively disenfranchised people in an integrated manner is critical.[6] Otherwise, the politically powerful water users will continue to prosper while deprivation continues among the poor. We return to this theme below.

Conclusion: Will the Poor Basin Resident Get Her Fair Share?

Continuities from apartheid to democracy: old paradigms in new bottles

The National Water Act introduced a new water management paradigm to support the restructuring of South African society as mandated by the constitution. Although many new ideas were introduced, we have also been struck by the high degree of continuity – mostly unconscious and denied if pointed out – in assumptions and concepts that may be impediments to achieving the equity goals, as they are hold-overs from an era with antithetical objectives. Van Koppen (2007) has also raised this point with regard to requirements that water investments must be 'economically viable' and even self-financing. Tapela (2005:5) argues that the emphasis on 'efficiency', user-pays principle and 'economic value' of water narrows the prospects of resource-poor, small-scale farmers.

This 'commoditization of water', rather similar to the current reliance on the market for acquiring land to implement land reform, is not conducive to encouraging smallholder farmers; rather, it further strengthens the hand of the large-scale users and weakens the case for reallocation to the poor. Further, in the current discourse, 'water scarcity' is redefined as an entirely physical phenomenon, not one that is largely socially and politically constructed (and can therefore be reconstructed, though not easily). By choosing to accommodate the large-scale water users and environmental requirements as a de facto high priority, it forces water reforms to deal at the margin.

The truth is that South Africa and the Olifants basin are not seriously constrained by an absolute physical scarcity of water; rather, the perceived 'scarcity' has been created by large allocations to commercial agriculture and mines, and now also to the ecological reserve, thus closing the door to other alternatives. But the discourse on this created situation of 'scarcity' is always – misleadingly – in terms of physical scarcity, thus avoiding assessment of other choices. Hence, the few attempts at scenario building, if they refer to reallocation at all, propose relatively small transfers from the rich to the poor, certainly potentially benefitting the poor while not threatening the rich but definitely not having much impact on equity. They assume the current status quo, i.e. continuing priority to large-scale sectors.

Another continuing, unexamined assumption is that, in agriculture, 'large is best'. The historical development of white agriculture in South Africa has led to large-scale, highly capitalized farms, now seen as inevitable: there is no vision for small- or medium-scale farmers, except as transitional to larger farms. Indeed, Lahiff (2007:11, 13) points out that explicit legal and policy restrictions against subdividing farms remain in place, based on a 1970 apartheid-era law 'inspired by the danger of ... blackening of the countryside'. Lahiff suggests the failure to subdivide is the single greatest contributor to the underperformance of land reform. It is based on the 'viable size' argument for maintaining white farmers' minimum incomes. Over time, the agrarian economy has been structured around the model of large-scale agriculture.

There is a hidden assumption of a trade-off between equity and productivity. However, small farms tend to be undercapitalized, with poor access to information and markets – lower water productivity is certainly not an inherent characteristic of small or medium-size farms, although total income from a small farm is lower. Therefore, official discussion revolves around how more of the large farms can become black owned, not whether there are more equitable alternatives.

As far back as 1977, South African water managers believed their approach was aligned with international standards, as documented at the Mar del Plata conference (van Koppen, 2007:36). Although the rhetoric emphasizes economic viability and user-pays principle, DWAF has continued to subsidize modern, large-scale white farms – the Lower Blyde Irrigation Board's new pipeline replacing a leaky canal was financed with a loan guarantee from DWAF (i.e. a subsidy) on a promise that 800 ha of additional land for previously disadvantaged farmers would also be included.[7]

The programmes to 'revitalize' small-scale irrigation in former homelands are also based on some old assumptions: that farmers are mostly men, and that small farms based on the old land allocations (1.28–5 ha) can be 'economically viable' for black families if only they have better technologies and better links to markets.[8] The Limpopo province is currently implementing a billion rand (US$130–200 million) revitalization programme. Initially designed to emphasize farmer empowerment, capacity building and community involvement, pressures to spend funds quickly led to a shift to promoting sophisticated technologies installed by commercial contractors with little beneficiary participation (de Lange, 2006:21–22; Denison and Manona, 2007:32–33, 35). It is unlikely that such a programme will make a substantial difference, as Tapela (2008) also concludes.

While DWAF is being substantially restructured, the main functional difference from the old department is the addition of forestry to its mandate: there has been no restructuring of water, land and agriculture into some kind of agrarian reform ministry, for example. Most literature has emphasized the break with past policies and paradigms, which in many respects is real, and South Africa deservedly receives much credit as an IWRM pioneer; however, even before 1994, South African water planners perceived themselves as pioneers in IWRM principles (van Koppen, 2007). It is important also to note the reality of continuity underpinning the new paradigm: it may be a new bottle but the contents are a mixture of old and new.

Institutional stagnation

While institutional reforms are stalled in the Olifants, there are many innovative experiments underway elsewhere, such as the estimated 200 catchment management forums. Therefore, it is a mistake to generalize to the entire country from this discussion – although it is equally wrong to claim that the Olifants findings are not relevant elsewhere. However, the evidence shows that transformation of irrigation boards to participatory and representative WUAs has stalled nationally. Promotion of new WUAs in small-scale schemes is proceeding slowly. In a few basins, catchment management agencies have been initiated, but in the Olifants the process was stopped when DWAF realized it was not leading to the kind of stakeholder-driven institution envisioned by the Water Act. Unfortunately, DWAF did not promote smallholder water user forums in the basin, to enable broader participation. The water allocation reform (WAR) programme itself is progressing slowly, partly because the disadvantages of the conversion of former rights to licences are becoming clearer. For example, it is simply impossible to issue credible licences to the thousands of small users.

One problem may be that DWAF is trying to do too many different and complex things simultaneously. Trying to achieve very difficult institutional reforms while also meeting stringent environmental standards, strengthening local government capacity and implementing major infrastructural projects, all while undergoing its own restructuring, is probably an impossible task for any organization. This is compounded by a more serious problem – the lack of an integrated approach across sectors and departments to institutional reform: land reform, agricultural services and mining are all under different departments. How can one achieve significant water equity unless the associated inequity in land is addressed simultaneously? How can local communities benefit from mines in their midst if they do not have a voice to demand a reasonable share of the benefits? This fragmentation may be the reason for DWAF's search for a 'balanced' approach – it has no choice.

Finally, the discourse on 'water scarcity' as a largely physical phenomenon has not helped. This socially created perception is rarely questioned and leads to claims that there are serious trade-offs between equity and productivity, that the options are limited and that satisfying downstream international and environmental demands while achieving real equity in benefits is impossible. This discourse has resulted in an inability to envisage alternative visions for the Olifants.

Potential for change under the democratic dispensation

The development trajectory of the Olifants basin simultaneously reflects the broader patterns of historical development in South Africa and the 'typical' pattern of basin development, where demand for water exceeds the available supply. The current incomplete and uncertain status of reforms represents a pattern characterizing most middle-income countries (for example, see Wester (2008) on Mexican reforms). All river basins are 'unique' in many respects, but there are also commonalities that provide grist for the science of river basin management.

The following are the most salient conclusions emerging from this study; they are discussed further below:

1. The Olifants is an extreme example of capture and development of natural resources, including water, for the benefit of a very small minority at the expense of the majority of inhabitants: it is a trajectory of water resources development initially for commercial agriculture, mining and energy, and more recently for industry and cities, now accompanied by concerns for environmental flows and availability of water for basic human needs.

2. Promulgation of a revolutionary water reform process after 1994, driven by constitutional and political imperatives, and expressed through the National Water Act of 1998, has not met expectations to date.

3. There are glaring contrasts among high expectations of using water as an instrument for poverty eradication and social reform, the cautious technocratic approach to implementation of reforms and disappointing outcomes to date.

4. A rhetorical and formal break with the past priority on development for the few has been accompanied by continuities that undermine reform objectives.

5. Opportunities for reducing poverty through achieving a higher degree of water equity and productivity do exist.

Within the international water management community, the NWA is rightly famous and is held up as a model. It is based on international 'best practices' such as Integrated Water Resources Management (IWRM) principles, democracy, meeting basic human water needs and prioritizing ecological requirements. Implementation of the NWA in the Olifants basin had begun even before it became law. There can be no doubting how seriously implementation is being pursued, or the professionalism of government departments, including DWAF. Nevertheless, progress has been slow.

The optimism about using water as a lever to achieve social and economic reforms was unrealistic for at least two reasons: (i) the cautious technocratic approach to implementation of water reforms; and, probably more salient, (ii) the lack of an integrated multi-departmental implementation.

DWAF wishes to achieve radical reforms without damaging the perceived stream of benefits from large-scale uses. Its officials usually work to 'perfect' policies and procedures in writing through consultation before any field testing is initiated. It has therefore been slow in establishing WUAs, transforming irrigation boards and implementing water reallocation. It has delayed the process of implementing the CMA out of well-placed fears that it would be captured by existing elites, but it has failed to promote proposed democratic grassroots forums. During this process, DWAF has seemed reluctant to try new ideas on a small scale to learn lessons before scaling up. Recently, it appears that DWAF has really been internalizing lessons learned, for example through its new initiatives on Water for Growth Development.

Another problem is the technocratic, as opposed to political, approach taken by DWAF. This reflects the technical expertise and mandate of the department. After the first Minister of Water Affairs (Professor Kader Asmal), the succeeding three ministers focused their attention primarily on delivering water supply and sanitation to the previously unserved population. This priority is understandable but may have been at the expense of actively supporting reforms.[9] Both the aborted CMA proposal process and the WAR programme have been left to technocrats, as if one can 'engineer' a satisfactory solution that provides water to new users while avoiding serious inconvenience to large-scale interests.

In fairness, it must be stated that the disappointing outcome of this cautious approach is largely a product of the lack of an integrated multi-departmental approach to reform – a higher-level political failure. Although DWAF has undertaken various efforts to establish coordinating committees with the Department of Agriculture, with mixed results, the problem is, to reiterate, a higher-level political failure. This is compounded by the efforts required to establish an entirely new local and provincial government structure to replace the pre-1994 territorial and institutional segregation. In hindsight, such an integrated approach might have directed attention to the root problem recognized in 1998 by the Minister of Water Affairs: the point is not 'dividing up so many buckets of water per person' but to produce and share equitably the maximum possible benefits.

The new South Africa is dramatically different from the old. There is now a remarkably open, democratic, inclusive and still idealistic political system. Nevertheless, as also noted by van Koppen (2007), one can also perceive striking continuities between the old and new regimes, suggesting a high degree of hidden 'path dependency'. Ideologically, ideas about the importance of the economy (cost recovery) have continued, even when accompanied by the reality of state subsidies. For example, the de Hoop dam will benefit large-scale mining firms most, with some 'trickle-down' to poor communities. While acknowledging substantial public investments for domestic water supply schemes for people in no position to cover the costs, these schemes are constructed to an entirely different standard (25 litres/person/day) than those in the wealthy cities. This seems similar to the old idea that the required landholding for a black farmer to be self-sufficient is smaller than for a white farmer. In the past, infrastructure was built to promote the interests of race-based (i.e. white) capitalists; today, with 'Black Economic Empowerment', a new black and white elite continues to receive extraordinary benefits. Water, like land, continues to be monopolized by a small group of privileged people, while the government continues its 'hydraulic mission', with priority for promoting large-scale interests (usually sweetened by reference to community benefits).

One lesson learned is that a single-factor or single-sector approach is inadequate. Providing a better water supply in the absence of other inputs is not enough for profitable agriculture. Similarly, hamstrung by legal impediments to subdividing farms, government has tried to allocate land to groups with little experience in agriculture and with insufficient institutional support. There has been insufficient examination of alternative futures for South African agriculture and water use.

It would be presumptuous for us to propose such alternative futures. However, we are prepared to offer the following ideas to stimulate thinking on this issue. In the short to medium term, government could adjust its investments to improve equity, productivity and well-being. Examples include large-scale implementation of household rainwater harvesting and other water infrastructure; a more bold approach to reallocating water from large-scale users to others; more effective technical, financial and institutional support for smallholder producers to enable them to increase their incomes in a sustainable way; and paying greater attention to ensuring that the benefits from large commercial water users such as mines are shared equitably with communities. Even the modest reallocation of water from large-scale commercial users to rural households is likely to have a useful impact on the well-being of poor rural people.

But for the longer term we believe a new agrarian vision is urgently required. A possible approach would be to commission a small group of eminent visionary people to articulate a set of alternative agrarian futures, including specific ideas on integrated implementation arrangements. The goal would be to achieve equitable land and water reforms that satisfy the needs and demands of rural and peri-urban people, recognition of women's roles in agriculture and small enterprises, provision of effective private and public support services to new farmers, and new models for wider sharing of benefits while minimizing local costs of mining mineral wealth. The commission's report can be used for widespread consultations on the alternatives, with strong political participation. These consultations would provide a platform for political leaders to move forward.

Acknowledgments

The Comprehensive Assessment of Water Management in Agriculture supported Douglas Merrey to prepare a synthesis of work done in the Olifants basin over the past decade (Merrey, 2007). This chapter draws from that work. Matthew McCartney has offered important comments on that synthesis work, which the authors greatly appreciate. Mike Muller, former Director General of DWAF, offered very critical observations on an earlier version of the paper and we are very grateful – even if we have not always accepted his perspective. We appreciate the comments of François Molle, Flip Wester and an anonymous reviewer. FANRPAN provided a support system to Doug Merrey to enable this chapter to be written, while both the Comprehensive Assessment and IWMI have supported Barbara van Koppen's work. The authors remain solely responsible for the contents of this chapter.

Notes

1 There are 19 officially designated 'Water Management Areas' in South Africa, which are intended to be river basin management units under the National Water Act of 1998.
2 The term 'white water economy' is taken from van Koppen (2007).
3 This quote and subsequent ones are taken from a selection of policy statements provided to us by Mr Mike Muller, former Director General of DWAF.
4 For lack of space we have not dealt with issues of water quality; however, there is increasing concern about its impacts on humans and wildlife; see, for example, the following report on crocodile deaths in the Olifants within the Kruger Park: www.int.iol.co.za/index.php?set_id=1&click_id=31&art_id=vn20080605055357280C518855, accessed 4 July 2008.
5 See Sullivan (2002) and Sullivan et al. (2002) for explanations of the WPI index calculation.
6 A point fully recognized by some officials, including B. Schreiner, but the institutional barriers to such integration are overwhelming.
7 Two years after the approval of this loan guarantee, it appeared the 'solution' was one or two large farms to be owned by black Africans under the government's Black Economic Empowerment (BEE) programme; BEE is increasingly controversial – critics perceive it as insufficiently broad based and therefore leading to changing the colour of the elite and not greater equity. Land claims have stalled this process. The new pipeline is currently operated profitably by the Rand Merchant Bank. We have no recent information on which to base further remarks.
8 Locally, small plots are often seen as acceptable because they enable more equitable land allocations, given the limited irrigated area available.
9 However, the 'Masibambane III' programme, co-financed by the European Union and other partners and recently launched by the DWAF Minister Hon. Lindiwe Hendricks, explicitly includes completion of departmental restructuring and promoting institutional reforms, as envisioned by the NWA (Water Wheel, 2008).

References

Ahmad, M.D., Magagula, T.F., Love, D., Kongo, V., Mul, M.L. and Kinoti, J. (2005) Estimating actual evapotranspiration through remote sensing techniques to improve agricultural water management: a case study in the transboundary Olifants catchment in the Limpopo basin, South Africa. Paper presented at the 6th WaterNet/WARFSA/GWP Annual Symposium, 1–4 November 2005, Ezulwini, Swaziland.

ARC (Agricultural Research Council), IWMI (International Water Management Institute) (2003) *Limpopo Basin Profile*. ARC–Institute for Soil, Climate and Water, ARC–Institute for Agricultural Engineering and IWMI, August 2003, for the Challenge Program on Water and Food. www.limpopo.arc.agric.za/profile.htm.

Backeberg, G.R. (2005) Water institutional reforms in South Africa. *Water Policy* 7(1), 107–123.

Basson, M.S. and Rossouw, J.D. (2003) *Olifants Water Management Area: Overview of Water Resources Availability and Utilization*. DWAF Report No. P WMA 04/000/00/0203. Final issue, September 2003. Department of Water Affairs and Forestry Directorate of Water Resources Planning and BKS (Pty) Ltd, Pretoria.

Bulpin, T.V. (1956) *Lost Trails of the Transvaal*. Stephan Phillips Africana Series, Somerset, South Africa.

Bundy, C. (1988) *The Rise and Fall of the South African Peasantry*, 2nd edn. David Philip, Cape Town and Johannesburg.

Carmo Vaz, A. (2000) Coping with floods – the experience of Mozambique. Paper presented at the first WARFSA/Waternet Symposium, Maputo, 1–2 November 2000.

Cullis, J. and van Koppen, B. (2007) *Applying the Gini Coefficient to Measure the Inequality of Water Use in the Olifants River Water Management Area, South Africa*. IWMI Research Report 113. IWMI, Colombo, Sri Lanka.

De Coning, C. (2006) Overview of the water policy process in South Africa. *Water Policy* 8, 505–528.

De Coning, C. and Sherwill, T. (2004) *An Assessment of the Water Policy Process in South Africa (1994 to 2003)*. Water Research Commission Report No. TT 232/04. Water Research Commission, Pretoria.

de Lange, M. (2004) Water policy and law review process in South Africa with a focus on the agricultural sector. In: Mollinga, P.P. and Bolding, A. (eds) *The Politics of Irrigation Reform: Contested Policy Formulation and Implementation in Asia, Africa, and Latin America*. Ashgate, Hants, UK, pp. 11–56.

de Lange, M. (2006) A literature study to support the implementation of Micro-AWM technologies in the SADC region. Paper produced for International Water Management Institute. Unpublished.

de Lange, M., Merrey, D.J., Levite, H. and Svendsen, M. (2005) Water resources planning and management in the Olifants basin of South Africa: past, present and future. In: Svendsen, M. (ed.) *Irrigation and River Basin Management: Options for Governance and Institutions*. CAB International, Wallingford, UK; International Water Management Institute, Colombo, Sri Lanka, pp. 145–168.

Delius, P. (1983) *The Land Belongs to Us: the Pedi Polity, the Boers, and the British in the Nineteenth-Century Transvaal*. Ravan Press, Johannesburg.

Denison, J. and Manona, S. (2007) *Principles, Approaches and Guidelines for the Participatory Revitalization of Smallholder Irrigation Schemes. Vol. 2: Concepts and Cases*. WRC Report No. TT 309/07. Water Research Commission and Arcus Gibb, Pretoria.

Department of Water Affairs (1986) *Management of the Water Resources of the Republic of South Africa*. Department of Water Affairs, Pretoria.

DWAF (Department of Water Affairs and Forestry) (1999) Comprehensive ecological reserve methodology. Available at: www.dwaf.gov.za/docs/Water%20Resource%20Protection%20Policy/river%20ecosystems/riv_sectionF_version10.doc.

DWAF (2004a) *Olifants Water Management Area: Internal Strategic Perspective*. Prepared by GMKS, Tlou and Matji on behalf of the Directorate, Water Resources Planning. DWAF Report No. P WMA 04/000/00/0304. Department of Water Affairs and Forestry, Pretoria.

DWAF (2004b) *National Water Resource Strategy*. First edition, September 2004. Department of Water Affairs and Forestry, Pretoria.

DWAF (2005) *A Draft Position Paper for Water Allocation Reform in South Africa: Towards a Framework for Water Allocation Planning*. Discussion document, January 2005, Directorate: Water Application. Department of Water Affairs and Forestry, Pretoria.

DWAF (2006) *Assignment to Develop and Test Methodologies for Determining Resource-specific General Authorizations under the National Water Act*. Prepared by Ninham Shand in association with Umvoto Africa and Synergistics Environmental Services. Directorate: Water Allocation, Department of Water Affairs and Forestry, Pretoria.

DWAF (2007) *Programme Guidelines for Intensive Family Food Production and Rainwater Harvesting*. Draft, March 2007. Department of Water Affairs and Forestry, Pretoria.

Earle, A., Goldin, J., Machiridza, R., Malzbender, D., Manzungu, E. and Mpho, T. (2006) *Indigenous and Institutional Profile: Limpopo River Basin*. IWMI Working Paper 112. International Water Management Institute, Colombo, Sri Lanka.

FAO (Food and Agriculture Organization of the United Nations) (2004) *Drought Impact Mitigation and Prevention in the Limpopo River Basin: a Situation Analysis*. Land and Water Discussion Paper 4. Prepared by the FAO Subregional Office for Southern and East Africa, Harare. FAO, Rome.

Faysse, N. (2004) *An Assessment of Small-scale Users' Inclusion in Large-scale Water Users Associations of South Africa*. IWMI Research Report 84. International Water Management Institute, Colombo, Sri Lanka.

Garduño, H. and Hinsch, M. (2005) *IWRM Implementation in South Africa: Redressing Past Inequities and Sustaining Development with a View to the Future*. World Bank Institute, Washington, DC.

Hope, R.A., Gowing, J.W. and Jewitt, G.P.W. (2008) The contested future of irrigation in African rural livelihoods – analysis from a water scarce catchment in South Africa. *Water Policy* 10, 173–192.

Kamara, A.B., van Koppen, B. and Magingxa, L. (2002) Economic viability of small-scale irrigation systems in the context of state withdrawal: the Arabie scheme in the Northern Province of South Africa. *Physics and Chemistry of the Earth* 27, 815–823.

King, J., Tharme, R.E. and de Villiers, M.S. (2000) *Environmental Flow Assessments for Rivers: Manual for the Building Block Methodology*. Water Research Commission Technology Transfer Report No. TT 131/00. Water Research Commission, Pretoria, South Africa.

Klarenberg, G. (2004) Fishing in troubled waters: two case studies of water quality management in sub-catchments of the Olifants basin, South Africa. Masters thesis in Law and Governance, Wageningen University, The Netherlands.

Lahiff, E. (2007) Land redistribution in South Africa: progress to date. Paper prepared for workshop on 'Land Redistribution in Africa: Towards a Common Vision'. www.sarpn.org/documents/d0002695/Land_Redistribution_South_Africa.pdf.

Le Roy, E. (2005) A study of the development of water resources in the Olifants catchment, South Africa: application of the WEAP model. MSc and Diploma thesis, Imperial College, London.

Lévite, H. (2003) Quantification of the inequities in water use in ex-homelands and ex-RSA in Olifants. Draft Discussion Note, 26 November 2003, International Water Management Institute, Colombo, Sri Lanka. Unpublished.

Lévite, H., van Koppen, B. and McCartney, M. (2003) The basin development trajectory of the Olifants basin, South Africa. Paper presented at the WARFSA/WaterNet Symposium, Gaborone, Botswana.

Ligthelm, M. (2001) Olifants water management area: catchment management agency establishment. In: Abernethy, C. (ed.) *Intersectoral Management of River Basins: Proceedings of an International Workshop on Integrated Water Management in Water-stressed River Basins in Developing Countries: Strategies for Poverty Alleviation and Agricultural Growth*. International Water Management Institute/Deutsche Stiftung für Internationale Entwicklung, Colombo, Sri Lanka, pp. 23–43.

Louw, M.D. and Palmer, C. (2001) *Olifants River Ecological Water Requirements Assessment. Ecological Management Class: Technical Input*. Final Report. Report PB 000-00-5499. DWAF, Pretoria.

Machethe, C.L., Mollel, N.M., Ayisi, K., Mashotala, M.B., Anin, D.D.K. and Vanasche, F. (2004) *Smallholder Irrigation and Agricultural Development in the Olifants River Basin of Limpopo Province: Management Transfer, Productivity, Profitability and Food Security Issues*. WRC Report No. 1050/1/04. Water Research Commission, Pretoria.

Magagula, T.F. and Sally, H. (2005) Water productivity in irrigated cultivated land in the Olifants basin in South Africa. Draft International Water Management Institute Research Report. Unpublished.

Magagula, T.F., van Koppen, B. and Sally, H. (2006) Water access and poverty in the Olifants basin: a spatial analysis of population distribution, poverty prevalence and trends. Paper presented at 7th Waternet/WARFSA/GWPSA Symposium, 1–3 November 2006, Lilongwe, Malawi.

McCartney, M. and Arranz, R. (2007) *Evaluation of Historic, Current and Future Water Demand in the Olifants River Catchment, South Africa*. IWMI Research Report 118. International Water Management Institute, Colombo, Sri Lanka.

McCartney, M.P., Yawson, D.K., Magagula, T.F. and Seshoka, J. (2004) *Hydrology and Water Resources Development in the Olifants River Catchment*. IWMI Working Paper. International Water Management Institute, Colombo, Sri Lanka.

Merrey, D. (2007) Balancing equity, productivity and sustainability in a water-scarce river basin: the case of the Olifants River basin in South Africa. Unpublished paper submitted to International Water Management Institute.

Merrey, D.J., Meinzen-Dick, R., Molinga, P. and Karar, E. (2007) Policy and institutional reform: the art of the possible. In: Molden, D. (ed.) *Water for Food, Water for Life: the Comprehensive Assessment of Water Management in Agriculture*. Earthscan, UK, pp. 193–232.

Molle, F. and Mollinga, P. (2003) Water poverty indicators: conceptual problems and policy issues. *Water Policy* 5, 529–544.

Mpahlele, R.E., Malakalaka, T.M. and Hedden-Dunkhorst, B. (2000) *Characteristics of Smallholder Irrigation Farming in South Africa: a Case Study of the Arabie–Olifants River Irrigation Scheme*. IWMI South Africa Working Paper. International Water Management Institute, University of the North, Colombo, Sri Lanka.

Muller, M. (2001) How national water policy is helping to achieve South Africa's development vision. In: Abernethy, C. (ed.) *Intersectoral Management of River Basins: Proceedings of an International Workshop on Integrated Water Management in Water-stressed River Basins in Developing Countries: Strategies for Poverty Alleviation and Agricultural Growth*. International Water Management Institute/Deutsche Stiftung für Internationale Entwicklung, Colombo, Sri Lanka, pp. 3–10.

Muller, M. (2007) Parish pump politics: the politics of water supply in South Africa. *Progress in Development Studies* 7, 33–45.

Post Uiterweer, N.C., Zwarteveen, M.Z., Veldwisch, G.J. and van Koppen, B. (2006) Redressing inequities through domestic water supply: a 'poor' example from Sekhukhune, South Africa. In: Perret, S., Farolfi, S. and Hassan, R. (eds) *Water Governance for Sustainable Development: Approaches and Lessons from Developing and Transitional Economies*. Earthscan, London, pp. 54–74.

Prasad, K., van Koppen, B. and Stryzepek, K. (2006) Equity and productivity assessment in the Olifants River basin, South Africa. *Natural Resources Forum* 30, 63–75.

Reader, J. (1998) *Africa: a Biography of a Continent*. Penguin Books, London.

Schreiner, B. and van Koppen, B. (2001) Catchment management agencies for poverty eradication in South Africa. Paper presented at 2nd WARFSA/WaterNet Symposium on Integrated Water Resources Management Theory, Practice, Cases, Cape Town, 30–31 October 2001.

Schreiner, B. and van Koppen, B. (2003) Policy and law for addressing poverty, race and gender in the water sector: the case of South Africa. *Water Policy* 5, 489–501.

Schreiner, B., van Koppen, B. and Khumbane, T. (2002) From bucket to basin: a new water management paradigm for poverty eradication and gender equity. In: Turton, A.R. and Henwood, R. (eds) *Hydropolitics in the Developing World: a Southern African Perspective*. Africa Water Issues Research Unit, Centre for International Political Studies, University of Pretoria, Pretoria. (CD-ROM produced by International Water Management Institute).

Schreiner, B., Mohapi, N. and van Koppen, B. (2004) Washing away poverty: water, democracy and gendered poverty eradication in South Africa. *Natural Resources Forum* 28 (3), 171–178.

Seetal, A.R. and Quibell, G. (2005) Water rights reform in South Africa. In: Bruns, B.R., Ringer, C. and Meinzen-Dick, R. (eds) *Water Rights Reform: Lessons for Institutional Design*. International Food Policy Research Institute, Washington, DC, pp. 153–166.

SEI (Stockholm Environment Institute) (2001) *WEAP: Water Evaluation and Planning System – User Guide*. Stockholm Environment Institute, Boston, Massachusetts.

Shah, T. and van Koppen, B. (2006) Is India ripe for integrated water resources management? Fitting water policy to national development context. *Economic and Political Weekly*, 5 August 2006, 3413–3421.

Shah, T., van Koppen, B., Merrey, D., de Lange, M. and Samad, M. (2002) *Institutional Alternatives in African Smallholder Irrigation: Lessons from International Experience with Irrigation Management Transfer*. IWMI Research Report 60. International Water Management Institute, Colombo.

Simpungwe, E. (2006) Water, stakeholders and common ground: challenges for multi-stakeholder platforms in water resource management in South Africa. PhD dissertation, Wageningen University, Wageningen, The Netherlands.

Stimie, C., Richters, E., Thompson, H., Perret, S., Matete, M., Abdallah, K., Kau, J. and Mulibana, E. (2001) *Hydro-institutional Mapping in the Steelpoort River Basin, South Africa*. IWMI Working Paper 17. International Water Management Institute, Colombo, Sri Lanka.

Sullivan, C. (2002) Calculating a water poverty index. *World Development* 30(7), 1195–1210.

Sullivan, C.A., Meigh, J.R. and Fediw, T.S. (2002) *Derivation and Testing of the Water Poverty Index Phase 1*. Centre for Ecology and Hydrology, Wallingford, UK.

Tapela, B.N. (2005) *Joint Ventures and Livelihoods in Emerging Small-scale Irrigation Schemes in Greater Sekhukhune District: Perspectives from Hereford*. Programme for Land and Agrarian Studies Research Report No. 21. School of Government, University of the Western Cape, Cape Town.

Tapela, B.N. (2008) Livelihoods in the wake of agricultural commercialization in South Africa's poverty nodes: insights from small-scale irrigation schemes in Limpopo province. *Development Southern Africa* 25(2), 181–198.

Terreblanche, S. (2002) *A History of Inequality in South Africa 1652–2002*. University of Natal Press, Pietermaritzburg.

Tharme, R.E. and King, J.M. (1998) *Development of the Building Block Methodology for Instream Flow Assessments and Supporting Research on the Effects of Different Magnitude Flows on Riverine Ecosystems*. Water Research Commission Report No. 576/1/98. Water Research Commission, Pretoria.

Thompson, L. (2001) *A History of South Africa*. Yale Note Bene Book. Yale University Press, New Haven and London.

Turton, A.R. (2003) The political aspects of institutional developments in the water sector: South Africa and its international river basins. Unpublished PhD dissertation. University of Pretoria, Pretoria.

Turton, A. and Meissner, R. (2002) The hydrosocial contract and its manifestation in society: a South African case study. In: Turton, A. and Henwood, R. (eds) *Hydropolitics in the Developing World: a Southern African Perspective*. African Water Issues Research Unit, University of Pretoria, Pretoria, pp. 37–60.

Turton, A.R., Meissner, R., Mampane, P.M. and Seremo, O. (2004) *A Hydropolitical History of South Africa's International River Basins*. Water Research Commission Report No. 1220/1/04. Water Research Commission, Pretoria.

van Heerden, P. (2004) *Estimating Water Requirements and Water Storage Requirements for Farms, Community and Backyard Gardens, and for Large Irrigation Systems: a User Manual for PLANWAT Version 1.2.3b and a CD-ROM to Install this Program*. IWMI Working Paper 82. International Water Management Institute, Colombo, Sri Lanka.

van Koppen, B. (2006) Basin development trajectory of the Olifants basin and implementation of the National Water Act. International Water Management Institute, Colombo, Sri Lanka. Unpublished report.

van Koppen, B. (2007) Institutional and legal lessons for redressing inequities from the past: the case of the Olifants Water Management Area, South Africa. Paper presented at the HELP Southern Symposium. CD HELP Southern Symposium. Help in Action. Local solutions to global water problems. Johannesburg, 4–9 November 2007.

van Koppen, B., Khumbane, T., de Lange, M. and Mohapi, N. (2006) Gender and agricultural productivity: implications for the revitalization of smallholder irrigation schemes program in Sekhukhune District, South Africa. In: Lahiri-Dutt, K. (ed.) *Fluid Bonds: Views on Gender and Water*. Stree, Calcutta, pp. 335–351.

van Veelen, M., Coleman, T., Thompson, H., Baker, T., Bowler, K., de Lange, M. and Sibuyi, I. (2002) Proposal for the establishment of a Catchment Management Agency for the Olifants Water Management Area. BKS Report 8184h. Draft submitted to Department of Water Affairs and Forestry by KBS (Pty) Ltd.

van Vuuren, A., Jansen, H., Jordaan, H., Van der Walt, E. and Van Jaarsveld, S. (2003) *Olifants Water Management Area: Water Resources Situation Assessment – Main Report*. DWAF Report No. P/04000/00/0101. Final Report, first issue (July 2003). Department of Water Affairs and Forestry, Directorate of Water Resources Planning, Pretoria.

Veldwisch, G. (2006) Local governance issues after irrigation management transfer: a case study from Limpopo province, South Africa. In: Perret, S., Farolfi, S. and Hassan, R. (eds) *Water Governance for Sustainable Development: Approaches and Lessons from Developing and Transitional Economies*. Earthscan, London, pp. 75–91.

Waalewijn, P., Wester, P. and Van Straaten, K. (2005) Transforming river basin management in South Africa: lessons from the Lower Komati River. *Water International* 30(2), 184–196.

Water Wheel (2008) 'Business unusual' gets donor nod. *Water Wheel* March/April 2008, Water Research Commission, Pretoria.

Wester, P. (2008) Shedding the waters: institutional change and water control in the Lerma–Chapala basin, Mexico. PhD dissertation, Wageningen University, The Netherlands.

Wester, P., Merrey, D.J. and de Lange, M. (2003) Boundaries of consent: stakeholder representation in river basin management in Mexico and South Africa. *World Development* 31(5), 797–812.

Ziervogel, G., Taylor, A., Thomalla, F., Takama, T. and Quinn, C. (2006) *Adapting to Climate, Water and Health Stresses: Insights from Sekhukhune, South Africa*. Stockholm Environment Institute, Stockholm.

4 From Half-full to Half-empty: the Hydraulic Mission and Water Overexploitation in the Lerma–Chapala Basin, Mexico

Philippus Wester,[1]* Eric Mollard,[2]** Paula Silva-Ochoa[3]*** and Sergio Vargas-Velázquez[4]****

[1]*Wageningen University, Wageningen, The Netherlands;* [2]*Institut de Recherche pour le Développement, Montpellier, France;* [3]*CH2MHILL, Water Business Group, San Diego, California, USA;* [4]*Instituto Mexicano de Tecnología del Agua, México; e-mails: *flip.wester@wur.nl; **eric.mollard@ird.fr; ***paula.silva@ch2m.com; ****kuirunhari@yahoo.com.mx*

Introduction

This chapter portrays the river basin trajectory of the Lerma–Chapala basin in central Mexico. It analyses the relationship between basin closure and the hydraulic mission, defined as the strong conviction that the state should develop hydraulic infrastructure to capture as much water as possible for human uses (Wester, 2008). In particular, it focuses on the role of the hydrocracy (hydraulic bureaucracy) in the creation of water overexploitation in the basin.

The Lerma–Chapala basin is in serious trouble, with water use at unsustainable levels and severe water pollution. Since the late 1970s, groundwater overexploitation has led to sustained declines in aquifer levels of 2 m/year on average, while surface water depletion has been close to, or has exceeded, annual river runoff in all but the wettest years. This was made possible by the drawing down of water stored in lakes and reservoirs. Twice in the 20th century (in 1955 and 2002), Lake Chapala, the downstream lake into which the Lerma River flows, nearly fell dry, losing more than 80% of its volume on both occasions. Between 2003 and 2008 above-average rainfall lessened the surface water crisis, with Lake Chapala recovering to above 80% of its storage capacity in September 2008, the highest level since 1979. While years of abundant rainfall can temporarily stop the overexploitation of surface water, the long-term consequences of water pollution and groundwater overexploitation are more dramatic and difficult to reverse. Tackling these three water crises requires addressing their interlinkages and the social mechanisms and institutional arrangements that govern water use.

The Lerma–Chapala basin provides a striking example of the complexities of water reforms in closed river basins, where consumptive water use is close to, or even exceeds, the level of renewable water availability (Keller *et al.*, 1996; Seckler, 1996). It is a basin in which many of the policy prescriptions emphasized in international water debates, such as irrigation management transfer (IMT) (Gorriz et al., 1995; Rap, 2006), integrated river basin management (IRBM) (Mestre, 1997; Wester et al., 2003) and increasing stakeholder participation in water management have been applied. Owing to the important economic and social interests linked to water in the densely populated and economically important Lerma–Chapala basin, it has served as a water policy testing ground for successive Mexican governments. Starting in the early 1990s, the federal government has enacted far-reaching water reforms (decentralization, participatory organizations, a new water law in 1992), accompanied by substantial funding for water treatment plants, support to water organizations, water-saving programmes and public-awareness campaigns. However, these efforts have not reversed environmental degradation in the basin nor led to a reduction in water use, and the three water crises remain dramatic today. This chapter explores why this is so, primarily focusing on surface water.

The next section introduces the basin and describes the process of basin closure. The following three sections provide a broad overview of the trajectory of the Lerma–Chapala basin, focusing on three periods (1500–1910, 1911–1980 and 1981 to the present). For each period, an analysis of the history of water development and the concomitant transformations in terms of water control and management are given. Conclusions are then drawn.

The Main Water Challenges in the Lerma–Chapala Basin

Physical setting of the Lerma–Chapala basin

The Lerma–Chapala basin is named after the Lerma River and the lake into which this river drains, Lake Chapala (see Fig. 4.1). When full, Lake Chapala discharges into the Santiago River, which flows in a north-westerly direction, to meet the Pacific after some 520 km. Since the early 1980s, very little water has flowed naturally from Lake Chapala to the Santiago River, due to dropping lake levels, and the Lerma–Chapala basin has, in effect, become a hydrologically closed basin. Lying between Mexico City and Guadalajara, the basin crosses five states (Querétaro, covering 5% of the basin, Guanajuato (44%), Michoacán (28%), México (10%) and Jalisco (13%)) and covers around 55,000 km^2, nearly 3% of Mexico's land area. Although the average annual runoff in the basin of 5513 Mm^3 (DOF, 2003) is only 1% of Mexico's total runoff, the basin is the source of water for 15% of Mexico's population (11 million in the basin and 2 million each in neighbouring Guadalajara and Mexico City). Located in central Mexico, the basin is an important agricultural and industrial area, containing around 13% of the area equipped for irrigation in the country and generating 9% of Mexico's gross national product (Wester et al., 2005).

Irrigated agriculture, covering some 795,000 ha, is the main water user in the basin. Eight irrigation districts (formerly state managed) cover around 285,000 ha, while some 16,000 farmer-managed or private irrigation systems (termed 'irrigation units' in Mexico) cover 510,000 ha. Twenty-seven reservoirs provide 235,000 ha in the irrigation districts with surface water, while around 1500 smaller reservoirs serve 180,000 ha in the irrigation units. An estimated 17,500 tube-wells provide around 380,000 ha in the basin with groundwater, of which 47,000 ha are located in irrigation districts (CNA/MW, 1999). The area actually irrigated between 1980 and 2001 is a matter of debate, with estimates ranging from 628,000 ha (CNA/MW, 1999) to more than a million ha (INE, 2003) per year.

Lake Chapala, with a length of 77 km and a maximum width of 23 km, is Mexico's largest natural lake. At maximum capacity the lake stores 8125 Mm^3 and covers an area of 1154 km^2 (Guzmán, 2003:110). When full, the average depth of the lake is 7.2 m, making it one of the world's largest shallow lakes. The shallow depth of the lake results in the loss of a large percentage of its storage to evaporation each year, with net evaporation of around 600 Mm^3 per year. Lake Chapala is highly valued by the inhabitants of Jalisco state, where the lake is situated, as well as by some 30,000 foreigners (mostly American retirees) living on its shores, and is a prime tourist destination. In addition, it provides Guadalajara, Mexico's second largest city, with 65% of its water supply.

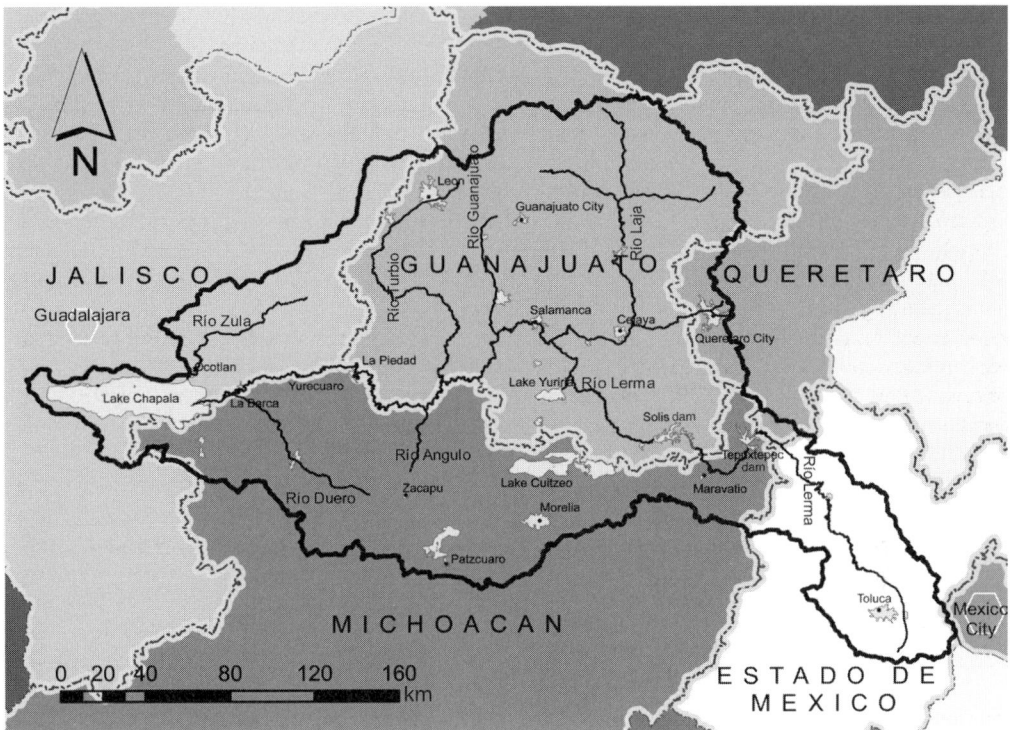

Fig. 4.1. States and rivers in the Lerma–Chapala basin.

Water overexploitation and Basin Closure

Since the early 1980s, surface water and groundwater in the basin have been overexploited. Although average rainfall from 1990 to 2001 (679 mm) was only 6% below the historical average (722 mm) (IMTA, 2002a), the amount of water depleted in the basin exceeded annual renewable water during this period, with no allocations for environmental flows. This was made possible by lowering the interannual stock of water stored in the basin's lakes, reservoirs and aquifers. Groundwater was overexploited, with declines in static aquifer levels of 1–5 m per year due to an estimated annual groundwater deficit of 1336 Mm^3 (IMTA, 2002a), while the consumptive use of surface water exceeded supply in all but the wettest years, nearly leading to the demise of Lake Chapala. Figure 4.2 presents the fluctuations in Lake Chapala's volume from 1934 to 2002, while Table 4.1 relates these fluctuations to developments in the basin. The section on Water Reforms and Water Transfers discusses how the lake fared after 2002.

Starting in 1945, water storage in the lake declined sharply, from an average of 6429 Mm^3 between 1935 and 1945 to 954 Mm^3 in July 1955, due to a prolonged drought combined with significant abstractions (750 Mm^3 per year on average) from the lake for hydroelectricity generation and irrigation (de P. Sandoval, 1994). During this period, around 214,000 ha were irrigated in the basin, mainly with surface water, and the constructed storage capacity in the basin was 1628 Mm^3. However, because of good rains towards the end of the 1950s, the lake recuperated, and storage averaged 7094 Mm^3 from 1959 to 1979.

In 1980, a second period of decline set in. By this time, constructed storage capacity in the basin had increased to 4499 Mm^3 and the average irrigated area had grown to around 680,000 ha, with a significant increase in groundwater irrigation. Although abstractions from the lake for hydropower generation had

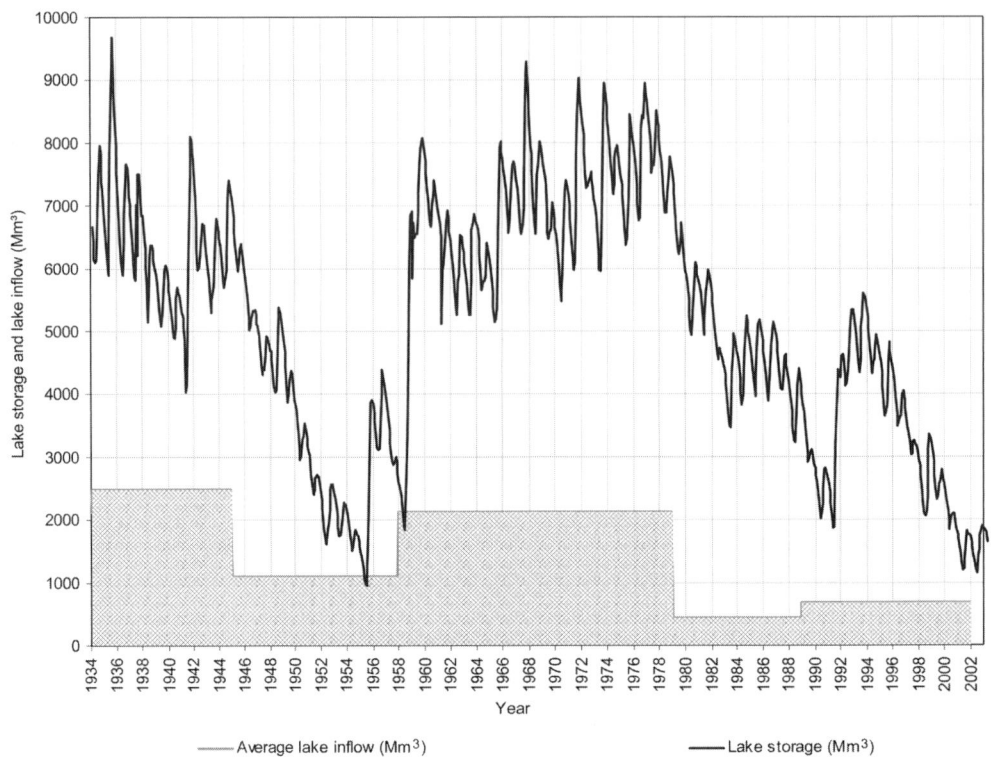

Fig. 4.2. Monthly Lake Chapala storage volumes and average inflows from 1934 to 2002.

Table 4.1. Overview of key water indicators in the Lerma–Chapala basin.

Period	Original (1934–1944)	Dry (1945–1957)	Wet (1958–1978)	Normal (1979–1988)	Latest (1989–2001)
Rainfall (mm/year)[a]	683	626	764	705	679
Inflow to Lake Chapala (Mm^3/year)[b]	2,485	1,085	2,127	429	677
Inhabitants (thousands of people)[c]	2,500 (1940)	3,000 (1950)	4,500 (1970)	8,700 (1990)	11,000 (2000)
Irrigated area (ha)[d]	155,000	214,000	508,000	675,000	689,000

Sources: [a]de P. Sandoval (1994) for all periods, except IMTA (2002a) for rainfall from 1989 to 2001; [b]de P. Sandoval (1994) up to 1988, BANDAS CD-ROMS for 1989 to 2001 (IMTA, 2002b); [c]de P. Sandoval (1994) for 1940, 1950, 1970. Census figures for 1990, 2000 from CNA/MW (1999); [d]Estimates of actual total irrigated area, averaged for the period, from CNA/MW (1999).

ceased, Guadalajara city started drawing large amounts of its urban water supply (between 200 and 400 Mm^3) directly from the lake. The combination of these factors and below-average rainfall (705 mm) resulted in declines in the lake's storage to around 2000 Mm^3 in 1990. After a good recuperation in the early 1990s, with lake storage reaching 5586 Mm^3 in October 1993 (68% of maximum storage), lake storage started declining again, dropping to

1145 Mm³ in June 2002 (14% of maximum storage), the lowest measured since 1955 (see Fig. 4.2).

Table 4.1 provides further details of the water situation in the basin, showing the sharp drop in inflows to Lake Chapala since 1979. While average rainfall from 1979 to 1988 was higher (705 mm) than from 1934 to 1944 (683 mm), the inflow to Lake Chapala was markedly lower (429 Mm³ versus 2485 Mm³). River inflow from 1989 to 2001 was slightly higher (677 Mm³), due to good rains in the early 1990s, but this was not enough to reverse the decline of Lake Chapala. Thus, the second period of lake decline was mainly due to the overextraction of water for urban use in Guadalajara and agricultural use both upstream and directly from the lake, and partly due to less rainfall. Between 1930 and 2000, the irrigated area in the basin increased fivefold, according to official statistics, and possibly by a factor of 7.5, while the population also increased fivefold during this period. The resulting levels of blue water depletion have made the basin very sensitive to variations in rainfall, with lower than average rainfall directly translating into reduced inflows to the lake. Between 1980 and 2001, the lake experienced a negative annual storage change of 191 Mm³ on average (IMTA, 2002a), but in years with above-average rainfall, such as 1991, the volume of the lake increased markedly.

To analyse Lerma–Chapala's trajectory, the hydraulic mission concept is used. Based on work by Reisner (1993) and Swyngedouw (1999), Wester defines the hydraulic mission as:

> the strong conviction that every drop of water flowing to the ocean is a waste and that the state should develop hydraulic infrastructure to capture as much water as possible for human uses. The carrier of this mission is the hydrocracy, which sets out to control nature and 'conquer the desert' by 'developing' water resources for the sake of progress and development.
>
> (Wester, 2008:10)

In Mexico, the hydraulic mission, the centralization of water development and the growth of the federal hydrocracy mutually reinforced one another and formed an important component of state formation in post-revolutionary Mexico. Three phases in the centralization of water resources development in Mexico can be identified: the birth of the hydraulic mission in the late 19th century, the rise of the hydraulic mission from the 1920s to the 1940s, and the heyday of the hydraulic mission from the 1950s to the 1970s. The following sections analyse these periods in the case of the Lerma–Chapala basin.

The Granary of Mexico: Water Development before the 1910 Revolution

Irrigation development in the Lerma–Chapala basin significantly expanded with the arrival of the Spaniards and the resulting colonization of the basin. The discovery of silver mines in Guanajuato in the 1550s led to the rapid settlement of the Bajío (a fertile valley in the basin covering most of Guanajuato, and parts of Querétaro and Michoacán) and the development of irrigated agriculture for wheat cultivation, mostly through private initiative and by monasteries (Murphy, 1986). The increasing demand for cereals by Mexico City led to the expansion of irrigation based on run-of-the-river irrigation schemes in the 17th and 18th centuries and the ingenious use of flood water through the construction of *cajas de agua* (embanked field ponds), primarily from tributaries of the Lerma River. This system consisted of interlinked and embanked fields of 5–200 ha each, filled in succession with flood water and with direct runoff from hills. These *cajas* (literally boxes) were drained in a staggered pattern after several months and then sown with wheat, while the larger *cajas* also stored water for supplementary irrigation. This form of controlled flooding was developed to a high degree of complexity in the Bajío (Sánchez, 2005). By the end of the colonial period, the basin's water resources were already intensively used, and by 1900 the run-of-the-river irrigation potential of the tributaries of the Lerma River had been largely developed, covering around 60,000 ha (SRH, 1953).

Towards the federalization of water allocation and development

The hydraulic mission started to gather force in Mexico towards the end of the 19th century, when the federal government began asserting its control over water both to promote commercial agriculture and to arbitrate in water allocation conflicts between *hacendados* (large landowners). Before then, irrigation and drinking water had largely been local affairs, although land and water rights were originally based on royal grants during the colonial period. The first 75 years of the 19th century were a period of turmoil and political unrest, with few new irrigation works in the basin. This changed in the last quarter of the 19th century, with attempts by *hacendados* to turn marshes and lakes into private property for land reclamation purposes.

The Porfirio Díaz regime (1876–1911), known as the Porfiriato, strongly supported private capital and foreign investment, and developed laws that led to extreme forms of land concentration. During the Porfiriato, the federal government established control over the country and focused on mining and railroad construction. An oligarchy of some 250 families, controlling 80% of the nation's land, handsomely profited from the increased production and trade, while an estimated 90–95% of rural households, forming 75% of Mexico's population, were landless according to the 1910 census (Hamilton, 1982). The extreme concentration of land ownership, with eight individuals holding 22.5 million ha in 1910, was a potent ingredient of the revolution that was to follow (Hamilton, 1982).

During the Porfiriato, the scale and number of hydraulic projects increased considerably, and the federal government started to play an active role in water development and the concessioning of water rights. In an excellent historical study, Aboites (1998) traces what he terms the federalization process in water affairs from 1888 to 1946. He indicates that, in the Mexican context, the term federalization refers to the process that led to the concentration of political and legal powers and faculties in the federal government, in short, centralization (Aboites, 1998). Before 1888, communities and municipalities administered water rights and water was controlled locally. This changed in 1888, when congress passed the *Ley General de Vías de Comunicación* (General Law on Communication Routes), which authorized the federal government to regulate the use of navigable and interstate rivers and specified that water concessions could only be issued by the federal government (Aboites, 1998). A decisive step in the federalization of water management was the amendment of Article 72 of the constitution in 1908, which placed rivers in the public domain. Based on this amendment, surface water as private property no longer existed and access to surface water was only possible through concessions issued by the federal government. Thus, in the space of 20 years, in legal terms, water in Mexico passed from being a local affair to falling in the public domain, administered by the federal government (Aboites, 1998).

Land reclamation projects in the Lerma–Chapala basin during the Porfiriato

Water development in the Lerma–Chapala basin during the Porfiriato mainly consisted of land reclamation, hydroelectricity projects and some irrigation development. These projects were undertaken by large landowners, sometimes in conjunction with foreign capital, and with an increasingly active involvement of the federal government in the funding and approval of these initiatives. The drainage of the Chapala and Zacapu marshes, and the proposals to drain the Lagunas de Lerma and the Cuitzeo and Yuriria lakes (see Fig. 4.1 for locations) stand out as examples of the land reclamation efforts (Wester, 2008). The expansion of run-of-the-river irrigation works on tributaries of the Lerma River also received attention, but the main incursion of the federal government in this area consisted of the formulation of river regulations.

The drainage of the Zacapu marsh (Ciénega de Zacapu), located in Michoacán near the headwaters of the Angulo River, is exemplary of how land reclamation projects were undertaken during the Porfiriato. As in other land reclamation projects, there was an important link between foreign capital, the federal bureaucracy and large *hacendados*. The Zacapu marsh, covering an area of around 150 km^2, was up to 8 m deep and surrounded by several haciendas and farming communities (Guzmán-Ávila, 2002). Eduardo Noriega, a *hacendado* and friend of Porfirio Díaz, obtained a concession from the federal government in 1900 to drain the marsh and construct a hydroelectricity plant near the exit of the marsh. As the Angulo was not navigable and did not form a boundary between two states and thus did not legally fall under federal jurisdiction, other *hacendados* challenged this concession, but to no avail. On the reclaimed land of 12,000 ha, Noriega developed an irrigation system, which started functioning in 1907, with a large loan from the federal government (Guzmán-Ávila, 2002).

The land reclamation fever rapidly spread throughout the basin during the Porfiriato, and various proposals were submitted to the federal government by *hacendados* to drain the Lagunas de Lerma and the Yuriria and Cuitzeo lakes. However, due to local opposition or struggles between *hacendados*, these works were not executed. A land reclamation project that was to have a lasting impact on Lake Chapala was the draining of the Ciénega de Chapala (Lake Chapala marsh). Until the late 19th century, Lake Chapala remained in its natural state, but this changed dramatically during the Porfiriato, as described below.

In 1894, a hydroelectricity plant, the first in Latin America and the second in the world, was constructed on the Santiago River at El Salto, some 60 km downstream of Lake Chapala, to provide Guadalajara with electricity. This plant received its water from Lake Chapala, which flowed into the Santiago River if the lake level was above *cota*[1] 95.00. The sill at the mouth of the Santiago River stopped the flow of water if the lake dropped below this level, while the form of the outlet to the Santiago River and the sediments deposited there by the Zula River, which joins the Santiago River just below Lake Chapala, restricted the amount of water leaving the lake above this level. This effectively blocked the outflow from the lake during the rainy season and could head up the water in the lake by 2–3 m. In one of the first studies on Lake Chapala, Miguel Quevedo y Zubieta shows that, on average, the lake reached *cota* 97.13 in the rainy season and would then fall to an average of *cota* 95.82 in the dry season, based on measured lake levels from 1896 to 1904 (Quevedo y Zubieta, 1906:18). As the average elevation of the Ciénega was *cota* 96.20, a large part of it would flood each year, depending on river inflows. When the Ciénega was flooded, Lake Chapala would reach a length of 100 km, a surface area of 1600 km^2 and would store around 9400 Mm^3 (de P. Sandoval, 1994:26).

During the dry season, when the lake dropped below *cota* 96.00, the little water that flowed into the Santiago was held up at the Poncitlán rapids. This led to the construction of a barrage at Poncitlán, completed in 1903, by which the level of Lake Chapala could be kept at *cota* 97.80. This made it possible to prolong high levels of storage in the lake, to be gradually released throughout the dry season for the El Salto hydroelectricity plant. However, it also entailed that the Ciénega de Chapala remained flooded longer. This led to complaints from *hacendados* with land in the Ciénega and motivated one of them, Manuel Cuesta-Gallardo, to develop plans to embank and drain the Ciénega de Chapala. He hired Luis P. Ballesteros to develop a plan for the reclamation and subsequent irrigation of the Ciénega, and in 1903 obtained a concession from the federal government to do so (Boehm, 1994). In 1905, work started on constructing embankments with a length of 95 km to separate the Ciénega from Lake Chapala, which was completed in 1910. A total area of 500 km^2 (50,000 ha) was cut off from the lake, reducing its storage capacity by some 1500 Mm^3 and leading to its current normal operating storage capacity of 7900 Mm^3 at *cota* 97.80 (Boehm, 1994).

Besides the land reclamation projects, the federal government became actively involved in drawing up river regulations. Based on the 1894 law, existing water rights had to be reconfirmed on rivers falling under federal jurisdiction, and the federal government had to approve new water concessions. Kroeber (1983) and Aboites (1998) provide a detailed account of how the Fifth Section of the *Secretaría de Fomento* drew up an increasing number of river regulations and how this led to increased federal control over water. In the Lerma–Chapala basin, the Laja River, a tributary of the Lerma in Guanajuato, provides an example of this process (Sánchez, 1999). In 1895, *hacendados* with colonial water rights on the Laja River requested that the federal government settle a water allocation dispute.

The federal government quickly established a commission to study the dispute, and in May 1897 decided that a complete study of the river was necessary to regulate all the water rights on the river. In 1901, the federal government enlarged the mandate of the study commission, to confirm and formalize all existing water rights and to conduct a full study of the river to verify if new water concessions could be awarded. Interestingly, the Laja was not a river falling under federal jurisdiction, but this did not prevent the Fifth Section from proposing a detailed river regulation in 1906 and establishing a permanent federal commission to inspect water withdrawals from the river. Although this was resisted by the haciendas drawing water from the Laja, the river was gradually brought under federal control (Sánchez, 1999).

This section has reviewed how the federal government increased its control over surface water during the Porfiriato. Through changes in the legal framework, the federal jurisdiction over rivers and lakes was expanded and the federal government became involved in confirming existing water rights and the formulation of river regulations. More importantly, large *hacendados* were granted concessions to drain lakes and to construct irrigation and hydroelectricity works, which frequently entailed the dispossession of previous water rights holders, primarily *campesinos* and *indigenas*, and also other *hacendados*. This oligarchic form of water resources development meant that the federal government itself did not construct water works, but rather supported a clique of *hacendados* with loans and water concession to do so. This changed after the Revolution of 1910–1920, as detailed below.

The Hydraulic Mission and the First Lake Chapala Crisis

The hydraulic mission of the hydrocracy and the bureaucratic–authoritarian state that developed in Mexico after the revolution of 1910–1920 strongly influenced water development in the Lerma–Chapala basin. The centralization of water development in Mexico accelerated in 1926 with the creation of the *Comisión Nacional de Irrigación* (CNI: National Irrigation Commission) and continued until the 1970s. These 50 years witnessed a large increase in the irrigated area in the Lerma–Chapala basin, intertwined with the formation and expansion of a strong hydrocracy with a keen sense of its hydraulic mission. The logo of the CNI and its successor, the *Secretaría de Recursos Hidráulicos* (SRH: Ministry of Hydraulic Resources), formed in 1946, contains the bold mission statement of Mexico's hydrocracy, namely *Por la Grandeza de México* (for the Greatness of Mexico). A more apt summary of the hydraulic mission is hard to come by.

The rise of the hydraulic mission: from oligarchic to revolutionary irrigation

The trend towards stronger federal control over water initiated under Porfirio Diaz's regime was consolidated in Article 27 of the 1917 Constitution. This article defined natural resources, including oil, land and surface water, as the inalienable property of the nation and established the *ejido* (common property) form of land tenure for the redistribution of land. Article 27 also established that the only way to gain access to surface water was through a concession granted by the federal government. Based on Article 27, the centralization of water management began in earnest in the 1920s, when President Calles launched a programme for the construction of large-scale irrigation districts and created the CNI as a semi-autonomous agency within the federal *Secretaría de Agricultura y Fomento* (SAyF: Ministry of Agriculture and Development). The CNI rapidly established itself as a competent hydrocracy and by 1935 was constructing 11 irrigation districts (IDs) throughout Mexico.

The CNI set out to develop 'revolutionary' irrigation systems, as opposed to the promotion of 'oligarchic' irrigation under the Porfiriato (Aboites, 1998). The revolutionary aspect initially consisted of using the construction of irrigation systems by the federal government to break up haciendas and colonize them with yeoman farmers, working and owning medium-sized irrigated farms (20–100 ha). The aim of the federal government was that this new rural middle class would gradually replace the large haciendas and would bring prosperity and stability to the countryside. Aboites (1998) has termed this 'revolutionary irrigation', as the post-revolutionary regime initially focused on using irrigation instead of land reforms to achieve the revolutionary promise of 'land and liberty', mainly in northern Mexico. With the more radical land reforms of the 1930s, attention shifted to supporting the *ejidos* (land reform communities) with irrigation works. In 1930, *ejidos* controlled only 15% of the land in irrigation districts, but by 1940 this had increased to 60% (Wionczek, 1982:370). Although the beneficiaries of the revolutionary irrigation policy were different, what remained the same was that the federal government led this social transformation process, by funding, designing and constructing the irrigation systems (Aboites, 1998). The management of the irrigation districts also became increasingly centralized from the 1930s onwards, although the water laws promulgated between 1926 and 1947 contained provisions for the creation of water boards to manage irrigation districts (Rap et al., 2004). However, the CNI frequently took control of the irrigation districts, as detailed below for the Lerma–Chapala basin.

Irrigation development in the Lerma–Chapala basin under the CNI

The following provides an overview of irrigation development in the Lerma–Chapala basin during the CNI era. Attention is mainly paid to the creation of the Alto Río Lerma Irrigation District (ARLID) in the Middle Lerma region, which was to become the largest irrigation district in the basin, and brief mention is made of developments in the Lower Lerma region. This brings out how the CNI increased its control over water in the basin and set in motion the process leading to water overexploitation.

Before the CNI started developing water resources in the basin, around 60,000 ha were already irrigated in the basin, with numerous run-of-the-river irrigation systems and *cajas de aguas* (SRH, 1953). Shortly after the CNI was formed, heavy rainfall in 1926 led to extensive flooding in the Lerma–Chapala basin. The CNI immediately focused its attention on the basin and formed two internal commissions to develop plans for the development of irrigation districts and hydroelectricity plants in the basin.

In their combined proposal, published in 1927, they recommended the construction of the Corrales dam on the Lerma River on the border of the Middle and Lower Lerma (see Fig. 4.3), to complement the Tepuxtepec dam, then under construction on the border of the Upper and Middle Lerma (Cuevas-Bulnes, 1941). The Corrales dam, with a planned storage capacity of between 750 and 1500 Mm^3, would serve to irrigate the lands of the Lower Lerma region, including the Ciénega de Chapala, and to generate hydroelectricity using the 150 m drop of the Zoró falls on the Lerma. They also recommended the construction of a new dam downstream of Tepuxtepec, to store more water for irrigation. It was estimated that 261,000 ha could be irrigated in the basin with surface water if these two new dams were built. Figure 4.3 presents the area currently irrigated in the basin and the main irrigation schemes and dams discussed in this chapter.

When the CNI presented its master plan, the construction of the Tepuxtepec dam had just started. In October 1926, a contract was signed between SAyF and the *Compañía de Luz y Fuerza del Suroeste de México* (Light and Power Company of Southwest Mexico), granting it an annual water concession of 750 Mm^3 for hydroelectricity generation and permission to construct the dam. The dam was completed in 1936, with a storage capacity of 370 Mm^3 (Santos-Salcedo, 1937). Between 1970 and 1973, the SRH elevated the dam's crest and increased its storage capacity to 585 Mm^3 (Garcia-Huerta, 2000).

After the construction of the Tepuxtepec dam, the amount of water flowing in the Lerma River increased during the winter season. This led to an increase in the irrigated area from some 36,000 ha in 1927 to some 46,575 ha in 1937 in the area that was to become the Alto Río Lerma irrigation district (Santos-Salcedo, 1937:160). This increase occurred mainly because the CNI had started rehabilitating the old run-of-the-river canals and constructing new ones on the Lerma River below the dam. In 1933, the CNI formed the Alto Río Lerma irrigation district, to fully develop the lands that could be irrigated with water from the Tepuxtepec dam. However, this created conflicts, and water users on already existing canals resisted the intrusion of the CNI. During

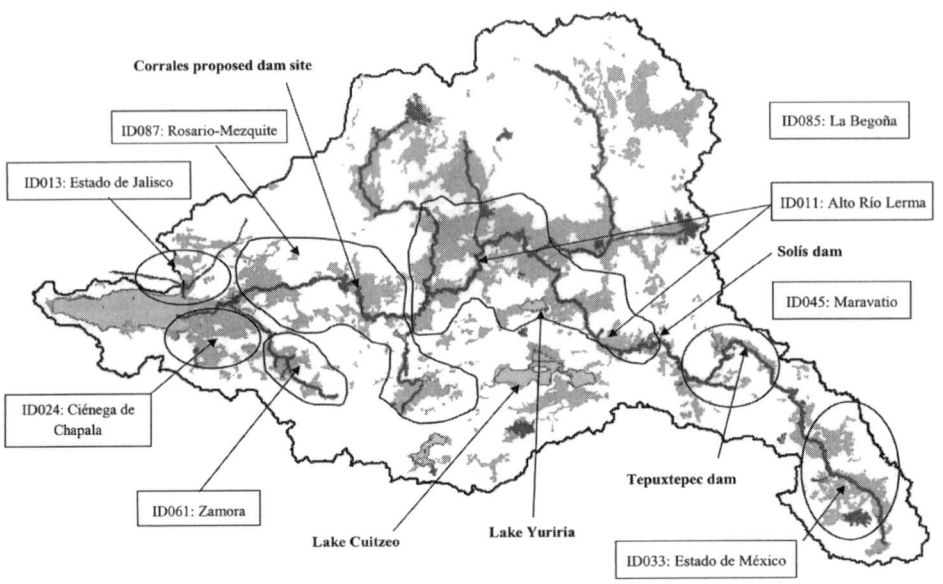

Fig. 4.3. Main dams and irrigation districts (IDs) in the Lerma–Chapala basin.

the 1920s, the *Dirección de Aguas* of SAyF had drawn up water distribution regulations for the run-of-the-river canals along the Lerma, including the canals of Acámbaro, Salvatierra, Valle de Santiago and Jaral de Progreso.

For these canals *Juntas de Aguas* (water boards) were established, based on the 1926 irrigation law, and the *Dirección de Aguas* attempted to regulate their water withdrawals by confirming existing water rights. In November 1933, an agreement was signed between the CNI and the *Dirección de Aguas*, in which control over all the irrigated areas from the Tepuxtepec dam to the city of Salamanca was passed to the CNI, to fall under the newly created Alto Río Lerma irrigation district. Through this agreement the CNI gained control over an irrigated area that until then had been managed locally for nearly 400 years. The increasing intrusion of the CNI led to protests from the existing *Juntas de Aguas*. Their protest was to cost them dearly. In February 1938, the CNI reacted by suspending all the *Juntas de Aguas* and taking over their responsibilities. It was not until the irrigation management transfer programme in the 1990s that these *Juntas de Aguas* were re-established, this time as water user associations (WUAs). Until then, the hydrocracy controlled the irrigation district.

While establishing its control over the run-of-the-river canals, the CNI also started work on the construction of the Solís dam, some 10 km upstream of Acámbaro in Guanajuato. The purpose of this dam was to improve flood control and store the water released (for hydro-electricity generation) from the Tepuxtepec dam for irrigation. Construction of the Solís dam, with a capacity of 800 Mm^3, started in 1939 and was completed in 1949. The CNI also built several large new canals to more than double the area under irrigation in ID011 to around 76,000 ha in 1946, up from 36,000 ha in 1927 (Wester, 2008). By 1940, the CNI had also developed plans for the further expansion of irrigation in the state of Guanajuato, including the Coria canal, to bring 25,000 ha under irrigation, and the Begoña dam on the Laja River, to irrigate some 18,000 ha. Owing to the first Lake Chapala crisis (see below) these works were delayed but were completed by the end of the 1970s.

A similar process occurred in the Lower Lerma region, where the CNI took control of the Ciénega de Chapala through the construction of irrigation and drainage works under the leadership of Ballesteros. Vargas-González (1993) provides a detailed account of how these developments interrelated with the redistribution of land in the area and how this led to increased federal control over the area. Ballesteros joined the CNI in 1926 as chief engineer of the Lower Lerma region and vigorously promoted the construction of the Corrales dam to increase the irrigated area in the Lower Lerma. In the end, the Corrales dam was not built, initially due to financial constraints and later because the proposed dam turned out to be sited on a geological fault. None the less, the water resources development plan presented by Ballesteros in 1927 was to guide developments in the basin until the late 1970s, and most of the works he and his CNI colleagues proposed in the 1930s were eventually constructed. This has led Pérez-Peña (2004) to speak of the 'Ballesteros school' in the development of the Lerma–Chapala basin, whose objective was the full utilization of the basin's water.

The above section has outlined how the CNI increased its role in water development in the Lerma–Chapala basin, by taking over the control of irrigation systems that had previously been managed locally, through both legal means and the construction of hydraulic infrastructure. In particular, the dissolution of the *Juntas de Aguas* in ID011 was a harbinger of the centralized water control that was to develop after the 1940s. The land reform partly helped the CNI to establish its control, but a stronger drive was its hydraulic mission to make good the promises of the revolution by developing 'revolutionary irrigation'. This mission was to reach its zenith between 1946 and 1976, with the creation of the SRH and the continued expansion of the irrigation frontier in the Lerma–Chapala basin.

The heyday of the hydraulic mission: river basin development and the SRH

During the 1940s, the concept of river basins as a unit of development started to gain force in Mexico, based on the Tennessee Valley Authority (TVA) model. During the election campaign of Miguel Alemán in 1946, the CNI lobbied the presidential candidate to initiate projects for regional development in various Mexican river basins and to form an overarching ministry of water resources. Directly after Alemán became president this happened, with the creation of the *Secretaría de Recursos Hidráulicos* (SRH: Ministry of Hydraulic Resources) in December 1946 to replace the CNI. The objective of the SRH was the comprehensive development of water resources and the concentration of the government's efforts in this field in a single organization.

Along with the concentration of water resources development in the SRH, river basin commissions were created by presidential decrees between 1947 and 1950 for several of Mexico's key basins, such as the Papaloapan, Tepalcatepec, Fuerte and Grijalva (Barkin and King, 1970). These commissions were to pursue comprehensive river basin development, based on the TVA model, but with the SRH minister as their president. The emphasis on comprehensive river basin development was to characterize the heyday of the hydraulic mission. From 1946 to 1976, the SRH vastly expanded its activities and mandate, with the river basin commissions serving to bypass state governments and other federal agencies. The SRH came to believe it was responsible for achieving 'the greatness of Mexico', not only through water resources development but also through regional development based on river basins. The hydraulic mission reached its zenith in the early 1970s with the passage of a new water law and the formulation of a national hydraulic plan.

In the Lerma–Chapala basin, the creation of the SRH coincided with the first Lake Chapala crisis, which lasted from 1945 to 1958. The following sections show how the hydraulic mission led to the 'overbuilding' of the basin, by reviewing the Lerma–Chapala–Santiago basin study commission created by the SRH in 1950, the controversies surrounding the first Lake Chapala crisis, and the continued expansion of the irrigation frontier in the 1960s and 1970s.

The Lerma–Chapala–Santiago study commission

In 1950, the SRH formed the Lerma–Chapala–Santiago basin study commission. This was strongly related to the first Lake Chapala crisis. In April 1947, the lake dropped below *cota* 95.15, at which point water no longer flowed to the Santiago River, for the first time since 1916. Hence, the three hydroelectricity plants on the Santiago, which depended on Lake Chapala, frequently had to stop operating. As these plants, owned by the *Nueva Compañía Eléctrica Chapala* (New Electricity Company of Chapala), were the only sources of electricity for Guadalajara, this led to strong demands from industrialists and the inhabitants of Guadalajara that the lake should be kept full by restricting irrigation in the basin. This led Orive-Alba, the SRH minister, to form a commission consisting of respected SRH engineers to study the problems of the basin. This commission set itself the task:

> ... to achieve a complete regularization of the existing water use systems [in the basin] and a better planning of those that can be realized in the future; arrive at a full understanding of the available water resources and their potential; and effectuate a more equitable water distribution in the basin through an adequate and combined operation [of existing infrastructure].
>
> (Vallejo-Ivens, 1963:5)

In a report published in December 1953, the commission set forth its recommendations for solving the lack of hydroelectricity and for fully utilizing the basin's water (SRH, 1953). The commission proposed the construction of a large hydroelectricity dam on the Santiago River, downstream of the confluence of several of its tributaries, to replace the plants that depended on Lake Chapala. It also recommended the construction of the Corrales dam on the Lerma River, with a storage capacity of 500 Mm^3, and the construction of the La Begoña dam on the Laja River, with a capacity of 180 Mm^3. Its other proposals consisted of plans to drain lakes throughout the basin to 'suppress unnecessary evaporation'. Thus, the commission recommended constructing a 20 km long and 6 m high embankment in Lake Chapala to reclaim 25,000 ha for agriculture. It also recommended draining Lake Cuitzeo by constructing a canal connecting it to the Lerma River, thus reclaiming 45,000 ha for agriculture, and draining Lake Yuriria to reclaim 7000 ha (SRH, 1953).

Although the execution of these plans would have a devastating effect on Lake Chapala, there was consensus in the commission on their desirability; the hydraulic mission was clearly in high gear. However, a contentious issue that the commission had to deal with was the sinking of deep tube-wells near the headwaters of the Lerma River to supply drinking water to Mexico City. In the 1940s, work started on canalizing the mountain streams feeding the Lerma and transferring this water to Mexico City through a tunnel. In addition to this transfer, it was proposed to sink deep tube-wells near the Lagunas de Lerma to augment the supply to Mexico City. The representative of the state of Mexico in the study commission strongly opposed this project (Santos, 2006). Guanajuato's representative also opposed the interbasin transfer, arguing it would have negative consequences for agriculture in Guanajuato. However, the government of the federal district persevered and succeeded in increasing the number of groundwater wells surrounding the Lerma wetlands. In the early 1950s, some 4 m^3/s (126 Mm^3/year) were transferred to Mexico City, increasing to 10 m^3/s (315 Mm^3/year) by the 1970s (Alba, 1988:163). These transfers affected the hydrologic cycle of the basin by sucking dry the Lerma River at its headwaters. After the interbasin transfer started, the Lagunas de Lerma and the wetlands of the upper Lerma quickly fell dry, to only partly fill during the rainy season. Another, even more contentious issue the study commission had to deal with was the sharp drop in the water levels in Lake Chapala. It had largely been created in 1950 to deal with this crisis, but, as the next section shows, in many ways its actions made the crisis worse.

The first Lake Chapala crisis (1945–1958)

From 1945 onward a period of lower than average rainfall (see Table 4.1), combined with extractions from Lake Chapala for hydroelectricity generation (520 Mm^3/year), resulted in the first Lake Chapala crisis. The response of the federal government to this crisis was strongly influenced by the hydraulic mission mind-set of the time and primarily consisted of efforts to secure the water supply of the hydroelectricity plants on the Santiago. As during the second Lake Chapala crisis (see Water Reforms and Water Transfers), the hydrocracy blamed the desiccation of the lake on the drought and the lake's high evaporation losses (de P. Sandoval, 1981). However, the extractions from the lake by the *Eléctrica Chapala* Company of some 520 Mm^3 a year, combined with 215 Mm^3 for irrigation, contributed strongly to the decline of the lake. Without these abstractions, the lake would not have fallen below *cota* 96.00 throughout the 1945–1958 period (de P. Sandoval, 1994). The efforts of the SRH and the Lerma–Chapala–Santiago study commission focused on ensuring these abstractions by a succession of hydraulic interventions in the lake. The majority of these works were planned and executed by the *Eléctrica Chapala* Company with authorizations from the SRH, while some were directly executed by the SRH. It is clear that the Lerma–Chapala–Santiago study commission, staffed by SRH hydrocrats, viewed Lake Chapala as an unaffordable luxury for Mexico and believed that its water should be used to the fullest extent possible.

A civil protest movement developed in Guadalajara during the first Lake Chapala crisis, just as it did decades later (see Water Reforms and Water Transfers), which went against the hydraulic mission of the SRH. Pérez-Peña (2004) provides a detailed account of the origin and activities of the *Comité de Defensa del Lago Chapala* (Committee for the Defence of Lake Chapala). This committee initially consisted of four people, with the author Ramón Rubín as its driving force, and was formed to protest against the 18 December 1953 presidential decree that authorized the Lerma–Chapala–Santiago commission to reduce the size of the lake's area by 25,000 ha. In January 1954, the committee sent an open letter to the president requesting the withdrawal of his decree. Throughout 1954, a range of academics, intellectuals and influential politicians joined the committee and pressured the Jalisco governor to stop the desiccation of the lake. Owing to pressure from the committee, the implementation of the presidential decree was stopped (and finally revoked in 1983). With the recovery of the lake in 1955, the activity of the committee lessened, and by 1958 it had faded away (Pérez-Peña, 2004).

Although the Lerma–Chapala–Santiago commission failed to construct a new embankment in Lake Chapala, it did sow the seeds for the second Lake Chapala crisis, by making the decision to use Lake Chapala for Guadalajara's water supply. In 1953, at the height of the first Lake Chapala crisis, the commission started work on developing the Atequiza–Las Pintas aqueduct to withdraw water from Lake Chapala for Guadalajara. The aqueduct's starting-point was the Ocotlán pumping station, which pumped water from Lake Chapala into the Santiago River, from where it flowed 40 km to the Atequiza canal. At the end of the Atequiza canal, water was pumped up 22 m to the newly dug Las Pintas canal (25 km long), which brought the water to the city's main water supply system. The initial capacity of this work was 1 m^3/s, but it was later increased to 9 m^3/s. The aqueduct entered into operation in 1956, although at that time the lake was nearly empty (de P. Sandoval, 1981).

In July 1955, the lake dropped to its lowest recorded level, namely *cota* 90.8 (954 Mm^3), resulting in a very erratic electricity supply to Guadalajara. However, very good rains in the autumn brought relief, and the lake recovered sufficiently to restart electricity production. By 1958, the lake had again dropped dangerously low, but another autumn of very good rainfall caused it to recover by nearly 5 m and the lake remained relatively full until 1979. The heavy rains of 1958 caused extensive flooding in the basin and serious damage to the Solís dam. As a result, between 1958 and 1982, the Solís dam was not filled to its full storage level but kept around 500 Mm^3. The water in excess of this storage was passed on to Lake Chapala until 1982, when the reconstruction of the Solís dam was completed.

Although the first Lake Chapala crisis had demonstrated that the basin had already reached its limits concerning water availability, the construction of new dams and the expansion of the irrigation frontier throughout the basin continued unabated during the 1960s and 1970s. Many of the works planned by the commission in 1953 were constructed by the SRH, and groundwater irrigation became increasingly important. The dam storage capacity in the basin more than doubled, from 1817 Mm^3 in 1959 to 3840 Mm^3 in 1979, the largest increase in the history of the basin (de P. Sandoval, 1994), while the irrigated area grew from 390,000 ha in 1960 to 640,000 ha in 1980, primarily in irrigation units (CNA/MW, 1999). The details of these developments will not be recounted here, but they clearly bear out that the hydrocracy took little heed of the warning of the first Lake Chapala crisis, but rather took it as an affirmation of its hydraulic mission to fully develop the water resources of the basin.

Water Reforms and Water Transfers: from Central Control to Negotiated Uncertainties

The drive by the federal government to mobilize ever more water through the construction of hydraulic infrastructure started to falter in the late 1970s, leading to the demise of the hydraulic mission in the 1980s and 1990s. In 1976, the river basin commissions were disbanded, and President López-Portillo merged the SRH with the Ministry of Agriculture to create the *Secretaría de Agricultura y Recursos Hidráulicos* (SARH: Ministry of Agriculture and Hydraulic Resources). This resulted in bureaucratic struggles and a politically expressed demand for renewed autonomy on the part of the hydrocrats, which they regained in January 1989, when the *Comisión Nacional del Agua* (CNA: National Water Commission) was created (Rap et al., 2004). Also, the focus on river basins was kept alive in the National Hydraulic Plan commission, where a group of water resource planners developed policy ideas on decentralized river basin management (Wester, 2008).

Although the 1960s and 1970s were the heyday of dam construction in the basin, with storage capacity more than doubling, the 1980s also saw some continued dam construction. The strengthening and raising of the Solís dam was important and was completed in 1982, which increased its storage capacity to 1200 Mm^3. Together with some minor dams, this increased storage capacity in the basin to 4499 Mm^3 by the end of the 1980s, which was nearly equivalent to the annual average surface water runoff in the basin. The elevation of the Solís dam coincided with the start of the second Lake Chapala crisis and was one of the contributing factors to the crisis, together with lower than average rainfall and the over-concessioning of surface water rights.

Another important development that affected Lake Chapala was that Guadalajara increased its withdrawals from the lake for its urban water supply. In the 1980s, a 42 km long pipe aqueduct was built to directly connect Lake Chapala with Guadalajara, fed by a pumping station with a capacity of 7.5 m^3/s on the shores of Lake Chapala. This aqueduct started functioning in 1992 and was intended to replace the Atequiza–Las Pintas aqueduct, constructed in the 1950s. However, Guadalajara continued to use both aqueducts and withdrew more than its annual concessioned volume of 240 Mm^3 from Lake Chapala. Guzmán (2003) estimates that Guadalajara withdraws around 450 Mm^3 from the lake each year, while an additional 130 Mm^3 are withdrawn from the lake for irrigation. These withdrawals are significant, as the average annual storage change in Lake Chapala from 1980 to 2001 was −191 Mm^3 (IMTA, 2002a). However, the Jalisco state government has consistently blamed the desiccation of Lake Chapala on excessive irrigation withdrawals upstream in Guanajuato and claims that it has reduced its withdrawals from Lake Chapala.

Concern about water quantity and quality in the Lerma–Chapala basin increased in the 1980s with the start of the second Lake Chapala crisis (1980–2002). The pace of institutional reforms increased after 1988, when the newly elected president of Mexico, Carlos Salinas, gave high priority to water issues (Rap et al., 2004). This materialized in the creation of the CNA in 1989, the transfer of government irrigation districts to users starting in 1989, and a new Water Law in 1992. These water reforms and larger political changes in Mexico in the 1990s, such as the transition to multi-party democracy and decentralization policies, led to a growing influence of new water actors in the basin, such as state water commissions, WUAs and environmental organizations. With the demise of the hydraulic mission and the rise of environmental issues, the demands and pressures on the hydrocracy changed fundamentally, from water supply development to water demand management.

This section analyses the attempts by the hydrocracy to deal with basin closure in the Lerma–Chapala basin in the 1990s and 2000s in this changed context, focusing on surface water allocation at basin level and groundwater regulation.

Attempts to bend down the water overexploitation curve

The main water management challenge in fully closed basins is bending down the water depletion curve. In the Lerma–Chapala basin, the hydrocracy made an attempt to bend the curve down in the 1990s by defining surface water allocation mechanisms at river basin level and by increasing the participation of state governments and, later on, of water users, in river basin management. In April 1989, the Mexican president and the governors of the five states in the basin signed a coordination agreement to improve river basin management and to 'rescue' Lake Chapala. The agreement contained commitments to modify water allocation mechanisms, to improve water quality, to increase water-use efficiency and to conserve the basin's ecosystems. In September 1989, a consultative council (CC) was formed to translate the agreement into action. Achievements of the CC include the formulation of a river basin master plan in 1993; a wastewater treatment programme, initiated in 1991; and a surface water allocation agreement, signed by the governors of the five basin states and the federal government in August 1991 (Mestre, 1997). However, these changes were carried out in a top-down manner, in which the political context considerably influenced how the policies were realized. This resulted in the exclusion of Lake Chapala as a 'water user' from the water allocation rules in the surface water allocation agreement (Wester et al., 2004).

The achievements of the CC led to the inclusion of an article in the 1992 Water Law on river basin councils (RBCs), defined as coordinating and consensus-building bodies between the CNA, federal, state and municipal governments, and water users. While responsibility for water management was retained by the CNA, the RBCs were conceived as important mechanisms for conflict resolution. The Lerma–Chapala CC became the Lerma–Chapala River Basin Council in January 1993. Currently, it consists of a governing board made up of the CNA Director, the five state governors and six representatives for water-use sectors (agriculture, fisheries, services, industry, livestock, urban). The RBC also includes a monitoring and evaluation group (MEG) and several specialized working groups. The MEG meets on a regular basis and is charged with preparing council meetings and applying the 1991 surface water allocation agreement (Wester et al., 2003).

In the Lerma–Chapala basin, surface water is allocated annually, based on concession titles and the surface water allocation agreement of August 1991. The concession titles set out the maximum volume that concession holders are entitled to, but the CNA may adjust the quantity that each user receives, based on water availability. The objective of the agreement was to save Lake Chapala, primarily to secure Guadalajara's domestic water supply. It sets out three allocation policies, namely *critical*, *average* and *abundant*, based on whether the volume of water in the lake is less than 3300 Mm^3, between 3300 and 6000 Mm^3, and more than 6000 Mm^3, respectively. For each allocation policy, formulas are used to calculate water allocations to the irrigation schemes in the basin, based on the surface runoff of the previous year. While no provisions for environmental flows were included in the agreement, the algorithms of the three allocation policies were designed to ensure sufficient carry-over storage in the basin's reservoirs. If adhered to, the modelling runs showed that this would generate sufficient spillage from reservoirs during the rainy season, and thus provide river inflows to Lake Chapala (Wester et al., 2005). However, a flaw of the agreement was that it was only based on rainfall data from 1950 to 1979, thus excluding the dry years in the 1940s and the 1980s. As a result, estimations of annual water availability, and hence water allocations, were too high, as become clear in the 1990s.

Since 1991, the MEG has met each year to apply the water allocation rules of the 1991 treaty, closely adhering to its provisions. According to CNA data, WUAs in the irrigation districts never used more water than allocated to them under the treaty (Wester et al., 2005). None the less, Lake Chapala's volume more than halved between 1994 and 2002. This led to intense debates in the RBC, with environmentalists and the Jalisco state government blaming the upstream irrigation districts in Guanajuato for using too much water. However, other contributing factors to the reduced inflows from the Lerma River to the lake are the following: CNA's weak control over surface water use in the small irrigation units, direct pumping from the river and Lake Chapala for irrigation, 10 years of lower than average rainfall, and reduced river base flows due to groundwater overexploitation (Wester, 2008). In addition, the 1991 treaty itself is partly at fault since it overestimated annual water availability and did not explicitly define environmental flows, which would have ensured a base flow in the Lerma River and hence inflows to Lake Chapala.

Water transfers and farmer initiatives to save water

Since 1999, political conflicts and negotiation processes surrounding the allocation of surface water have dominated the Lerma–Chapala RBC. Although stakeholder participation in water management has been broadly accepted in Mexico, the relationships between social and government actors are strongly influenced by a long tradition of concentration of political and decision-making power at the federal level (Vargas and Mollard, 2005). Negotiations in the past were common, albeit with the federal authority as the central actor, commanding patronage and corporatist relationships. The traditional coalition between farmers (to obtain subsidies from the government), the administration (dependent on politicians, also at the local level) and elected representatives (to avoid unrest in their states) continues to be strong, alongside stakeholder participation, decentralization and multi-party elections in Mexico. Currently, the decentralization of water management to river basins entails the creation of different spaces for social participation, which changes conflict-solving and negotiation practices.

In November 1999, because of critically low lake levels, and under pressure from Jalisco to secure Guadalajara's water supply, the CNA transferred 200 Mm^3 from the Solis dam, the main water source of the largest irrigation district in the basin, to Lake Chapala. This was the first time that surface water was physically transferred from the agriculture sector to the urban and environmental sectors under the 1991 treaty. A second transfer of 270 Mm^3 followed in November 2001, as lake levels continued to decline. These water transfers were met with staunch resistance from farmers, mostly from the middle of the basin, and undermined the legitimacy of the RBC. Farmers felt that their water was being stolen, as they received no compensation and because the 1991 treaty did not outline procedures for water transfers. In contrast, environmentalists and the Jalisco state government argued that much more water had to be transferred to save the lake, as around 10 Mm^3 were needed to raise the lake level by 1 cm. This led many in Jalisco to refer to the water transfers as 'aspirins' for the lake's headaches, with the media calling for much stronger medicine to cure the lake.

Before 1999, none of the WUA leaders in the Alto Río Lerma irrigation district were actively involved in the RBC. However, the water transfers galvanized these leaders to act. In May 2000, the presidents of WUAs from Jalisco, Guanajuato and Michoacán met one another for the first time to discuss ways to strengthen their position in the RBC. Until then, WUAs had only dealt with the CNA, and there were no horizontal linkages between WUAs from different irrigation districts. In 2001, the WUAs established a new working group in the RBC, under the leadership of the representative for agricultural water use on the RBC. Until the end of 2002, this *Grupo de Trabajo Especializado en Planeación Agrícola Integral* (GTEPAI: Specialized Working Group on Integral Agricultural Planning) attempted to strengthen the negotiation position of irrigators in the RBC. A central element of GTEPAI's strategy was to show that the irrigation sector was serious about saving water and hence a credible negotiation partner. The cooperation of government agencies, agro-industries and producers under the GTEPAI initiative resulted in a change in cropping patterns during the winter season of 2001/02. Throughout the basin, GTEPAI facilitated the conversion from wheat (four irrigation turns) to barley (three irrigation turns) on 47,000 ha. This resulted in a record production of barley, reduced imports for breweries, and claimed water savings of 60 Mm^3 (Paters, 2004). While GTEPAI improved farmer representation and participation in the RBC, its efforts to save water went unrecognized by the other members of the RBC.

While the farmer representatives took the lead, the threat of civil disobedience by farmers decreased. However, in November 2002, when the CNA decided that a third water transfer of 280 Mm^3 was to take place during the summer of 2003, tensions increased and farmers warned that they would occupy the Solis dam to prevent the transfer. Simultaneously, the representative of agricultural water use on the RBC was pressured to resign from the RBC during the MEG meeting in November 2002. The disappointment of farmer representatives and others involved with GTEPAI was such that they decided to dissolve the GTEPAI and to revert to interest group politics.

During the summer of 2003, unexpected heavy rains coincided with the third water transfer, causing floods in many parts of the basin. Instead of being accused of stealing irrigation water from farmers, the CNA was blamed for aggravating flooding through the water transfer. Although the very good rains of 2003 led to a spectacular recovery of Lake Chapala, with stored volumes jumping from 1330 Mm^3 in June 2003 to 4250 Mm^3 in January 2004 (see Fig. 4.4), this did not cool down tempers, as Jalisco wanted a full lake and had secured CNA's support for this. In November 2003, the Jalisco representative on the RBC again demanded the transfer of water from upstream dams to Lake Chapala, fuelling the anger of farmer representatives and further straining the relationship with Guanajuato. None the less, the CNA announced that 205 Mm^3 would be transferred, representing 50% of the unallocated water stored in the basin's reservoirs, and on 27 November 2003 opened the Solís dam. However, the CNA denied that this was a transfer, arguing that it was necessary for the hydraulic security of the Solís dam. The WUAs in the Middle Lerma did not buy

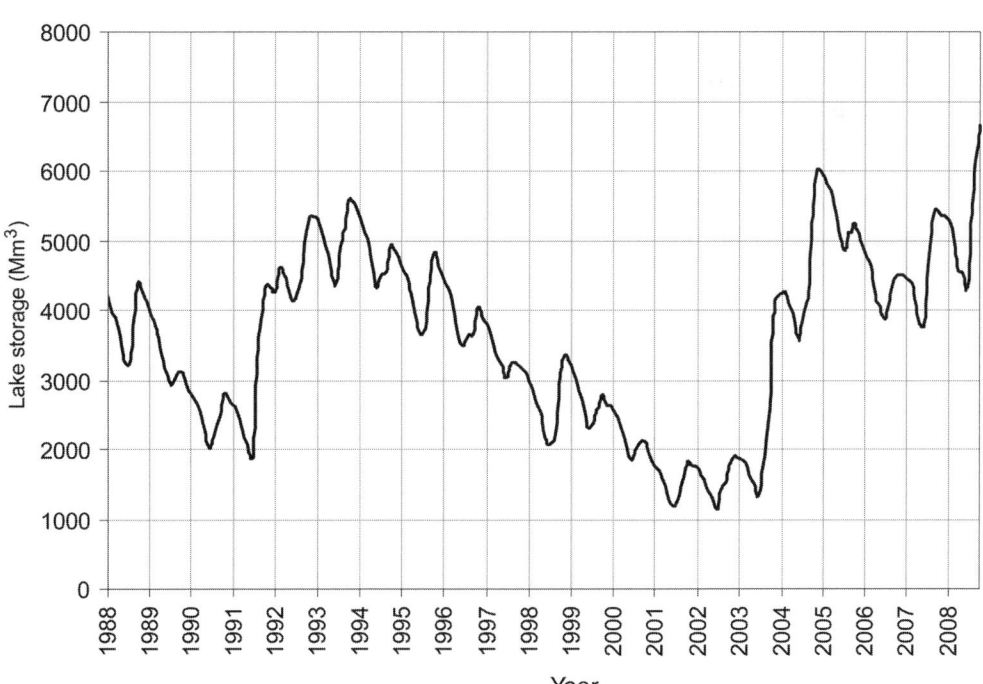

Fig. 4.4. Monthly Lake Chapala storage volumes from January 1988 to October 2008.

into this excuse and, for the first time, took the issue to court on 12 December 2003. The judge of the Celaya district court ruled in favour of the farmers and ordered that the transfer be stopped. However, by the time the judge forbade the transfer, the water had already flowed, with 174 Mm^3 reaching the lake.

Under pressure from Jalisco, 955 Mm^3 was transferred from reservoirs in the basin to Lake Chapala between 1999 and 2004, of which 817 Mm^3 arrived (Dau-Flores and Aparicio-Mijares, 2006). Although these water transfers were insufficient to 'save' the lake and could be seen as an instance of symbol politics, they did have consequences. First, around 100,000 ha could have been irrigated with this 'excess' water. The reduced allocations to the irrigation districts negatively affected farmers' livelihoods, the larger agricultural economy and the performance of the WUAs that depended for their income solely on irrigation service fees. In addition, the leadership of the WUAs was severely questioned by water users because of the lack of water for irrigation, although there was water available. Second, Jalisco could claim that it was saving the lake, as without the transfers Lake Chapala would have dropped to 746 Mm^3 in July 2002, 208 Mm^3 less than the lowest level in 1955 (Dau-Flores and Aparicio-Mijares, 2006:68). Third, the CNA reaffirmed its position as the central decision maker in the basin, although the transfers damaged its legitimacy and reputation. Last, farmer representatives became actively involved in negotiations at the river basin level and developed an initiative to switch to less water-demanding crops.

Renegotiating the surface water allocation agreement

Throughout this period, a parallel process was underway to revise the 1991 water allocation agreement. In this process, the controversies and conflicts in the basin came together, such as the conflict between agricultural interests and those defending the lake (environmentalists and Guadalajara/Jalisco state), the decentralization struggles between the CNA and the states in the basin, and the clash between a technocratic approach to allocating water and a negotiated agreement approach. In 1999, the members of the RBC decided to revise the agreement, as it was clear that it was not rescuing Lake Chapala. This was attributed to weaknesses in the 1991 agreement, including an overestimation of water availability in the basin, an underestimation of the area under irrigation and the lack of mechanisms to control the clandestine use of water (Güitrón, 2005). In 1999 and 2000, detailed hydrological studies were carried out by a consultant hired by the CNA to develop a new model for calculating surface runoff, without this leading to major changes in the water allocation agreement.

In March 2002, the Jalisco representative on the RBC requested a full revision of the 1991 allocation agreement, leading to the creation of a new working group, called the *Grupo de Ordenamiento y Distribución* (GOD: Ordering and Distribution Group). This group consisted of the CNA, government officials of the five states in the basin and consultants hired by Jalisco and Guanajuato. To develop consensus in this group, it was felt necessary to contract a 'neutral' outsider to execute the hydrological studies and develop a new water allocation model. Thus, it was decided to contract IMTA (Instituto, Mexicana de Tecnología del Agua), Mexico's water research institute. This proved to be important, as IMTA became a mediator and provided the negotiation parties with updated and revised hydrological data and water allocation scenarios (IMTA, 2002a). Until the end of 2003, little progress was made in the negotiations, although the detailed studies and their discussion in the RBC did lead to a new consensus on hydrological data and the design of the water allocation model.

Behind the scenes, the revision of the surface water agreement became linked to negotiations surrounding the construction of two new dams in the Santiago basin, both located in Jalisco. The Arcediano dam on the Santiago River is to provide Guadalajara with water, so that the city can stop withdrawing water from Lake Chapala. The second dam will be located on a tributary of the Santiago River, and will provide León, the largest city in Guanajuato, with water. However, to receive this water Guanajuato must guarantee that it will allow the return flows from León to flow to Lake Chapala. The discussions on the financing of these dams became increasingly linked to the water allocation negotiations, to such an extent that political brokerage at high levels was needed to reach a simultaneous deal on both issues. In early 2004, President Fox made the allocation of federal funds to the construction of these two dams conditional to the signing of a new water allocation agreement (Campillo, 2004).

Thus, the last phase of the negotiations was entered into under a charged political atmosphere. At an RBC meeting held in May 2004, the CNA regional office presented an 'optimized' water allocation scenario that did not include the need for water transfers. Instead, it was proposed that the volume stored in the reservoirs of the basin would not exceed their normal storage capacity, by keeping the emergency flood storage empty. Hence, any excess storage water would be discharged to Lake Chapala. The 'optimized' allocation scenario also showed that, irrespective of Lake Chapala's volume, farmers would always receive at least 50% of their concessioned volume. The good rains of 2004, with Lake Chapala reaching 75% of its capacity in November, helped pave the way for the signing of a new surface water allocation covenant in December 2004. The revised agreement entails further reductions in allocations to irrigation if water levels in Lake Chapala are low, but it does not explicitly contain provisions for environmental flows. The resistance of farmer representatives to the new covenant decreased after the presentation of the 'optimized' water allocation scenario, and after the inclusion of an article in the covenant that it could be revised each year. The pressure exerted by the Mexican president and the issue linkage with the construction of new dams were also important elements that led to the signing of the new covenant. However, without the good rains of 2003 and 2004 the story would have been quite different, and it remains to be seen how well the new water allocation covenant will function when the next dry period occurs.

The invisible water crisis: groundwater overexploitation

A more pressing issue than surface water allocation in the Lerma–Chapala basin is the serious overdraft of the basin's aquifers, estimated at 1336 Mm^3 per year (IMTA, 2002a). The situation in the Middle Lerma region is particularly acute, with extractions exceeding recharge by 40% (CEAG, 2006). As some 380,000 ha in the basin are irrigated with groundwater, and industrial and domestic uses depend almost entirely on groundwater, the long-term consequences of continued groundwater overexploitation overshadow those of Lake Chapala drying up. However, efforts to reduce groundwater extractions have yielded few results to date.

In 1993, the Lerma–Chapala RBC signed a coordination agreement to regulate groundwater extraction in the basin, but progress on the ground has been limited (CNA, 1993). The weak control of the CNA over groundwater extractions and the high social and political costs of reducing groundwater exploitation are primary obstacles. Although the constitution mandates the federal government to intervene in overexploited aquifers by placing them under *veda* (prohibition), thereby prohibiting the sinking of new wells without permission from the federal government, the experience with *vedas* has been disappointing (Arreguín, 1998). For example, the number of wells in Guanajuato alone increased from approximately 2000 in 1958 to 16,500 in 1997, although the drilling of new wells in the whole state was already forbidden in 1983 (Guerrero, 2000).

Based on the recognition that *vedas* had not worked and to counter the continued depletion of groundwater in the basin, the CNA started promoting the formation of *Comités Técnicos de Aguas Subterráneas* (COTAS: Technical Committees for Groundwater) in selected aquifers in the Lerma–Chapala basin in 1995 (Wester, 2008). Through the establishment of COTAS, the CNA sought to organize aquifer users, with the aim of establishing mutual agreements for reversing groundwater depletion. Based on developments in the state of Guanajuato, where the *Comisión Estatal de Agua de Guanajuato* (CEAG: Guanajuato State Water Commission) enthusiastically promoted the creation of COTAS (Guerrero, 2000; Wester, 2008), the structure of the COTAS has been defined at the national level in the rules and regulations for RBCs (CNA, 2000). In these rules, the COTAS are defined as water user organizations, whose membership consists of all the water users of an aquifer. They are to serve as mechanisms for reaching agreement on aquifer management, taking into consideration the needs of the sectors using groundwater (CNA, 2000).

As with the RBC, government has played an active role in forming and promoting the COTAS but with a much larger involvement of state governments. In the state of Guanajuato, 14 COTAS (of which 11 fall in the Lerma–Chapala basin) have been formed with the financial, logistical and technical support of CEAG (Hoogesteger, 2004; Sandoval, 2004). While CEAG has encouraged the COTAS to set their own agenda, it has retained an important influence on the COTAS. Because agriculture is the major groundwater consumer, most of the discussions in the COTAS in Guanajuato revolve around increasing irrigation efficiencies and reducing water use by the agriculture sector.

On paper, COTAS are platforms where all the users of an aquifer meet to reach agreements on aquifer management. However, user participation has been quite low, notwithstanding attempts by the state water commissions to involve as many stakeholders as possible. In part, this is due to a lack of reliable information on the owners of pumps in an aquifer and the lack of infrastructure and human resources on the part of the COTAS, making it difficult to summon all the users. Hence, during the formative stage of the COTAS only well-known people were invited to participate (Wester, 2008). In the majority of cases, the representatives of the agriculture sector in the COTAS are commercial farmers or agro-industrialists. This procedure, which has not brought together all the pumpers in an aquifer but rather builds on a small group of leaders who are not necessarily representative, has hamstrung the effectiveness of the COTAS. Although nearly all stakeholders agree that the situation is grave, this has not yet translated into a multi-stakeholder process to reach a negotiated agreement on reductions in groundwater extractions. Hence, the overall impact of the COTAS has been minimal. None has yet devised mechanisms to significantly reduce groundwater extractions, and the tough issue of how to reach agreement on an across-the-board reduction in pumping has not yet been broached.

Furthermore, many participants and staff of the COTAS and CEAG have become frustrated because the COTAS have little power to make a real difference in groundwater extractions. This is because they have no faculties to control groundwater extractions and have to rely on the goodwill of users and other institutions, particularly the CNA. As the CNA is the only government agency that can issue pumping permits, and is responsible for the enforcement of aquifer regulations, groundwater users are keen to maintain good relations with the CNA. In addition, the CNA has taken a back seat in the COTAS, and has emphatically not given them a mandate, thus sending the message to groundwater users that the COTAS are irrelevant. The CEAG has continued to promote the COTAS, in the hope that it can wrestle some control over groundwater away from the CNA. However, as long as the CNA continues to give preference to the lucrative business of legalizing 'irregular' pumps instead of throwing its weight behind the COTAS, the chances of a negotiated agreement on reductions in groundwater extractions are bleak.

Conclusions

This chapter shows how the hydraulic mission, embedded in the various manifestations of the hydrocracy in Mexico, led to the 'overbuilding' of the Lerma–Chapala basin and the concomitant overexploitation of water. The trajectory of the Lerma–Chapala basin is comparable to that of many other closing river basins, starting with small-scale, local water management, and then progressing to large dams and irrigation schemes funded, built and operated by the state. Technology development has been an important driving force of the hydraulic mission, as without reinforced concrete and hydrocarbon-fuelled machinery most of the large hydraulic works could not have been constructed. Other important drivers were the availability of labour and capital, which were frequently constraining factors in the history of the Lerma–Chapala basin. The specifics of how the Lerma–Chapala basin was overbuilt have been detailed above, and have led to the current challenges the basin is facing, such as environmental degradation, overexploitation of water, increasing social conflicts and the need for all involved actors to develop new ways to negotiate their way out of basin closure.

The closure of the Lerma–Chapala basin is a combination of increasing human pressures on water, the overconcessioning of water rights, and rainfall fluctuations. However, the creation of water overexploitation in the basin was not inevitable or an automatic process, but the outcome of the hydraulic mission of the federal government's hydrocracy. In its efforts to 'develop' the basin, the hydrocracy was strongly supported by state governments and water users to achieve the fullest utilization of water for the greatness of Mexico. The conviction that every drop of water evaporating from Lake Chapala is a 'waste' is still strong today among farmers and hydrocrats; it partly explains the lack of concerted efforts to reduce consumptive water use in irrigated agriculture. If Lake Chapala had not been the main source of water for Guadalajara and an important tourist destination, it is doubtful whether the state of Jalisco would have made an effort to 'rescue' the lake.

Another important finding presented in this chapter is the role of water abstractions from Lake Chapala. It is probable that the first and second Lake Chapala crises would not have occurred if no abstractions from the lake had taken place. This is an important point as, throughout the years, hydrocrats have argued that the cyclical declines in Lake Chapala were due to years of drought. While years of less rainfall obviously lead to lower inflows to the lake, the yearly abstraction of 520 Mm^3 from the lake during the 1940s and 1950s for hydroelectricity generation were an important cause of the first Lake Chapala crisis. The relatively wet period in the 1960s and 1970s made it possible for the hydrocracy to execute the water infrastructure development plans it had formulated since the 1930s. In particular, the elevation of the crest of the Solís dam in 1982 was important, as this increased the storage capacity in the Middle Lerma region. However, irrigation is not fully to blame for the second Lake Chapala crisis. From 1980 to 2001, the overall negative annual storage change of the lake was 191 Mm^3, while withdrawals from the lake for Guadalajara's water supply were at least 240 Mm^3 and possibly as high as 450 Mm^3 per year. Without these withdrawals the lake would not have declined.

The presence of Lake Chapala at the downstream end of the Lerma–Chapala basin poses special challenges for water management in the basin. This revolves around the extent of fluctuations in the lake's volume that are regarded as acceptable. Before the hydraulic interventions of the 20th century, high lake levels resulted in outflows discharging to the Santiago River. The hydraulic modifications of Lake Chapala and the construction of dams upstream largely cancelled these outflows and, depending on rainfall levels, resulted in the retraction or expansion of the lake's volume. The above-average rainfall between 2003 and 2008 led to a good recovery of the lake, showing how sensitive it is to variations in rainfall. In effect, it has temporarily reopened the basin from a surface water perspective. With a lake that is so sensitive to rainfall variations, the determination of the range of acceptable variations in its volume is subjective and its quantification raises political difficulties. In years with lower rainfall, farmers need more water while there is less water available, leading to reduced inflows to the lake. To stop the lake from falling below critical levels, water needs to be transferred from dams precisely when farmers need it most. This calls for the design of compensation mechanisms for farmers to forgo irrigation in dry years, but this option has not yet been considered in the Lerma–Chapala basin.

The key finding of this chapter is how difficult it is to reduce consumptive water use in closed basins, even if a range of water reforms are attempted and serious efforts are made to arrive at negotiated agreements on surface water allocation mechanisms. The three responses to river basin closure identified by Molle (2003), namely allocation, conservation and supply augmentation, are clearly in evidence in the Lerma–Chapala basin. Part of the answer as to why it is so difficult to reduce consumptive water use is because of the 'overbuilding' of the basin and the hydro-social-networks (Wester, 2008) constituted around, and by, the hydraulic infrastructure in the basin. The construction of hydraulic infrastructure tends to ensure that water is withdrawn from the hydrological cycle into the hydro-social cycle, thereby creating constituencies dependent on water for their livelihoods. For example, the widespread hydraulic modifications to Lake Chapala changed it from a natural lake into a managed storage reservoir, on which Guadalajara depends for its urban water supply. The political and economic repercussions are such that it is very difficult to reduce withdrawals from the lake, while the existence of the Chapala–Guadalajara aqueduct provides 'easy' water, which precludes attempts to increase water delivery efficiencies in the city. Similarly, the dams, irrigation canals and tube wells constructed in the basin have led to the development of numerous hydro-social-networks that are bent on continuing the abstraction of water for irrigation. Left to their own devices, these hydro-social-networks will continue withdrawing more water than is sustainable.

Note

1 The depth of Lake Chapala is measured with a locally defined benchmark, originally called the *acotación* (elevation mark) and later the *cota* (benchmark). This benchmark was established around 1897, with *cota* 100 defined as the bottom of the keystone of the sixth arch of the bridge over the Santiago in Ocotlán (destroyed in 1965 when a new bridge was built). This elevation of this point was later determined to be 1526.80 m above sea level. At present, the lake's normal maximum operating level is at *cota* 97.80, while at around *cota* 90.00 it is nearly empty.

References

Aboites, L. (1998) *El Agua de la Nación. Una Historia Política de México (1888–1946)*. CIESAS, Mexico City.
Alba, C. (1988) The rise and fall of a Mexican regional planning institution: Plan Lerma. In: Quarles van Ufford, P., Kruijt, D. and Downing, T. (eds) *The Hidden Crisis in Development: Development Bureaucracies*. Free University Press, Amsterdam, pp. 159–174.
Arreguín, J. (1998) *Aportes a la Historia de la Geohidrología en México, 1890–1995*. CIESAS and Asociación Geohidrológica Mexicana, Mexico City.
Barkin, D. and King, T. (1970) *Regional Economic Development: the River Basin Approach in Mexico*. Cambridge University Press, London.
Boehm, B. (1994) La desecación de la ciénaga de Chapala y las comunidades indígenas: el triunfo de la modernización en la época porfiriana. In: Viqueira-Landa, C. and Medina-Mora, L.T. (eds) *Sistemas Hidráulicos, Modernización de la Agricultura y Migración*. El Colegio Mexiquense and Universidad Iberoamericana, Mexico City, pp. 339–384.
Campillo, B. (2004) El conflicto por agua en la cuenca Lerma–Chapala. In: Blazquez-Graf, N. and Cabrera-López, P. (eds) *Jornades Anuales de Investigación, 2004*. Universidad Nacional Autónoma de México, Mexico City, pp. 201–217.
CEAG (Comisión Estatal del Agua de Guanajuato) (2006) *Memoria Institucional 2000–2006 de la Comisión Estatal del Agua de Guanajuato (CEAG)*. Comisión Estatal del Agua de Guanajuato, Guanajuato City, Mexico.
CNA (Comisión Nacional del Agua) (1993) *Acuerdo de Coordinación que Celebran el Ejecutivo Federal y los Ejecutivos de los Estados de Guanajuato, Jalisco, México, Michoacán y Querétaro con el Objeto de Realizar un Programa de Coordinación Especial que Permitirá Reglamentar el Uso, Explotación y Aprovechamiento de las Aguas Subterráneas de la Cuenca Lerma–Chapala*. Comisión Nacional del Agua, Mexico City.
CNA (2000) *Reglas de Organización y Funcionamiento de los Consejos de Cuenca*. Comisión Nacional del Agua, Mexico City.
CNA/MW (Montgomery Watson) (1999) *Proyecto Lineamientos Estratégicos para el Desarrollo Hidráulico de la Región Lerma-Santiago-Pacífico. Diagnostico Regional*. Comisión Nacional del Agua/Montgomery Watson, Guadalajara, Mexico.
Cuevas-Bulnes, L. (1941) *Memoria del Distrito de Riego de 'El Bajo Río Lerma' Jalisco y Michoacán*. Comisión Nacional de Irrigación, Mexico City.

Dau-Flores, E. and Aparicio-Mijares, F.J. (2006) *Acciones para la Recuperación Ambiental de la Cuenca Lerma-Chapala*. Comisión Estatal de Agua y Saneamiento de Jalisco, Guadalajara, Mexico.
de P. Sandoval, F. (1981) *Obras Sucesos y Fantasías en el Lago de Chapala*. Gobierno del Estado de Jalisco, Secretaria General de Gobierno, Unidad Editorial, Guadalajara, Mexico.
de P. Sandoval, F. (1994) *Pasado y Futuro del Lago de Chapala*. Gobierno del Estado de Jalisco, Secretaria General de Gobierno, Unidad Editorial, Guadalajara, Mexico.
DOF (Diario Oficial) (2003) Acuerdo por el que se dan a conocer las denominaciones y la ubicación geográfica de las diecinueve cuencas localizadas en la zona hidrológica denominada Río Lerma Chapala, así como la disponibilidad media anual de las aguas superficiales en las cuencas que comprende dicha zona hidrológica. *Diario Oficial de la Federación* 601(11), 2–11.
Garcia-Huerta, M.L. (2000) Irrigación y política: historia del distrito de riego num. 11 del Alto Río Lerma 1926–1978. BA thesis, Universidad Autónoma del Estado de México, Toluca, Mexico.
Gorriz, C.M., Subramanian, A. and Simas, J. (1995) *Irrigation Management Transfer in Mexico: Process and Progress*. World Bank Technical Paper No. 292. World Bank, Washington, DC.
Guerrero, V. (2000) Towards a new water management practice: experiences and proposals from Guanajuato state for a participatory and decentralized water management structure in Mexico. *International Journal of Water Resources Development* 16(4), 571–588.
Güitrón, A. (2005) Modelación matemática en la construcción de consensos para la gestión integrada del agua en la cuenca Lerma-Chapala. In: Vargas, S. and Mollard, E. (eds) *Los Retos del Agua en al Cuenca Lerma–Chapala*. Instituto Mexicano de Tecnología del Agua/Institut de Recherche pour le Développement, Jiutepec, Mexico, pp. 25–44.
Guzmán, M. (2003) *Chapala. Una Crisis Programada*. H. Congreso de la Unión, Mexico City.
Guzmán-Ávila, J.N. (2002) Las disputas por las aguas del Río Angulo en Zacapu, 1890–1926. In: Ávila-García. P. (ed.) *Agua, Cultura y Sociedad en México*. El Colegio de Michoacán, Zamora, Michoacán, Mexico, pp. 137–148.
Hamilton, N. (1982) *The Limits of State Autonomy: Post-Revolutionary Mexico*. Princeton University Press, Princeton.
Hoogesteger, J. (2004) The underground: understanding the failure of institutional responses to reduce groundwater exploitation in Guanajuato. MSc thesis, Wageningen University, Wageningen, The Netherlands.
IMTA (Instituto Mexicano de Tecnología del Agua) (2002a) *Revisión y Adecuación del Modelo Dinámico de la Cuenca Lerma-Chapala y Aplicación de Diversas Políticas de Operación y Manejo Integrado del Agua*. Informe final Proyecto TH-0240. Instituto Mexicano de Tecnología del Agua, Jiutepec, Mexico.
IMTA (2002b) *BANDAS: Banco Nacional de Datos de Aguas Superficiales. Hidrometría y Sedimentos hasta 2002*. Eight volume set of CD-ROMS. Instituto Mexicano de Tecnología del Agua, Jiutepec, Mexico.
INE (Instituto Nacional de Ecología) (2003) *Diagnóstico Bio-físico y Socio-económico de la Cuenca Lerma–Chapala*. Instituto Nacional de Ecología, Mexico City.
Keller, A., Keller, J. and Seckler, D. (1996) *Integrated Water Resource Systems: Theory and Policy Implications*. IIMI Research Report 3. International Irrigation Management Institute, Colombo, Sri Lanka.
Kroeber, C.B. (1983) *Man, Land, and Water: Mexico's Farmlands Irrigation Policies 1885–1911*. University of California Press, Berkeley and Los Angeles.
Mestre, E. (1997) Integrated approach to river basin management: Lerma–Chapala case study – attributions and experiences in water management in México. *Water International* 22(3), 140–152.
Molle, F. (2003) *Development Trajectories of River Basins: a Conceptual Framework*. IWMI Research Report 72. International Water Management Institute, Colombo, Sri Lanka.
Murphy, M.E. (1986) *Irrigation in the Bajío Region of Colonial Mexico*. Westview Press, Boulder and London.
Paters, H. (2004) Water and agriculture in the Lerma–Chapala basin in central Mexico: farmers' efforts to manage decentralization and save surface water. MSc thesis. Wageningen University, Wageningen, The Netherlands.
Pérez-Peña, O. (2004) Chapala, un Lago que Refleja un País: Politica Ambiental, Accion Ciudadana y Desarrollo en la Cuenca Lerma Chapala Santiago. PhD dissertation, Universidad de Guadalajara, Guadalajara, Mexico.
Quevedo y Zubieta, M. (1906) *La Cuestión del Lago de Chapala. Dictamen Presentado al Sr. Ministro de Fomento sobre el Aprovechamiento de las Aguas del Lago de Chapala*. Talleres de Tipografía y Fotograbado P. Rodríguez, Mexico City.
Rap, E. (2006) The success of a policy model: irrigation management transfer in Mexico. *Journal of Development Studies* 42(8), 1301–1324.

Rap, E., Wester, P. and Pérez-Prado, L.N. (2004) The politics of creating commitment: irrigation reforms and the reconstitution of the hydraulic bureaucracy in Mexico. In: Mollinga, P.P. and Bolding, A. (eds) *The Politics of Irrigation Reform: Contested Policy Formulation and Implementation in Asia, Africa and Latin America*. Ashgate, Aldershot, UK, pp. 57–94.

Reisner, M. (1993) *Cadillac Desert: the American West and its Disappearing Water* (revised and updated). Penguin Books, New York.

Sánchez, M. (1999) Sin querer queriendo: los primeros pasos del dominio federal sobre las aguas de un río en México. *Relaciones. Estudios de Historia y Sociedad* 80 (Otoño 1999), 69–98.

Sánchez, M. (2005) *'El Mejor de los Títulos': Riego, Organización y Administración de Recursos Hidráulicos en el Bajío Mexicano*. El Colegio de Michoacán, Zamora, Michoacán, Mexico.

Sandoval, R. (2004) A participatory approach to integrated aquifer management: the case of Guanajuato State, Mexico. *Hydrogeology Journal* 12(1), 6–13.

Santos, I. (2006) Los afanes y las obras. La Comisión Lerma–Chapala–Santiago (1950–1970). *Boletín del Archivo Histórico del Agua* 11(34), 29–38 (septiembre–diciembre).

Santos-Salcedo, J. (1937) Memoria del sistema nacional de riego Núm. 11 Alto Río Lerma, Gto. *Irrigación en México* 14(4–6), 156–170.

Seckler, D. (1996) *The New Era of Water Resources Management: from 'Dry' to 'Wet' Water Savings*. IIMI Research Report 1. International Irrigation Management Institute, Colombo, Sri Lanka.

SRH (1953) *Comisión Lerma–Chapala–Santiago*. SRH, Mexico City.

Swyngedouw, E. (1999) Modernity and hybridity: nature, regeneracionísmo, and the production of the Spanish waterscape, 1890–1930. *Annals of the Association of American Geographers* 89(3), 443–465.

Vallejo-Ivens, F. (1963) Origen, finalidades y resultados, hasta 1963, de la Comisión del Sistema Lerma–Chapala–Santiago. In: *Memoria de Tecer Seminario Latino-Americano de Irrigación del 17 al 29 de febrero de 1964. Tomo 1: Generalidades del Seminario, Grandes Proyectos e Investigaciones*. Secretaría de Recursos Hidráulicos, Mexico.

Vargas, S. and Mollard, E. (2005) Contradicciones entre las perspectivas ambientales de los agricultores y la defensa de sus intereses en al cuenca Lerma-Chapala. In: Vargas, S. and Mollard, E. (eds) *Problemas Socio-ambientales y Experiencias Organizativas en las Cuencas de México*. Instituto Mexicano de Tecnología del Agua/Institut de recherche pour de développement, Jiutepec, Mexico, pp. 64–82.

Vargas-González. P. (1993) *Lealtades de la Sumisión. Caciquismo: Poder Local y Regional en la Ciénega de Chapala, Michoacán*. El Colegio de Michoacán, Zamora, Mexico.

Wester, P. (2008) Shedding the waters: institutional change and water control in the Lerma–Chapala basin, Mexico. PhD dissertation, Wageningen University, Wageningen, The Netherlands.

Wester, P., Merrey, D.J. and de Lange, M. (2003) Boundaries of consent: stakeholder representation in river basin management in Mexico and South Africa. *World Development* 31(5), 797–812.

Wester, P., Vargas, S. and Mollard, E. (2004) Negociación y conflicto por el agua superficial en al Cuenca Lerma–Chapala: actores, estrategias, alternativas y perspectivas (1990-2004). On the CD-Rom *Agricultura, Industria y Ciudad. Pasado y Presente. III Encuentro de Investigadores del Agua en la Cuenca Lerma-Chapala-Santiago*, Chapala, Jalisco, 6–8 October 2004. El Colegio de Michoacán-Universidad de Guadalajara, Zamora, Mexico.

Wester, P., Scott, C.A. and Burton, M. (2005) River basin closure and institutional change in Mexico's Lerma–Chapala basin. In: Svendsen, M. (ed.) *Irrigation and River Basin Management: Options for Governance and Institutions*. CAB International, Wallingford, UK, pp. 125–144.

Wionczek, M.S. (1982) La aportación de la política hidráulica entre 1925 y 1970 a la actual crisis agrícola Mexicana. *Comercio Exterior* 32(4), 394–409.

5 Managing the Yellow River: Continuity and Change

David Pietz[1]* and Mark Giordano[2]**
[1]*Washington State University, Pullman, Washington, USA;*
[2]*International Water Management Institute, Colombo, Sri Lanka;
e-mails: *pietz@wsu.edu; **mark.giordano@cgiar.org*

Introduction

In 1997, the Yellow River dried up, 750 km from its mouth in the Bohai Sea, triggering significant comment and concern both within and beyond China. In China, this drying-up elicited a broad response in print and broadcast media about the environmental consequences of rapid economic development. At the same time, the state directed a range of scientific and technical organizations to focus research on the causes of water depletion in the Yellow River basin. Internationally, the general issue of water scarcity in north China prompted speculation about China's future ability to feed itself and the consequent impact on global grain markets (Brown and Halweil, 1998).

As suggested by the dramatic photographs of the desiccated river bed, the protagonist in this contemporary drama was indeed the Yellow River. One explanation for the vigorous domestic and international response to the drying-up lies in the tangible economic importance of the Yellow River to the North China Plain – the 'breadbasket' of China. The problems with the Yellow River suggested the profound impact that resource scarcity could have on China's continuing transformation to a global economic power. An additional explanation for the outcry generated by the drying up of the Yellow River was cultural. In the historical memory of past and contemporary Chinese, the Yellow River is the 'mother river' – the river that sustained the growth of Chinese civilization. To witness this river fail to reach the sea was to conjure up a host of negative images about the Chinese and China.

The goal of this chapter is to elucidate the contemporary relevance and importance of the Yellow River by exploring the trajectory of its historical development. This historical trajectory includes the trends over time in the physical development of the river's water resources, including traditional river-control practices. Just as importantly, it also includes the evolution of traditional values and symbols related to water and the river. As a result, this chapter, perhaps to a greater degree than any other in this volume, devotes substantial space to understanding the relevance of historic cultural antecedents to current issues. The physical and cultural aspects together help explain contemporary approaches to hydraulic management of the Yellow River basin and the options that Chinese society and basin managers have for the future.

Physical Geography of the Yellow River Basin

Most descriptions of the Yellow River's geography commence with a recitation of facts. For example, the Yellow River begins in the Qinghai–Tibetan Plateau of Qinghai province, from where it flows across eight other provinces and autonomous regions, before emptying into the Yellow Sea north of the Shandong peninsula (Fig. 5.1). With a length of over 5400 km, the Yellow River is the second longest in China and the tenth longest in the world, and drains an area larger than France. The basin contains approximately 9% of China's population and 17% of its agricultural area. While such static figures may be of passing interest, it is a deeper understanding of variation in the Yellow River basin's physical geography that is necessary if one wishes to understand the issues which both the Chinese government and basin residents face in their daily efforts to use, manage and protect the river. For accomplishing this formidable task, and for analysis, the river is often divided into its three main reaches.

Upper reach

The upper reach of the Yellow River drains just over half of the total basin area and extends from the river's origin in the Bayenkela mountains to the Hekouzhen gauging station downstream from the city of Baotou. On the Qinghai–Tibetan Plateau, where the Yellow River begins, steep rock slopes, low evaporation and high moisture retention produce runoff coefficients estimated to range from 30 to 50% (Greer, 1979; World Bank, 1993). This, combined with relatively high precipitation levels, results in this westernmost region of the upper reach contributing 56% of the entire river's total runoff by the point of the Lanzhou gauging station (YRCC, 2002b). As the river moves northward from there into the Ningxia/Inner Mongolian plains and the Gobi desert, potential evaporation rises to levels several times that of precipitation. The spatial variation in flow contribution within the upper reach is further exacerbated by human usage patterns. In the most western regions of the upper reach, relatively low population densities, agricultural development and industrializa-

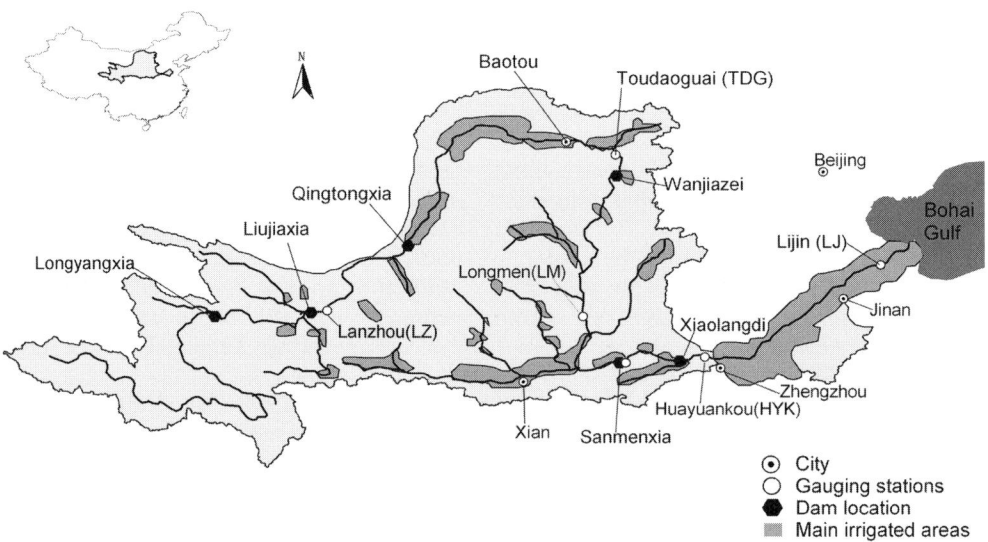

Fig. 5.1. The Yellow River basin.

tion limit *in situ* usage. As the river moves northward from Lanzhou, the agricultural population, with its long history of irrigation, and a growing industrial base substantially increase water withdrawals.

Middle reach

The middle reach, covering 46% of the basin area and providing virtually all of the remaining runoff, begins at the Hekouzhen gauging station (YRCC, 2002a). The middle reach of the Yellow River plays a significant role in basin water balances and availability for human use for two reasons. First, the reach includes some of the Yellow River's major tributaries, such as the Fen and the Wei, which contribute substantially to the total flow. Second, as the river begins its 'great bend' to the south, it cuts through the Loess Plateau and its potentially fertile but highly erodible loess soils. These soils enter the main stem and its tributaries as massive quantities of silt, resulting in average sediment concentrations unprecedented among major waterways and giving both the river and the sea into which it flows, their common 'Yellow' names (Milliman and Meade, 1983).

Sediment levels in the Yellow River are caused, in part, by such natural factors as the erodibility of the loess soils already mentioned, low average precipitation (which retards the growth of soil-stabilizing vegetation); and an increase in the gradient and power of the Yellow River as it passes through the most erodible zone. However, these levels are clearly exacerbated by anthropogenic factors, many of which have been in place for centuries or millennia (Ronan, 1995). While there is debate on the degree to which the Loess Plateau was 'naturally' forested, it seems clear that as early as the Qin and Han dynasties, large areas of land had been deforested for fuelwood and agricultural expansion, a factor believed to have contributed to increased erosion and, perhaps, regional desiccation (Menzies, 1995). Whatever the cause, the long-standing nature of the sedimentation phenomenon can be seen in the Chinese use of the phrase 'when the [Yellow] river runs clear' to mean 'never'. As will be described later, control of the potentially devastating Yellow River floods, which are greatly exacerbated by the high sediment loads generated in the middle reach, has formed a central theme in Chinese water management and politics for at least 3000 years. In addition, control of sedimentation to reduce the severity and frequency of flooding, accomplished through flushing, is now estimated to require about 25% of the total Yellow River flow and so is a major factor in current utilization of basin water.

Lower reach

The lower reach of the Yellow River commences at the apex of the natural basin in Taohuayu near the city of Zhengzhou and forms one of the most unique river segments in the world. Here, the sediment transported from the middle reach begins to settle as the river spills onto the flat North China Plain, producing a consistent aggradation of the bed and a naturally meandering and unstable channel (Ren and Walker, 1998). This instability has, in fact, been so severe that the Yellow River has had six major channel changes over the past 3500 years, in which the outlet to the sea has shifted 400 km from one side of the Shandong peninsula to the other (Greer, 1979). These massive shifts in the river channel, as well as more frequent smaller movements, have clearly caused problems for the millions of people who have attempted to farm the fertile alluvial soils of the lower reach. In response, successive river managers down the millennia have constructed levees along the banks of the Yellow River in an attempt to stabilize the main channel. While such structures may hold the channel in the short term, their success depends on consistently raising levee walls as sediment elevates the level of the channel constrained within.

Over time, the process of raising levees has contributed to a 'suspended' river, in which the channel bottom is above ground level, sometimes by more than 10 m (Leung, 1996). This raising of the channel above the level of the neighbouring countryside has clear implications for the severity of flooding when the levees inevitability fail in their function. In addition, the elevated bed alters the meaning of the Yellow River basin concept. With the channel above ground level, the surrounding landscape cannot drain into the river nor can tributaries enter it. This essentially means that the river 'basin' becomes a narrow corridor no wider than the few kilometres' breadth of the embanked channel. With almost no inflow, the contribution of the lower reach is limited to only 3% of the total runoff. While much of the sediment is deposited in the lower reach, approximately half has historically reached the river's outlet to the sea. These large deposits have, until recently at least, caused the river's delta to expand outward, creating substantial new farmlands (Ren and Walker, 1998).

Extra-basin issues including the south–north transfer

While the above discussion focused on the current geographical boundaries of the Yellow River basin, it is important to note that these boundaries, particularly in the lower reach, have changed, and may again change, over time. As mentioned, the high sediment load of the Yellow River makes the channel very unstable in the lower reach, where the topography is extremely flat. When the Yellow River's channel shifts, typically after a flood event or through human intervention, it connects hydrologically with either the Hai River system to the north or the Huai River system to the south, resulting in an expansion of basin boundaries across various portions of the North China Plain. The last time such a change occurred was in 1938, when the Yellow River's south dyke was purposefully breached at Huayuankou to block an advance of the Japanese army. The river was returned to its present course by engineering means in 1947 (Todd, 1949). The imposition of the Grand Canal, which runs perpendicular to the generally east-to-west-flowing rivers of eastern China, and which essentially links all basins from Hangzhou north to Tianjin, further complicates the strict definition of basin boundaries in the lower reach.

Another problem confusing the understanding of the Yellow River basin boundaries is the lack of congruence between the geographical extent of the basin as commonly delineated and the relevant hydrological units. For example, in the lower reach of the basin, seepage from the suspended main stem of the river recharges groundwater aquifers in both the Hai and Huai basins, where it is extracted for crop production. Additional water is also transferred out of the basin for industrial and domestic use, especially to the cities of Jinan, Qingdao and Tianjin. Of potentially greater significance for the future is the planned construction of the 'south waters north' engineering schemes, which may eventually transfer large amounts of water from the Yangtze River basin into the Yellow River, further marring the relevance of the geographical definition of the Yellow River basin (Biswas et al., 1983).

Water and Governance in Chinese History

This and the following section explore how the state, during the late Imperial, early Republican, Nationalist and People's Republic periods sought to manage the Yellow River in central China. They identify multiple meanings of water in general and the Yellow River in particular during the *longue durée* (an approach to the study of history, giving priority to long-term structures over events) of Chinese history and examine how these meanings shaped 20th-century efforts to control the Yellow River. Despite fundamental differences in political form among the various Chinese state-building projects of the 20th century, each state was fundamentally driven by similar modernizing assumptions, and each sought to selectively draw upon multiple historical meanings of the Yellow River and water in similar ways.

As reflected in the official histories written during the Imperial period, the origin of Chinese civilization is directly connected to water. One of the first renderings of this creation myth comes from the *Annals of History* (Shiji), written by Sima Qian (circa 145–90 BC) during the Han dynasty. Yü the Great, reputed to be the founder of China's first dynasty, is credited with draining the great north-central plain by digging discrete channels to lead the water of the Huai, Yellow, Yangtze and Wei rivers to the sea. The ordering of these waterways, collectively known as the 'four great rivers' (*sidu*), was attributed to Yü by most of the great historical writers of Imperial China

(Wang, 1987). The work of Yü the Great led to the development of sedentary agriculture and gave rise to a state that promoted agricultural pursuits and was sustained by appropriating a portion of the agricultural surplus. In sum, Yü the Great was responsible for the development of the cradle of Chinese civilization. The creation tale not only helped to legitimize the veneration of agriculture by later Confucian states, but also continued to inspire Chinese water-control endeavours throughout the Imperial period and beyond (Levenson and Schurman, 1969).

Complementing the connection of the birth of Chinese civilization with water were other systems of early Chinese thought that arose during the period of the 'Hundred Schools' in the late Zhou, early Warring States period (circa 500 BC). Adherents of the Naturalist school of thought, which developed during this period, sought to explain nature on the basis of the complementary cosmic principles of yin and yang (Fairbank et al., 1989). Yang represents the male, light, hot and active qualities, while yin represents the forces of femaleness, darkness, coldness and passivity. These opposing elements, however, represent complementary forces that comprise nature.

The Naturalists also stressed the basic concept of the 'five elements' to explain the composition of nature. The five elements – fire, earth, metal, water and wood – came to represent a pre-science, which was used in combination with other cosmic correlations, including numerology and astrology, to formulate calendars and to form the foundation of geomancy (fengshui). The point here is that the view of water as articulated by the Naturalists (female, dark, passive) complemented the creation myth surrounding Yü the Great. The connection centred on the qualities of femaleness as giver of life, or that which was responsible for the birth of a civilization. At the same time, we can see an affinity between the passive, dark qualities of water as described by the Naturalists and the historical sanction that manipulating water gave to Yü the Great.

Taoism was another major Chinese philosophical movement with direct connection to water. The meaning of water in philosophical Taoism represents an alternative to the creation myth surrounding Yü the Great and the concepts of the Naturalists. Water is perhaps the supreme moral example of the stricture to find harmony with 'the way' (tao) through the principle of wu-wei, or do-nothingness. Left to its own accord, water finds its harmony with the way by effortlessly following the contours of the land. Water as an object of contemplation intending to reveal moral truths informed much of China's cultural production during the Imperial period. Viewed by Taoists as something to be admired rather than controlled, mountains and other features of the natural landscape were rendered in poetry, painting and gardens as places of contemplation, where it was possible to connect with the ultimate realities of nature and to escape worldly concerns. The quietude of unaltered landscapes was a recurring poetic and philosophical theme during the Imperial period (Murphey, 1967; Greer, 1979).

Certainly, this sort of cultural expression was produced by, and for the benefit of, the literati, and it is precisely these people who were the face of the predominant socio-political system in imperial China – Confucianism. Indeed, Confucianism and Taoism share a fundamental similarity in their respective view of the unity of heaven, earth and man. There is a long tradition of Confucian-trained members of the bureaucratic class absenting themselves on occasion from their administrative duties and seeking a more contemplative life in nature.

Despite the strength of the Taoist traditions regarding nature, and the expression that this was often given in cultural production, it is equally true that nature in Imperial China was altered in a massive way. Deforestation of upstream regions supported, expanded and intensified agricultural pursuits necessary to support expanding populations. Farmers viewed water as a means of supporting these pursuits, and as something which needed solutions for managing both dry (irrigation) and wet (flood control) periods. Imperial states, in turn, through the medium of the administrative bureaucracy, viewed water as a means of promoting agriculture, thereby increasing expropriation of the agricultural surplus to expand and sustain the empire.

Development and Management of the Yellow River Basin

For most of the Imperial period, the Imperial Chinese state expended considerable resources in controlling the water of the Yellow River. One focus was the early development of irrigation. An additional focus was the construction of an extensive canal system connecting the Huai with the Yellow and Yangtze river valleys to facilitate the transport of the agricultural surplus to capital regions. The building of these canals created a complex matrix of waterways involving the lower Yellow plains. Complicating water controls were the periodic shifts of the Yellow River. Throughout the Imperial period, state priorities remained centred on maintaining the system of canals that provided the artery of grain tribute transport to northern capitals.

Canal transport and irrigation became intimately tied to the growth of Imperial power. Canal transport, developed within the context of warfare, served the formation of political power. Irrigation sustained agricultural development, which, in turn, expanded revenue for the political centre. Thus, the importance of water spawned a need to create an administrative organization to develop and maintain large canal and irrigation systems. Although the degree to which the central government was involved in local irrigation projects was in fact limited, 20th-century sinologists such as Wittfogel (1957) correctly identified the importance of effective water management to maintaining the state and the empire during the Imperial period. The pattern for subsequent water administration was established during the Han dynasty (206BC–220AD). In the Imperial capital, *dushui* (the office of the Director of Water Conservancy), under the Ministry of Public Works, was created as a planning and coordinating organization for the management of all river basins in China. At the same time, responsibility for labour recruitment and construction was delegated to local administrative units (Greer, 1979). The central challenge to successful water management during the Han dynasty, and later, was the ability to coordinate the efforts of the centre and the locality.

The imperial period

Governments in the early Imperial period persistently faced a cycle of water management issues: heavy dependence on water development for irrigation and grain transport led to a breakdown of hydraulic conditions when central authority waned, which in turn mandated large expenditures to restore stability. Managing this cycle required central capacity to undertake large-scale engineering projects. Indeed, throughout the Imperial era, rulers repeatedly viewed the regulation of water as providing legitimacy to rule. The historical precedent was Yü the Great, who claimed the right to rule based on his success in regulating water during the prehistorical period. Indeed, official dynastic histories esteemed the rule of individual rulers or their dynastic houses by claiming the legitimate historical mantle of Yü the Great. Such was the legitimizing rhetoric of Ming (1368–1644), who administered Yellow-conservancy projects in the mid-15th century.

Throughout the Ming and Qing (1368–1911) dynasties, Yellow River policy was guided by two differing principles: (i) diverting the flow of the Yellow River to the sea through different channels; and (ii) increasing the scouring capacity of the Yellow River by *shu shui gong sha* (confining the river between high dykes). Although these schemes were alternately adopted, they were guided by the singular goal of protecting grain transport (Huang, 1986). The debate between those advocating each of the two main engineering approaches was couched in moral terms. This debate was between a 'Confucian approach', which sought to regulate the behaviour of waterways through human action (i.e. digging channels to divert flows), and a 'Taoist approach', which sought benefit through the natural quality of water (i.e. allowing the natural forces of water to wash away silt) (Wu and Fan, 1993).

The struggle waged by the Ming court to regulate the Yellow River reveals several points. With overall management premised on safeguarding canal transport, the options available to management officials were limited. The two alternatives within this context, 'dividing the flow of the Yellow River' and 'utilizing a single flow to scour', remained the normative approaches to the management of the Yellow River well into the 20th century. In 1578, important additions to these fundamental approaches were proposed, including the construction of retention basins in upstream segments of the Yellow River to regulate flows in periods of heavy runoff. However, these plans were abandoned. One potential explanation is the fractured nature of administrative authority over waterways in central China. Competing bureaucratic units during the Ming dynasty, such as the Grand Canal Commission, Board of Public Works and provincial organizations, exerted pressures not always complementary to one another. The Qing dynasty, however, established the view that complete centralized control over the Yellow River was critical. The creation of the Yellow River Administration (YRA) in the early Qing dynasty (circa 1700) was the institutional expression of this sentiment.

The YRA was created in the early Qing period and headed by a director general appointed by the central government. With offices in Jining (Shandong province), the YRA served as a planning and coordinating organization for the lower Yellow River basin, the Grand Canal, and the lower Huai River valley. The functional goal of the YRA was to maintain grain transport from the south. As such, the YRA was essentially an adjunct of the Grain Transport Administration, as its primary function was to prevent flooding in the lower Yellow and Huai rivers, which would endanger the smooth functioning of the Grand Canal (Pietz, 2002). The historical importance of the YRA was that it was the first administrative organization in China to consider basin-wide issues, even though its actual operation was restricted to the lower Yellow River basin. Thus, when basin-wide river management gained currency in the early and mid-20th century in North America and Europe, China already had institutional experience with basin governance concepts.

In 1855, the Yellow River yet again changed course. The river breached its banks in Henan and adopted a northerly course, running through Shandong province to the sea. By this time, much of the grain tribute to the capital Beijing was transported by ocean. But the shift of the Yellow River rendered any transport via the Grand Canal hopelessly inefficient and expensive. Thus the immediate rationale for central control of the Yellow River, namely maintenance of the canal system, was lost. As a consequence, the YRA was abolished in 1856. The removal of central management of Yellow River control ultimately left local and provincial institutions responsible for water management in their immediate locales. The general collapse of Qing provincial and local government institutions, mirroring the deterioration of central capacity, meant that Yellow River management languished. By the end of the dynasty in 1911, water-control structures along the Yellow River, particularly in the lower reaches, were collapsing.

Basin development and management during the early 20th century

The period between 1855 and 1927 represented an important transformation in Yellow River management. The shift of the Yellow River in 1855 triggered the withdrawal of state patronage over water management, although there were attempts during the last years of the Qing and the early years of the Republican period to reconstitute centralized control. By the so-called Warlord period (1915–1926) the fundamental collapse of central political authority in China precluded any functioning of centralized water administration. Still, reformers among China's political elites retained the ideal of centralized control – realizing the reformulation of centralized management during the 1930s, in the Nationalist period.

With the nominal reunification of the country by the Nationalist Party after 1927, the new government embarked on an ambitious 'reconstruction' campaign to promote national strength. Consistent with Imperial patterns, Chiang Kai-shek and the Nationalist government immediately sought sanction to rule by 'ordering the waters' of the empire.

Coupled with this traditional concern of stabilizing the agricultural economy, the Nationalist government's state-building efforts were heavily influenced by the trend toward growing state capacity in many countries during the mid-20th century. The Nationalists re-established centralized institutions to manage the water on the North China Plain. In 1933, the national government established *Huanghe shuili weiyuanhui* (the Yellow River Water Conservation Commission, or YRWCC) having, in 1932, organized the National Economic Commission (NEC), whose purpose was to promote modern industrial growth by improvements in agricultural production and marketing. The formation of the NEC and its goals were familiar patterns engendered by the worldwide economic depression. As a supra-bureaucratic economic planning and coordinating body, the NEC had a number of analogues in different countries suffering from the worldwide depression, as state intervention in the economy was deemed necessary to optimize allocation and utilization of resources. The NEC underwent a series of reorganizations in mid-1933, which gave it broad jurisdiction over water conservancy and other reconstruction activities aimed at reviving the agricultural infrastructure (Anon., n.d.).

Another significant change in water management during the late 19th and early 20th centuries was the potential of water to serve modern industrial development. Although the specific goal was indeed industrial development, the more instrumentalist view of water serving state-sponsored economic growth (i.e. agricultural growth) during the Imperial period provided the basic assumption. Although small, China's modern economic sector experienced sustained growth in the late 19th and early 20th centuries. Several prominent Chinese industrialists in the early 20th century advocated active water management policies to promote cotton production and effective water transport to and from industrial enterprises centred in the Yangtze River delta region.

A third important development during the early Republican period, which established a pattern that would largely be consistent throughout 20th-century Yellow River management, was the introduction of modern hydraulic science into China. Initially introduced by foreign technical experts, a strong nationalistic tendency soon served to impel the development of native talent. Based on European and American models, engineering training institutes were founded that trained Chinese students in fundamental engineering practices, such as surveying. One example is the Hehai Engineering Institute (presently HeHai University) in Nanjing, founded by Zhang Jian, whose students would come to provide a cadre of well-trained technicians in the years to come.

The development of a cadre of hydraulic engineering and technical professionals during the first several decades of Republican China reflected increasing levels of technical education during this period (Strauss, 1998). Technical personnel in positions of policy planning included members of the commission itself, as well as directors of the Engineering Office and senior engineers. These individuals all received advanced training in engineering in the USA or Europe. Most of the engineering personnel at both the low and mid-levels received training in their specialties from the growing number of engineering departments at colleges and universities in China. In 1935, there were a total of 37 institutions of higher education offering degrees in civil and other fields of engineering (Huang, 1986). Included in this number were institutions such as the Qinghua University and *Shuili gongcheng zhuanmen xuexiao* (the former Water Conservancy Training Institution) in Nanjing, which became part of *Guoli zhongyang daxue* (National Central University).

The last broad development of Yellow River management during the early to mid-20th century was the pattern of developing foreign partners in water management. This development, however, reflected the troubled relationship that China had with the USA and European powers. In some ways, the power of the traditional role of water and the cultural significance of the Yellow River in China also mitigated the success of international cooperation. An early effort was led in 1914 by the American Red Cross, which attempted to secure an agreement for a loan to pursue an aggressive water management scheme on the North China Plain. Ultimately, the plan failed because of problems related to leadership of the project and over differing conceptions about the technical approaches to water management in China. The Chinese leader of the project, Zhang Jian, suggested that the American chief engineer simply did not understand the special nature of China's water and traditional methods of dealing with it. This sensitivity to the special nature of China's water and a certain reverence to past Chinese accomplishments in managing water continued to be an undercurrent even as China intensified these sorts of transnational cooperative efforts over the next decades (Pietz, 2006).

Transnational cooperation continued to develop during the Nationalist period. In early 1931, the government invited the directors of the League of Nations' Economic and Financial Section and its Communications and Transit Section to visit China to advise on reconstruction projects (National Economic Council, 1934). In addition, the Board of Trustees of the Returned British Boxer Indemnity Commission designated that 66% of the money from the British Boxer Indemnity be returned to China to assist water conservancy projects. Finally, the United Nations Relief and Rehabilitation Administration sponsored Yellow River management operations following the end of World War II. In all, the record of international cooperation in Yellow River management during the 20th century was spotty. But China's pattern of seeking these partnerships suggests a general trend in the internationalization of China's water management.

The ability of the Nationalist government to realize its Yellow River conservancy plans during the 1930s was conditioned by difficulty in controlling resources at the local level. In other words, it could organize and plan but it struggled to build. Several projects were completed but on a smaller scale and beyond schedules originally envisioned. This was primarily due to inadequate labour conscription and the inability to enforce work discipline. The government tried campaigns of moral suasion and the dispatch of Nationalist troops to ensure compliance with its goals, but projects were persistently obstructed by the inability to mobilize conscripted labour.

Basin development since 1949

Yellow River management was in a shambles by the time of the Communist victory in 1949. In large measure, difficult conditions in the lower portions of the valley were severely aggravated by Chiang Kai-shek's order to destroy the southern dykes of the Yellow River near Huayuankou in 1937. This decision was intended to slow the advance of Japanese troops from the north. The massive flood towards the lower Huai River valley indeed brought a pause to the Japanese invasion, but the longer-term consequences were to destroy much of the conservancy works that the Nationalist government had managed to build during the prior decade. Although there were some projects initiated after the end of the Pacific War in 1945, the state of the Yellow River was indeed precarious when Mao Zedong led the communists to power in 1949.

The developments described above during the Nationalist period, namely centralization, modern industrial development, introduction of modern science and technology, and international cooperation in water management, suggest that hydraulic engineering during this period was increasingly reflective of standards and practices that prevailed in the industrialized countries of the time. One need only look to the institutional model of river management in China during the Nationalist period (the Tennessee Valley Authority) to get an understanding of the types of 'mega-project' that China was moving towards. Does the history of Yellow River conservancy under the Chinese Communist Party after 1949 suggest continuities with these trends? The answer is yes for much of the post-1949 period. Beginning in 1958, however, with the onset of the Great Leap Forward, China modified this orientation towards the grand project by introducing small-scale projects that emphasized local administration, mass mobilization, a celebration of traditional notions of water conservancy (i.e. a certain anti-modernism) and self-reliance. Thus, after 1958 there was a dual character to Yellow River engineering: mega-projects combined with small-scale installations.

Looking back at such diverse approaches to Yellow River engineering, one is certainly tempted to come to some conclusion as to which paradigm best served the goals of river management. The problem, of course, is defining these goals. There were multiple goals, and respective goals, it was argued, could be best achieved by different approaches. The purpose of the following examination of Yellow River engineering after 1949 is not to evaluate differing approaches to river management but is, instead, intended to delineate areas of continuity and change. One significant difference in Yellow River management effort after 1949 was the degree of local political control attained by the new government, and hence the ability to sufficiently mobilize labour for conservancy projects. In other important respects, however, the decade after 1949 reflected broad continuities and discontinuities with earlier Yellow River management efforts. Institutional structure, modern technology and international cooperation were all issues that would be at the centre of fierce debates over the Yellow River.

Institutional structure: centralization and decentralization

One of the key policy debates after 1949 was over the institutional structure of the Yellow River control. In its most distilled manner, the debate was over whether water management could best be pursued with a centralized structure. Immediately after 1949, the government of the People's Republic of China (PRC) had, by and large, assumed the institutional structure of the Yellow River Conservancy Commission (YRCC, the successor to the YRWCC), as it had been established in 1946 during the Nationalist era.

The first large-scale water management plan adopted by the government after 1949 was focused on the Huai, not the Yellow, River. This plan clearly signalled the degree to which water management immediately after 1949 would be centrally planned and financed. Begun in 1950, the plan called for the creation of nine upstream reservoirs, strengthening dykes in the middle and lower reaches, and improving the storage (Hongze Lake) and drainage capacity in the lower portions of the river. State expenditures for the Huai River project during the 1950s were high. Between 1949 and 1952, state spending on the Huai River scheme was 64% of all government expenditures on river management in China (Vermeer, 1977). Water officials felt that immediately rectifying the Huai River was critical to addressing long-term social and political disruption in the valley.

The Huai River plan provided the basic blueprint for the Yellow River plan adopted by the government. In 1955, the Technical and Economic Plan for Yellow River Comprehensive Utilization was submitted to the state council by the YRCC. This was probably the first ever comprehensive development plan for the basin, and focused on power generation in the upper reach, flood control in the middle reach and irrigation downstream. The ambitious plan, approved by the First People's Assembly in July 1955, envisioned, among other items, the construction of an astounding 46 large dams on the Yellow River's main stem (Greer, 1979). It is interesting to note that, probably because of Soviet influence and aid, the water-engineering efforts in the early 1950s were *relatively* capital intensive rather than labour intensive, as had traditionally been the case in Chinese water development (Chi, 1965). At the basin level, the YRCC was responsible to the Ministry of Water Conservancy and was the representative of centralized control over the breadth of the basin. Although labour mobilization remained the responsibility of provincial and sub-provincial institutions, the Yellow River Commission held overall coordinating functions over technical elements of the engineering plans.

Beginning in 1958, however, water management administration experienced a strong trend toward decentralization. Corresponding with the communalization push, administration and spending on Yellow River projects increasingly became the responsibility of provincial and local governments or the communes. This shift from central to local control was influenced by several factors: incorporation of small projects alongside large ones, the increasing labour element of overall project design and execution, and the primacy given to local irrigation projects that were more suited to local control (Wu and Fan, 1993).

Science and technology: modern hydraulic engineering and mass mobilization

Behind the plans of the early People's Republic of China for the development of the Yellow River basin was a strong belief in the ability of human ingenuity to overcome nature. This belief emanated from the tremendous pride and euphoria following the defeat of Japan, victory in the Chinese Civil War and the establishment of 'New China', and the success in stopping the advance of US and UN forces in the Korean peninsula. If the Chinese people could defeat feudalism and imperialism, why would not it also be possible to conquer the Yellow River? Why would it not be possible to use the will of the people to make the river 'run clear' for the first time in history? The then commissioner of the YRCC, Yang Huayun, presented such visions during a field trip to the Yellow River by Chairman Mao through a promise: the Yellow River would be made peaceful for at least 300 years through the construction of the planned large dams. A somewhat more realistic assessment of the potential to control the river is attributed to Mao in his suggestion that the Yellow River problems could be 'well handled' although not necessarily fully resolved; in this respect, the actions of the government were to follow the ambitious plans.

An example of the resolve to develop the river is seen in the name of the first major irrigation project under the new development plans, the People's Victory Canal, located in Henan province. This project, which still provides the name to a brand of cigarettes, was designed to divert Yellow River water by gravity to irrigate almost 100,000 ha of farmland (Zhang and Shangshi, 1987). Signalling the symbolic and real significance of such undertakings, Chairmen Mao visited the project in October 1952, when he officially opened its diversion gates. Irrigation and dam construction continued through the late 1950s under the slogan 'big diversion, big irrigation'. However, the primary means to complete projects shifted from capital to labour, probably in large part due to the withdrawal of Soviet aid. In fact, the decision made in 1957 to 'depend on the masses' and rely more on local capital in water construction projects can be seen in some ways as the beginning of the nationally disastrous Great Leap Forward, which began in 1958.

Although voluntarism was a critical element of the regime's ruling psychology, science and technology were still valorized during the decade of the 1950s. During the first period, the ambitious Yellow River engineering plans were, in part, predicated on data and plans gathered and formulated by the technical staff of the Nationalist government's YRCC. Although the number of technical specialists throughout China was limited, large numbers of such experts were heavily recruited by the new government's YRCC after 1949 to participate in some of the nation's premier projects (Vermeer, 1977). So, by the mid-1950s, newly minted technical experts from a growing number of technical institutions in China joined with experts who had received their training and work experience during the Nationalist period and were, together, vital participants in the conceptualization of the Yellow River engineering scheme.

The orientation towards technical expertise and notions of modern hydraulic practices came under attack with the onset of the Great Leap Forward policies in 1958. As an auxiliary to the rectification campaigns such as the Anti-Rightist Movement, which saw the discrediting of many water conservancy technical experts and the move towards greater local administration of water control projects, these projects themselves increasingly became conceptualized and executed by subunits of the People's Communes (usually the production brigade). The mantra became cheaper, quicker, better, etc., as Yellow River conservancy projects were the result of local initiative designed to meet local problems. The ideal was indeed not to conform to the abstract notions of modern hydraulic practices, but projects were designed to fill practical needs and were to be executed through the sheer power of the human will, that is to say by a massive mobilization of labour.

International cooperation and self-reliance

The pattern of seeking international technical and financial assistance established during the Nationalist period was continued during the first decade of the PRC. After 1949, however, American, Dutch and German engineers were replaced by technical experts from the Soviet Union. Indeed, up to the onset of the Great Leap Forward, all water conservancy projects in China were advised by Soviet engineers.

Perhaps the best-known example of Soviet technical cooperation was the construction of the Sanmenxia dam (1958–1960). The

Sanmenxia reservoir was created behind the first significant dam in history to be built on the main stem of the Yellow River. However, because of the failure of the Soviet engineers to appreciate the nature of the sediment load in the river and the Chinese enthusiasm of the period to carry the project forward, the dam was woefully unsuited and the reservoir was silted within only a few years of construction. This, in turn, caused the waters of the Yellow River to back up into the Wei River basin, where they inundated land and threatened the ancient city of Xian with flooding. The failure of Sanmenxia, the similar failure of early irrigation projects and the famine which occurred in the aftermath of the Great Leap Forward were shocks to the leadership of the People's Republic in Beijing as well as to the YRCC (Greer, 1979; Becker, 1998). Together, these events caused a new sense of realism in policy and dampened the enthusiasm for pure engineering solutions to development problems and programmes. Better effort was made to understand the role of sediment in reservoir operations; dam construction plans were modified; and the number of new reservoirs to be constructed was reduced. Drainage development and irrigation system rehabilitation were also begun, and farmers were slowly re-convinced of the potential value of irrigation construction.

Soviet advisors packed up and returned to the Soviet Union by 1960. Beneath the mantra of self-sufficiency after 1960, Yellow River management was to be guided by the inspiration of the masses. The Cultural Revolution, which lasted from 1966 to 1976, brought political chaos to China, including the Yellow River basin. Somewhat surprisingly, the moderately revised development plans of the 1950s, and heavy government investment in the basin, continued despite the chaos, without substantial debate (Stone, 1998). Giant power-generating reservoirs were constructed in the upper basin; a soil-conservation campaign created new terraced fields on the Loess Plateau of the middle reach; and irrigation diversions were substantially expanded in the lower reach, especially in Shandong and Henan provinces. Meanwhile, village-based water management systems, including canal maintenance and water allocation between neighbouring villages, were shaped in the basin, although they were structured based on the political overtones of the time.

The Contemporary Setting: Change and Response

With the death of Mao Zedong in 1976, Deng Xiaoping came to power and helped to introduce a wide-ranging set of reforms that swept through China in the 1980s (Meisner, 1999; Naughton, 2003). The commune system that had been established in villages was abolished and a rural household responsibility system moved production decisions and power towards individual farmers (Ash, 1988). Government planning and control became more decentralized and, as also occurred in the agriculture sector, public investment in the water sector declined. Environmental awareness later started to grow and a more politically liberal atmosphere allowed people to review past basin strategies and lessons. In 1984, the state council approved the Second Yellow River Basin Plan, which listed soil-erosion control in the middle reach as the most important policy objective, as opposed to power generation and flood control, as had been emphasized in the 1954 plan.

Changing political economy

Following these changes, the late 1980s and early 1990s saw the arrival of a new water era for China. In the Yellow River, this was reflected in two ways. First, the rule of law was given added relevance. Second, economic growth placed increasing demand on water resources, in both quantitative and qualitative terms. Together, these and other factors caused fundamental changes in both perceptions of appropriate water policy and management, and, increasingly, in water management practice.

The major legal landmark for water policy was the 1988 Water Law, which provided the basic framework and principles for water management in the 1990s. This was followed by related legislation, including the Water Pollution Prevention and Control Law, the Soil and Water Conservation Law, and the Flood Control Law. A large body of additional administrative rules and ministerial regulations related to water were also passed, along with a number of other laws at least indirectly related to water.

This move towards legalism took place at a time of dynamic economic growth and structural change, which began in the early 1980s. Increasing liberalization of markets and foreign investment helped to sustain rapid economic growth. Industrial output increased dramatically. Increasing agricultural labour productivity and de facto and de jure changes in residency rules freed people from the farms and allowed rapid urbanization. While population growth has slowed, expansion continues and, importantly, rising affluence has caused dietary changes which favour meats and contribute to massive growth in feed grain use, with concomitant increases in crop water demand.

New challenges for the river

The key factors driving Yellow River management in the new era are thus not water itself but rather the larger economic and social environment, which has shifted pressure and focus. While flood control is still important, water stress is now probably the number one issue for most basin authorities and residents. How water stress rose in prominence can be seen by looking at three factors: a decline in water supplies, an increase in demand and a growing awareness of environmental water needs.

On the supply side, runoff substantially decreased in the 1990s, as shown in Fig. 5.2. One question is whether the decline is caused by secular declines in long-term precipitation levels brought about, perhaps, by global climatic change. As a similar, but apparently less severe, dry spell to that which occurred in the 1990s also occurred from 1922 to 1932, it is suspected by some that the Yellow River is now at the tail-end of a 70-year cycle, and that rainfall levels and river flows will therefore begin climbing in the near future. However, the figure graphically shows that the runoff decline is not a phenomenon of only the 1990s, but that other factors must also be at work. Possibilities include changes in land use, which have altered rainfall/runoff ratios (Zhu et al., 2004), and increased irrigation (Yang et al., 2004), including groundwater irrigation, perhaps in part as a response to declining surface supplies. Although a slowing of the problem is evident in the early 21st century, consistent with near average rainfall (YRCC, 2007), it is debatable whether this is evidence of a turnaround. There is no question, however, that the reduced runoff has contributed to supply constraints.

Even if runoff levels do increase, they might well be offset by decreases in effective supply due to pollution. Water pollution, in general, has been called the number one environmental issue in China (Jun, 2004). For the Yellow River, the declining state of water quality is exemplified in Fig. 5.3, which shows changes

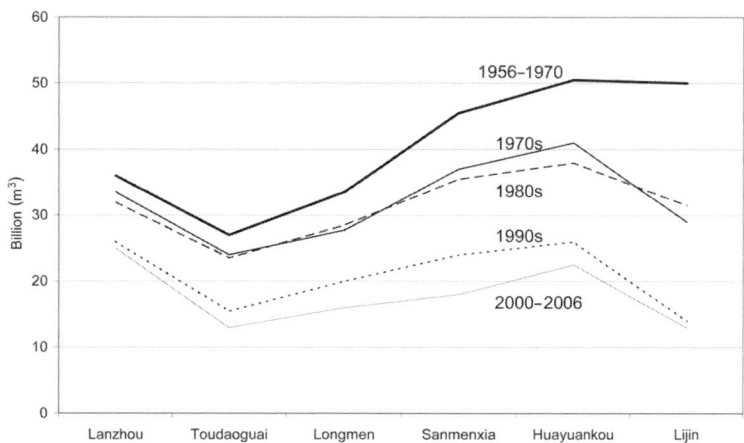

Fig. 5.2. Yellow River runoff, 1956–2006. Source: YRCC 2002b, 2007.

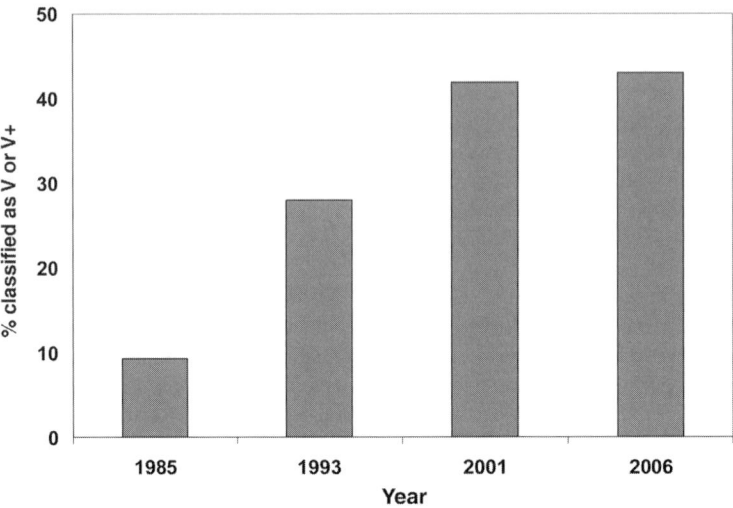

Fig. 5.3. Severely polluted length of the Yellow River (% classified as class V or V+). Source: Yellow River Water Resources Bulletin, YRCC. www.yellowriver.gov.cn/other/hhgb/

in percentages of the river's length classified under the Chinese system to be in the lowest-quality grade (V) or even worse (V+) – levels unsuitable for most direct human use. Nearly half the river now falls into one of these categories, and the Yellow River is now perhaps the second most polluted river in China.

One major pollutant source is industrial and domestic waste discharged into the Yellow River's main stem and tributaries. While there is substantial discharge from all provinces, Shaanxi contributes over one-quarter of the total, and the Wei River tributary contributes the largest share, almost 30% of the basin total. Two other important pollution sources are the unmeasured discharge from rural township and village enterprises (TVEs) and non-point pollution sources from agriculture. Beginning in the 1980s, TVEs developed rapidly throughout China and have often been allowed to remain out of compliance from wastewater laws and regulations because of their limited technology and financial levels, difficulty in monitoring their discharge, and the general trend in decentralization of economic control and management. From the early 1980s to the mid-1990s, farmers substantially increased their use of fertilizers and pesticides, with the result that a considerable fraction of residues now enters the river with return flow from irrigation.

On the demand side, total use (depletion) has increased only somewhat over the past one and a half decades (Table 5.1), in large part because there is little additional water to develop. However, there has been substantial change in the geography of use, with upstream regions consuming more and downstream regions less. Sectorally, there have also been moderate reductions in agricultural use, more than offset by dramatic growth in industrial and domestic depletion. Partially in response to declining surface supplies and increasing demand, groundwater pumping has also increased dramatically since the late 1980s. Available data from 1980 to 2002 show that groundwater abstraction increased by 5.1 billion m^3 billion, or 61%, reaching 13.5 m^3. However, since groundwater data are notoriously difficult to collect, especially for agriculture, where most use occurs, it is possible that actual use is even higher than the figures suggest (Wang et al., 2007a). In fact, the lower Yellow River basin is part of a now-infamous groundwater drawdown problem in the North China Plain, which has been suggested to be a threat to a substantial part of China's future food supply (Foster and Chilton, 2003). Even using formally collected statistics for the most recent period available (Table 5.2), combined surface water and groundwater depletion is now equal to nearly 80% of total withdrawals, which are themselves equal to nearly 90% of annually renewable water resources.

The outcome of declining supplies and increasing demand has already been the seasonal desiccation of portions of the Yellow River, discussed at the beginning of this chapter. From 1995 to 1998, there was no flow in the lower reach for some 120 days each year, and in some cases flow ended over 700 km from the sea, failing even to reach Shandong province. This cut-off inflow has important repercussions to basin function for three reasons. First, it obviously limits the availability of surface water for human use in downstream provinces and, less obviously, reduces groundwater recharge in the lower reach (because of the raised channel, discussed further below, this impact may be outside formal basin boundaries). Second, it negates the competence of the river to carry its heavy sediment load to the sea, potentially resulting in a more rapidly aggrading and flood-prone channel than would otherwise exist (although low flows also tend to be associated with lower sediment loads). Third, it has clear consequences for the ecology of the downstream areas and, in particular, for the Yellow River delta and coastal fisheries. The reduction in flow, coupled with success in flood control in the past five decades, has caused a retreat of the delta shoreline, intrusion of salt water, and increased salinity and lowering sea water temperature in the Bohai estuary. Further complicating matters, the Shengli petroleum field, the second largest petroleum oil source in China, is located in the delta and competes with the trickling river flow for environmental needs.

Table 5.1. Yellow River water depletion (billion m^3) by sector and reach, 1988–1992 and 2002–2004. Source: Cai, 2006.

Years	Reach	Total	Agriculture	Industrial	Domestic
1988–1992	Upper	13.11	12.38	0.51	0.22
	Middle	5.44	4.77	0.38	0.28
	Lower	12.18	11.24	0.55	0.38
	Basin	30.72	28.39	1.45	0.89
2002–2004	Upper	17.54	15.71	1.42	0.41
	Middle	5.71	4.16	0.97	0.58
	Lower	8.44	7.04	0.82	0.58
	Basin	31.69	26.91	3.21	1.57
Difference	Upper	34%	27%	179%	84%
	Middle	5%	–13%	155%	108%
	Lower	–31%	–37%	49%	54%
	Basin	3%	–5%	121%	77%

Table 5.2. Yellow River resources, withdrawal and depletion (billion m^3), 2004–2006.

Annual water resources		55.5
Withdrawal		
	Total	48.9
	Surface water	35.3
	Groundwater	13.5
Depletion		
	Total	38.2
	Surface water	28.6
	Groundwater	9.5

Since the 1998 strengthening of the 1987 Water Allocation Scheme and the operationalization of the Xiaolangdi dam, discussed below, the YRCC has managed to nominally end absolute flow cut-off, an important accomplishment. Even so, it is now clearly established that environmental water demands have not been adequately included in existing allocation schemes. According to basin managers, the primary environmental water use in the Yellow River is for sediment flushing to control potentially devastating floods, and it has been estimated that this would require about one-quarter of the Yellow River's flow (Zhu et al., 2004). The special challenge of flood control in the lower reach is caused when sediment transported from the middle reach begins to settle as the river spills on to the flat North China plain, producing a naturally meandering and unstable channel (Ren and Walker, 1998). It is calculated that roughly 1 trillion t of sediment enter the Yellow River each year. Of these, 400 million t are calculated to be captured by two large reservoirs and various irrigation diversions, 100 million t are believed to settle within the lower reach, and an additional 100 million t are flushed to the sea through dry-season minimum flow. To flush the remaining 400 million t, an environmental water requirement of 14 billion m^3 (3.5 billion m^3 of water per 100 million t of sand), which is more than one-quarter of the recent flow, is currently estimated to be necessary (Giordano et al., 2004).

To control the impact of that sediment which is not flushed, successive river managers over millennia have constructed levees to contain the Yellow River. While such structures may hold the channel in the short term, their success depends on continually raising the levee walls as new sediment elevates the level of the channel constrained within. Over time, the process of levee raising has contributed to a 'suspended' river, in which the channel bottom is above ground level, sometimes by more than 10 m (see Fig. 5.4). Since the founding of the People's Republic, the levees have held, but obviously the levee-raising solution cannot continue indefinitely. The current comprehensive flood management plan comprises a range of interrelated strategies. These include extensive soil and water conservation programmes in the upper and middle river reaches (particularly in the Loess Plateau); the construction of multi-purpose reservoirs; adjustment and strengthening of levees in the lower river reach; the development and improvement of flood-retention basins; the implementation of development and building controls in flood-prone areas; and planning measures, such as the relocation of families presently living in areas of high flood risk, such as the inner flood plain (Giordano et al., 2004).

In the more 'traditional' sense of ecological use, Chinese scientists, and the Chinese in general, increasingly recognize the environmental services that high-quality water flow brings. In the case of the Yellow River, these are largely discussed in terms of flow maintenance for biodiversity protection and sustenance of wetlands and fisheries at the mouth of the river, and for dilution and degradation of human-introduced pollutants. That concepts of environmental flows and values have changed is evident in the water-utilization accounts provided by the YRCC. The environment as a user of water was first included in basin water accounts as recently as 2004. While the most recent figures place environmental use at only 2% of total depletion, a more realistic figure would be likely to approach one-third of annual flow (Zhu et al., 2004).

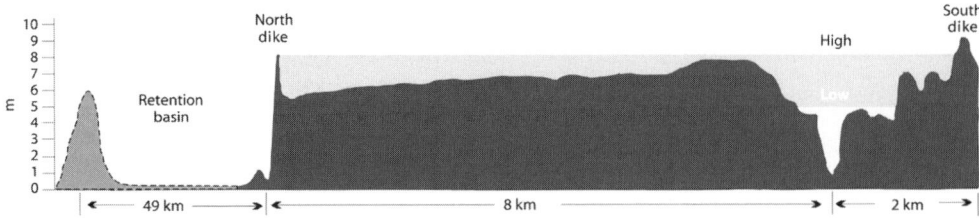

Fig. 5.4. Schematic representation of a cross-section of the Yellow River.[1] Source: after Ronan, 1995.

Institutional response

With effective supply decrease, increases in demand from traditional users and growing recognition of environmental needs, the Yellow River Basin is now effectively closed in most senses of the term. As a result, there is a clear need for water policy to shift away from a singular emphasis on flood control and resource development, and towards comprehensive basin management strategies. Such a new direction in thinking was, in fact, reflected in Article 1 of the 1988 Water Law, which stated that the document was 'formulated for the rational development, utilization, economization and protection of water resources, for the prevention and control of water disasters and for the realization of sustainable utilization of water resources in order to meet the needs in national economic and social development'. In other words, water management in China in the 1990s, harkening back to the Tang dynasty edicts, was officially going to take a more comprehensive approach, which would include concepts of economic value and trade-offs, resource protection and sustainable development, among others.

To carry out such changes in management, however, would require a movement in institutional structures. While the YRCC was already ostensibly serving as the river basin authority, in practice its powers for basin management and planning were limited and unclear. However, the changes in thinking brought about in part by the 1988 Water Law slowly began to be reflected in the management mandate of the YRCC. For example, in 1997, the state council approved the 'Outline of Yellow River Harnessing and Development', which, though still calling for the construction of 36 additional large dams, began addressing the issues of comprehensive utilization of the basin water resources. In 1998, the state council, the Ministry of Water Resources and the National Planning Committee issued the 'Yellow River Available Water Annual Allocation and Main Course Regulating Scheme' and the 'Management Details of Yellow River Water Regulating', leading the way to the first basin-wide, main-course flow regulation, which began the following year.

Perhaps more fundamentally, the Ministry of Water Resources brought forward ideas for the conceptual transformation of water resource development and management in China, from engineering-dominated approaches to approaches based on demand management and the value of water resources (a shift from emphasis on *gondchengshuili,* engineering water benefits, to *ziranshuli,* broader water resources benefits) (Boxer, 2001). Following this shift, concepts such as water pricing, water rights and water markets were further discussed and tested, and are now beginning to have an impact on water management across China, including the Yellow River basin.

Changing mechanisms and adaptation

The overarching changes in institutional structures and approaches brought new mechanisms through which water users have to, or choose to, use the resource. Following from the water-resource-based approach and the overarching change in political economy, calls for the use of water pricing as a mechanism to regulate use have now become almost universal in official discussions of water policy change. While the meaning and impact of water pricing in China, and elsewhere, are contested, the use of water pricing as a policy tool is at least premised on the assumption that it will provide incentives for farmers, the largest water user group, or, in practice, their direct water suppliers, to reduce water use and increase efficiency (Lohmar et al., 2007). A confounding issue, however, is that it is farmers who have benefitted least from China's economic growth, and increasing rural incomes is now also a major policy goal. Thus the government is struggling with ways in which pricing can be used as a tool for water savings and investment, while at the same time protecting or improving farmer welfare. As a result, water price increases are being discussed in terms of broader agricultural reform policies, which include reductions in rural taxation rates and new rural investments.

Often connected to water pricing reform is the establishment of water user associations (WUAs). As with pricing, devolution of at least some irrigation management control to local levels fits in with the overall push in China towards market principles, as well as with 'global' trends in water management paradigms. This is evidenced in the large involvement of international organizations in the funding of Chinese projects to create and support WUAs in the Yellow River and elsewhere. In theory, WUAs place management closer to the actual uses and therefore improve service and provide a mechanism for both fee collection and, therefore, sustained investment in operations and maintenance (Lohmar et al., 2007). This is expected to result in better long-term use of water, as well as improved farmer outcomes.

In practice, the utility of water pricing and WUAs as efficiency- and livelihood-enhancing tools is still the subject of debate. For example, it has been suggested that, given the low level of current prices, the level of increase needed to induce demand response may not be politically feasible, and the initial result of pricing may thus simply be one of a welfare transfer away from farmers without associated changes in water-use levels or practices (Yang et al., 2003; Barnett et al., 2006). Some empirical analyses have shown that this is not necessarily the case (Huang et al., 2006; Liao et al., 2007); however, even these analyses highlighted the incompatibility of agricultural water prices with rural poverty-alleviation goals. A second issue, perhaps especially important in the Yellow River's lower reach and the associated basins of the North China Plain, is that direct water pricing can, at present, only be applied to state-controlled surface water supplies, not to privately accessed groundwater. Some of the implications as related to water use are discussed further below.

In addition to direct effects on water-use decisions, increased prices and irrigation management reform are also hoped to provide indirect incentives for the adoption of water-saving technologies. There is, in fact, evidence since the 1980s of increasing use of such technologies, including field levelling, plastic sheeting, canal lining and sprinkler irrigation (Blanke et al., 2007). However, adoption still seems to be confined mostly to low-cost options appropriate for individual household use only. It has also been suggested that, even in the face of increasing scarcity, the water-related incentives for water users and managers to adopt most technologies are still simply too low.

To address this issue, new approaches are being sought. For example, there is at least one ongoing experiment with large-scale 'water trading', in which industry invests in agricultural water-savings technology, and other farmer benefits, in exchange for access to the water saved. This experiment is taking place between farmers in the Hetao irrigation district in Inner Mongolia (the largest in the Yellow River basin), and in the downstream industry near Baotou city.

There is also evidence that, even without sufficient incentives to adopt water-saving technologies, farmers are adapting to changing water and market circumstances in other ways. For example, as formal surface water allocations have declined, farmers have switched from low- to high-value crops, a phenomenon made profitable by the rising demand for vegetables, fruits and meat in growing cities, or by changing farming practices (as highlighted by Moya et al., 2004, in the Yangtze basin).

There is, however, a question on the extent to which these responses to planned (e.g. pricing) and unplanned (e.g. declining surface deliveries) actions result in real water savings. For example, reduction in the agricultural application of surface irrigation can, in some cases, simply reduce groundwater recharge, recharge that would later have been pumped and used again elsewhere. Kendy (2003) and Kendy et al. (2003) have highlighted this outcome for an area of the North China Plain, where virtually all annually renewable water is used (depleted) and groundwater tables are falling with agricultural and urban expansion. As Kendy et al. (2003) show, while water might be used and reused more wisely, bringing a balance between water supply and demand can only come from reduced use. With almost no water reaching the sea, it could be argued that the same holds true for the Yellow River in general.

Engineering not forgotten

Changing institutional structures and options for individual response to the new water challenges in the Yellow River have been closely connected with China's evolving political economy over the past quarter century. But China has, of course, long been famous for the use of large-scale engineering as a tool for water management. Thus it should come as no surprise that engineering solutions still form a large part of official efforts to manage the Yellow River, even in the new environment. These continuing engineering efforts can be put into three general categories – landscape change, water control and water mobilization.

In terms of landscape change, perhaps the most important is related to the Loess Plateau in the Yellow River's middle reach. Large-scale engineering efforts to transform the landscape of the Loess Plateau began in the 1950s and have included sediment-retention dams, revegetation and strip farming. Perhaps the most visually stunning means, which highlights the true magnitude of the input and the impact on the land surface, has been the creation of terraces on the steeply sloping gullies, easily visible with the naked eye even from commercial flights. While the early efforts at transformation of the plateau were couched in terms of agricultural output increases, they are now promoted on the basis of sediment reduction and poverty alleviation. By the turn of the 21st century, somewhat more than one-third of the farmland in the most erodible areas was considered to have been brought basically under control.

Related at least in part to engineering efforts at sediment control has been the continued construction of large-scale dams for water control. Most prominent of these is the recently completed Xiaolangdi dam, located in the lower middle reach, the largest dam on the Yellow River and second in China only to the Three Gorges. While a multi-purpose project, the dam's most heralded feature is its possibly unique system of tunnels and underground powerhouses, which make it possible to flush sediment through the creation of controlled floods. While the dam has been financed in part with foreign funds and constructed with the involvement of foreign engineers, it was built with a thoroughly Chinese understanding of the Yellow River's problems, showing that, since Sanmenxia, much has been learned in terms of both engineering skill and the management of international relations. In fact, the dam has been considered a major success and has even managed to avoid the criticism by international NGOs levelled against many other large-scale water-engineering projects in China. This may, in part, be because an international environmental expert panel was included in the project, perhaps a first for such a significant project in China (Gunaratnam et al., 2002).

Beyond Xiaolangdi, at least two dozen additional dam projects on the Yellow River and its tributaries are still planned. However, swamping any of these projects in terms of scale and impact, and certainly in controversy, is the effort to mobilize water in the south–north water-transfer scheme. While formally started late in 2002, the scheme was initially conceptualized in the 1950s (Greer, 1979) to move 50 billion m^3 of water, approximately the annual flow of the Yellow River, from the Yangtze basin in the south to the Yellow River and the North China Plain. If completed as in present plans, the south–north transfer will involve three routes, known by their relative geographic position – eastern, middle and western. The eastern and middle routes cross the Yellow River, before delivering most of their planned water further north. The western route would transfer water directly into the Yellow River. Because of the costs per unit of water moved, the diversion can only be justified on the basis of domestic and industrial demand. None the less, it can still be argued that agriculture is an indirect beneficiary, since the new water availability would reduce pressure on diversions from agriculture (Berkoff, 2003). In terms of direct impact on the Yellow River itself, the outcomes are not clear. Most of the planned transfers through the eastern and middle routes will be used outside the basin. The transfers from the western route would increase Yellow River flows directly, with the greatest benefit to provinces in the middle reach. However, as this route is the most costly and difficult to build, it is not clear whether it will ever be constructed.

While the south–north transfer is in many senses a classic engineering project of the hydraulic mission era, it is being justified on economic grounds. In fact, firms are expected to buy and market the water. Thus, even what might in the past have been thought of as a pure engineering endeavour now also has the flavour of the new economic environment.

Old tensions revisited and continuing transformation

The closure of the Yellow River basin has come at a time of, and in large part because of, larger economic and political change within China. The resulting management challenge brings to light again an age-old governance tension in China on the balance between central and local power. In essence, the necessary shift toward basin-scale management considerations implies a role for central authority, even if with a broader range of social input in decision making. At the same time, economic liberalization, even with 'Chinese characteristics', implies decentralized authority and the use of individual-oriented market incentives to drive resource use and conservation.

The potential conflict this can cause for water management is evidenced in the dichotomy in the authority and decisions between surface water and groundwater use. Allocation of surface water in the Yellow River remains the mandate of the YRCC and, with recent improvements in bureaucratic operation, monitoring ability and engineering control, it has been able to manage allocations between provinces reasonably well, even in the face of growing scarcity.[2] The end of Yellow River flow cuts is partial evidence. However, rapid growth in groundwater use over the last one or two decades (Wang et al., 2007b), along with the growth of private tube-well ownership (Wang et al., 2005) since 1979, has weakened the meaning of that control. For example, Molden et al. (2007) have shown that farmers in the Zhanghe irrigation district of the Yellow River's lower reach responded to declining surface water allotments by switching to self-supplied groundwater. The overall water result was not so much a change in the volume of water used, as was intended by the allocation reduction, but rather a change in the source of that use. The options and choices of individuals in effect nullified the ability of the YRCC. This is a conflict likely to surface in other areas as well. While it is not yet clear where the final balance of power will lie or how legal and regulatory change, and enforcement, will help to take the best from each approach, the history of adaption in the Yellow River to date suggests that solutions will be found.

Conclusions and Implications

To many an observer, the events reflected in the post-1949 history of Yellow River management may indeed suggest much that was novel, and much that was unprecedented, in Chinese history. It is our argument, however, that to look at this period in such a discrete manner is to neglect important historical continuities that can be viewed as an entire 20th-century effort to devise some type of political and social system to replace the Imperial system that fell in 1911. Much of this 20th-century effort was informed by the values and images of water and the Yellow River, as these evolved during the Imperial period. Although an examination of Yellow River conservancy certainly reflects broad and often bitter disagreement about institutional arrangements, China's role in the world and state–society relations, from the perspective of the post-Mao period there remain important continuities with patterns that were initiated and developed during the past. Despite fundamental differences in political form among the various Chinese state-building projects of the 20th century, each state was fundamentally driven by similar modernizing assumptions, and each sought to draw selectively upon multiple historical meanings of the Yellow River and water in similar ways.

Since the fall of the Imperial system in 1911, China has sought to reconstruct a state system able to ensure national survival and to pursue the goals of renewed wealth and power. Lasting for much of the last century, China's search for political form has expressed itself in experiments from one end of the 20th-century political spectrum to the other: representative democracy, warlordism (decentralization), quasi-fascism, communism and, most recently, capitalism with, what the government calls, Chinese characteristics. Transcending all these political–economic boundaries was water. More to the point, a major consideration of each successive state-building effort in the 20th century was how to effectively manage water to serve the goals of nation building and modernization. During the 20th century, every Chinese state sought to address the hydraulic breakdown on the North China Plain that had occurred during the late Qing period. The

Republican government after 1911, the Nationalist state after 1927, and the communist government after 1949 all sought to assume the historical legitimacy conferred by effectively regulating the Yellow River water. Although there were fundamental differences in political ideologies and organization during each political experiment during the 20th century, there were historical themes that transcended these boundaries. For example, the quest to establish a vigorous modern national identity among the peoples of the empire was a goal, transcending political–economic divides, of virtually every Chinese elite.

Water management in the 20th century was also informed by fundamental assumptions and goals that cut across the traditional political boundaries. Several pan-20th-century developments included faith in: (i) administrative centralization; (ii) modern industrial development; (iii) modern science and technology; and (iv) transnational cooperation. In turn, many of these assumptions and goals were informed, or promoted, by selective views of water that existed in the Imperial period. Traditional views of water, such as the politically legitimizing role of 'ordering the waters', centralized water management and the entire Confucian notion of active manipulation of water to serve the broader goals of statecraft, were never far below the surface, and infuse contemporary China's predilection for an activist government role in managing water on the North China Plain. The Confucian traditions that premise good government on the ability to 'control the waters' continue to animate the tendency within the YRCC to promote engineering solutions to water-scarcity issues. One need only offer the South to North Water Diversion Project as the latest supporting evidence of this bias. In contrast, a growing sensibility in China of environmental degradation has spawned a nascent environmental movement, which has promoted non-engineering approaches (e.g. conservation) to water issues. In the rhetoric of this movement, one clearly sees an implicit, and occasionally explicit, re-emergence of an aesthetic that is informed by traditional Taoist sensibilities. The continuing existence of these sensibilities is likely to mean that China has the capacity to be flexible in its management policies – able to execute shifts from engineering and non-engineering approaches by selectively calling upon historical and philosophical sanction.

The historical tension between centralized control and local autonomy continues to define the challenge of managing water in contemporary China. The imperatives of economic reform have entailed a significant devolution of central administrative power in China since 1978. Water planners recognize the historical lesson of effective central presence in managing the Yellow River, but efforts to successfully mediate local and regional interests have been difficult. Negotiating and enforcing water allocation compacts between provinces continues to be a major challenge. Below the provincial level, local governments are caught between serving central mandates and local constituents. By and large, pollution and groundwater exploitation continue to increase under the pressures of local economic development. This historical and contemporary tension between centre and locality will continue to define China's attempt to implement a national water strategy well into the future.

Since 1978, the YRCC has deepened commitments to internationalization that emerged during the 20th century. Although periods like the Great Leap Forward have witnessed water management premised on local initiative and local technical knowledge, the current patterns of internationalization are the consequence of the state's promotion of modern science and technology. Indeed, much of the content of international technical exchange and capital was embedded in the context of engineering solutions adopted by the state, and state involvement in scientific, technical and financial networks has also introduced the range of experiences, engineering and otherwise, that nations and regions have undergone in water management.[3] Similarly, the emphasis on market justifications for both water investment and management is largely premised on international practices. Indeed, one might suggest that with the historical emphasis on 'ordering the waters' in China, coupled with China's current commitment to international experience, we may see a certain synthesis of tradition and contemporary approaches to Yellow River management, which may well represent models for other regions of the world.

In the more immediate realm, the entire context of the Yellow River basin's closure has intensified the competing interests over water resources since the well-publicized 'shock' of the basin drying up in 1997. At the very centre of China's attempt to formulate institutional arrangements and responses lie the fundamental tensions arising from expanding urban populations, burgeoning industrial production and consumer demands for greater food resources. The trajectory for the Yellow River basin in the context of water scarcity will include adjustments in utilization, allocation and institutional responses, all shaped by the historical context of river and water management outlined above.

Notes

1 Just above the railway bridge linking modern Zhengzhou with Ximxiang, i.e. just west of the old Bian canal.
2 Under the system, the YRCC controls all key surface water reservoirs and surface water abstraction points and assigns use quotas on behalf of the central government to each basin province and autonomous region, plus Hebei and Tianjin. The quotas are adjusted proportionally each year, based on expected water availability. However, the system is more nuanced than this simple explanation suggests and provides opportunities for negotiation and adjustment, based on immediate conditions. See Zhu (2006) for additional details.
3 For an example of such commitments note the series of International Yellow River Symposiums held since 2000.

References

Anon. (n.d.) Tongyi shuili xingzheng ji shiye banfa dagang, and tongyi shuili xingzheng shiye jinxing banfa. *Number Two Historical Archives* (Nanjing, China) 44, 77.
Ash, R.E. (1988) The evolution of agricultural policy. *The China Quarterly* 116, 529–555.
Barnett, J., Webber, M., Wang, M., Finlayson, B. and Dickinson, D. (2006) Ten key questions about the management of water in the Yellow River basin. *Environmental Management* 38(2), 179–188.
Becker, J. (1998) *Hungry Ghosts: Mao's Secret Famine.* The Free Press, New York.
Berkoff, J. (2003) China: the south–north water transfer project – is it justified? *Water Policy* 5(1), 1–28.
Biswas, A.K., Dakang, Z., Nickum, J.E. and Liu, C. (eds) (1983) *Long-distance Water Transfer: a Chinese Case Study and International Experiences.* United Nations University, New York.
Blanke, A., Rozelle, S., Lohmar, B., Wang, J. and Huang, J. (2007) Water saving technology and saving water in China. *Agricultural Water Management* 87(2), 139–150.
Boxer, B. (2001) Contradictions and challenges in China's water policy development. *Water International* 26(3), 335–341.
Brown, L. and Halweil, B. (1998) China's water shortage could shake world food security. *World Watch Magazine* 11(4), 10–21.
Cai, X. (2006) *Water Stress, Water Transfer and Social Equity in Northern China: Implications for Policy Reforms.* Human Development Report Office Occasional Paper. United Nations Development Programme, New York.
Chi, W. (1965) Water conservancy in Communist China. *The China Quarterly* 23, 37–54.
Fairbank, J., Reischauer, E. and Craig, A. (1989) *East Asia: Tradition and Transformation.* Houghton Mifflin, Boston.
Foster, S. and Chilton, P. (2003) Groundwater: the processes and global significance of aquifer degradation. *Philosophical Transactions of the Royal Society London, B* 358, 1957–1972.
Giordano, M., Zhu, Z., Cai, X., Hong, S., Zhang, X. and Xue, Y. (2004) *Water Management in the Yellow River Basin: Background, Current Critical Issues and Future Research Needs.* Comprehensive Assessment Research Report 3. Comprehensive Assessment Secretariat, Colombo, Sri Lanka.
Greer, C. (1979) *Water Management in the Yellow River Basin of China.* University of Texas Press, Austin and London.
Gunaratnam, D., Xie, Q. and Ludwig, H. (2002) The International Environmental Expert Panel for major dam/reservoir projects: the Yellow River, China. *The Environmentalist* 22(4), 333–343.
Huang, L. (1986) Huaihe liuyude shuili shiye. Master's thesis. National Taiwan Normal University.

Huang, Q., Rozelle, S., Howitt, R., Wang, J. and Huang J. (2006) Irrigation water pricing policy in China. Paper presented at the IAAE Preconference, Water/China Joint Session 12 August 2006, Gold Coast, Australia www.montana.edu/econ/seminar/Archive/irrwaterpricepolicychina.pdf

Jun, M. (2004) *China's Water Crisis*. Eastbridge Books, Norwalk, Connecticut.

Kendy, E. (2003) The false promise of sustainable pumping rates. *Ground Water* 41(1), 2–4.

Kendy, E., Molden, D.J., Steenhuis, T.S. and Liu, C.M. (2003) *Policies Drain the North China Plain: Agricultural Policy and Groundwater Depletion in Luancheng County, 1949–2000*. Research Report 71. International Water Management Institute, Colombo, Sri Lanka.

Leung, G.Y. (1996) Reclamation and sediment control in the middle Yellow River valley. *Water International* 21, 12–19.

Levenson, J. and Schurman, F. (1969) *China: an Interpretive History, from the Beginnings to the Fall of Han*. University of California Press, Berkeley.

Liao, Y., Giordano, M. and de Fraiture, C. (2007) An empirical analysis of the impacts of irrigation pricing reforms in China. *Water Policy* 9 (Suppl. 1), 45–60.

Lohmar, B., Huang Q., Lei, B. and Gao Z. (2007) Water pricing policies and recent reforms in China: the conflict between conservation and other policy goals. In: Molle, F. and Berkoff, J. (eds) *Irrigation Water Pricing: the Gap between Theory and Practice*. Comprehensive Assessment of Water Management in Agriculture Series, No. 4. CAB International, Wallingford, UK, pp. 277–294.

Meisner, M. (1999) *Mao's China and After: a History of the People's Republic*. Simon and Schuster, New York.

Menzies, N.K. (1995) Forestry. In: Needham, J. (ed.) *Science and Civilization in China*, Vol. 6, part III. Cambridge University Press, Cambridge, pp. 543–667.

Milliman, J.D. and Meade, R.H. (1983) World-wide delivery of river sediment to the oceans. *Journal of Geology* 91(1), 1–21.

Molden, D., Bin, D., Loeve, R., Barker, R. and Tuoung, T. (2007) Agricultural water productivity and savings: policy lessons from two diverse sites in China. *Water Policy* 9 (Suppl. 1), 29–44.

Moya, P., Hong, L., Dawe, D. and Chen, C. (2004) The impact of on-farm water saving irrigation techniques on rice productivity and profitability in Zhanghe irrigation system, Hubei, China. *Paddy and Water Environment* 2(4), 207–215.

Murphey, R. (1967) Man and nature in China. *Modern Asian Studies* 1(4), 313–333.

National Economic Council (1934) *Number Two Historical Archives (Nanjing, China)* 44, 78.

Naughton, B. (2003) *Growing Out of the Plan: Chinese Economic Reform 1979–1993*. Cambridge University Press, Cambridge, UK.

Pietz, D. (2002) *Engineering the State: the Huai River and Reconstruction in Nationalist China, 1927–37*. Routledge, New York.

Pietz, D. (2006) Controlling the waters in twentieth-century China. In: Tvedt, T. and Jakobsson, E. (eds) *A History of Water: Water Control and River Biographies*. I.B. Taurus, London, pp. 92–119.

Ren, M. and Walker, H.J. (1998) Environmental consequences of human activity on the Yellow River and its delta, China. *Physical Geography* 19(5), 421–432.

Ronan, C.A. (1995) *The Shorter Science and Civilization in China: an Abridgement of Joseph Needham's Original Text*. Cambridge University Press, Cambridge, UK.

Stone, B. (1998) Developments in agricultural technology. *The China Quarterly* 116, 767–822.

Strauss, J. (1998) *Strong Institutions in Weak Polities: State Building in China, 1927–1940*. Studies on Contemporary China. Clarendon Press, Oxford.

Todd, O.J. (1949) The Yellow River reharnessed. *Geographical Review* 39(1), 38–56.

Vermeer, E. (1977) *Water Conservancy and Irrigation in China*. Leiden University Press, Leiden.

Wang, J., Huang, J. and Rozelle, S. (2005) Evolution of tubewell ownership and production in the North China Plain. *The Australian Journal of Agricultural and Resource Economics* 49(2), 177–195.

Wang, J., Huang, J., Blanke, A., Huang, Q. and Rozelle, S. (2007a) The development, challenges and management of groundwater in rural China. In: Giordano, M. and Villholth K. (eds) *The Agricultural Groundwater Revolution: Opportunities and Threats to Development*. CAB International, Wallingford, pp. 37–62.

Wang, J., Huang, J., Rozelle, S., Huang, Q. and Blanke, A. (2007b) Agriculture and groundwater development in northern China: trends, institutional responses, and policy options. *Water Policy* 9 (Suppl. 1), 61–74.

Wang, Z. (1987) *Huaihe liuyou zhili zongshu*. Bengbu: Shuili dianlibu zhihuai weiyuanhui, Bengbu

Wittfogel, K. (1957) *Oriental Despotism: a Comparative Study of Total Power*. Yale University Press, New Haven.

World Bank (1993) *China: Yellow River Basin Investment Planning Study*. Report No. 11146-CHA, June 30, 1993. World Bank, Washington, DC.

Wu, R. and Fan, C. (1993) Huaihe xiayoude honlao zaihai taolun. *Jianhuai shuilishi lunwenji*. Zhonggui shuili xuehui shuilishi yanjiuhui, Beijing

Yang, D., Li C., Hu, H., Lei, Z., Yang, S., Kusuda, T., Koike, T. and Musiake, K. (2004) Analysis of water resources variability in the Yellow River of China during the last half century using historical data. *Water Resources Research* 40 W06502, doi:10.1029/2003WR002763.

Yang, H., Zhang, X. and Zehnder, A. (2003) Water scarcity, pricing mechanism and institutional reform in northern China irrigated agriculture. *Agricultural Water Management* 61(2), 143–161.

YRCC (Yellow River Conservancy Commission) (2002a) Yellow River Basin Planning. www.yrcc.gov.cn/. March, 2002 (in Chinese).

YRCC (2002b) Information made available during meetings between the YRCC and the International Water Management Institute. Zhengzhou, China, September–October, 2002.

YRCC (2007) Water Resources Bulletin. www.yellowriver.gov.cn/other/hhgb/

Zhang, Z. and Shangshi, D. (1987) The development of irrigation in China. *Water International* 12(2), 46–52.

Zhu, Z., Giordano, M., Cai, X. and Molden, D. (2004) The Yellow River basin: water accounting, water accounts and current issues. *Water International* 29(1), 2–10.

Zhu, Q. (2006) *Preliminary Assessment on the Impacts of Unified Water Regulation in the Yellow River*. Chinese Ministry of Water Resources. Beijing. www.mwr.gov.cn/english1/20060110/20060110104326 MDXBUU.pdf

6 The Colorado River: What Prospect for 'a River No More'?

Douglas J. Kenney
University of Colorado, Boulder, Colorado, USA;
e-mail: douglas.kenney@colorado.edu

Introduction

The Colorado River of the American Southwest is among the most studied, contested and valued rivers in the world, annually providing water and electricity to roughly 30 million residents, generating 11.5 billion kWh of hydroelectricity, and irrigating more than 3 million acres (1.2 million ha) of crops (Adler, 2007). This is remarkable in many ways, not least of which being the observation that, just 150 years ago, Lieutenant Joseph C. Ives (1861:110) concluded his exploration of the basin with this remarkably misguided assessment:

> The region last explored is, of course, altogether valueless. It can be approached only from the south, and after entering it there is nothing to do but to leave. Ours has been the first, and will doubtless be the last, party of whites to visit this profitless locality. It seems intended by nature that the Colorado River, along the greater portion of its lonely and majestic way, shall be forever unvisited and undisturbed.

How does a river change from being 'altogether valueless' to becoming critically important in, roughly, the span of two human lifetimes? The answer lies not so much with the river itself, or even in the lands drained by the river, but in how human ingenuity and institutions have shaped how value is created and measured. The combination of an arid, sunny climate with abundant lands having good soils would, without irrigation, indeed be only of limited human value. But irrigation – aptly deemed 'reclamation' in the American West – has transformed the region, first for the benefit of farming, and more recently for booming sunbelt cities such as Las Vegas, Phoenix, Los Angeles and Denver. As part of this transformation, the jagged mountains, massive canyons and vast deserts that once made the region inhospitable are now viewed as amenities worthy of reverence and protection. It is a region, and a history, full of contrasts and paradoxes, with a future being shaped by a continuous stream of newcomers, including 37 million visitors annually to Las Vegas and 5 million to the Grand Canyon, and welcoming nearly one million new permanent residents annually to the seven Colorado River states.

Given the rate of change in the Colorado River basin, it is difficult to predict the future with any confidence, especially since an unwelcome new era is emerging: an era of limits. It is increasingly unrealistic to accommodate new demands in the basin simply by drawing on unused supplies, as users already exist to utilize every drop of the Colorado; the river has not consistently reached the ocean for decades. Rather, meeting new, mostly urban, demands requires actions that resonate through the water community in some way: for example, drawing on surplus flows in wet years, transferring water from agricultural to urban users in normal years, and tapping reservoir storage in dry years. This last scenario has been particularly evident in recent years; reservoirs that were 90% full in 2000 were less than half their capacity by 2004.[1] While much of this decline can be rightly attributed to the onset of drought (particularly severe in 2002), other conspirators have been population growth and the corresponding expansion of the water infrastructure to serve these new populations. From 1920 to 1990, the population of the Colorado River basin states increased more than sevenfold, giving way to an even more explosive growth in the 1990s, when four basin states (Nevada, Arizona, Colorado and Utah) led the USA in percentage population growth, while another (California) led in terms of absolute population growth (Census Bureau, 2001; Grand Canyon Trust, 2005).[2] In 2004, one senior official estimated that the size of the population relying on water from the Colorado River had increased by 26% in the past decade (Griles, 2004). Also impressive is population growth in the final reaches of the river, across the border in Sonora and Baja, Mexico. While drought conditions may end at any time, rapid population growth is expected to continue, and, additionally, the wealth of recent research suggests that climatic change will hit this region harder than most – reducing streamflows anywhere from 11 to 45% by 2100 (Christensen and Lettenmaier, 2006; Hoerling and Eischeid, 2007).[3] This is the backdrop against which irrigation, urbanization and environmentalism are now colliding, all within the context of laws, customs and values shaped over a remarkably short time-frame.

Physical and Environmental Setting

The Colorado River is primarily fed by snowmelt originating high in the Rocky Mountains of Colorado and Wyoming. Every spring and summer, this water races downhill in a generally south-west direction, pulling in tributaries from New Mexico and Utah to form the main channel slicing through arid lands in Arizona, Nevada, California and a small section of Mexico (Fig. 6.1) (for general summaries, see Carothers and Brown, 1991; Pontius, 1997; Gleick et al., 2002; Project Wet, 2005). Many maps of the Colorado show the 632,000 km^2 basin as ending at the US–Mexico border – undoubtedly a politically motivated decision, but actually not terribly inaccurate as over 95% of the basin is in the USA. The overwhelming majority of management decisions and engineering works are located in the USA, and the river ends soon after crossing the international border, disappearing completely in most years into waiting fields before it can reach its natural terminus at the Colorado River delta along the Gulf of California.

One of the few qualities of the Colorado River that is not on a grand scale is the flow of the river. For legal reasons (discussed later), main-stem Colorado River flows[4] are reported at Lee Ferry (or adjacent to Lee Ferry), the midpoint of the river just downstream of the Glen Canyon dam (see Fig. 6.1). Gauging records are interpreted with respect to known upstream patterns of water storage and consumption to estimate the natural (i.e. unaltered) flow. The total annual natural flow of the river at this point averages approximately 15 million acre-feet (MAF)[5] (roughly 18 billion m^3).[6] None the less, while not among the top 20 US rivers in terms of flow volume, the Colorado is still an impressive and welcome asset in what is primarily an arid basin. Much of the lower basin, home to the most productive agriculture, receives only 100 mm of precipitation annually. An ambitious programme of hydraulic engineering has taken full advantage of these modest and highly variable flows (see Fig. 6.2). Along its course, the river is now harnessed by roughly two dozen significant storage and diversion projects, most notably the Glen Canyon dam (forming Lake Powell) and the Hoover dam (forming Lake Mead), bracketing both ends of the region's signature natural attraction, the Grand Canyon. Water storage facilities on the Colorado River can hold roughly 4 full years of flow, a tremendous asset in terms of water supply management, but achieved at the expense of transforming the river from an unpredictable and sediment-heavy, warm-water stream to an elaborate plumbing system of relatively clear and cold water, flowing in highly predictable (and tempered) patterns – described by Fradkin (1981) as 'a river no more'.

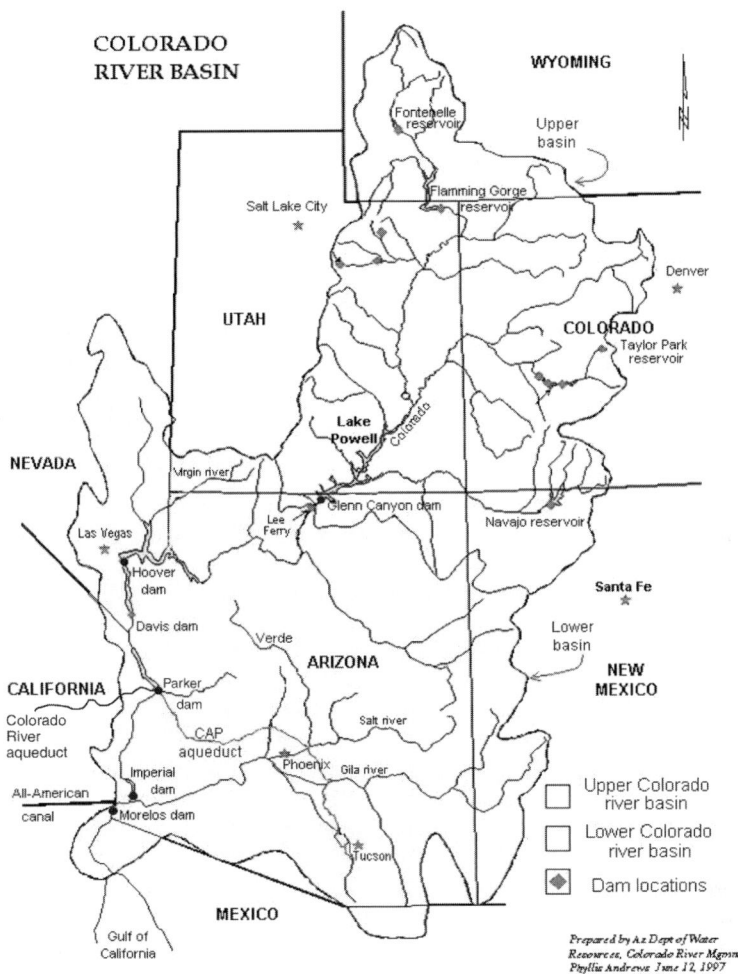

Fig. 6.1. The Colorado River basin (modified from the original).

The environmental consequences of this modified hydrograph are felt throughout both the basin and the local ecosystems, with native fish species providing perhaps the best indicator of the environmental costs of river development. The construction of water infrastructure, particularly the Hoover and Glen Canyon dams, has created an environment where non-native species have displaced most native species; four remaining native fish species (humpback chub, razorback sucker, bonytail chub and Colorado pike minnow) are listed as endangered (Carothers and Brown, 1991; Adler, 2007). Of particular salience has been the removal of both sediment from the river by the storage reservoirs and water from the system by out-of-basin exports. Many of the major users of Colorado River water – including those in southern California, Colorado's Front Range, central Utah, and the Rio Grande valley in New Mexico – are located outside the Colorado hydrologic basin. The ecological impact of the resulting changes to the volume, timing, temperature and chemical composition (especially the enhanced salinity) of flows is further compounded by the introduction of exotic species, including trout (for the cold-water fisheries), horses and burros, tamarisk (aka salt cedar), and plant and animal species associated with farming and ranching (Adler, 2007).

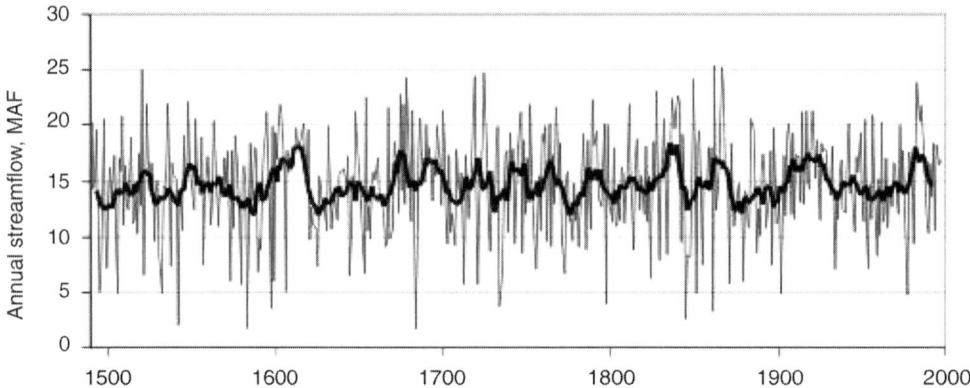

Fig. 6.2. Reconstructed natural Colorado River flows (at Lees Ferry) (Courtesy Jeff Lucas and Connie Woodhouse). (Dark line indicates 10-year averages.)

Environmental restoration programmes in both basins – the Upper Colorado River Endangered Fish Recovery Program and the Lower Colorado Multi-Species Habitat Conservation Plan – exist to coordinate mitigation but, ironically, both efforts are explicit in allowing still additional river development and consumption. No ecosystem is more threatened by this accumulation of storage and diversion facilities than the Colorado River delta, primarily located in Mexico. Diminished flows due to upstream consumption, including long time periods during the initial filling of the Mead and Powell lakes, have starved the delta of flows, reducing the area covered by wetlands to less than a tenth of its original 728,000 hectares (Glennon and Culp, 2002). The delta now survives on roughly 1% of the river's natural flow, this water originating mostly as agricultural return flows and occasional reservoir spills – such as the El Niño-inspired floods of the early 1980s (Fig. 6.3). Given the increasing water demands, likely decreased flows due to climate change and currently low storage levels, major reservoir spills may never recur (Gertner, 2007). Current efforts to improve the efficiency of upstream water-delivery systems threaten further reductions in flow.[7]

An Institutional History of the Colorado River Basin

The institutional arrangements of the Colorado River basin have evolved over several decades of conflict and compromise. Most histories of the basin focus on the evolution of the so-called 'Law of the River', a collection of federal and state laws and court decisions that, collectively, apportion the flow of the river among the seven basin states and Mexico (e.g. see Lochhead, 2001, 2003). However, while the Law of the River is undoubtedly important and is central to understanding both the basin's past and future, it is only one component of the overall institutional framework. There are many political, social, cultural and environmental factors which not only fill out the legal skeleton provided by the Law of the River but also frequently articulate a competing set of values. The result is that the modern institutional arrangements of the Colorado River are bifurcated, and the primary source of this bifurcation is paradigmatic. Specifically, the institution features an odd balance of a 'private commodity' paradigm, featuring an emphasis on water development and the rights of individual rights-holders, with a 'public value' paradigm, emphasizing resource protection, value pluralism and democratic (i.e. collective and participatory) decision making. Not surprisingly, given their inherent incompatibility, these paradigms did not evolve simultaneously or in a coordinated manner, but evolved rather sequentially and incrementally. It is against this backdrop that new institutional arrangements are now being sought, pushed by the harsh reality of a limited water supply but constrained by the lack of a coherent vision regarding the appropriate goals of water management.

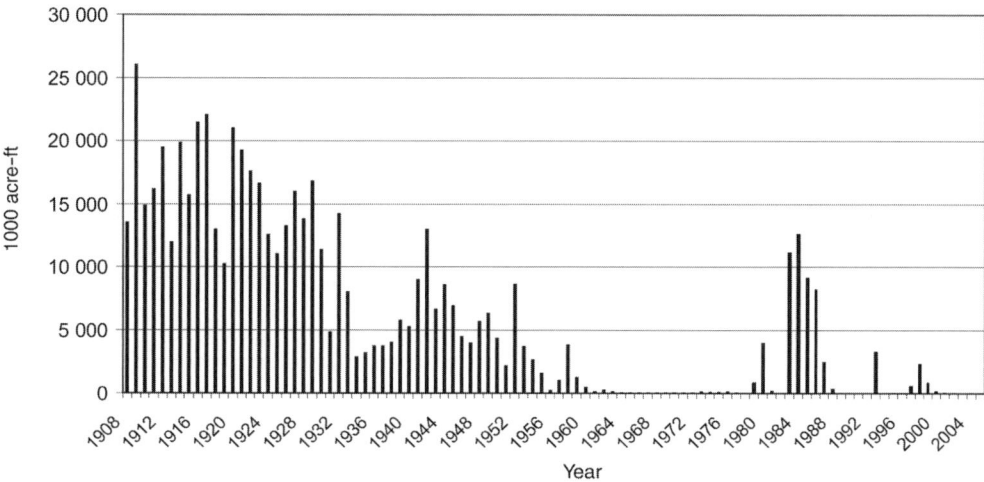

Fig. 6.3. Colorado River flows to the delta (adapted from data compiled by Kevin Wheeler).

In what follows, the institutional history of the Colorado River is reviewed in eras defined by these two dominant paradigms, focusing primarily on the major portion of the basin that lies within the seven US Colorado River states. In contrast to a traditional Law of the River history, which begins with the Colorado River Compact of 1922, this review begins with the arrival of the first Europeans in this part of the New World, as this provides the origins of the private commodity paradigm, which still largely shapes the institutional arrangements in the Colorado. In these early decades, the conflicts between countries, and, later, US states, for the bounties of the Colorado occurred within this dominant paradigm. Conflicts thus typically did not feature fundamental disagreements regarding values or ideologies but were primarily distributive in nature – i.e. each party wanted to secure as much of the river's benefits as possible – and were focused on issues of apportionment, development and consumption, while systematically devaluing non-monetary, public and systemic values of the river.

Evolution and reign of the private commodity paradigm

Early exploration and settlement

The origins of the region's private commodity paradigm can be traced back to the post-Columbian era of European expansion into the New World. The first wave of European explorers in the 1530s comprised the Spanish conquistadors, most prominently Francisco Vasquez de Coronado, who led the ultimately unsuccessful search for the mythical Seven Cities of Cibola, thought to contain mineral riches similar to those in the Inca Empire of Peru and the Aztec Empire in Mexico (Waters, 1946; DeVoto, 1952; Brandon, 1990). Finding no gold, these excursions ultimately gave way in the 1600s to Spanish missionary *entradas*, aimed at bringing Christianity to the region. Much like the conquistadors, the missionaries greatly improved the geographic knowledge of the lower Colorado basin but were otherwise unsuccessful, as the padres could claim few souls and only one mission (San Xavier, near modern-day Tucson, Arizona) survived after missionary efforts were abandoned in 1781. By the 1800s, the English and French had replaced the Spanish as the major European influences in the region, this time concentrated in the upper basin. Like the Spanish earlier, these were not immigrants looking for homesteads but were entrepreneurs looking to extract wealth – in this case, beaver skins for the European hat industry (Waters, 1946; DeVoto, 1952).

By the 1840s, the fur industry was in decline, but global forces were still shaping events in the Colorado River basin. As Waters (1946:185) writes:

Across all Europe – in France, Austria, Germany, Hungary, Italy – geysers of unrest broke out. In an unparalleled outpouring of human emotion the tide swept over Europe, and kings ran before it in terror. All of South and Central America rose in revolt against their Spanish masters, establishing their independence. In North America, Mexico broke free from Spain and then the Republic of Texas from Mexico. The United States, declaring war against Mexico, took most of the Colorado River basin including what was to become Nevada, Utah, California and most of Arizona, New Mexico, Colorado and Wyoming.

Soon, almost the entire Colorado River basin became the legal domain of the USA, with the obvious exception being the failure to acknowledge the sovereign rights of the indigenous peoples (known as Indians or Native Americans). Dozens of tribes are indigenous to the region, including Apaches, Navajos, Hopis, Zunis and Utes. Beginning with the conquistadors, each wave of Anglo settlement occurred with little regard to native peoples, cultures and rights, a tradition that improved only marginally under US control, as wars and treaties forced great reductions in territories under tribal control. Addressing the so-called 'Indian problem', however, was insufficient by itself to stimulate Anglo settlement of the basin, and if the USA had learned anything from the European competition for the New World, it was that the key to holding land was promoting settlement (DeVoto, 1952). Given that settlement of arid territories is innately tied to water management, water policy thus became a tool of national security and national economic development.

It was in this context that gold deposits were first discovered in the West, prompting the California Gold Rush of 1849, followed a decade later by similar gold rushes in Colorado and Arizona (Waters, 1946). Succeeding where the conquistadors had failed over 300 years earlier, thousands of entrepreneurs flooded into the region from across the globe in search of mineral wealth. Eventually, the mining 'boom towns' evolved more diversified economies or went bust as mineral reserves were exhausted or spread too thin among competing miners, but the legacy of the boom on water resources has endured, largely due to the evolution in the mining camps of the prior appropriation doctrine of water allocation, since adopted and practised in all of the Colorado River states (and beyond) (Pisani, 1992).

Four elements of prior appropriation are particularly noteworthy (Tarlock et al., 2002; Kenney, 2005). First, unlike the riparian doctrine practised in the eastern USA, water rights established under prior appropriation are not linked in any way to land ownership, thereby ensuring that western development was not limited to stream corridors but can, instead, reach wherever the combined forces of engineering and economics can provide water services. Second, water rights established through prior appropriation are limited to legally recognized 'beneficial uses', which until recently only included industrial, agricultural, municipal and domestic uses, while excluding most environmental uses. Third, prior appropriation water rights are a form of private property right, which can be bought and sold with relatively few restrictions, the primary one being that no transfer can be permitted that 'injures' other legally established prior appropriation rights. Fourth, and most significantly, the prior appropriation doctrine is based on the tenet of priority and, specifically, the notion that the first person to beneficially use a water source should, in perpetuity, retain the right to continue to use the same volume of water (and for the same uses) every year.

Perhaps the best way to understand prior appropriation is to consider how a 'call on the river' works. A 'call' is the term used to describe a situation when insufficient water is available in a given year to satisfy the needs of all parties with recognized water rights. The origin of these rights can be traced back to the initial settlement of the region and the first uses of water for recognized purposes. Over time, an inventory of these uses was developed, and each 'right' was recorded with respect to the location of use, the amount of use, the purpose of use and the first date of use. While the details vary somewhat among the western states, each generally established a water management agency to record and monitor the exercise of these rights, with these efforts organized at sub-state scales defined by the major river basins. A call is most likely to occur in a drought, and begins when a water rights-holder complains to the state agency about the unavailability of water. To satisfy the call, the administrator orders some users to completely cease diversions, beginning with the most junior (the youngest rights), followed by the second most junior, and so on, until the available supply again matches the volume of the remaining rights. Note that this is not a system based on sharing or proportional cutbacks; junior water rights are cut off in their entirety, one by one, until the remaining rights-holders can use their rights in their entirety. In practice, this can be highly complex, as seniors and juniors are scattered throughout a basin, in different reaches and sub-basins. A particularly challenging situation arises when the most senior users are far downstream, as this requires the upstream juniors (perhaps in a different sub-basin) to allow water to flow past their diversion structures to ensure that the downstream senior is satisfied. Administering these programmes is a challenge to legal institutions, engineering systems and social systems, but provides the benefit of encouraging and protecting early investments in water projects (Kenney, 2005).

The priority concept not only provided a strong incentive to rapid settlement but also enshrined the key elements of the private commodity paradigm – i.e. the notion that water is an economic commodity which should be privately owned and manipulated for the benefit of entrepreneurial capitalism. It is worth noting that this approach to water allocation and management differs significantly from what was observed in many of the first agrarian settlements in the West, particularly the Mormon communities that sprang up in Utah in the late 1840s, the Hispanic *acequia* communities of northern New Mexico, or any of the Native American communities (Waters, 1946; Maass and Anderson, 1978). These communities all featured collective or centralized control of water resources, an approach strongly endorsed by western visionary John Wesley Powell. Powell – best remembered for his exploration of the Colorado River in 1869 – was one of the first men to openly question the logic of the private commodity paradigm, instead arguing for small communal societies nourished by the careful and sustainable utilization of the region's limited natural resources (Powell, 1890; Stegner, 1953). Powell's well-reasoned argument in favour of moderation and community control was widely ignored.

Following the US Civil War of the 1860s, a large and restless eastern population was ready to heed Horace Greeley's famous advice and head west, and did so at the urging of a national government that provided a variety of homesteading programmes designed to promote an agrarian West, a popular national goal (Pisani, 1992). Many homesteaders soon discovered, however, that the small land allotments (often just 160 acres, or 65 ha), lacking reliable water supplies, were simply not suited to farming. It is estimated that two-thirds of all homesteaders failed, often leading to the consolidation of land in the hands of banks and other 'empire builders', who found large tracts well suited to low-density ranching (Stegner, 1953). Where agrarian communities flourished – particularly in pockets of California, Arizona, Utah and Colorado – it was because of their location along perennial streams that were well suited to the construction of water storage and diversion works. If agrarian settlements were to take hold on a large scale, then water development on a large scale seemed the obvious answer.

Apportionment and lower basin development

By the early 1900s, it was apparent that the dream of an agrarian West – viewed by the progressive national government as more ideologically desirable than mining or ranching economies – would require development of the West's large river systems, particularly the Colorado River. The Reclamation Act of 1902 was thus enacted to bring the financial and technical resources of the federal government to task, initially under a funding mechanism designed to recoup costs from project beneficiaries, but eventually evolving into a programme of blatant subsidies and political favours (Worster, 1985; Reisner, 1986; Wahl, 1989). Many of the initial targets of the federal reclamation programme were in the lower Colorado River basin, where fertile soils, long growing seasons and favourable topography provided an ideal opportunity for large-scale irrigation, if only the flow of the river – once termed a 'natural menace' by the Bureau of Reclamation

(USBR, 1946) – could be controlled by upstream storage. Existing irrigation developments along the Palo Verde, Yuma, Imperial and Mexicali valleys (in the Arizona–California–Mexico border region) had not only already demonstrated the potential for irrigation but had also shown the vulnerability of these operations to flooding and siltation.

Large-scale river development could not proceed, however, until an understanding was reached regarding the legal apportionment of the river's flow among the seven US states and Mexico. Owing to political unrest in Mexico and a reluctance of water interests in the USA to acknowledge any obligation by the upstream nation to maintain flows to Mexico, it was quickly decided that an apportionment was needed just between the states of the upper basin (Colorado, New Mexico, Utah and Wyoming) and the lower basin (Arizona, California and Nevada) (Hundley, 1975). Despite the fact that the prior appropriation system was already in effect (intrastate) in each of the seven US Colorado River states, it was argued by the upper basin that this approach would not be equitable at the interstate scale, given that the lower basin was being settled at a much faster rate. The upper basin states thus wanted a permanent reservation of water for their use (regardless of when that use would eventually occur), and unless they got this, they would use all means necessary to block any apportionment and, more importantly, any of the desired lower basin developments – particularly the Hoover dam. Thus, the seeds of a very hard-fought compromise were sown, and a new institutional mechanism – the interstate compact – was unveiled to produce the Colorado River Compact of 1922, the first of nearly two dozen water allocation compacts now in existence in the American West (Hundley, 1975; Tyler, 2003).

As case-specific solutions to interstate water allocation disputes, each compact is unique, but the Colorado River Compact is particularly unusual, in that it features an apportionment of specific, long-term (decadal) volumes of water rather than annual percentages or standards requiring the maintenance of a constant minimum flow rate at the state line. The key element of the compact is found in Article III(d), which requires the states of the upper basin to release 75 MAF of water every 10 years past Lee Ferry (see Fig. 6.1) to the lower basin (or an annual average of 7.5 MAF), which seemed a modest burden, given that the annual flow of the river was estimated at this time to at least exceed 16 MAF and perhaps to be as high as 20–22 MAF (Hundley, 1975). The roughly two decades of gauging data available suggested an average flow of 16.8 MAF. However, as shown earlier in Fig. 6.2, this estimate has proven to be highly flawed, as gauging records and tree-ring studies both suggest the long-term flow of the river is approximately 15 MAF (Woodhouse et al., 2006).

This error can potentially work to the disadvantage of the upper basin states, given the downstream release requirement. In a manner very analogous to a call on a prior appropriation regime, in an extended dry period, if satisfying the lower basin delivery obligation meant insufficient water remained to serve upper basin users, then those users would presumably be prevented from diverting and using the water as it flowed through these headwaters states. This situation has never happened, in part due to two protections provided to the upper basin. First, the compact's 10-year accounting method allows reduced deliveries in dry years, as long as they are offset by higher deliveries in wet years (within any 10-year period). Second, as discussed later, a major storage reservoir (Lake Powell) now exists just upstream of the delivery point, allowing the upper basin to maintain steady downstream deliveries even when faced with highly variable inflows – at least as long as water remains in storage. This capability has been exploited to create a hydropower-focused water management regime that keeps releases relatively constant, which ironically eliminates much of the flexibility inherent in the 10-year accounting method.

The compact was ratified[8] as part of the federal Boulder Canyon Project Act of 1928, which authorized the Boulder dam, renamed the Hoover dam, and the All-American canal, so named since it would divert water from the river to agricultural users in southern California, in a structure that would not cross over the international line (unlike an existing canal, which was being used by both Mexican and American interests). It also provided an interstate apportionment among the lower basin states of 4.4 MAF to California, 2.8 MAF to Arizona and 0.3 MAF to Nevada.[9] This element of the Boulder Canyon Project Act has been the subject of considerable litigation, mostly resolved in Arizona v. California (1963), but has survived intact. With these provisions in place, construction of the Hoover dam (along the Arizona–Nevada border) was completed by 1935 at a cost of US$49 million (in 1935 prices) and at least 96 lives. The project has dramatically reduced the flood danger downstream, while providing over 26 MAF of storage capacity (in Lake Mead) and 2000 megawatts of hydropower capacity. Soon thereafter, in 1941, the Parker dam was built downstream on the river (along the Arizona–California border), to provide a diversion point for the Colorado River aqueduct, which provides municipal and industrial water to southern California cities (Fig. 6.1). As seen with the other lower basin projects, the Parker dam was fraught with controversy, with Arizona unsuccessfully using both litigation and the Arizona National Guard in a futile attempt to slow California's use of the river (Mann, 1963).

The apportionment of the Colorado River was completed in the 1940s in two separate actions. First, a 1944 Treaty with Mexico (Mexican Water Treaty of 1944) apportioned a minimum of 1.5 MAF/year (roughly 10% of the river's natural flow) to be delivered at the international border. This is water in addition to the 7.5 MAF allocated annually to both the upper and lower basins, and thus increased the overall annual apportionment of the river to 16.5 MAF. Initial discussions with Mexico in 1910 had been based on a potentially equal division of flows at the border, an arrangement that had disintegrated by 1923 to the point where the USA suggested it was not obligated to provide any delivery (based on the infamous but ultimately insignificant Harmon Doctrine) (Hundley, 1966). The deal enacted was, thus, yet another hard-fought compromise and was tied to another apportionment decision regarding the shared Rio Grande River, where Mexico has the strategic advantage of being the upstream party on the critical reach (Hundley, 1966).

The second apportionment decision of the decade came in the Upper Colorado River Basin Compact of 1948, which apportions the upper basin share among the four states as follows: 51.75% to Colorado, 23% to Utah, 14% to Wyoming and 11.25% to New Mexico. Percentages are used since the amount of water reserved for the upper basin is theoretically 7.5 MAF/year, but due to the flawed flow assumptions used in the Colorado River Compact and the new delivery obligation promised to Mexico – both of which must be satisfied before the upper basin can take its apportionment – it is widely assumed that the flows available to the upper basin may not consistently exceed 6 MAF (Tipton and Kalmbach, 1965).[10] This compact also featured the establishment of an Upper Colorado River Commission to monitor consumption levels and, if necessary, interpret and enforce complex rules for sharing upper basin shortages. This has never been necessary; upper basin consumption has never exceeded 4 MAF/year (see Table 6.1). Exactly how the Upper Colorado River Commission would calculate and enforce shortages among the four states remains to be seen, especially since no curtailment of upper basin uses is likely to be initiated by the commission until legal ambiguities regarding the full Colorado River Compact are first addressed. The rules of the upper basin compact generally call upon each state to curtail water uses in proportion to levels of use in the preceding years, although exactly how this would be implemented by state agencies within each state is a further source of uncertainty. Given recent drought conditions, several upper basin states have initiated these discussions.

Omissions in the apportionment scheme

Before moving forward with a discussion of upper basin and Arizona water development, still nested within the private commodity paradigm, it is worth noting that the seven-state and international apportionment of the Colorado River, as completed in 1948, left many issues unresolved for future generations. The apportionment framework is not only based on flawed flow assumptions and ambiguities about how future shortages would be handled, but also contains several notable substantive omissions. Many of these omissions have not been fully addressed as yet, with progress delayed for

Table 6.1. Colorado River main-stem consumption and deliveries to Mexico (thousand acre-feet).

	1975	1980	1985	1990	1995	2000	2005
Upper basin (UB)							
Colorado	1,789	1,754	1,993	2,102	1,711	2,383	1,856
New Mexico	293	424	393	362	387	337	466
Utah	616	670	759	784	792	774	853
Wyoming	278	353	351	520	436	421	405
UB total	3,001	3,220	3,541	3,803	3,366	3,953	3,618
Lower basin (LB)							
Arizona	1,208	1,035	1,032	2,117	2,029	2,643	2,429
California	4,937	4,680	4,710	5,163	4,837	5,258	4,344
Nevada	154	228	373	311	350	450	292
LB Total	6,299	5,943	6,115	7,591	7,216	8,351	7,065
Evaporation	2,093	2,063	1,841	1,598	1,703	2,102	1,360
Total USA consumption	11,393	11,226	11,497	12,992	12,285	14,406	12,043
Delivered to Mexico	1,656	6,143	13,396	1,676	1,838	2,145	1,725

Note: UB totals include minor deliveries in north-eastern Arizona (not shown). Data for 2005 are provisional; evaporation losses, in particular, are very rough estimates. During the current drought, inflows have been approximately 62% of the 30-year average in 2000, 59% in 2001, 25% in 2002, 51% in 2003, 49% in 2004, 105% in 2005, 71% in 2006, and 68% in 2007; 2008 was expected to be an average or above-average year. Data are compiled from the Bureau of Reclamation statistics, primarily the *Consumptive Use and Losses* reports and *Decree Accounting* statements.

decades until crises and changes in the paradigm provided a more conducive policy-making environment. Four of these omissions include Indian water rights, environmental flows, groundwater and water quality.

The basic apportionment is nearly silent on the issue of Native American (Indian) water needs, with the exception of language in Article VII of the Colorado River Compact – later repeated in many subsequent compacts – stating that 'nothing in this compact shall be construed as affecting the obligations of the United States of America to Indian tribes'. This language was inspired by the landmark Winters decision in 1908 (Winters v. United States, 1908), which established as precedent the federal responsibility to provide tribes relegated to reservations with the water resources needed to sustain these new tribal homelands. Translating this principle into actual water management in the Colorado River basin is an ongoing process, subject to considerable debate and litigation, especially in the lower basin, where the vast majority of the basin's large reservations are located. Arizona, in particular, features several tribes with Colorado River rights of great seniority, as these rights are defined as originating with the dates of the Indian treaties or the establishment of reservations, actions that typically took place before widespread homesteading by Anglos. Additionally, these rights can be quite large, as they have since been defined as the amount of water that would be needed to irrigate all the 'practicably irrigable acreage' within the reservation.[11] By some estimates, large reservations – such as the Navajo reservation in north-eastern Arizona – could conceivably be awarded the entire flow of the Colorado River under this calculus. Politically, this outcome is unacceptable to the non-Indians that would be displaced, so the 'solution' has been to withhold from tribes the financial resources needed to develop water projects until they agree to settlements that dramatically scale-back the size of their rights (Burton, 1991; Thorson et al., 2006).

While the ethics of this approach are certainly debatable, the effectiveness is undeniable; many Navajos, for example, still do not have potable domestic water supplies in their communities. In contrast, several tribes have negotiated settlements tied to the Central Arizona Project (discussed in the following section), which now delivers approximately 0.55 MAF annually (about one-third of project capacity) to tribal lands in central Arizona.[12]

Another largely unresolved issue is the need for environmental flows. As suggested earlier in the discussion of the Colorado River delta, the reservation of water for environmental flows was not explicitly provided for in either compact or in the treaty, with the exception that each jurisdiction retains great latitude in how apportioned water is used internally. States can, theoretically, reserve a component of flow for environmental needs, but the incentive to do so is limited by the lack of any assurance that other states would follow suit and, more importantly, by the evolution of water allocation rules during an era and paradigm where environmental protection took a back seat to water development. In the Colorado basin (as in many other places), protecting the environment was seen as something that could wait until the basic sustenance needs of homesteading populations could be assured. As discussed later, this era did not arrive in this basin until the 1970s.

Groundwater is also not mentioned in the apportionment scheme, a common (and often problematic) omission in western water compacts generally, but one that has thus far been tolerable in this case, since the centrepiece of the Law of the River is the requirement to deliver a fixed volume of surface water at a given point (Lee Ferry) and, subsequently, the apportionment of that surface water to three states (and eventually Mexico) downstream. From the standpoint of the overall basin, how groundwater is managed upstream is largely irrelevant as long as the delivery obligation is satisfied. Similarly, groundwater use in the lower basin is an important issue – overdrafting in Arizona is a chronic problem – but is largely outside the scope of the Law of the River, which has been interpreted by the courts as not applying to lower basin tributaries. Groundwater law is extremely complex and non-uniform across (and sometimes within) the basin states, with most regimes awarding rights based on either priority (as done with surface water) or land ownership, or some combination thereof (Bryner and Purcell, 2003).

Finally, water quality is also omitted from the apportionment scheme, which has primarily been an issue due to the accumulation of salts as the river moves downstream. This is a result of natural processes and human activities, including out-of-basin imports of freshwater in the upper basin, saline irrigation return flows and evaporation from reservoirs. At one point in the 1960s, excessive salt in the river resulted in a brief international incident with Mexico, which convincingly argued that its apportionment could not be satisfied with water too salty for irrigation. In response, the treaty was modified in 1973 to reflect this understanding, and an ongoing remediation programme was established under the Colorado River Basin Salinity Control Act of 1974 (Holburt, 1975; Adler, 2007).

Upper basin and Arizona development

With the completion of the basic basin-wide apportionment through the Mexican Treaty and Upper Basin Compact, and given the economic boom that followed the end of World War II, the states of the upper basin mobilized to pursue their share of federal water development funds. Arizona was also now in line for water projects, having seen the futility in spending decades unsuccessfully fighting Californian projects. In fact, the first of the big post-war project proposals was for the Central Arizona Project (CAP), a vast aqueduct that can convey approximately 1.5 MAF of water from the main stem (on the Arizona–California border) to interior regions, including the cities of Phoenix and Tucson, traversing over 541 km and 732 m in elevation. The project was designed to ease groundwater overdrafting problems throughout the state. Included in the CAP proposal were dams at Bridge (or Hualapai) and, later, Marble canyons, bracketing Grand Canyon National Park, to provide the hydropower (and the hydropower revenues) necessary to support the project in terms of both electricity (for pumping) and economic subsidies for the intended market of both agricultural and municipal users (Terrell, 1965a).

This idea of using 'cash register' hydroelectric dams to subsidize water deliveries was eagerly embraced by upper basin users, who sought to implement the concept on their own forthcoming projects.

While the economic and environmental merits of the CAP were debated in Congress, the upper basin pursued projects, first gaining resumption of work on the Colorado–Big Thompson Project[13] (initiated in 1938 but delayed by World War II), and then initiating congressional consideration of the multi-faceted Colorado River Storage Project (CRSP). After initial discussions, it was determined that the CRSP would consist of five cash register dams and 15 'participating projects' (i.e. regional irrigation systems), and would use the new economics proposed in the still-pending CAP bills to achieve what the General Accounting Office has since calculated as a 100% subsidy for the participating projects – truly a stunning fall for a programme that still claims to be fee based, but only a slightly larger subsidy than the system-wide reclamation project average.[14] In Congress, the CRSP bill enjoyed the support of the upper basin states and Arizona, but was opposed by a coalition of southern California water interests, fiscal conservatives and environmentalists (Terrell, 1965b).

The emergence of environmentalism as a political force in Colorado River politics was largely a new phenomenon, foreshadowing the eventual emergence of the public values paradigm. At issue in the CRSP bill was the proposal to build the Echo Park dam inside the Dinosaur National Monument (along the Utah–Colorado border). Ultimately, securing passage of the Colorado River Storage Project Act of 1956 meant abandoning the Echo Park dam proposal in exchange for an enlarged project at Glen Canyon – a Faustian bargain that is now widely regretted among environmental interests, due to the submergence of the spectacular canyons that characterize the Glen Canyon region (Terrell, 1965b). The dams authorized by CRSP provide nearly 34 MAF of storage capacity in four major units – Glen Canyon on the Colorado River in Arizona, Flaming Gorge on the Green River in Utah, Navajo on the San Juan River in New Mexico, and the Curecanti (now the Aspinall) Unit on the Gunnison River in Colorado. Eleven participating projects were also authorized to use the stored water, a great irony to many, given that the US Department of Agriculture was actively working elsewhere in the country at this time to take 40 million acres out of production to ease national crop surpluses (Terrell, 1965b).

Still additional projects in the upper basin (and elsewhere) were authorized in 1968 when the CAP legislation was finally enacted. As seen in the CRSP process, the passage of the Colorado River Basin Project Act meant abandoning the environmentally controversial 'Grand Canyon dams', this time traded for a massive coal-fired power plant (the Navajo Generating Station), which ironically impedes visibility of the canyon spared from the dam builders. Perhaps more than any other example, the coalition building and deal making associated with the act embodies the distributive politics epitomized by western water conflicts, as Arizona got its long-desired CAP only by conceding to California a junior water priority for Colorado River flows serving the project, and adding language authorizing projects in Nevada (the Southern Nevada Supply Project), Utah (re-authorization of the Dixie Project and provisional authorization of the Uintah Unit of the Central Utah Project), New Mexico (authorization of Hooker dam or alternative), and Colorado (authorization of the Dolores, Dallas Creek, San Miguel, West Divide, and Animas–La Plata Projects) (Ingram, 1990).[15] Overall, the Colorado River Basin Project Act legislation features a palpable lack of internal consistency or financial integrity, and marks the high water mark for the private commodity paradigm.

The era of the public values paradigm

The successful efforts to block the Echo Park and the Grand Canyon dams were the precursors of a larger movement which fundamentally altered the legal, political and ideological foundations of the Colorado River. Until this point, the battles for the Colorado River, while heated and protracted, were among parties that viewed the resource through a common lens, emphasizing development, entrepreneurialism and private control. Sustaining the political viability of this paradigm required strict adherence to three related myths: (i) the economic argument that the federal reclamation programme pays for itself in user fees, a claim that is more than true for the multipurpose dams but only rarely a reality for the irrigation projects; (ii) the notion that these efforts worked to the benefit of the family farmer and other individual entrepreneurs, when in reality the benefits largely accrued to empire-builders such as banks, railroads and corporate agriculture; and (iii) the notion that the economic benefits of water development were so vast and fundamental as to render any concern over ecological impacts, the loss of environmental services, or the deterioration of other instream values as inconsequential (Fradkin, 1981; Reisner, 1986). Adhering to these now discredited myths fuelled numerous political careers and widespread economic development, and undoubtedly helped achieve the national goal of western settlement, but it also created something heretofore missing from the region: an urban constituency drawn to the aesthetic and environmental amenities of the region, supportive of public lands and other collective resources, and emphasizing quality of life over return on investment. It is more than a little ironic that aggressive water development activities in the West have created the infrastructure necessary to support approximately 55 million residents in the Colorado River basin states – up from 4 million just a century earlier (see Fig. 6.4) – and the subsequent rise of an 'ethic of place' (Wilkinson, 1990), based primarily on a public values paradigm.

The federal environmental movement

Efforts to reconcile these two competing worldviews take place in several arenas. One of the most controversial has been the evolution of federal environmental policy. Unlike the conservation movement of the early 1900s and the associated focus on the scientific utilization of natural resources (Hays, 1959), modern environmentalism has a strong preservationist ethic, which questions the underlying logic of utilitarianism, and also has a strong urban, aesthetic and public-health orientation (Paehlke, 1989). These threads run through several national laws enacted in the late 1960s and early 1970s, including, among others, the Wild and Scenic Rivers Act of 1968, the National Environmental Policy Act of 1969, the Clean Water Act of 1972, and the Endangered Species Act of 1973 (Rasband *et al.*, 2004). These acts, all applicable in the Colorado River basin, are forceful articulations of preservation, moderation and deliberative decision making, and all feature new opportunities for citizens to participate in decision making through both formal decision-making processes and a rapidly growing variety of ad hoc collaborative efforts (Kenney *et al.*, 2000). Of particular salience in

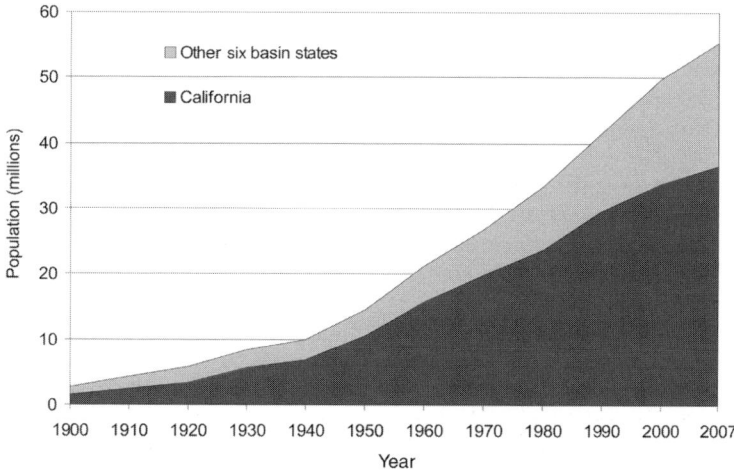

Fig. 6.4. Population growth in the Colorado River basin states (1900–2007). (Courtesy Brad Udall.)

the basin has been the Endangered Species Act, which effectively blocks new developments found to jeopardize the continued existence of threatened and endangered species, and which has forced many operational modifications to existing projects. Federal legislation enacted in this era and focusing on public lands management also articulates similar public values principles, a notable observation given that over half the Colorado River basin is federal public lands – a figure that jumps to almost three-quarters if tribal lands are included.

Still notably absent from this body of federal environmental legislation are rules requiring, as a matter of course, the reservation of water instream for environmental flows. Unless necessary in a given river stretch to protect an endangered species or to sustain the purposes of a federally reserved area (e.g. a waterfall associated with a national park), federal laws generally defer to the tradition in state water law of allowing water users to consume rivers in their entirety. Western states now provide some mechanisms for devoting water rights to instream flows, but these tend to be very limited in scope, often relying on water rights that are junior to traditional consumptive users (Gillilan and Brown, 1997). To the extent that rivers in arid regions of the American West retain some perennial flows, the cause is often the presence of senior water rights-holders downstream, which precludes some upstream (junior) diversions, or, on a larger scale, the existence of interstate compacts that require the maintenance of specified flow levels downstream. Since most demands on the Colorado River are in the lower reaches of the river, both legal requirements and economic patterns ensure that water flows remain relatively high (compared with unaltered flows) until reaching major diversion structures, mostly in California and Arizona. What is not maintained, however, are the peak flows needed to sustain the geomorphology and habitat characteristics required by native species. Major environmental restoration programmes in the upper basin, in the Grand Canyon reach (of the lower basin), and proposed efforts in the delta, for example, are all based around the desire to restore periodic peak flows, a goal that often runs counter to the purpose of constructing and operating water-storage reservoirs (Adler, 2007). To the extent that progress is made on these environmental issues, it usually takes the form of reservoir operational changes, including well-publicized (but very isolated and temporary) flood releases from the Glen Canyon dam. The actual removal of dams has been discussed, but is not an idea that has taken root in the Colorado basin.

In addition to substantive changes in water management, federal environmental laws also reshape the governance landscape. A strong theme running through most modern environmental legislation is a distrust of federal natural resource agencies, especially those accustomed to producing natural resource commodities. As a result, agency decision-making processes were reformed to be more specified and transparent than ever, with public participation, benefit–cost studies and environmental assessments as required elements, and with abundant opportunities for judicial review of decisions. Additionally, many natural resources agencies at all levels of government have found it increasingly worthwhile to work collaboratively with groups of public and private stakeholders on a variety of natural resource issues. The so-called 'watershed initiatives' are one expression of this phenomenon, mostly of the 1990s (Kenney et al., 2000). These groups have been much more active in the small watersheds of the Pacific Northwest than those of the Colorado basin, and have found much more success dealing with water-quality issues than the water-supply disputes that characterize the more arid regions of the West, including the Colorado River basin, where the seniority concept is often viewed as an impediment to collaborative problem solving. None the less, they are one additional element of the Colorado's evolving institutional framework, encouraging a greater consideration of environmental and other public values as part of water management.

These changes in law and governance, combined with the demographic transformation of the region associated with its sudden urbanization, have presented a particular challenge for the region's primary dam builder and traditional enabler of the private commodity paradigm: the Bureau of Reclamation. A reorganization and temporary name change to the Water and Power Resources Service

(1979–1982) was one attempt to publicly embrace an evolving focus from water development to management. Similarly, the agency's need to rethink its constituency was perhaps firstly and most clearly articulated in its *Assessment '87* report, in which it noted:

> As irrigated agriculture becomes a smaller part of its mission, the Bureau needs to identify all of its constituencies. At the same time, however, it must assure agricultural interests that they are not being abandoned where there is a legitimate need for a continuing Federal presence. By working with new constituencies in potential partner arrangements, the Bureau can make an easier transition to an effective resource management organization.
> (USBR, 1987)

Although still an agency dominated by water resource engineering, by most measures the Bureau of Reclamation has been successful in evolving its mandate to include substantial foci on water-system efficiency, environmental mitigation, conflict resolution and urban water issues. A similar evolution has taken place in the other branches of the federal government. In Congress, key natural resource committees, once routinely dominated by powerful western defenders of reclamation programmes, now often feature members sceptical of (if not openly hostile to) environmentally and economically unsound reclamation programmes that are blatantly contradictory to the values expressed by their increasingly urban constituencies. Also, since the federal environmental movement, support for additional subsidized western irrigation projects has been spotty at best among most presidential administrations, first, and perhaps most famously, demonstrated by President Carter's 'hit list' of reclamation projects unveiled in the late 1970s, followed soon after by President Reagan's much less-publicized, but ultimately more effective, efforts to discourage questionable projects by the use of less-generous federal cost-sharing requirements (Reisner, 1986). To be politically viable, modern federal reclamation projects typically need to be small, feature extensive environmental mitigation elements, and be tied to Indian water rights settlements, such as the Animas–La Plata Project, nearing completion in south-western Colorado (Pollack and McElroy, 2001).

States, markets and the evolving role of agriculture

Although the Colorado River states have enacted several state laws consistent with the public values paradigm, the level of activity has generally trailed that of the federal government, perhaps in part due to the very fact that federal programmes now effectively cover issues of pollution and species protection, and also due to the observation that the state's role in water issues has generally been limited to administering prior appropriation rights, established, in most cases, decades before the modern environmental movement. Layering public interest protections and new efficiency standards on top of already established rights is a difficult task, which most states have been reluctant to tackle; rather, the more common focus is on establishing modest instream flow programmes (within the framework of priority rights) and adding terms to newly established or modified rights (Kenney, 2001). Of particular concern are rights transferred from one user to another – often in the modern era from agricultural to urban users. Outside some so-called 'water banking' activities, the legal transfer of water rights between Colorado River states is nearly non-existent and remains a highly delicate topic, but market-based water transfers within states are commonplace, and are the primary tool used to adapt the allocation of water in this region transitioning from rural to urban.[16]

The growing frequency of water transfers in the western states says a lot about the past, present and future of irrigated agriculture, although the message is far from clear (MacDonnell, 1999). Despite the emergence of several large cities highly dependent upon Colorado River flows (e.g. Las Vegas, Los Angeles, San Diego, Phoenix, Tucson, Denver, Albuquerque, Salt Lake City), the greater part – probably more than two-thirds – of Colorado River flows are still used in agriculture.[17] The most productive areas are in southern California and western Arizona, which produce roughly 80% of the winter vegetables of the USA (Project Wet, 2005). In the upper basin, much of the agricultural activity is focused on producing cattle feed; it has been argued that cattle are the single largest consumer of Colorado

River water (Fradkin, 1981). Thus, while the political might and economic importance of the agricultural sector have declined significantly, agriculture is still an important player in Colorado River water issues. Increasingly, agriculture plays two, largely contradictory, roles in western water issues: first, as a 'water source' for cities wishing to purchase rights to sustain ongoing population growth; and second, as a cultural and aesthetic amenity that urban dwellers often wish to sustain. Similarly, the viewpoint of irrigators towards water markets features two seemingly incongruent threads: first, that water markets provide an essential revenue stream for financially strapped or retiring farmers; and second, that the collective impact of markets can be a detrimental force undermining the viability of rural communities (Howe et al., 1990). Not surprisingly, western state legislators are frequently caught in a dilemma of trying to streamline water transfers (to increase the efficiency and utility of transfers) while trying to ensure that transfers offer protection to third parties and public interests, typically defined to include rural communities dependent on farming economies and, less frequently, on environmental resources (National Research Council, 1992).

Living with Limits: a New Era for the Colorado?

The challenge

For several decades, water demands on the Colorado River have roughly matched the full available yield of the river, with most consumption happening in the last third of the basin. According to records provided by the US Bureau of Reclamation, from 1996 to 2000 (prior to the current drought), annual water consumption (depletion) averaged approximately 15.5 MAF: 8.0 in the lower basin, 3.7 MAF in the upper basin, 1.8 MAF in Mexico, and 2 MAF lost through reservoir evaporation (USBR, 2004).[18] Table 6.1 provides additional statistics on patterns of water consumption at 5-year intervals (not averages). Particularly noteworthy in Table 6.1 is the rise in demand throughout the 1980s and 1990s and, conversely, the sharp decline (evident by 2005) after the onset of aggressive drought-coping measures. Figures provided for Mexico are for deliveries, not consumption, although in most years the two values are comparable, given the tradition of full use in the basin.

Notwithstanding the important long-term challenges of finding water for environmental restoration and for some Indian communities with unresolved water rights claims, in most other respects, this tradition of full use is not inherently problematic, as long as the least reliable component of water yield is only used as a supplemental supply (ideally for low-valued uses) and not as the baseline supply supporting urban growth. Unfortunately, this is not the situation in many pockets of the basin, as rural uses generally precede urban uses (and thus rank higher within states' prior-appropriation systems). This is an unusual situation, but it is one that can be remedied. As noted above, state water laws provide an important mechanism to reallocate water (and the risk of shortages) through voluntary agricultural to urban water transfers, ranging in form from the dozens of small transactions occurring each year along Colorado's Front Range to the massive deals in southern California that have weaned urban areas off surplus flows (i.e. flows in excess of the state's apportionment) through complex conservation and transfer arrangements with major irrigation districts. But, ultimately, the efficacy of this strategy for managing water supply risk in particular locales in the Colorado River basin is shaped and limited by the larger interstate rules of water allocation codified in the Law of the River and, perhaps more importantly, by the realization that the overarching challenge in the basin is to acknowledge and live within the limits of the river. This challenge has a particularly complex flavour in the Colorado River basin due to the river's overallocation.

In theory, the Law of the River provides the framework within which water budgets can be established and shortages allocated, if necessary, between the Colorado River states and Mexico. However, as noted earlier, the apportionment found in the Law of the River is flawed in many ways, as it annually allocates 16.5 MAF (7.5 MAF for each basin and 1.5 for Mexico) from a river that yields, at best, 15 MAF. The fact that the Colorado River is overallocated has been widely understood for many decades but has become more difficult to ignore as urban growth results in larger (and firmer) water demands and as drought conditions have gripped the basin. Several trends suggest this situation could worsen; population growth, climatic change and energy development all suggest further stress on water resources. Faced with these pressures, states such as Colorado and Arizona, which historically have not used their full apportionments, continue to pursue additional development and consumption of the river. To not do so would ease stress on the river but only by imposing burdens (limits) on their own residents for a situation that others have primarily created and benefit from, and from which the Law of the River is supposed to provide protection. Somewhat ironically, this expansion of use has become more realistic as problems of overuse have forced California to scale back its use to its legal apportionment (from 5.2 to 4.4 MAF/year).[19] But the calculus remains unchanged: if all states pursue plans that target consumption at the level of their legal apportionments, and if those apportionments are collectively more than the river provides, then the situation is inherently unsustainable. This reality is particularly troublesome in an era of climatic change; even a modest 10% reduction in flows would provide a tremendous challenge to the regional water budget.

Solutions?

The twin forces of drought and growing demands, and the net impact of declining reservoir storage (see Fig. 6.5), prompted the federal government in 2005 to warn the states that they needed to develop a plan for sharing shortages or the federal government would do so independently. Ironically, despite all the nuanced language in the Law of the River, there had always been much ambiguity in how shortages in the lower basin should be handled. While the Upper Basin Compact provides some rules and establishes a commission to calculate and enforce shortages in that part of the basin, the legislation apportioning lower basin shares does not explicitly address the allocation of potential shortages and does not establish a commission to address the issue. The Supreme Court in the Arizona v. California (1963) litigation appointed the Secretary of the US Department of Interior to make these decisions when necessary, and, in 2005, the Secretary made it clear that her preference was to ratify a scheme developed by the states rather than to impose her own solution.[20] For the states, this was a formidable political challenge, as no state official wanted to agree to a reduction of its apportionment or to any change in the management of reservoirs or water accounting that modified the reliability of that apportionment. Political careers in the

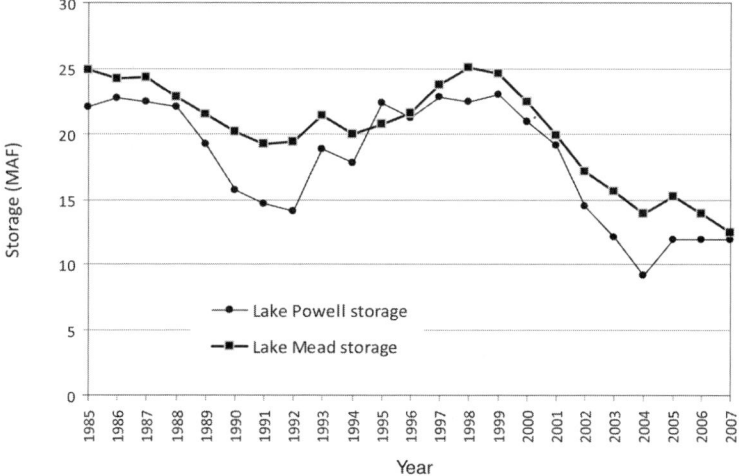

Fig. 6.5. Storage in Lakes Powell and Mead, 1985–2007.

American West have been historically built on the ability of leaders to obtain more water (Reisner, 1986; Ingram, 1990). Voluntarily agreeing to take less could be viewed publicly as failure and even as immoral, as the 'rights-based' tradition of water law in the West makes it very difficult to consider compromise or sharing (Wolf, 2005). The situation was, at best, a zero-sum game and explained why resolving the problem had been deferred for decades.

Through an elaborate planning and decision-making process centred around a document known as the Environmental Impact Statement (EIS), the states and federal government in 2007 concluded a contentious negotiation modifying reservoir operations (for Lakes Powell and Mead) and specifying rules for sharing shortages in the lower basin (USBR, 2007). The new rules call for water storage to be balanced more equally between the two main reservoirs, and prescribe a schedule of lower basin curtailments should storage in Lake Mead fall below specific elevations. Following the political compromise made back in 1968, which subordinated the water right of the Central Arizona Project to other lower basin users, it is the CAP that will bear the brunt of shortages. As before, the Secretary of the Interior is empowered to administer the programme and retains sole decision-making authority should water levels drop below the levels described in the shortage-sharing schedule. Although many issues about apportionment and shortage sharing remain, these new rules address the most pressing omissions in the legal framework.

The reservoir operations and shortage-sharing rules were the most debated elements in the EIS process; however, the new rules also address mechanisms (and incentives) for supply augmentation and conservation (USBR, 2007). These elements may be the linchpins to future progress, as including these elements allows the states to maintain the goal of additional development and use of the river, and transform the politics back to a positive-sum situation. In the past, the key to positive-sum bargaining in the basin was to expand the available benefits (i.e. water and power) through new storage and conveyance facilities, and by excluding public value proponents from decision making. Today, the situation is more complex, as far fewer opportunities exist for increasing yield through new storage, and environmental interests are an entrenched stakeholder, empowered by both law and public sentiment. The result has been the emergence of an unusually rich suite of strategies for increasing yields and avoiding (overcoming) limits, highlighted by efforts to eliminate reservoir spills (and associated overdeliveries to Mexico), marketing of water salvaged through conservation programmes, the eradication of water-loving tamarisk and Russian olive trees, weather modification (i.e. cloud seeding), desalination, the proposed importation of water from neighbouring basins, and compensated fallowing of agricultural land.[21]

Each of the augmentation and conservation strategies raises a host of difficult legal and political issues; by comparison, the engineering and economic challenges are almost inconsequential. One emerging issue is best expressed as the 'efficiency paradox', which refers to the observation that 'inefficiencies' associated with leaky canals, reservoir spills, inefficient irrigation practices and other system losses are often the primary source of water for valued environmental resources, such as the Colorado River delta, the Salton Sea (in southern California) and many other sites of high ecological importance. If these interests are considered – i.e. if the paradigm of decision making is broadened to include environmental values – then these efforts are not truly an augmentation strategy offering mutual benefits but are merely a zero-sum reallocation from public environmental interests to water users. Thus, while not as obvious as a debate over a new dam, this movement toward 'conservation and augmentation' strategies on the Colorado River is none the less another paradigmatic conflict and brings into question whether the full meaning of limits, restraint and sustainability will ever take hold in this basin.

As seen in intrastate water politics, the role of agriculture is also a prominent consideration in the future of regional (interstate) water management. For example, California's recent efforts to scale back its overall consumption to its legal apportionment has primarily been achieved through the reallocation of water from agricultural to urban users, with damages to agricultural interests offset by cash payments (these are voluntary transactions) and by efficiency programmes that allow most farming operations to continue with less consumption (but with less recharge of the Mexicali aquifer used in Mexico and less runoff for regional sinks, such as the Salton Sea, which is a critical habitat for migratory waterfowl). Agricultural interests in California and the other lower basin states are also implicated by the emerging ICS (Intentionally Created Surplus) programme, which allows water saved through 'extraordinary' conservation, efficiency projects, land fallowing and river augmentation to be transferred to other, mostly urban, users (USBR, 2007). Notwithstanding the environmental issues associated with the efficiency paradox and the hesitancy of regional leaders to embrace concepts of limits and sustainability, these ICS efforts offer many benefits to cities struggling to serve growing populations and farmers looking to stabilize (or even augment) revenues while responding to concerns about the high level of water use in agriculture.

Concluding Thoughts

The Colorado River of the south-western USA remains one of the world's most intriguing natural resources, valued as a critical water supply in an arid and suddenly populous region, and a source of natural beauty and grandeur few other rivers can match. It is also one of the world's most overstressed rivers, burdened by high expectations and by an institutional framework lacking in vision, coherence and sound assumptions about what is, and what should be, available to the community of farmers, cities and other water interests. Once immersed in these institutional issues, it is difficult to be optimistic about the river's future, particularly as growth and climatic change further challenge traditional management solutions, and regional (basin-wide) forums of planning and action are largely non-existent. Many organizations – including the Upper Colorado River Commission – exist with an interest in particular Colorado River issues and subregions, but there remains no river basin organization within which to study, consider and facilitate fundamental change in the basin. This institutional deficiency has been noted by several authors, who argue that the establishment of a basin-wide commission would be a valuable first step in framing, debating and ultimately addressing the issues in the Colorado basin that transcend the interests and authorities of any given state or interest group (e.g. see Kenney, 1995; Morrison et al., 1996; Getches, 1997). The basin states have not been receptive to these proposals, in part due to concerns about establishing mechanisms that may increase the influence of Indians, Mexico, the federal government or environmental interests in basin politics.

Ultimately, a new way of doing business will need to emerge in the basin – either incrementally or in a dramatic rush, perhaps triggered by empty reservoirs – and regardless of what that 'new way' looks like, it seems certain that few interests will be transformed as fundamentally as the agriculture sector. Even today, in a service area of over 30 million residents and a period of water stress, agriculture still consumes the greater part of the Colorado River water, often for uses that, in economic terms, are of low value. Ironically, this is perhaps the best long-term hope for this basin, as this provides an opportunity for market-based water reallocations, which could sustain cities and the most profitable farms for several decades. Agricultural to urban water reallocations are already seen throughout the basin, especially in southern California, and are finally emerging at a larger regional scale in the lower basin, through water-banking schemes and, potentially, the emerging ICS programme.

Water marketing, however, while probably more ecologically benign than the efficiency projects, comes with several hidden costs. Disentangle markets from legal constraints, and economic subsidies and the cities, industrial users and some instream uses (particularly hydropower) would find ample supplies; some farmers would enjoy needed revenue; and the highest-valued agriculture, particularly for fruits and vegetables, would continue uninterrupted for decades as lower-value feed crops were first phased out. Probably fairing less well would be non-market and public values (e.g. environmental resources) and rural communities dependent on lost farming economies. Additionally, the promise of the Colorado River Compact would be lost – i.e. the idea that a certain amount of water should be reserved for each region of the basin, in perpetuity, to support local lives and lifestyles, regardless of whether they were economically competitive with those in other regions. If not for this arrangement, farmers in Wyoming, for example, would never be competitive for water with casinos in Las Vegas. Perhaps that is fine; at the least, it is explicit in identifying that trade-offs need to be made if the region is ever to live within its means. That, after all, seems to be the biggest omission in the current arrangements, and in the current discussions on how to move forward. What should Colorado River allocation, management and use look like, given inherent limits in water supply and the imperative to consider traditionally excluded parties – the environment, tribes, Mexico – better in decisions? If history is a guide, then this is a question that is likely to exceed the capabilities of existing institutional decision-making forums, political leaders and paradigms. There is work to be done.

Notes

1. These statistics are compiled from data recorded by the US Bureau of Reclamation: http://www.usbr.gov/uc/water/crsp/cs/gcd.html.
2. Population statistics are compiled by the US Census Bureau and distributed online at www.census.gov.
3. These figures come from recent studies using the general circulation models (GCMs) associated with the fourth Intergovernmental Panel on Climate Change (IPCC) assessment. A summary of these and other relevant studies is provided in Appendix U of USBR, (2007).
4. Lower basin tributaries are much smaller, perhaps 2–3 MAF (million acre-feet) but are, more importantly, legally considered as outside the apportionment and management scheme of the Colorado River.
5. Estimating long-term natural (i.e. unaltered) streamflows at Lee Ferry is an inexact science, coloured by technical and political complications. Generally, these efforts fall into two general categories: those based on actual stream gauges (usually beginning in 1906) and those based on tree-ring reconstructions (which can go back as far as the year 762) (see www.colorado.edu/resources/paleo/lees/). Estimates based on actual stream gauges are primarily offered by the Upper Colorado River Commission and by the US Bureau of Reclamation, and usually fall in the range of 15.1–15.3 MAF/year (e.g. see UCRC, 2004; USBR, 2006). Slight differences generally reflect how many of the recent drought years are included in the analysis. Those based on tree-ring reconstructions suggest a lower long-term average. For example, the landmark study by Stockton and Jacoby (1976) suggested an average as low as 13.4 MAF/year. More recent reconstructions from 1490 to 1997 by Woodhouse et al. (2006) and from 762 to 2005 by Meko et al. (2007) suggest an annual value of 14.7 MAF.
6. Water volume in the western USA is measured in acre-feet. One million acre-feet (MAF) = 1.233 billion m3. Throughout the rest of this chapter, the MAF unit is used exclusively, despite its unfamiliarity outside the western USA, as the flow and apportionment numbers expressed in MAF units have great familiarity and significance in the region, and are of a convenient scale.
7. Of particular concern are efforts to line the All-American canal to reduce cross-border seepage and to construct a Drop 2 reservoir to catch main-stem overdeliveries to Mexico (with most of the 'conserved' water going to San Diego and Las Vegas). On the Colorado, seepage, reservoir spills and other 'inefficiencies' are often an important source of water for environmental resources. In most cases, water managers are under no obligation to continue these flows, and face powerful incentives to capture this water to serve growing human demands.
8. In order to take full effect, a compact must be signed by the negotiators, ratified by the legislatures of each of the participating states, and then be ratified by the federal government. The Colorado River compact was signed by the states in 1922, but was not officially ratified until it was accepted (ratified) by Congress in the 1928 legislation. The process was highly unusual in that Congressional ratification occurred before Arizona ratified the agreement, which did not occur until 1944. The delay, in large part, could be traced to a long-standing dispute between Arizona and California, which was not resolved until the conclusion of the Arizona v. California litigation many years later.
9. As noted later, rules for allocating shortages were not established until 2007.
10. The 6 MAF value is produced by subtracting 7.5 MAF (the lower basin apportionment) and 1.5 MAF (the Mexican apportionment) from a likely average yield of 15 MAF. It is only if the river's yield is 16.5 MAF or higher, as originally believed, that the upper basin receives the full apportionment of 7.5 MAF. The most controversial part of this analysis is the treatment of the

Mexican apportionment, which is to be reduced in some proportional (but otherwise unspecified) way to uses by the USA in a drought crisis. Since the Mexican obligation is a relatively small amount of water, any interpretation does not invalidate the observation that the upper basin is the primary entity harmed by the overallocation of flows.

11 Many of the key elements of tribal water rights in the Colorado River basin were established as part of the Arizona v. California (1963) litigation, which established the 'practically irrigable acreage' standard for measuring rights, reiterated the great seniority of these rights and quantified rights for five lower main-stem tribes at over 900,000 acre-feet. Since tribal water rights are subtracted from the apportionments of the states in which they are located, there is a zero-sum competition for Colorado River flows among Indians and non-Indians within each basin state.

12 CAP allocations are listed at http://www.cap-az.com/docs/SubcontractStatusReport_03_13_08.pdf.

13 As the name implies, the Colorado–Big Thompson Project diverts water from the Colorado River main stem in western Colorado to the Big Thompson River in eastern Colorado, using the Adams tunnel to avoid the necessity of pumping water over the continental divide. The exceedingly complex project, completed in 1956, exports roughly 260,000 acre-feet/year to a mix of agricultural and municipal interests along Colorado's Front Range (Tyler, 1992).

14 Overall, the General Accounting Office (GAO, 1981, 1996) and Water Resources Council (1975) estimate federal irrigation project subsidies in the range of 82–98%.

15 This type of political behaviour is often called logrolling, and occurs when legislators from various jurisdictions all agree to support each other's proposed projects in their home districts. In this way, a project with only local appeal can gain the support of a broad base of legislators.

16 It is worth noting that Nevada has been the primary entity promoting interstate water transfer mechanisms, such as the water banks and the intentionally created surplus (ICS) programme (discussed later), as it is the only basin state that already uses its full apportionment exclusively for municipal uses (e.g. Las Vegas), and is thus very limited in its ability to support urban growth based on water transfers from agriculture.

17 Compiling water and water-use statistics in the Colorado River basin is notoriously difficult for many reasons, including the separation of administrative responsibilities between the upper and lower basins, and the differing traditions regarding the inclusion (or exclusion) of tributaries and the accounting of water (and water uses) once exported from the hydrologic basin. Additionally, patterns of water use can change significantly year to year; figures are updated frequently, and there is rarely agreement on any single set of statistics as being 'official' or formally accepted. With these caveats, the best available data come from the *Consumptive Uses and Losses Reports* issued by the Bureau of Reclamation (see www.usbr.gov/uc/library/envdocs/reports/crs/crsul.html). Unfortunately, these reports are not very useful for tracking main-stem use of Colorado River water in lower basin agriculture, which is shifting rapidly – particularly in southern California. Statistics for the upper basin suggests that agricultural land area and water consumption have both increased by about 10%, from the 1981–1985 to the 1996–2000 period, comprising in both periods about 68% of all upper basin consumption. These values have probably dropped in recent years due to drought conditions.

18 During the current drought, this total level of use has been reduced by efforts in California to scale back overuse (to its legal apportionment), by a reduction in the amount of spills and overdeliveries to Mexico, and through reduced evaporation from reservoirs that are currently at unusually low levels. Collectively, these efforts have re-balanced the system-wide water budget at least temporarily, assuming average yields – a condition that has existed in only one year between 2000 and 2007.

19 The so-called '4.4 Plan' is implemented as part of the Quantification Settlement Agreement (QSA) and is described in *The Colorado River Water Delivery Agreement* (text available at www.saltonsea.water.ca.gov/docs/crqsa/crwda.pdf).

20 As part of the litigation, a Special Master employed by the court suggested that lower basin shortages be apportioned in ratios matching the apportionment; thus, California's share of reductions would be 4.4/7.5, Arizona's 2.8/7.5 and Nevada's 0.3/7.5. The court rejected this approach as being overly rigid.

21 An inventory of augmentation options was recently compiled in research commissioned by the Southern Nevada Water Authority and is summarized at: www.snwa.com/assets/pdf/augmentation_summary.pdf.

References

Adler, R.W. (2007) Restoring the environment and restoring democracy: lessons from the Colorado River. *Virginia Environmental Law Journal* 25(1), 55–104.

Brandon, W. (1990) *Quivira: Europeans in the Region of the Santa Fe Trail, 1540–1820*. Ohio University Press, Athens, Ohio.

Bryner, G. and Purcell, E. (2003) *Groundwater Law Sourcebook of the Western United States*. Natural Resources Law Center, Boulder, Colorado.

Burton, L. (1991) *Indian Water Rights and the Limits of Law*. University of Kansas Press, Lawrence, Kansas.

Carothers, S.W. and Brown, B.T. (1991) *The Colorado River through the Grand Canyon: Natural History and Human Change*. University of Arizona Press, Tucson, Arizona.

Census Bureau (2001) *Population Change and Distribution: Census 2000 Brief*. C2kbr/01-2. US Department of Commerce, Economics and Statistics Administration, Washington, DC.

Christensen, N. and Lettenmaier, D.P. (2006) A multimodel ensemble approach to assessment of climate change impacts on the hydrology and water resources of the Colorado River Basin. *Hydrology and Earth System Sciences Discussion* 3, 1–44.

DeVoto, B. (1952) *The Course of Empire*. Houghton Mifflin Company, Boston, Massachusetts.

Fradkin, P.L. (1981) *A River No More: the Colorado River and the West*. The University of Arizona Press, Tucson, Arizona.

GAO (General Accounting Office) (1981) Federal charges for irrigation projects reviewed do not cover costs. Report No. PAD-81-07

GAO (1996) Bureau of Reclamation: information on allocation and repayment of costs of constructing water projects. GAO/RCED-96-109.

Gertner, J. (2007) The future is drying up. 21 October 2007. www.nytimes.com/2007/10/21/magazine/21water-t.html?_r=1&oref=slogin

Getches, D.H. (1997) Colorado River governance: sharing federal authority as an incentive to create a new institution. *University of Colorado Law Review* 68(3), 573.

Gillilan, D.C. and Brown, T.C. (1997) *Instream Flow Protection: Seeking a Balance in Western Water Use*. Island Press, Washington, DC.

Gleick, P.H., Cooley, H., Katz, D. and Lee, E. (2002) *The World's Water, 2002–2003*. Island Press, Washington, DC.

Glennon, R.J. and Culp, P.W. (2002) The last green lagoon: how and why the Bush administration should save the Colorado River delta. *Ecology Law Quarterly* 28(4), 903–907.

Grand Canyon Trust (2005) *The Colorado – a River at Risk: Coping with Drought in the Colorado River Basin, 2005–2006*. Flagstaff, Arizona.

Griles, S.J. (2004) Building on success: facing the challenges ahead. Address to the Colorado River water users association. www.crwua.org/news/norton2004.html (accessed 17 December 2008).

Hays, S.P. (1959) *Conservation and the Gospel of Efficiency: the Progressive Conservation Movement 1890–1920*. Harvard University Press, Cambridge, Massachusetts.

Hoerling, M. and Eischeid, J. (2007) Past peak water in the southwest. *Southwest Hydrology* 6(1), 18–19, 35.

Holburt, M.B. (1975) International problems on the Colorado River. *Natural Resources Journal* 15(1), 11–26.

Howe, C.W., Lazo, J.K. and Weber, K.R. (1990) The economic impacts of agriculture-to-urban water transfers on the area of origin: a case study of the Arkansas River valley in Colorado. *American Journal of Agricultural Economics* 72(5), 1200.

Hundley, N. Jr (1966) *Dividing the Waters: a Century of Controversy between the United States and Mexico*. University of California Press, Berkeley, California.

Hundley, N. Jr (1975) *Water and the West: the Colorado River Compact and the Politics of Water in the American West*. University of California Press, Berkeley, California.

Ingram, H. (1990) *Water Politics: Continuity and Change*. University of New Mexico Press, Albuquerque, New Mexico.

Ives, J.C. (Lieutenant, US Corps of Topographical Engineers) (1861) *Report Upon the Colorado River of the West, Explored in 1857 and 1858* (1st Session, 1861). Government Printing Office, Washington, DC.

Kenney, D.S. (1995) Institutional options for the Colorado River. *Water Resources Bulletin* 31(5), 837–850.

Kenney, D.S. (2001) *Two Decades of Water Law and Policy Reform: Conference Report*. Natural Resources Law Center, Boulder, Colorado.

Kenney, D.S. (2005) Prior appropriation and water rights reform in the western United States. In: Bruns, B.R.,

Ringler, C. and Meinzen-Dick, R. (eds) *Water Rights Reform: Lessons for Institutional Design.* International Food Policy Research Institute, Washington, DC, pp. 167–182.

Kenney, D.S., McAllister, S.T., Caile, W.H. and Peckham, J.S. (2000) *The New Watershed Source Book.* Natural Resources Law Center, Boulder, Colorado.

Lochhead, J.S. (2001) An upper basin perspective on California's claims to water from the Colorado River, Part 1: the Law of the River. *University of Denver Water Law Review* 4(2), 290.

Lochhead, J.S. (2003) An upper basin perspective on California's claims to water from the Colorado River, Part II: the development, implementation and collapse of California's plan to live within its basic apportionment. *University of Denver Water Law Review* 6(2), 318.

Maass, A. and Anderson, R.L. (1978) *... and the Desert Shall Rejoice.* The MIT Press, Cambridge, Massachusetts.

MacDonnell, L.J. (1999) *From Reclamation to Sustainability: Water, Agriculture, and the Environment in the American West.* University Press of Colorado, Boulder, Colorado.

Mann, D. (1963) *The Politics of Water in Arizona.* University of Arizona Press, Tucson, Arizona.

Meko, D.M., Woodhouse, C.A., Baisan, C.H., Knight, T., Lukas, J.J., Hughes, M.K. and Salzer, M.W. (2007) Medieval drought in the upper Colorado River basin. *Geophysical Research Letters* 34, L10705, 24 May.

Morrison, J.I., Postel, S.L. and Gleick, P.H. (1996) *The Sustainable Use of Water in the Lower Colorado River Basin.* Pacific Institute for Studies in Development, Environment and Security, Oakland, California.

National Research Council (1992) *Water Transfers in the West: Efficiency, Equity, and the Environment.* Committee on Water Management. National Academies Press, Washington, DC.

Paehlke, R.C. (1989) *Environmentalism and the Future of Progressive Politics.* Yale University Press, New Haven, Connecticut.

Pisani, D.J. (1992) *To Reclaim a Divided West: Water, Law, and Public Policy, 1848–1902.* University of New Mexico Press, Albuquerque, New Mexico.

Pollack, S.M. and McElroy, S.B. (2001) A-LP lite: a compromise project that fulfills the United States' Trust responsibility in an environmentally responsible manner. *Natural Resources Journal* 41(3), 639–642.

Pontius, D. (1997) *Colorado River Basin Study.* Western Water Policy Review Advisory Commission, Denver, Colorado.

Powell, J.W. (1890) Institutions for arid lands. *The Century* XL (May to October), 111–116.

Project Wet (2005) *Discover a Watershed: the Colorado.* Project Wet International Foundation, Bozeman, Montana.

Rasband, J., Salzman, J. and Squillace, M. (2004) *Natural Resources Law and Policy.* Foundation Press, Stamford, Connecticut.

Reisner, M. (1986) *Cadillac Desert: the American West and its Disappearing Water.* Viking Penguin, New York.

Stegner, W. (1953) *Beyond the Hundredth Meridian: John Wesley Powell and the Second Opening of the West.* University of Nebraska Press, Lincoln, Nebraska.

Stockton, C.W. and Jacoby, G.C. (1976) Long-term surface water supply and streamflow trends in the upper Colorado River basin based on tree-ring analyses. *Lake Powell Research Project Bulletin* 18, 1–70.

Tarlock, A.D., Corbridge, J.M. and Getches, D.H. (2002) *Water Resource Management: a Casebook in Law and Public Policy.* Foundation Press, New York.

Terrell, J.U. (1965a) *War for the Colorado River. Volume 1: the California–Arizona Controversy.* The Arthur H. Clark Company, Glendale, California.

Terrell, J.U. (1965b) *War for the Colorado River. Volume 2: Above Lee's Ferry – the Upper Basin.* The Arthur H. Clark Company, Glendale, California.

Thorson, J.E., Britton, S. and Colby, B.G. (2006) *Tribal Water Rights: Essays in Contemporary Law, Policy, and Economics.* University of Arizona Press, Tucson, Arizona.

Tipton and Kalmbach, Inc. (1965) Water supplies of the Colorado River available for use by the states of the upper division and for use from the main stem by the states of Arizona, California and Nevada in the lower basin. Prepared for the Upper Colorado River Commission, Denver, Colorado.

Tyler, D. (1992) *The Last Water Hole in the West: the Colorado–Big Thompson Project and the Northern Colorado Water Conservancy District.* University of Colorado Press, Niwot, Colorado.

Tyler, D. (2003) *Silver Fox of the Rockies: Delphus E. Carpenter and Western Water Compacts.* University of Oklahoma Press, Norman, Oklahoma.

UCRC (Upper Colorado River Commission) (2004) *Fifty-sixth Annual Report of the Upper Colorado River Commission.* Salt Lake City, Utah.

USBR (US Bureau of Reclamation) (1946) The Colorado River: a natural menace becomes a national resource. Draft of March, 1946. United States Department of the Interior, Washington, DC.
USBR (1987) *Assessment '87 … a New Direction for the Bureau of Reclamation.* United States Department of the Interior, Washington, DC.
USBR (2004) Colorado River system consumptive uses and losses report, 1996–2000. United States Department of the Interior, Washington, DC.
USBR (2006) 2006 hydrologic determination (draft). Report issued by the Secretary of Interior, US Department of Interior, as prepared by USBR, as part of analysis of water availability from Navajo reservoir and the upper Colorado River basin for use in New Mexico. United States Department of the Interior, Washington, DC.
USBR (2007) *Colorado River Interim Guidelines for Lower Basin Shortages and Coordinated Operations for Lake Powell and Mead.* Final Environmental Impact Statement (FEIS). US Department of the Interior, Washington, DC.
Wahl, R.W. (1989) *Markets for Federal Water: Subsidies, Property Rights, and the Bureau of Reclamation.* Resources for the Future, Washington, DC.
Water Resources Council (1975) *Options for Cost Sharing, Part 5a: Implementation and OM&R Cost Sharing for Federal and Federally Assisted Water and Related Land Programs.* Water Resources Council, Washington, DC.
Waters, F. (1946) *The Colorado.* Holt, Rinehart and Winston, New York.
Wilkinson, C.F. (1990) Toward an ethic of place. In: Udall, S.L., Nelson Limerick, P., Wilkinson, C.F., Volkman, J.M. and Kittredge, W. *Beyond the Mythic West.* Western Governors' Association, Gibbs-Smith Publisher, Layton, Utah, pp. 71–94.
Wolf, A. (2005) Transboundary water conflicts and cooperation. In: Kenney, D.S. (ed.) *In Search of Sustainable Water Management.* Edward Elgar Publishing, Cheltenham, UK.
Woodhouse, C.A., Gray, S.T and Meko, D.M. (2006) Updated streamflow reconstructions for the upper Colorado River basin. *Water Resources Research* 42(5), W05415, doi:10.1029/2005WR004455.
Worster, D. (1985) *Rivers of Empire: Water, Aridity, and the Growth of the American West.* Pantheon Books, New York.

Legal Documents: Cases, Statutes, Compacts, Treaties and Agreements

Arizona v. California, 373 U.S. 546 (1963).
Boulder Canyon Project Act of 1928, 43 U.S.C.A. § 617.
Clean Water Act of 1972 (Federal Water Pollution Control Act Amendments), 86 Stat. 816
Colorado River Basin Project Act of 1968, 43 U.S.C.A. § 1521.
Colorado River Basin Salinity Control Act of 1974, 43 U.S.C.A. § 1571.
Colorado River Compact of 1922, C.R.S.A. §37-61-101 (45 U.S. Stat. 1057).
Colorado River Storage Project Act of 1956, 43 U.S.C.A. § 620.
Endangered Species Act of 1973, 87 Stat. 884.
National Environmental Policy Act of 1969, 83 Stat. 852.
Treaty between the United States of America and Mexico respecting utilization of waters of the Colorado and Tijuana Rivers and of the Rio Grande, Feb. 3, 1944, U.S.-Mexico, 59 Stat. 1219. (Mexican Water Treaty of 1944).
Upper Colorado River Basin Compact of 1948, C.R.S.A. §37-62-101 (63 Stat. 31 (1949)).
Wild and Scenic Rivers Act of 1968, 82 Stat. 906.
Winters v. United States, 207 U.S. 564 (1908).

7 Sharing Scarce Resources in a Mediterranean River Basin: Wadi Merguellil in Central Tunisia

Patrick Le Goulven,[1]* Christian Leduc,[2]** Mohamed Salah Bachta[3]*** and Jean-Christophe Poussin[2]****

[1]*Institut de Recherche pour le Développement, UMR G-EAU (Gestion de l'Eau, Acteurs, Usages), Quito, Ecuador;* [2]*IRD, UMR G-EAU, Montpellier, France;* [3]*Institut National Agronomique de Tunisie, Tunis Mahrajène, Tunisia; E-mail: *patrick.legoulven@ird.fr; **christian.leduc@ird.fr; ***bachta.medsalah@inat.agrinet.tn; ****jean-christophe.poussin@ird.fr*

Introduction

Owing to its geographical location, Tunisia is under the influence of a Mediterranean climate in its northern part and a Saharan climate in the southern part. This climatic discontinuity results in a strong variability in water availability and defines three agro-climatic zones: (i) the northern area, with its forestry and agricultural vocation, which includes the Medjerda (the only permanent river) and provides 82% of the country's surface water; (ii) the semi-arid centre, initially a wide rangeland with large plantations, characterized by violent and sporadic runoff; and (iii) the southern area (62% of the country), where settlements are concentrated around water sources (springs, oases) and where people live on extensive pastoralism. Tunisia has many aquifers, storing 720 Mm^3 each year in the northern and central areas and 1250 Mm^3 in the south of the country (DGRE, 1995).

In spite of this contrasting geo-climatic situation, Tunisia has always found ways to make the best out of limited resources, in particular during Carthaginian, Roman and Arabic times, when the country was known for its urban development and flourishing agricultural production. Transfers of water have been implemented since antiquity and water shortages are an old challenge, which Tunisia has managed through its extensive traditional know-how (Treyer, 2002). With a population of approximately 10 million and the availability of water resources below 500 m^3/capita/year, Tunisia has been able to meet the needs of its various economic sectors, even during severe droughts: coverage of drinking water supply reaches 100% in cities and more than 80% in rural areas, without rationing, even in periods of shortage.

This has been achieved through policies defined in the 1970s, when the Tunisian government built works to develop and regulate water resources, transferring water from the hinterland to the coastal areas, out of a concern for equity and economic development. This strategy equipped the country with an extensive water infrastructure, comprising 29 large dams, 200 tanks, 766 lake-reservoirs, more than 3000 boreholes and 151,000 wells. This ensured the satisfaction of agricultural needs (80% of the overall consumption) and

© CAB International 2009. *River Basin Trajectories: Societies, Environments and Development* (eds F. Molle and P. Wester)

allowed the development of mass tourism along the coast, an activity characterized by a seasonal demand for good-quality water.

The central area has not been directly impacted by the growth of tourism but it has undergone changes through its relationship with the coastal area (called the 'Sahel') in terms of labour migration, water transfers and emergence of new markets for agricultural produce. Kairouan, the main town in this area, is located above an aquifer which collects the water of three river basins (Zeroud, Merguellil, Nebhana) draining the Tunisian central highlands. These basins were first closed by dams (constructed between 1965 and 1989) designed to protect Kairouan from exceptional floods. The Merguellil basin was also the target of various soil and water conservation works from the 1960s onwards, which formed a part of successive regional development plans and water resources management policies. However, water users and other stakeholders played a very small part in these strategies – which were planned and implemented by the central administration – and these are nowadays increasingly questioned and challenged.

The Merguellil basin provides an ideal case study to analyse the effect of the progressive establishment of water infrastructure, its use by various segments of the population and their impact on the spatial and social distribution of water resources. In spite of costly investments, the water tables of the aquifers of the upper basin and of the Kairouan plain are dropping at an increasing rate. The Merguellil basin also provides the opportunity to examine the modes of governance, as well as the economic and regulatory tools which might assist in the control of access to water resources.

Characteristics of the Merguellil Basin

Environmental context

Located in semi-arid central Tunisia, the Merguellil basin is one of the three main river basins of the southern side of the Tunisian ridge, flowing into the Kairouan plain. The upper part of the Merguellil basin, upstream of the El Haouareb dam, has a surface area of 1180 km² and is delimited by a succession of *djebels* (mountainous ridges). The El Haouareb dam itself is anchored in two lateral *djebels* (Aïn El Rhorab and El Haouareb). The lower Merguellil basin, limited to the north by Djebel Cherichira and to the south by Draa Affane, is part of the large Kairouan plain, which covers 3000 km² (Fig. 7.1).

The altitude in the upper basin varies between 200 and 1200 m, with 33% of the area between 200 and 400 m, 36% between 400 and 600 m, 20% between 600 and 800 m and 11% higher than 800 m. In the lower basin, the altitude of the plain decreases steadily from 200 to 80 m in Kairouan. The mountain ranges in the upstream basin consist of sedimentary deposits, with a large predominance of limestone (sometimes dolomite). They may be covered by other deposits (sand, sandstone, sandy clay) from the Miocene epoch, especially in the El Ala area.

Soil texture varies from clay to sand. In the upper basin, the main soil types include shallow soils over a calcareous crust and deep soils over sandstone. The highest parts of the basin have forests and scrubs (Kesra forest). Overgrazing and land clearing have greatly damaged large areas, and natural vegetation has been replaced by species of lesser interest with regard to both economic use and protection against erosion.

Many forms of erosion can be observed in the Merguellil basin. Among them, erosion in gullies prevails, not only in sandy areas with poor vegetation but also in some clayey areas. Soil erosion is obviously higher in areas with steep slopes (sometimes over 12%). According to a recent study, arable lands threatened by soil erosion total 670 km².

The mean temperature is 19.2°C in Kairouan (a minimum of 10.7°C in January and a maximum of 38.6°C in August). Winter is cool in the north-west of the basin and temperate elsewhere. In the upper basin, temperatures are below 10°C between December and February (e.g. 5.7°C in Makthar in January) and around 25°C in July and August. The relative humidity varies between 70 and 55% in winter and between 40 and 55% in summer. Between May and August, the climate is very dry. The mean annual potential

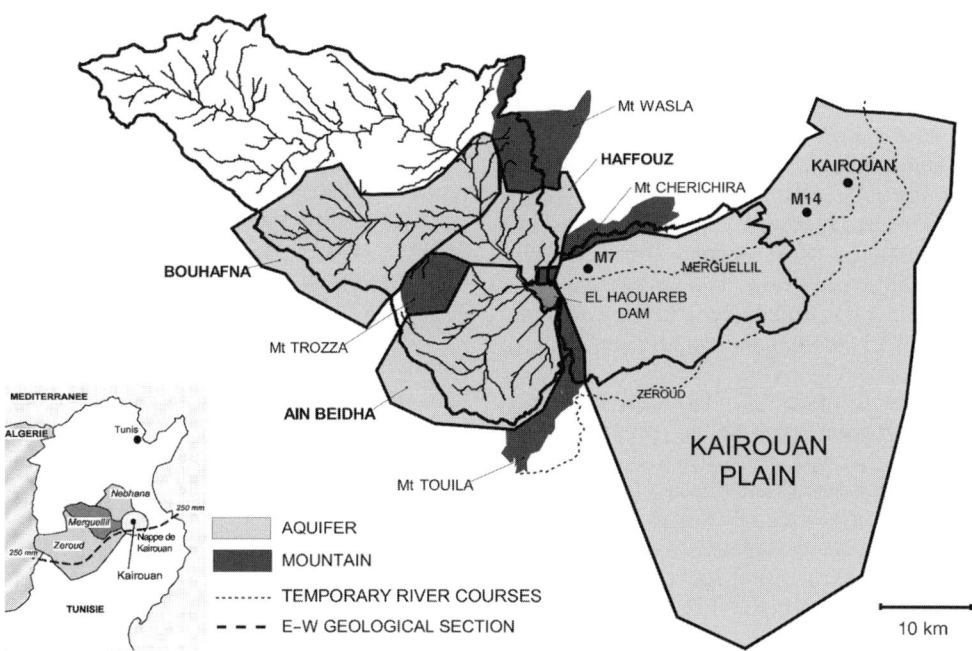

Fig. 7.1. Location of the study area, limits of the upper and lower sub-basins and of the different aquifers (M7 and M14 are piezometers referred to in Fig. 7.3).

evapotranspiration (Penman) is close to 1600 mm in Kairouan and decreases with altitude.

Rainfall

Rainfall measurements in and around the Merguellil basin are very variable in terms of record period, completeness, reliability and representativeness. The first measurements started before 1900 but these are very few. Information on rainfall became relatively abundant only after 1970. The mean annual rainfall is about 300 mm in the plain and increases up to 510 mm in the upper part of the basin, with a gradient of about 20 mm per 100 m of altitude. The two rainiest months are October and March. As is common in semi-arid areas, rainfall varies widely in time and space: since 1925, extreme values measured in Kairouan were 703 mm in 1969/70 and 108 mm in 1950/51. Analysis of rainfall series shows a slight decrease in yearly values between 1976 and 1989, but no trend that could be statistically considered as being beyond normal variability.

Surface runoff

The total length of Wadi Merguellil down to Kairouan is about 90 km. The Merguellil basin is endoreic (with no outlet to the sea) and its rivers have sporadic flows, which may be very violent. This ephemeral regime is a fundamental characteristic of the regional hydrology: about 80% of the annual flow is produced in 12 days. Before the construction of the El Haouareb dam, the largest floods of Wadi Merguellil reached the El Kelbia *sebkha*, a large salt lake located close to the sea, which often dried up. The smaller floods used to vanish in the Kairouan plain by both evaporation and infiltration to the aquifer.

Surface runoff is observed through a network of five stations covering different sub-basins. The El Haouareb reservoir represents a good hydrological 'integrator' of the whole upper basin, as witnessed by the dam water level (Fig. 7.2).

For the period 1989–2005, the mean annual flow of Wadi Merguellil, estimated at the El Haouareb dam, was 17 Mm^3, with a minimum of 2.5 Mm^3 in 2000–2001 and a maximum of 37.6 Mm^3 in 2004–2005. These values can be compared with the exceptional flood of autumn 1969, estimated at about 175 Mm^3, with a peak flow of over 3000 m^3/s, resulting in a severe inundation of the Kairouan plain, with high human and material losses (Bouzaïane and Lafforgue, 1986).

Outflow from the El Haouareb reservoir consists of evaporation (25%), pumping and releases (12%), and uncontrolled infiltration to the karst aquifer (63%). Because of this exceptional karstic loss, dam releases are very limited and the dam has completely dried up several times. The dam has never spilt and the highest water level was reached in February 2006.

Groundwater

Three small, interconnected aquifers (Aïn Beidha, BouHafna, Haffouz–Cherichira) can be found in the lower part of the upper Merguellil basin (Fig. 7.1). Depending on place and time, they interact with the drainage network in both directions: springs supplying rivers or floods recharging alluvium and linked aquifers.

The Kairouan plain aquifer represents a much larger water storage because of its horizontal extent and a thickness of up to 800 m of alluvium and colluvium (Nazoumou, 2004). Water table levels are regularly measured in more than 100 piezometers (the oldest measurements date back 40 years). Some level recorders complement these monthly measurements, as well as physical and chemical field measurements and geochemical and isotopic analyses. Changes in the plain water table levels reflect the variability of recharge (e.g. a rise of up to 10 m after the 1969 exceptional flood) and the ever-increasing rate of pumping.

Water resources and their changes with time

The Merguellil basin has been under an ever-growing human pressure for 40 years. This pressure has taken different forms:

- The 1969 catastrophic flood led to the construction of the Sidi Saad dam in the Zeroud valley in 1981 and of the El Haouareb dam in the Merguellil valley in 1989 (and to the north, the Nebhana dam, built in 1965). This significantly increased evaporation and reduced infiltration in the Kairouan plain. Part of the reservoirs' water is pumped to supply public irrigation schemes downstream of the dam.

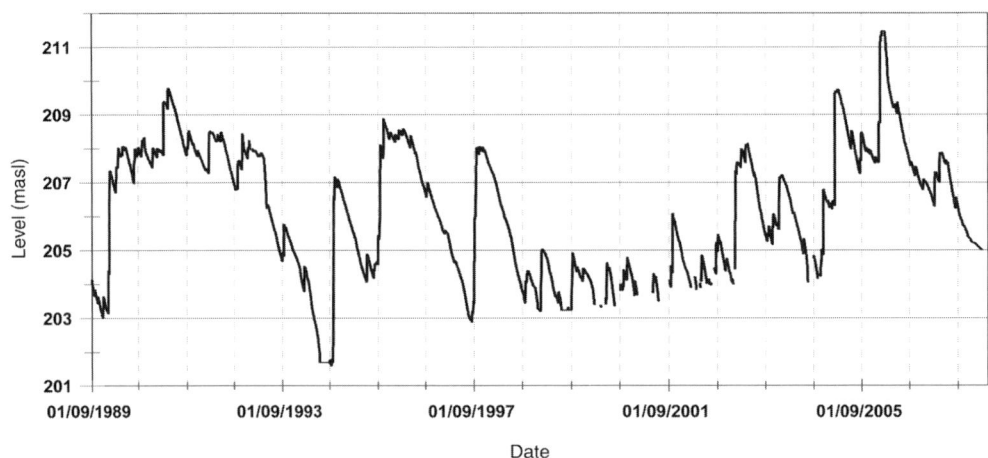

Fig. 7.2. Water level of the El Haouareb dam (because of siltation, the lowest levels, recorded in 1994, 1997–2005 and 2008, correspond to a complete drying up of the dam).

- The Kairouan plain aquifer was previously fed by the rapid infiltration of flood water, which was the major component of its water budget, and by lateral groundwater inflows from adjoining aquifers. Since 1989, the El Haouareb dam has stopped most of the Merguellil flow (dam releases have represented only 6% of the dam water) and the plain aquifer is now recharged by the horizontal transfer through a karstic system that mixes water from the Aïn Beidha aquifer and from the dam reservoir. Isotopic analyses (Ben Ammar et al., 2006) showed that releases from the dam (i.e. post-1989) have not flowed further than the first 7 km downstream of the dam. Present groundwater dynamics are largely driven by pumping for irrigation use (Fig. 7.3).

In the upper basin, the overexploitation of the BouHafna aquifer and the subsequent drop in groundwater levels were believed to be responsible for the declining base flow observed in Wadi Merguellil (Kingumbi, 2006). An alternative explanation links this decrease in river flow to the expansion of soil and water conservation (SWC) works in the upstream basin (Lacombe, 2007).

To reduce siltation in the reservoirs of the three large dams in the region, SWC works have been implemented in each upper basin. In the Merguellil basin in particular, they currently consist of 25,000 ha of contour-ridged terraces, 45 small tanks and five larger lakes. Presently, more than 20% of the upper basin area has been affected by conservation works.

Another important change in water resources in the Merguellil basin is the development of pumping from both surface water and groundwater. The first public irrigation schemes were implemented in the 1970s. During the 1980s, private wells were subsidized and their number rapidly increased, from 100 in the 1960s to about 5000 at present. Abstraction of groundwater is not only very intense in the Kairouan plain but also occurs at a lower rate in the upstream part of the Merguellil basin. At the same time, the increase in population and expansion of water supply networks has led to much larger withdrawals from aquifers, upstream and downstream. Moreover, the export of water to urban areas and tourist activities along the coast are other major factors contributing to the present overexploitation of aquifers.

Development and Settlement of the Basin through History

An ancient history characterized by invasions

From prehistoric times, Tunisia has taken part in Mediterranean agrarian civilizations founded

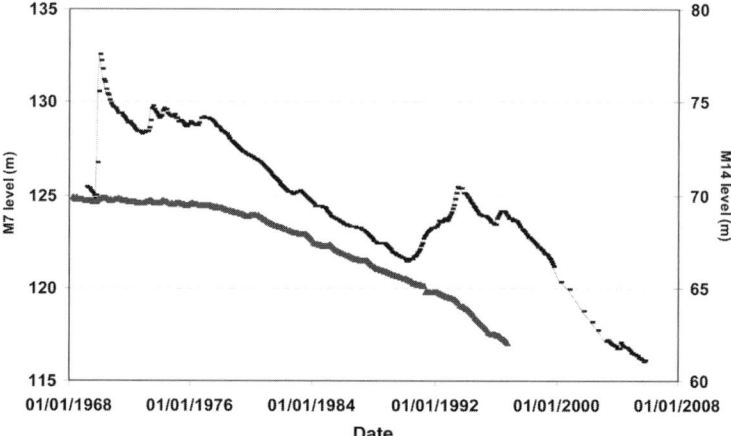

Fig. 7.3. Long-term changes in the water table level of the Kairouan plain: upstream piezometer M7 (upper curve) recorded the main regional events (climate and pumping), while downstream piezometer M14 (lower curve) seems to be affected by irrigation pumping only.

on cereals (maize, barley) and small cattle (small ox, grey donkey, sheep). The common olive tree and the carob tree were known by the first inhabitants of the country, the Berbers, along with the wild vine, the fig tree and the almond tree, which are thought to be indigenous species. But it was the Phoenician colonization, around 800 BC, which brought the techniques of arboriculture, and multiplied orchards and gardens, as testified by the famous treatise written by Magon during the 3rd century BC.

Rome colonized the northern part of Tunisia in 146 BC, and the centre and the south later on. Its presence until the 5th century AD led to a refining of Carthaginian farming techniques and the development of very sophisticated irrigation schemes (Géroudet, 2004a). Tunisian maize and olive oil were essential condiments in Roman food. In the same way, the techniques of breeding and craft industry developed: the orange-red, sigillated ceramics produced in the heart of the area were found in the whole Mediterranean basin as of the end of 2nd century AD.

In the Merguellil region, Romans populated the mountainous zones in particular (Kesra, Makthar, El Alaa, Djebel Ousselet) and developed olive trees. They were grown at higher altitudes, in sandy soils, with low densities (five to six trees/ha) and were surrounded by small, low walls to ensure good water supply (Géroudet, 2004a). The Romans had a good command of many hydraulic techniques and constructed many cisterns, tanks and aqueducts. Nowadays, farmers still collect the fruits of 'Roman olive trees'.

Between the fall of the Roman and Byzantine empires and the French colonization, the countryside was regularly plundered by successive invasions by tribes coming from Arabia (in particular starting from the 7th century), and from the south of Egypt (Beni Hillel in 1051). This latter tribe devastated the south and the heart of the country, destroyed the city of Kairouan and the countryside – cutting down part of the old Roman olive trees – and destroyed water-harvesting structures. Fertile areas were thus abandoned and sedentary populations hid in the mountains: livestock breeding replaced agriculture. There was no notable improvement of agriculture under the Ottoman Empire, when the heavy taxes imposed on peasants limited agricultural development (Géroudet, 2004b).

Founded in 670 AD by Oqba Ibn Nafaa, the city of Kairouan is located on a rich alluvial plain and close to a *sebkha*, which was used as a water reserve and pasture for the horses and camels of armies. One century later, the city was connected by underground conduits to large reservoirs located a few kilometres apart (Qsar Al Mâ) and supplied by neighbouring wadis (Mahfoudh et al., 2004). The Aghlabides (800–909 AD) enlarged these reservoirs, and Kairouan was, by then, the intellectual and political centre of the Maghreb.

Spatial settlement during the 18th and 19th centuries

Historical records from the 18th century reveal a tribe-based society where individuals only exist as a member of a tribe. The history of its founder, of his migrations and then of his descendants belongs to the history of the tribe and makes it possible to legitimize the occupation of a particular piece of land. The extension of the territory occupied by these tribes was not measured and natural topographic elements provided only landmarks. Boundaries were social: the territory of the tribe ended where that of another tribe started. Certain tribes have an affiliation to a *marabout* (religious leader); others have neither a founder nor a history of migration and this is the case with the Berber tribes.

Livelihoods combined a concentration of agricultural production on limited areas where water was accessible with an extensive exploitation of dispersed pastoral resources (Genin et al., 2006). Rangelands were collective and croplands were at the disposal of all, protected against foreign invasions by the whole tribe. Rules for cultivated land, gardens and water use varied according to groups and natural conditions. Since Roman times the local population has acquired a good technical know-how concerning the control of stream flows and the enhancing of infiltration using small-scale land and water conservation works.

In such a contrasting environment, the inventoried techniques are very diverse: benches (*tabias*) and field terraces on sloppy land, works in the wadis to slow down the flow (*jessours*), small tanks to store water in summer, canals to divert and spread flood flow (*mgoud*), water collectors with controlled and directed flow (*meskat*), groundwater tapping, and distribution by irrigation channels or *seguias*.

The rudimentary aspect of these works was compensated for by a sophisticated social management of water based on the *seguia*, which is used as a dividing, conveyance and final distribution channel, regulated by elaborate and precise customary rights. The same social group selected the site and the characteristics of the works, and organized their maintenance and their exploitation.

Conditions of access to water structured these rural tribes into groups of owners (water and land), and beneficiaries without property rights and generally marginalized, who provided labour for construction, maintenance and management of these infrastructures. The cohabitation of these two social groups was defined by a set of rules referring to water property, its distribution and maintenance of the works. These rules were accepted by all and specified the statutes and the roles of each individual in the tribe. Even if water often remained 'the friend of the powerful' (Bedoucha, 1987), there was a coherence between technical tools and management goals, which ensured the overall performance of the system.

French protectorate (1881–1956) and evolution of societies

More centred on the north and the Sahel, the French presence in the Merguellil area was limited to only two large properties: the El Haouareb farm (3700 ha) and the Ousseltia-Pichon farm (8000 ha). Their contribution remained technical in nature, with the development of strictly rainfed arboriculture. However, colonization modified the basin landscape, as tractors and modern ploughs partially destroyed the *fesguias*,[1] the *seguias*, the terraces and the small stone-wall reservoirs which obstructed passage, thus increasing erosion.

Tribal structures changed little by little. Generalization of schooling made the prevailing social order anachronistic, with younger generations rarely agreeing to return to agricultural work. Demographic growth caused strong pressure on natural resources, and successive intergenerational divisions resulted in strategies of both agricultural intensification and land expansion through clearing, with a consequent clear degradation of natural resources.

These external intrusions often came with radical technological innovations, which generally remained the prerogative of certain classes within rural societies, leading to a form of dualistic agriculture. This implied notable changes in the social relationships regarding property, access to resources and their exploitation towards productive ends. The reproduction of rural societies was affected by processes of adjustment and the end of a social model based on access to water governed by rules defined and enforced by local communities.

Independence and the development of large-scale hydraulics works

Post-independent Tunisia sought to structure space and sedentarize its population through the development of hydro-agricultural schemes. Collective decision making would be entrusted to the government and the administration, and would thus free individuals from the tribal structure (Bachta *et al.*, 2005; Bachta and Zaibet, 2006). The state built a large hydraulic infrastructure to capture, transfer and allocate water resources. This intervention can be broken down into three phases (Feuillette *et al.*, 1998), followed by a period of reflection.

Technical investments

The construction of large infrastructural works was based on a logic of better distributing water resources between areas and of multiple uses in order to accelerate the development of the country, by increasing agricultural production in the northern area (considered as the 'breadbasket of the country'), then by transferring the surplus towards coastal areas to feed the main centres of population, the tourist industry and key zones of high-value irrigated agriculture.

Legal and incentive measures

In parallel, authorities created legal regulations (the Water Code) which transformed the legal status of the resource (public, inalienable and imprescriptible), established new rights of use and entrusted their implementation to the Ministry of Agriculture. Centralized institutions were created to implement national strategies at a regional level: sectoral general directorates, regional development offices, and a national company (SONEDE) for domestic water supply and sanitation (WS&S).

Technical investments were encouraged by incentives to intensify water use and by allowing easy and cheap access to the resource to priority uses: to ensure water supply and sanitation to all, to stabilize rural incomes in order to limit rural outmigration, to ensure food security and to develop both export agriculture (citrus fruits) and tourism.

The economy phase

Mobilization of water resources slowed down because the cost of construction increased. Authorities strengthened regulation and state control (including the collection of fees) and also improved technical management of existing hydraulics works (interconnections between works, multiple uses, optimization of allocation). Water demand is high and diversified but there is yet no way of calling into question the use made of the resource. The difficulties encountered in the collection of operation and maintenance (O&M) fees in public irrigation schemes contributed to the extinction of the development offices in 1986 and to their replacement by the Regional Commission for Agricultural Development (CRDA) and local associations of collective interest (AIC) institutions, whose attributions were defined by amendments to the Water Code in 1987:

- AICs are endowed with a legal status and are created on the initiative of the users or the administration. A yearly management agreement must be signed with the administration. AICs cover the O&M costs of the hydraulic infrastructure put under their responsibility. The revenue comes from the contributions of the members, subsidies from the CRDA, and the sale of water.
- The CRDA is a structure created by the 1989 decentralization law and established

at the level of the *gouvernorat*.[2] It is a pluridisciplinary structure composed of sections (*arrondissements*), which represent most national general directorates. Supervision of the AIC is one of the missions of the CRDA that receives the fees for the use of water. The water resources section is a water police, which establishes fines and lawsuits and transmits them to the police chief for processing by a civil jurisdiction.

Forthcoming shortages and ad hoc policies

At the end of the 1990s, various national studies forecast a substantial discrepancy between water demand and water supply by around 2010. This led Tunisia to strengthen its control of water demand, using technical tools supported by complementary economic, legal and institutional measures:

- Pricing policy: the price of agricultural water doubled between 1989 and 1996. Pricing of domestic water follows a block-tariff, consumption by hotels being charged at the higher block level.
- Incentives for water savings: laws and decrees were promulgated to encourage irrigators to save water; localized irrigation was subsidized up to 60%; and the state rehabilitated infrastructure in public schemes.
- Reinforcement of collective management: AICs were turned into groups of collective interest (GIC), with extended roles and responsibilities. The authorities hoped to achieve financial disengagement and the collection of fees through the AICs.
- Small and medium hydraulic works: in 1995, the legislation on SWC defined various types of work to be implemented: *tabias* and vegetated benches on the top of the slopes to fight erosion and to increase water infiltration into the soil; short-lifespan tanks with no technical management to control siltation of larger dams (1000 planned, more than 400 built in 2000); larger tanks linked to human uses (203 planned, more than 100 completed); and water structures to infiltrate water or to spread flood water. These small works offer a better distribution and allocation of water

resources in the basin. They should increase storage capacity by using underground storage, slow down erosion and protect larger reservoirs by trapping part of the sediments. However, they decrease the flow to downstream dams, and the management of the basin becomes more complex since it must integrate a whole set of superimposed water management schemes.

Impacts of public policies on the Merguellil basin

From 1956 to 1962, reafforestation works and construction of *tabias* were carried out to help solve unemployment problems. From 1962 to 1973, a development plan for the upper basin was designed under USAID auspices but its implementation – without local participation – was not a success. From 1974 to 1980, due to opposition from peasants, SWC works were carried out only on state land and forests.

In 1989, the El Haouareb dam was built, just before Wadi Merguellil reaches the Kairouan plain. This dam, oversized to protect Kairouan from extreme floods (such as that of 1969), supplies the El Haouareb irrigation scheme and recharges the downstream aquifer through continuous seepage. The rights to use the aquifer are assigned by the state, which established 15 public irrigation schemes supplied with groundwater wells as part of its policy to settle nomadic groups.

From 1990 onwards, the SWC directorate set up a decennial strategy (1991–2000), focused primarily on the treatment of hill slopes and the construction of small tanks (Fig. 7.4). Currently, 17% of the basin is occupied by terraces made out of dry stone or by terraces with total flow retention (*tabias*), in particular in the very fragile Zebes and Haffouz sub-basins (Ben Mansour, 2006). The upper basin also includes about 30 small tanks (with a capacity lower than 0.5 Mm^3), built and managed by the SWC directorate, which store 2.5 Mm^3/year on average, and five larger tanks, built by the dam directorate and managed by the SWC directorate. With a storage capacity above 1 Mm^3, these tanks receive annual average contributions of 2.8 Mm^3. These reservoirs were initially built to trap sediments but authorities later tried to select reservoir sites close to the population and exploitable land.

The development of water infrastructure in

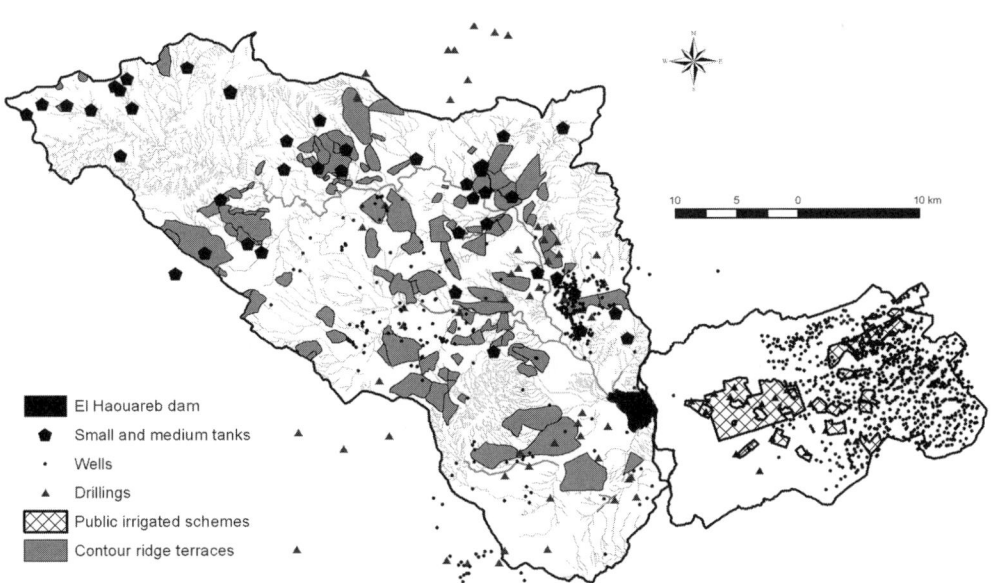

Fig. 7.4. Water infrastructure in the Merguellil basin.

the Merguellil basin illustrates the impact of successive public policies well, with the notable exception of groundwater wells, which have proliferated since 1974 in both upstream and downstream aquifers, despite official prohibition. The wells are deepened by a local manual technique (*forage à bras*) as the water table drops, without intervention of the CRDA water police because authorities prefer to turn a blind eye to these practices and to encourage regional agricultural development.

Present Agriculture and Water Uses in the Merguellil Basin

The Merguellil basin overlaps with seven administrative districts (*delegations*) belonging to two *gouvernorats* (Siliana and Kairouan). In 1994, the population in the study area totalled 102,600, 85% of which resided in the *gouvernorat* of Kairouan (Géroudet, 2004c). The pattern of settlement between delegations is almost identical in the censuses of 1974, 1984 and 1994, except for the population of Chébika, which almost doubled between 1974 and 1994 (Fig. 7.5). However, the last census, in 2004, showed an inversion in the demographic trends, which had been characterized up to that point by a regular increase in population. A decrease in the remote rural population of the basin (approximately 85% of the total) is now expected.

Deep wells tapping several aquifers in both the upper and lower parts of the basin ensure supply of drinking water to this population. Small water supply schemes supplying isolated communities are managed locally, while large ones are managed by the SONEDE national company. However, more than 80% of the water pumped for domestic consumption is exported out of the basin area towards the large cities on the coast. Withdrawals for domestic use represent more than half of the withdrawals in the upper basin and less than one-third in the lower basin. Industrial use is marginal (less than 2% of the total).

We focus hereafter on agricultural water uses. Small-scale farming prevails in the basin (55% of farms are under 5 ha, 81% less than 10 ha). In the majority of cases, division of land at inheritance remains oral and farmers thus do not have ownership titles, which prevents them from getting bank loans. Many farmers

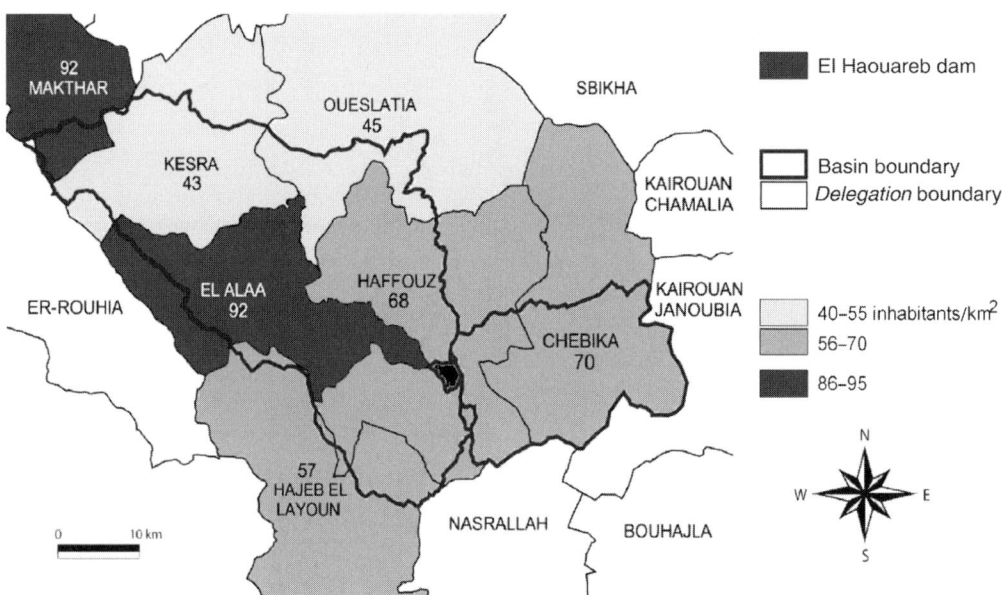

Fig. 7.5. Population density per *delegation* in 1994 (administrative district).

are trying to continue farming but on landholdings that are too small to be economically viable and hence either engage in pluri-activity or try to migrate.

Agricultural development of the basin upstream and assessment of water uses

An investigation of 5045 farm units in Haffouz district, carried out in 2002 by the CRDA of Kairouan, identified eight main cropping systems, including irrigated systems (arboriculture and olive trees, olive trees alone, cereals, winter vegetables, summer vegetables) and rainfed systems (arboriculture and olive trees, olive trees, cereals). This survey also identified a typology of farming systems that can be considered as representative of the upper basin.

Types of farms and crops in Haffouz district in 2002

Farms are divided into seven types according to their cropping patterns. The first four types are based on dry farming and include types T1 (farms cultivating mainly olive and almond trees); T2 (farms cultivating cereals with a large proportion of fallow and rangeland); and T3 and T4 (both cultivating mainly cereals and olive trees but with different average areas: 45.2 ha for T3 and 6.8 ha for T4). The last three types refer to irrigated cropping: T5 (irrigated vegetable cropping in rotation with olive trees and orchards); T6 (irrigated cereals); and T7 (irrigated olive trees and orchards).

The first four types make up about 90% of the farms in the district (T1 and T4 alone total 80% of farms), while farms based on irrigated crops are very few: types T5, T6 and T7 represent only 12% of the farms and are mainly found in the sectors (sub-subdistricts) of Haffouz, Khit El Oued and Aïn Beidha (Fig. 7.6). Type T7 includes most of the irrigated farms, which are concentrated in only a few *douars* (settlements). The analysis clearly shows a strong spatial heterogeneity of farming systems, related to strong differences in access to irrigation water.

The major part of the agricultural area is cultivated with rainfed crops (cereals and olive trees). The extent of fallow lands, linked to the mode of rainfed (dry) farming, explains the low cropping intensity, between 57% and 98%, with an average of 73%. In most sectors, irrigated crops make up less than 10% of the agricultural area, except in Haffouz, where they correspond to nearly 40% of the cropped area. Vegetables and irrigated cereals are cultivated in rotation with olive and almond trees.

Agricultural uses of water in the upper basin are little developed: irrigated crops cover only 2700 ha out of 33,000 ha of cultivated land. Perennial crops and olive, almond and apricot trees cover 1700 ha, while summer vegetables are planted on less than 400 ha. Distribution and types of uses depend on access to water. Irrigation with surface water (by pumping from Wadi Merguellil in particular) is very unpredictable in summer. Aquifers are very localized and the drilling of wells less convenient than in the plain downstream because of the relief.

Assessment of flows

Using studies on the El Haouareb dam (Kingumbi, 1999) and calculations made on small tanks (Lacombe, 2007), a first assessment of surface water and green (soil) water in the upper area can be made for the 2000–2004 period (Fig. 7.7):

- Out of 409 Mm^3 of rainfall, 175 Mm^3 (43%) are lost by evaporation, including 41% by evapotranspiration of the natural vegetation and 2% from small dams and the El Haouareb dam.
- The larger part of rainwater (89%) is stored as green water: 48% is consumed by cultivated areas, rangelands and forests, on a total area of 605 km^2, and 41% by the natural vegetation.
- Runoff water accounts for only 11% of rainfall; once evaporation in the dam is deducted the quantity of water which can be used for productive purposes amounts only to 7.8% of the basin inflow.

This shows the paramount importance of dryland farming, which uses the soil storage capacity (600 Mm^3 for the basin, based on a reserve of 100 mm) and whose storage efficiency (88% for daily rainfall under 15 mm) is important (Dridi, 2000). If rainfed cereal production in the basin, which produces, on

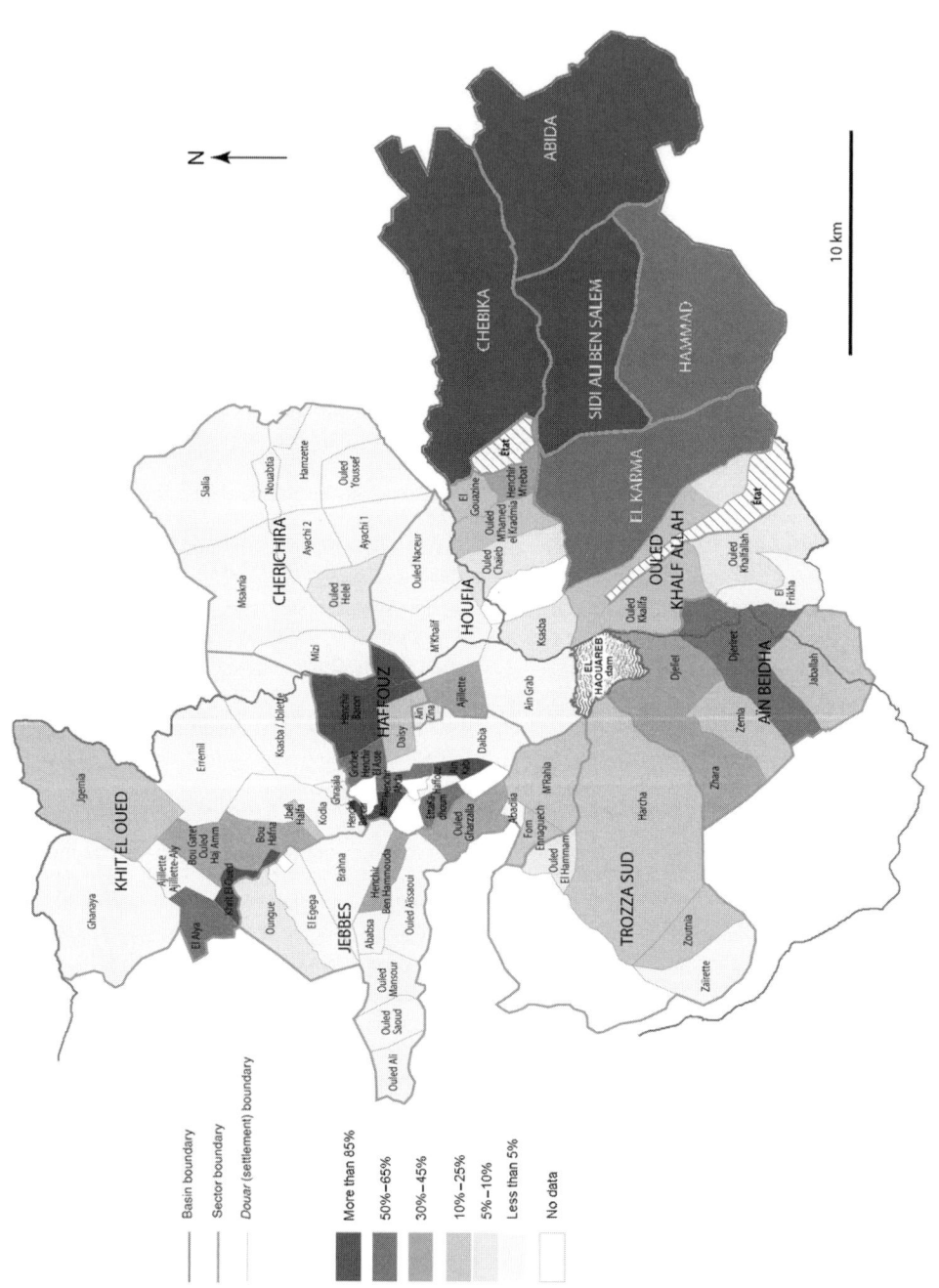

Fig. 7.6. Density of irrigated crops in 2002 in the central part of the Merguellil basin.

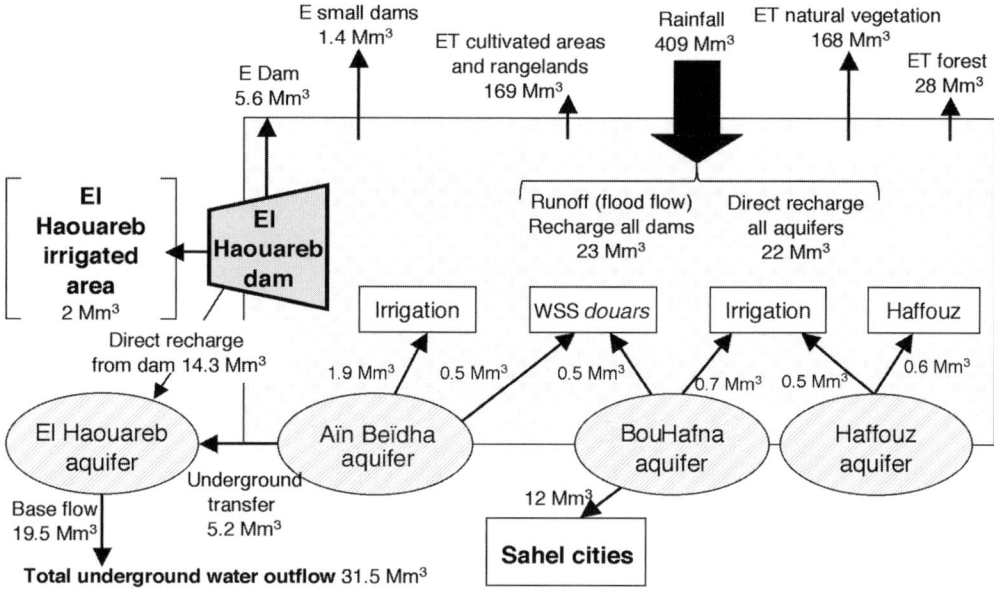

Fig. 7.7. Assessment of average flows of the upstream zone of Merguellil basin (2000–2004) (E, evaporation; ET, evapotranspiration; WSS, water supply and sanitation; *douars*, settlements).

average, 2.6 t/ha, were to be replaced by an equivalent production under irrigation, an area of 6500 ha (at 4 t/ha) would be needed, which would require 14 Mm^3 of blue (renewable runoff and groundwater) water, which is 44% of the amount exported today. This gives an idea of the interest in seeking drought-resistant varieties (Luc, 2005).

Agricultural development and water use downstream of the El Haouareb dam

Types of farms and cropping patterns

An exhaustive farm inventory and a first classification resulted in the identification of eight main farming systems:

- The first three types mostly combine rainfed agriculture with livestock (more than 1.5 sheep/ha) and are characterized by their dominant crops: olive trees (type T1), cereals (T2), and cereals intercropped with olive trees (T3). When water is available in farm types T1 and T2, summer vegetables are irrigated either in association with olive trees (T1) or alone (T2).

- Types T4 and T6 have a cropping pattern made up of approximately 20% of rainfed crops, exclusively of cereals in T4, and 80% of irrigated crops. T4 farms grow olive trees with cereals and intercropped irrigated vegetables; T6 farms have the same irrigated crops but without olive trees.

- Types T5, T7 and T8 are exclusively irrigating. T5 is primarily made up of summer vegetables cropping in full field; T7 is primarily olive trees intercropped with some summer vegetables, and T8 shows an important proportion of fruit-bearing orchards, alone or together with olive trees.

The proportion of irrigated crops varies from 24% in the Houfia sector to 88% in the Chebika sector. The proportion of summer vegetables, alone or intercropped with olive trees, varies between 11% of the cropped area in the Houfia sector and nearly 40% in the Ouled Khalf Allah sector.

With the exception of the Houfia sector, which distinguishes itself by its strong proportion of rainfed crops, agricultural development is rather homogeneous in the plain downstream of the El Harouareb dam. All sectors have access to irrigation water, either through public schemes or through private wells and boreholes pumping water from the Kairouan aquifer.

This results in a cropping pattern that includes 70% of irrigated crops, with 30% devoted to summer vegetables (melons, watermelons, peppers and tomatoes). This cultivation of summer vegetables is the mainstay of irrigation development because it yields handsome revenues. Its development is associated with the adoption of drip-irrigation, which is subsidized by the state at the level of 60% of capital costs for small farmers. This irrigation technique is also very labour saving, and associated fertigation allows farmers to strongly increase yields and therefore incomes.

Farmers can increase their areas cultivated with summer vegetables, but these crops are very risky and sensitive to market variations and vagaries. For example, prices of melons and watermelons are divided by three between the first early productions and the main production season, approximately one month later.

Assessments of flows

In Table 7.1, the accounting of green water considers the whole area of the main plain, excluding 27,350 ha of *djebels* (Table 7.1).

In the lower basin, more than 60% of rainfall is consumed by crops (Fig. 7.8). For non-cultivated areas, the overall consumption of rainwater is estimated at 25.2 Mm³. Volumes abstracted from the Kairouan aquifer for municipal and industrial uses represent a total of 15 Mm³ (values given by the Kairouan CRDA).

Only the contributions from rainfall and the dam are measured values, while other variables are estimated. Urban abstraction represents almost half of agricultural use. Since water is

Table 7.1. Assessment of green water consumption according to rainfall.

Year	Dry	Average	Wet
Total contribution of rainfall in Mm³	49.2	82.3	99.8
Total green water consumed in Mm³	30.5	52.2	60.1
Total green water consumed (% rain)	62.0	63.4	60.2

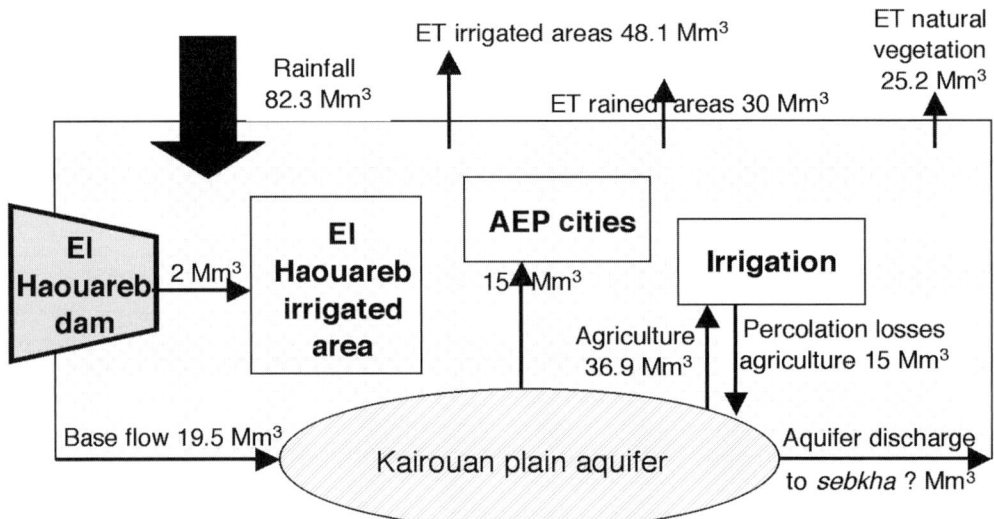

Fig. 7.8. Assessment of average flows in the lower Merguelli basin (1994–2003) (ET, evapotranspiration; AEP, water supply project; *sebkha*, salt lake).

exported and there is no return flow to the aquifer, this amounts to a net loss for the zone. The main inflow is rainfall and groundwater flows from the upper basin; the contribution of the dam through releases of surface water is very limited. Observations of aquifer levels confirm the imbalance between inflow and outflow and point to a shortfall of 17.4 Mm^3, with agriculture as the main cause for this imbalance (net consumption of 21.9 Mm^3) (Luc, 2005).

Competition for water between agriculture and other activities is very strong in the plain, but all sectors do not face the same constraints. Drinking water supply is a priority according to the Water Code, and abstraction is supposed to be done only through authorized and controlled boreholes. Agricultural use in public schemes is also based on controlled groundwater abstraction, but the administration has, in fact, very little control over private wells. These wells are deepened in order to follow the decline of the aquifer and have spread out in the area, despite renewed prohibition. They came along with changes in agricultural practices through the introduction of melons and watermelons, both of which ensure a handsome income to farmers.

Assessment of flows in the entire basin

Most of the agricultural production in the upper basin is based on green water and its contribution is also very substantial in the plain. In such a context, it is interesting to increase the volume of water stored in the ground and to make better use of it. For blue water at a global level, withdrawals for municipal and industrial use (27 Mm^3) and irrigation (34 Mm^3) are of the same order of magnitude and almost exclusively withdrawn from aquifers that are overexploited. The replacement of dryland farming by irrigated crops would jeopardize the balance of the aquifers and affect the export of drinking water to the Sahel zone.

Assessment of the lower basin is more complex. The aquifer drops but its functioning is poorly known. The basin being closed, with no natural outflow except the *sebkha* and the sea in case of extraordinary events, all the blue water available is currently mobilized and depleted. To build further tanks and dams in the upper basin would be tantamount to a re-appropriation and redistribution of existing (and already used) resources. The only water which is not yet fully mobilized is green water. Because of farmers' aversion to risk, production of dry farming is limited, even in wet years. Thus, substantial progress is possible by developing crop varieties with better resistance to drought and stress and by improving cropping techniques, for example dry farming and mulching.

The very significant amount of water exported for domestic use induces a real competition for water and could place farmers in a difficult situation in case of shortages, since drinking water is a national priority.

Impacts and Valorization of Installations

After reviewing the status of main crops in the basin, we now turn to the assessment of the impact of water-harvesting structures in the upper basin. This is done through the comparison of the water flows before (1989–1996) and after (1997–2005) the implementation of SWC works.

Impacts of installations on the allocation of water

The analysis of runoff–rainfall relationships showed that the expansion of SWC infrastructure on 21% of the upper basin area between 1989–1996 and 1997–2005 induced a drop of 41–50% in the runoff to downstream areas (Lacombe, 2007). This confirms the efficiency of water-harvesting techniques (Nasri, 2007) and their capacity to reduce runoff towards the dam. But this, of course, comes at the detriment of the users of the Kairouan aquifer, who have seen water tables dropping by between 0.25 and 1 m/year since the 1980s (Feuillette, 2001).

Assessments of the tabias

Dridi (2000) showed that soil depth seldom exceeds 1 m. Below this, limestone or the rock substratum stops or modifies the percolation of water. The recharge of aquifers by *tabias* is probably non-existent, and the water stored in the unsaturated zone is taken up again by evapotranspiration (Favreau et al., 2001). By collecting surface water charged with sediments, the *tabias* create wet and fertile lands close to the ditches (Fig. 7.9) and increase soil moisture downstream of the bunds. Roose (2002) showed that olive trees make the best use of the *tabias* in central Tunisia, with a line of trees planted downstream of the bund and two upstream.

In the upper catchment, areas provided with *tabias* are cereal fields or rangelands for cattle (Dridi, 2000; HAR, 2003). Areas already cultivated are generally not equipped with *tabias* because they do not allow the passage of bulldozers. Today, the crops for which *tabias* are likely to be further developed are cereals. Mechergui (2000) showed that by increasing storage of green water by more than 20%, *tabias* can increase yields from 1 t/ha to 1.5 t/ha. But these potential benefits are compensated for by losses in cultivated area and in output due to increased difficulties in treating and harvesting the product. On the basin scale, the agronomic benefits derived from *tabias* can thus be considered as virtually nil (Lacombe, 2007), although their impact on erosion is probably positive.

Assessment of tanks

In the upper basin, 47% of the 46 tanks are exploited by 270 farmers, who irrigate 669 ha (vegetables (2%), olive trees (69%), almond trees (14%), and other trees (15%)), which corresponds to an average of 5.9 farmers/tank and 12.5 farmers/exploited tank. In 63% of the cases, water is abstracted by engine-driven pumps, and in the remaining cases irrigation is done by gravity, water sometimes being transported by cisterns.

The yield output of olive trees irrigated by tanks is low (4.5 kg/tree), compared with yields in the plain (25 kg/tree, and 10 kg/tree in non-irrigated situations; Feuillette, 2001). Limiting factors are related to conditions of shallow soils and a harsher climate at a higher altitude, which increase the likelihood of harmful frosts. Irrigation of olive trees is especially useful in periods of drought, to save young seedlings. Consequently, the agronomic valorization of tanks cannot be assessed only in terms of volumes withdrawn and intended for irrigation. Volumes of water applied are low, in particular for the almond and the olive trees, and the use of cisterns is constraining and expensive.

Global assessment of the upper catchment before and after implementation of SWC

Over the 1989–1996 period, the average flow of the Merguellil was 20 Mm^3/year at the level of the El Haouareb dam (Fig. 7.10, left part). Over the 1997–2005 period, only 70% of this flow reached the dam. Water balance studies carried out on tanks showed that the latter captured a sixth of the total volume harvested by all structures, the remainder being collected by *tabias*. This has resulted in a new distribution of collected flows (Fig. 7.10, right part).

In this assessment, the *tabias* were not forested, in accordance with the situation observed in 2005. In addition, frequent passage of livestock on the bunds is responsible for early degradation of *tabias*. At the basin level,

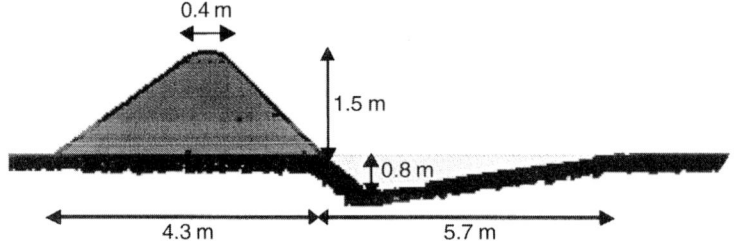

Fig. 7.9. Cross-section of anti-erosive bunds (*tabia* system).

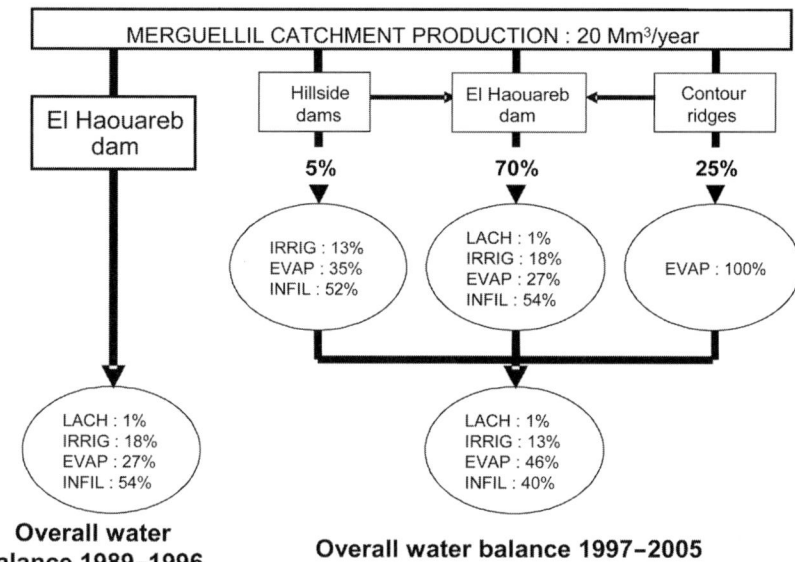

Fig. 7.10. Change in flow distribution in the Merguellil basin (1989–1996 versus 1997–2005) (LACH: release; IRRI: irrigated agriculture; EVAP: evaporation; INFIL: infiltration).

most of the water harvested is transformed into evapotranspiration, without any significant agronomic benefit. As for tanks, it is assumed that percolation losses are passed on to the aquifers, where they can be exploited through pumping. This assumption is based on the studies by Grünberger et al. (2004), who found evidence of a transfer from the El Gouazine tank to the alluvial aquifers of downstream valley bottoms.

Even accounting for the inaccuracy of these assessments, it remains clear that water-harvesting structures result in reductions of infiltrated and abstracted flows of 14 and 5 percentage units, respectively, and an increase in evaporation flows from 2 to 46%, lost for any use. On the basis of an annual production of about 20 Mm³/year in the basin area, upstream SWC works therefore induce a loss of 3.8 Mm³/year of water to evaporation. This volume would make it possible to irrigate between 374 and 910 ha of vegetable cultivation with two harvests, in summer and winter.

If all the bunds were forested, as an ideal way to make use of collected water, one would obtain an increase of 25% in the agricultural value compared with the current situation, and a reduction of only 6% of evaporation compared with the situation without *tabias*. But this assumption is based on the planting of 1,310,000 olive trees in three lines separated by 10 m and with enough roots to exploit all of the unsaturated zone.

Water productivity in the upper and lower basins

The econometric analysis carried out by Albouchi (2006) on the upper and lower parts of the Merguellil basin for the 1994–2003 period examined the use of four factors mobilized in the aggregate production of the basin:

- Land: if the agricultural area is larger in the upper zone, irrigation is less developed (7% of cultivated land, against 26% in the lower basin). If water is not used efficiently, continuous expansion will increase pressure on water, in particular in the lower part. Intensification of farming systems remains limited in the two zones, with a cropping intensity of 68% upstream, against 87% downstream.

- Water: water consumption strongly increased in the two zones, from 5.5 to 9 Mm^3 in the upper basin and from 13.5 to 34 Mm^3 in the lower basin, i.e. increases of 65% and 152%, respectively. In contrast, we can observe a drop in water consumption per hectare on irrigated land in the downstream part, but withdrawals are still at 3859 m^3/ha, against 2315 m^3/ha in the upper basin. This lower application of water upstream reflects a still 'extensive' and recent irrigation, but farmers are increasingly attracted by crops with high added value. Downstream irrigators are already specialized in such products and, rather, seek to save water in order to expand their plots.
- Labour: agricultural employment, including available family labour and occasional hired labour, increased by approximately 13% in the two zones because of the expansion of irrigated areas and of the limited mechanization of agriculture. Increases in employment and per hectare labour input are higher in the lower basin (twice the values for the upper basin), which can be explained by the intensity of labour demand in vegetable cultivation. Farms in the upper part are characterized by lesser integration into the market, and resort to the abundant family labour and seasonal migrations to the lower basin, when occasional labour may amount to 60% of all labour requirements.
- Capital: capital is limited to the use of variable inputs; collected data evidenced a more intensive use of capital in the lower part, 203 Tunisian Dinars (DT)/ha versus 82 DT/ha in the upper part, which reflects the differential level of intensification. Production of cereals and vegetables is higher downstream (12,550 t and 56,280 t, against 12,240 t and 7740 t upstream), due to the irrigation of the former (maize, barley) and to a better technical control of the latter. However, tree and animal production are still higher upstream, because of the importance of olive trees and because all herds (sheep, goats, cattle) are larger.

The net income is understandably higher in the downstream part, with a 10-year observation average value of DT 12.3 million. In the upstream part, this income only totalled DT 2.5 million and was even negative during the first 4 years of observation, showing that the land and family labour factors were not valued at market prices. This can explain the migration of young people from the upper to the lower part of the basin, where irrigation allows for the remuneration of all factors and the accumulation of profit.

The price of irrigation water varies between 0.08 and 0.12 DT/m^3 over the whole zone of study, while average water productivity is 0.18 DT/m^3 upstream and 0.39 DT/m^3 downstream. This shows the existence of rents in the two zones, particularly downstream, where revenue is accumulated in the value of land. The analysis also indicates that there is considerable potential to improve the economic efficiency of the two zones and to increase the generated overall income. A first estimate of the determinants of efficiency discards factors like specialization, integration into the market, availability of fodder or the degree of technical assistance, which are statistically non-significant. The effect of water saving is significant but weak, which can be explained by a poor command of these recently introduced techniques and by a tendency to expand irrigated areas rather than to intensify cultivation on them (Bachta et al., 2000). Access to credit has a strong positive effect on technical, allocative and economic efficiency; according to some farmers, the lack of financing is much more constraining than the lack of water. Producers having access to credit generally have better information on the prices of products and of production factors, and are more motivated to adopt new technologies. Finally, small-scale farming negatively influences both technical and economic efficiency, which suggests that a land policy designed to prevent or limit the fragmentation of farms would be desirable.

According to the analysis, continued implementation of SWC works in the upper basin without attendant public support would lead to a drop of overall gross product in the basin. Further construction would deprive the most productive downstream zone of part of its water supply and would accentuate pressure on the aquifer. This loss can be mitigated by increasing the efficiency of the upper basin (land reform, access to credit), but none of the scenarios considered is able to reverse the conclusions of the analysis.

Prospective Analysis

Possible or planned changes

There is often confusion between hydrological variability, which is naturally very high in the Mediterranean region, and climate change. Until now, no statistically significant change in rainfall has been observed at the Mediterranean scale (Cudennec et al., 2007), although depending on authors and methods, opposite conclusions were proposed in countries such as Spain and Italy. An overall worsening of rainfall in the Mediterranean basin remains hypothetical. At the Tunisian scale, the study of 43 long data sets of rainfall (Sakiss et al., 1994) did not find any consistent trend. Changes that could be identified at an even smaller scale, such as central Tunisia (e.g. Kingumbi et al., 2005), have to be put in the context of this overall stability.

Present global circulation models do not represent the Mediterranean climate with satisfying accuracy, especially with regard to the extreme rainfall variability in time and space. Whatever the future climate in Tunisia, climatic changes in other parts of the world may also affect the study area in indirect ways. The recent worldwide rise in cereal prices, for example, partly linked to droughts in North America and Australia, led to a quick and unexpected increase in the Tunisian fixed prices.

The previous sections showed that the most important and rapid changes in regional water resources are consequences of human activities on the environment, and the induced redistribution of water in time and space, between blue water and green water, groundwater and surface water, and upstream and downstream parts of the basin. New conservation works planned by the authorities consist of three dams and bench terraces over 30,000 ha, which will further decrease the river flow, the inflow to the El Haouareb dam and, consequently, the recharge of the Kairouan plain aquifer.

Concerning the uses of water, Tunisia is characterized by a recent fast demographic growth (population has doubled since 1970 but is now levelling off), a concentration of the population in the coastal urban centres and a seasonal peak in the flux of foreign tourists (6 million/year). The increases in population and in the standard of living, and improved coverage of water supply in rural areas, have resulted in a continuous rise in domestic water demand. At the local scale, the Merguellil basin experienced a continuous demographic growth up to 1994, but the 2004 census showed a more complex evolution in recent years. While Kairouan and its surroundings continue to grow, an exodus of the rural population towards towns close to Chébikah and Haffouz is observed. This trend, caused by social and economic problems, is likely to continue in the next decades, according to the pattern widely observed in the Mediterranean basin.

These local and regional trends will increase the demand for water transfers towards urban and tourist centres of the Sahel, which already amount to 12 Mm^3/year, as well as withdrawals for the local population and industries (presently 15 Mm^3/year). But the water demand for agriculture is, and will remain, the biggest problem. With the integration of Tunisia into the world agricultural markets, the growing demand from local and tourist populations, vegetable farming – a very water- and labour-intensive activity – should continue to expand, together with plantations of new fruit species in demand on the world market (almonds, apples, etc.).

The future of SWC works

The SWC works of the upper basin are not completely satisfactory: a large part of the harvested water is lost by evaporation and very few new water uses have developed. In theory, the planting of 1.3 million olive trees would allow full use of the harvested water, but this number should be doubled to make the *tabias*, planned up to 2011, profitable. This hypothesis appears quite unrealistic, in particular because they require regular maintenance. In the absence of maintenance, they are damaged by breaches, which automatically cause other breaches in other *tabias* downstream. This succession of ruptures accentuates the runoff concentration and leads to the formation of gullies. Erosion in areas of SWC works without maintenance is then even worse than in natural conditions. Studies on small tanks in El Gouazine showed that, 8 years after their construction, these SWC works had already lost 60% of their storage capacity. The runoff coefficient and the sediment transport went back to their initial values before the SWC works (Nasri et al., 2004).

Under such conditions, there is little likelihood that the remaining population of the upper basin will undertake the maintenance of plantations and conservation works, especially when people were not consulted during their construction, or even opposed them, thus forcing authorities to build them on public land.

Small tanks were initially established to trap sediments and to protect the El Haouareb dam. They fulfil their role properly since their average lifespan is about 20–25 years. Currently, only some of the largest and well-located tanks are exploited in an intensive way. In general, there is no collective management, with each irrigator using pump sets subsidized by the government to abstract water from the tank. Intensification is limited because the risk of water shortage is high, especially during drought years. Setting up water user associations is not expected to change this situation, especially given the fast sedimentation of the reservoirs.

It is often said that global warming will increase the frequency of extreme events, in particular droughts lasting several years, with more irregular rainfall. If such changes occur, they would be detrimental to the lifespan of SWC works (via breaching, faster filling) and to their usefulness, because ditches are not able to store water to bridge rain events. However, the El Haouareb dam would be of greater interest in stopping exceptional floods and reassuring rural and urban populations downstream in the Kairouan plain. It is large enough to undergo sedimentation and subsequent reduction in storage for many years.

One can thus imagine the future with an extensive use of most of the basin upstream (rainfed farming of olive trees and cereals, grazing lands) and an intensive exploitation of land endowed with groundwater resources, on an area that would guarantee water supply for a sequence of drought years. In addition to its protective role against flooding of the city, the El Haouareb dam would only be used to transfer water to the Kairouan aquifer. The management of water in the zone would be, in fact, essentially limited to the management of groundwater.

Groundwater management

Dynamics of the upper basin aquifers (Haffouz, Aïn Beidha, BouHafna) are insufficiently known (recharge, connections), but the first models available emphasize their fragility. They are exploited by private wells, without any control on withdrawals, and also supply the Sahel region with domestic water. The Kairouan plain aquifer supplies about 15 well-based public schemes and thousands of private wells, manually deepened if necessary when the water table drops. Although formally a protected area, the Kairouan aquifer remains, in fact, a free-access collective resource: restrictive regulations are not respected and hand-dug boreholes are always more numerous, even in public schemes. Farmers attempt to free themselves from the rigidity of water turns and directly access groundwater to cultivate melon and watermelon with drip-irrigation. Law enforcement is difficult because the water police function is entrusted to the institution in charge of regional development, which chooses to preserve social peace in an area struck by unemployment rather than to safeguard the long-term durability of the system.

Water shortages are not imminent, given the thickness of the aquifer, but the continuous drop in the groundwater level will make access to water more and more difficult for small producers; those located at the aquifer margins will see their resource disappear. In addition, the deeper layers of the aquifer have lower transmissivity and a higher mineral content. Water quality will most likely deteriorate in the long term.

Aware of the importance of aquifers, Tunisian authorities explore various recharge techniques, even if this will probably not be sufficient to offset overdraft. The El Haouareb dam loses most of its water through karstic cracks, which constitute an unexpected but interesting process of recharge of the Kairouan plain aquifer. This saves water from evaporation and smoothes inter-annual climatic fluctuations (the storage capacity of the Kairouan plain aquifer is much higher than the dam's, by more than an order of magnitude). The rest of the dam water is pumped for irrigation and also evaporates. Keeping water in the dam for satisfying the irrigation demand is a short-term approach, but is not pertinent at a regional scale. A more proactive management would include dam releases in spring, in order to save a part of the dam water presently lost by evaporation (five times higher in July than in January). According to our calculations, about half of the released volumes would be gained by the Kairouan plain aquifer: since 1989, 24 Mm^3 could have been saved, which is greater than the mean annual groundwater inflow to the Kairouan aquifer from the dam and the upstream Aïn Beidha aquifer over the same period. Obviously, such an improvement would only bring a limited amount of 'new' water into the regional water budget, but this is the most efficient solution in a region where water resources are scarce and already overexploited.

Alternative solutions should also be searched for. Urban and tourist centres of the Sahel coast could turn to desalination of seawater for domestic supply, as in several other Mediterranean areas. The price of desalination strongly dropped with the development of reverse osmosis (between €0.5 and €1.00 per m^3). For instance, the Balearic Islands and the Canaries resort to desalination to meet the demand of the tourist season, as does Andalusia, which runs several plants in order to avoid conflicts between tourist and agricultural activities.

The overexploitation of the aquifers in the basin cannot be stopped without an effective management of agricultural demand and its acceptance by the population. An agent-based simulation model was used to test various management policies (Feuillette et al., 2003) by representing the interactions between water supply and demand, investment strategies of farmers and their decisions to dig wells, and interactions between farmers. A first scenario simulates the impact of adoption of drip-irrigation systems, as advocated by the authorities. The model revealed an increase in farm income but a very clear rise in pumping and in the number of wells. Ground-truthing showed that the farmers make a benefit after 1 year only because the drip system is subsidized by the government by up to 60%. This enables farmers to grow melons and watermelons, sold at a good price to the hotels on the coast, 100 km away. In the following years, farmers try to increase production by all means: extension of the irrigated area, renting of additional land, increase in the power of pumps, digging of new wells, etc. The shift to micro-irrigation, without attendant measures, would compound the overexploitation of the aquifer, i.e. the exact opposite of the expected outcome. Additional simulations showed that the combination of drip-irrigation and a very high pricing of water when used beyond a given quota gives the best results: reasonable profit-making by farmers through a better valorization of water combined with a substantial reduction in total withdrawals.

To enforce regulation, the administration can try to control current individual dynamics by strengthening control: electrification of all the pump sets, systematic installation of three-phased electric meters and political support to the national electric company to recover fees. Such proposals are obviously difficult to implement. The administration can also initiate a process of turning management responsibilities over to water user associations at the borehole level. Faysse (2001) showed the collective benefit resulting from the application of flexible rules by an association of irrigators. Concerning deep-well-based irrigation schemes, management by associations with increasingly broad attributions, and the disengagement of the state at the CRDA level, gradually become the rule. However, the adoption of a more participatory management could be temporarily hindered by difficulties in identifying supportive interest groups. The support of users, who have long been kept away from management, will probably come up with the awareness of the fragility of the aquifer, regarded until now as an inexhaustible resource crossed by large underground rivers.

Conclusion

Located in the heart of Tunisia in a semi-arid zone, the Merguellil basin belonged, until the

19th century, to a tribe-based society. Agricultural modes of production typical of this Mediterranean environment combined a concentration of investments on the limited areas where water was accessible with an extensive exploitation of scattered pastoral resources.

During the French protectorate, the arrival of a few colonists induced a first modification of land use, but the most important changes in land and water resources development policies happened after independence, when the appearance of the basin was deeply modified: closing of the basin by the El Haouareb dam in 1989, and implementation of water-harvesting structures and small tanks since the 1980s.

The Merguellil basin is typical of the problems faced in Tunisia and in the Mediterranean basin in general: limited water resources; intermittent flows; strong increase in, and diversification of, demand; strong human-induced hydrological changes; competition between declining upstream rural societies and a more dynamic urban/tourist downstream, or coastal, zone; and very localized uses of overexploited aquifers.

The soil and water conservation (SWC) works in the Merguellil basin, planned at the national level and implemented without consultation or participation of the users, are consequently very poorly exploited. These works result in a reduction of infiltration and abstraction and in a substantial increase in losses by evaporation. By reducing surface runoff, they also decrease the inflow to the El Haouareb dam and, consequently, the recharge of the Kairouan aquifer, which is used in a very productive way.

The SWC works are likely to deteriorate with time because the administration has not planned any maintenance and the population hardly feels concerned. In the near future, management of aquifers will be the central issue, because of their pivotal role in supporting intensification of irrigated production and supplying cities. The remainder of the basin will probably remain confined to extensive farming: rangelands and rainfed farming of cereals and olive trees.

To restore a balanced use of the aquifers without impacting the development of the area too much, the administration will have to implement demand-management policies, an uphill battle with very few successes recorded worldwide.

Acknowledgments

We wish to thank all those who took part in the work undertaken on the Merguellil basin, within the framework of the MERGUSIE programme, and in particular the many students of Tunisian and French universities. Thank you all those who elaborated the various thematic syntheses within the framework of a convention with IWMI: their documents were used as a support of this chapter.

Notes

1 *Fesguias* are big public water tanks with a capacity that can reach hundreds of cubic metres.
2 *Gouvernorat*, or *wilaya* in Arabic: administrative region directed by a governor appointed by the president. There are 24 *gouvernorats* in Tunisia.

References

Albouchi, L. (2006) Gestion de l'eau en Tunisie: d'une période de mobilisation à une politique de réallocation de la ressource selon sa valorization économique. Cas du bassin versant du Merguellil, Tunisie Centrale. PhD thesis, University Montpellier I, Montpellier, France.

Bachta, M.S. and Zaibet, L. (2006) Espaces agraires et environnement en Tunisie: la permanence des cadres spatiaux au cours du 20ème siècle. Séminaire sur l'avenir de l'agriculture en Méditerranée, projet UE Wademed, Cahors, France, 6th November 2006.

Bachta, M.S., Le Goulven, P., Le Grusse, P. and Luc J.P. (2000) Environnement institutionnel et relations physiques pour une gestion intégrée de l'eau dans le milieu semi-aride méditerranéen. Le cas tunisien. Séminaire International Montpellier 2000, "Hydrologie des Régions Méditerranéennes". Montpellier, France, 11–13 Octobre 2000. *PHI-V/ Documents Techniques en Hydrologie*, 51, 177–186.

Bachta, M.S., Zaibet, L. and Albouchi, L. (2005) *Impact Assessment of Water Resources Development in the Merguellil Basin: Kairouan, Tunisia*. Report for the Comprehensive Assessment of Water Management in Agriculture. International Water Management Institute, Colombo, Sri Lanka.

Bedoucha, G. (1987) *L'Eau, l'Amie du Puissant. Une Communauté Oasienne du Sud Tunisien*. Editions des Archives Contemporaines, Paris.

Ben Ammar, S., Zouari, K., Leduc, C. and M'Barek, J. (2006) Caractérisation isotopique de la relation entre barrage et nappe dans le bassin du Merguellil (plaine de Kairouan, Tunisie centrale). *Hydrological Sciences Journal* 51(2), 272–284.

Ben Mansour, H. (2006) *Dynamiques des Ouvrages de Conservation des Eaux et des Sols dans le Bassin Amont du Merguellil*. Report for the Comprehensive Assessment of Water Management in Agriculture. International Water Management Institute, Colombo, Sri Lanka.

Bouzaïane, S. and Lafforgue, A. (1986) *Monographie Hydrologique des Oueds Zeroud et Merguellil*. Technical report. Directorate Générale des Ressources en Eau/Institut Française de Recherche Scientifique pour le Développement en Coopération, Tunis, Tunisia.

Cudennec, C., Leduc, C. and Koutsoyiannis, D. (2007) Dryland hydrology in Mediterranean regions – a review. *Hydrological Sciences Journal* 52(6), 1077–1087.

DGRE (Direction Générale des Ressources en Eau) (1995) *Situation de l'Exploitation des Nappes Phréatiques*. Technical report. Ministère de l'Agriculture, Tunisia.

Dridi, B. (2000) Impact des aménagements sur la disponibilité des eaux de surface dans le bassin versant du Merguellil (Tunisie centrale). PhD thesis, University of Strasbourg, France.

Favreau G., Leduc C. and Schroeter, P. (2001) Reply to comment on 'Long-term rise in a Sahelian water-table: the Continental Terminal in South-West Niger' by Leduc, C., Favreau, G., Schroeter, P. (2001). *Journal of Hydrology* 243, 43–54; *Journal of Hydrology* 255(1–4), 263–265.

Faysse, N. (2001) L'influence des règles collectives d'allocation de l'eau sur les choix stratégiques des agriculteurs. Des petits périmètres irrigués tunisiens aux prélèvements en rivière dans le bassin de l'Adour. PhD thesis, University of Paris, X Nanterre, Paris.

Feuillette, S. (2001) Vers une gestion de la demande sur une nappe en accès libre: exploration des interactions ressources usages par les systèmes multi-agents. Application à la nappe de Kairouan, Tunisie Centrale. PhD thesis, University of Montpellier II, Montpellier.

Feuillette, S., Le Goulven, P. and Bachta, M.S. (1998) Les pouvoirs législatifs, règlementaires et juridiques en Tunisie confrontés à la gestion des nappes souterraines. Colloque SFER « L'irrigation et la gestion collective de la ressource en eau en France et dans le monde », Montpellier, 19–20th November 1998.

Feuillette, S., Bousquet, F. and Le Goulven, P. (2003) SINUSE: a multi-agent model to negotiate water demand management on a free access water table. *Environmental Modelling & Software* 18(5), 413–427.

Genin, D., Guillaume, H., Ouessar, M., Ouled Belgacem, A., Romagny, B., Sghaïer, M. and Taamallah, H. (2006) *Entre Désertification et Développement, la Jeffara Tunisienne*. IRD–Cérès Editions–IRA, Paris.

Géroudet, C. (2004a) *Histoire du Peuplement et Histoire Agraire*. Report for the Comprehensive Assessment of Water Management in Agriculture. International Water Management Institute, Colombo, Sri Lanka. www.iwmi.cgiar.org/assessment/files/word/projectdocuments/merguellil/partie1.pdf

Géroudet, C. (2004b) *Politiques Agricoles et Structures Foncières*. Report for the Comprehensive Assessment of Water Management in Agriculture. International Water Management Institute, Colombo, Sri Lanka. www.iwmi.cgiar.org/assessment/files/word/projectdocuments/merguellil/partie2.pdf

Géroudet, C. (2004c) *Evolution Démographique de 1966 à Nos Jours*. Report for the Comprehensive Assessment of Water Management in Agriculture. International Water Management Institute, Colombo, Sri Lanka. www.iwmi.cgiar.org/assessment/files/word/projectdocuments/merguellil/partie3a.pdf

Grünberger, O., Montoroi, J.P. and Nasri, S. (2004) Quantification of water exchange between a hill reservoir and groundwater using hydrological and isotopic modelling. El Gouazine, Tunisia. *C.R. Geoscience* 336, 1453–1462.

HAR (Société d'hydraulique et d'aménagement rural) (2003) *Etude de Planification des Aménagements CES dans le Gouvernorat de Silinana. Phase 2*. Rapport Final Pour la Direction Générale de l'Aménagement et de la Conservation des Terres Agricoles (DG/ACTA). Ministère de l'Agriculture, de l'Environnement et des Ressources Hydrauliques, Tunis, Tunisia.

Kingumbi, A. (1999) *Bilan et Modélisation de la Retenue du Barrage d'El Haouareb*. Mémoire de DEA en Modélisation Hydraulique et Environnement. Ecole Nationale d'Ingénieurs de Tunis, Tunis, Tunisia.

Kingumbi, A. (2006) Modélisation hydrologique d'un bassin affecté par des changements d'occupation. Cas du Merguellil en Tunisie centrale. PhD thesis, University of Tunis, El Manar, Tunisia.

Kingumbi, A., Bargaoui, Z. and Hubert, P. (2005) Investigation of the rainfall variability in the central part of Tunisia. *Hydrological Sciences Journal* 50(3), 493–508.

Lacombe, G. (2007) Evolution et usages de la ressource en eau dans un bassin versant aménagé semi-aride. Le cas du Merguellil en Tunisie Centrale. PhD thesis, University of Montpellier 2, Montpellier.

Luc, J.P. (2005) *Blue and Green Water and Water Accounting: the Merguellil River Basin.* Report for the Comprehensive Assessment of Water Management in Agriculture. International Water Management Institute/Institut de recherche pour de développement, Tunisia. www.iwmi.cgiar.org/assessment/files_new/research_projects/river_basin_development_and_management/bluegreenwater_rapport_luc.pdf

Mahfoudh, F., Baccouch, S. and Yazidi, B. (2004) *L'Histoire de l'Eau et des Installations Hydrauliques dans le Bassin de Kairouan.* Report for the Comprehensive Assessment of Water Management in Agriculture. International Water Management Institute, Colombo, Sri Lanka. http://www.iwmi.cgiar.org/assessment/files/word/projectdocuments/merguellil/histoire eau kairouan.pdf

Mechergui, M. (2000) La petite hydraulique et son impact sur la vie du paysan, les eaux de ruissellement, la conservation en eau et en sol et les ressources en eau vers l'aval dans un bassin versant: cas de deux bassins versants de Siliana et du Kef en Tunisie. Conférence: Relations terre-eau dans les bassins versants ruraux, Conférence électronique, 18 septembre au 27 octobre 2000. FAO, Rome, www.fao.org/AG/agL/watershed/watershed/papers/papercas/paperfr/case18fr.pdf

Nasri, S. (2007) Caractéristiques et impacts hydrologiques de banquettes en cascade sur un versant semi-aride en Tunisie centrale. *Hydrological Sciences Journal* 52(6), 1134–1145.

Nasri, S., Albergel, J., Berndtsson, R. and Lamachère, J.M. (2004) Impact des banquettes sur le ruissellement d'un petit bassin versant. *Revue des Sciences de l'Eau* 17(2), 265–289.

Nazoumou, Y. (2004) *Analyse du Réservoir Aquifère de la Plaine de Kairouan.* Report for the Comprehensive Assessment of Water Management in Agriculture. International Water Management Institute, Colombo, Sri Lanka.

Roose, E. (2002) Banquettes mécaniques et techniques traditionnelles de GCES pour la zone méditerranéenne semi-aride de Tunisie. In: Roose, E., Sabir, M. and De Noni, G. (eds) *Techniques Traditionnelles de GCES en Milieu Méditerranéen.* Institut de recherche pour le développement, Rabat, Morocco, pp. 155–168.

Sakiss, N., Ennabli, N., Slimani, M. and Baccour, H. (1994) *La Pluviométrie en Tunisie a-t-elle Changé Depuis 2000 ans?* Institut National de la Météorologie & Institut National Agronomique de Tunisie, Tunis, Tunisia.

Treyer, S. (2002) *Analyse des Stratégies et Prospectives de l'eau en Tunisie. Rapport I: Monographie de L'eau en Tunisie.* Report for 'Plan Bleu'. Plan Bleu, Foundation Sophia Antipolis, Antibes, France.

8 Water Competition, Variability and River Basin Governance: a Critical Analysis of the Great Ruaha River, Tanzania

Bruce A. Lankford,[1]* Siza Tumbo[2]** and Kossa Rajabu[3]

[1] University of East Anglia, Norwich, UK; [2] Sokoine University of Agriculture, Morogoro, Tanzania; [3] WWF-Tanzania, Rujewa, Tanzania (deceased); e-mails: *b.lankford@uea.ac.uk; **sdt116@suanet.ac.tz

Introduction

This chapter analyses historical irrigation and river basin developments and narratives to demonstrate particular dimensions of water competition in the Great Ruaha River basin in southern Tanzania. Alongside this, we identify three interrelated scalar and emergent dynamic behaviours revealed as a part of basin development. These 'systems' behaviours relate to the growth and coalescing of areas of smallholder irrigated farms since the late 1950s. The three concepts are termed 'parageoplasia',[1] 'non-equilibrium behaviour' and 'share modification'. These insights provide additional layers to the ideas captured in Molle's (2003) conceptual framework for river basin development, specifically on the demand–supply equation, where we bring additional thinking to his allocation 'third way' and on the nature of basin development. While exploring the broad narrative of growth in water demand, we explore further dimensions arising from a highly variable inter-/intra-annual water availability, which affects the distribution of water and impacts of this growth curve, as informed by a sub-Saharan environment.

As well as explaining the concept's terms, we argue that the ideas revealed by this case study might have application to smallholder irrigation elsewhere in savannah agro-ecologies in Africa. The chapter explores how this analysis leads to new insights – particularly in relation to adaptation to climatic change expressed through increased variability of rainfall and river flow (Milly et al., 2008).

Context

The allocation and equity of division of water between sectors in certain kinds of basins is particularly difficult when rapid growth in one sector establishes a basin-wide potential towards disequilibrium. The term disequilibrium is used in the rangelands' ecological sense (Sulllivan and Rohde, 2002), pertaining to dramatic changes in inputs such that a medium-term, predictable resource offtake from a climax ecology is denied. Explored in more detail by Lankford and Beale (2007), basin disequilibrium occurs because of external and internal perturbations of water catchments and linked interconnections between upstream and downstream water-use systems. Externally derived perturbations arise via a variable water supply, expressed through climate and weather, bringing inter- and intra-seasonal fluxes of drought and wetness, potentially further exacerbated by climatic change. Internal perturbations occur due to feedback connections between linked sectors or systems where water abstraction and depletion occur – particularly in the irrigation sector, where depleted quantities are both large and highly variable inter- an intra-seasonally. Both types of perturbations pose problems for the management of river basins, particularly the 'equilibrium' expectation that the quality and quantity of water are either only mildly varying or predictable or both, and can be managed accordingly.

Unrealized or unfounded expectations about the slow and/or predictable behaviour and development of basins in turn generate challenges for dividing water between sectors such as rural and urban areas, industries that use water, agriculture and tourism. While many of these flux-related issues are relevant to water governance institutions globally, problems are particularly acute in semi-arid developing countries in Africa, where a particular type of water resource instability exists. This environment should be contrasted with the characteristics of temperate, humid flood-plain river basins of richer developed nations, shown on the left in Fig. 8.1. Typically, in northern Europe, greater stability and predictability are conferred by natural means (temperate/oceanic rainfall patterns, use of groundwater aquifers and low daily evaporation rates) and artificial means (river-training works, storage, piped reticulation, prediction and hydrological information via a network of monitoring stations). This supply-side predictability and stability allows society to monitor rising demands and therefore determines the 'sustainable' gross abstraction of water and hence environmental headroom (Carnell et al., 1999). A regulatory approach to water, providing water rights to users, is achievable under such circumstances. Such a situation is further mollified by the fact that the underlying economy is not irrigation based (the UK uses 2% of fresh water for irrigation (Weatherhead, 2007)) and can invest in less water-intensive activities (e.g. light industry or service sectors), thereby reducing the demand for water.

However, the right side of Fig. 8.1 shows that instability in semi-arid Africa arises from the interplay of combined natural and institutional factors: high climate variability; minimal natural and artificial storage buffering; direct

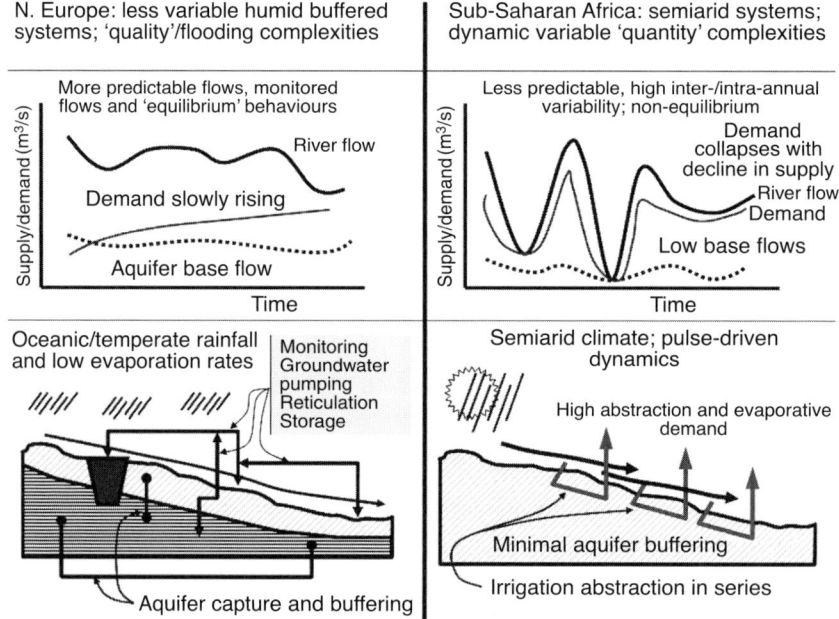

Fig. 8.1. The characterization of equilibrium and non-equilibrium river basin systems.

and immediate access to water for agriculture, often fed by gravity in a series of intakes; and significant abstraction and depletion rates, arising from high evaporative demands coupled with water spreading for irrigation. Here we observe a particular characteristic of such environments, where actual use follows supply closely, sometimes up to 100% of what is available. As water supply declines during the dry season or drought, so does usage, often over several orders of magnitude – in other words, daily demand for one area might vary from 5–10 m^3/s in the main rainfall season to 0.50–0.1 m^3/s in the dry season. In this environment, demand is a function of livelihoods that are immediately dependent on natural resources, with few options for switching to an economy that is less reliant on water. In addition to the large area of potentially irrigable land, this is one reason why potential demand is so high and why usage closely follows supply. Moreover, river flow and rainfall monitoring networks and mechanisms for mitigating or sharing varying and declining resources tend to be weak (Donkor, 2003), which undermines both transparency and predictive and risk-based responses. In these conditions, a normative regulatory approach to river basin management is much more problematic.

Added to this comparative analysis are the three key solutions to managing water sufficiency – supply, demand and allocation (share) management – each taking a part and role during river basin development. As rivers close, and when the fixes of supply-side infrastructural development become increasingly expensive, attention turns to issues of demand management, water conservation and water allocation (Molle, 2003; Molden et al., 2005). Consequently, governments and NGOs, as well as the academic community, seek new and innovative understandings of the governance of demand management and of the means to share limited but varying amounts of water between users. This chapter explores the trajectory of the Ruaha River basin and stresses the challenging specificities of sub-Saharan African environments.

The growth of smallholder irrigated farms in the Usangu plains, in the Great Ruaha basin, from approximately 1000 ha in 1960 to an area of between 20,000 and 40,000 ha in the present day, with an associated rise in water competition, provides three insights on river basin systems and, as a consequence, new entry points for the refinement of irrigation and river basin management. These ideas illustrate related, but separable, issues that inform systems policy. Brief descriptions are given below, and illustrated in Table 8.1 and Figs 8.2 and 8.3. The chapter explores the ideas and their implications for river basin management in greater detail; captured in Fig. 8.3, they illuminate other possibilities related to, and building upon, the S-shaped model of basin development.

Parageoplasia

This term applies to non-local externalities created by upstream water depletion in a river

Table 8.1. Three basin behaviours observed in southern Tanzania.

Idea	Observation	Resulting from	Outcomes	Policy implication
Parageoplastic behaviour	Exported aridity downstream with specific timing, quantity and quality dimensions	Increased area of dry- and/or wet-season irrigation upstream	Altered behaviours and outcomes downstream	Discern parageoplastic links followed by basin or local solutions
Non-equilibrium systems and behaviour	Fluctuating area of wet-season irrigation between upper and lower limits	Climatic and weather variability leading to changes in rainfall and runoff amounts	Supply–demand equation non-linear, complicating allocation	Rethink irrigation planning methods to allow abstraction to mirror runoff flux
Share modification	Uneven proportional division of varying river flows between sectors	Poorly conceived irrigation design and installation of irrigation intakes	Changing inequity of supply between sectors	Remodel or refit irrigation intakes to improve proportional division of river flow

174   B. Lankford et al.

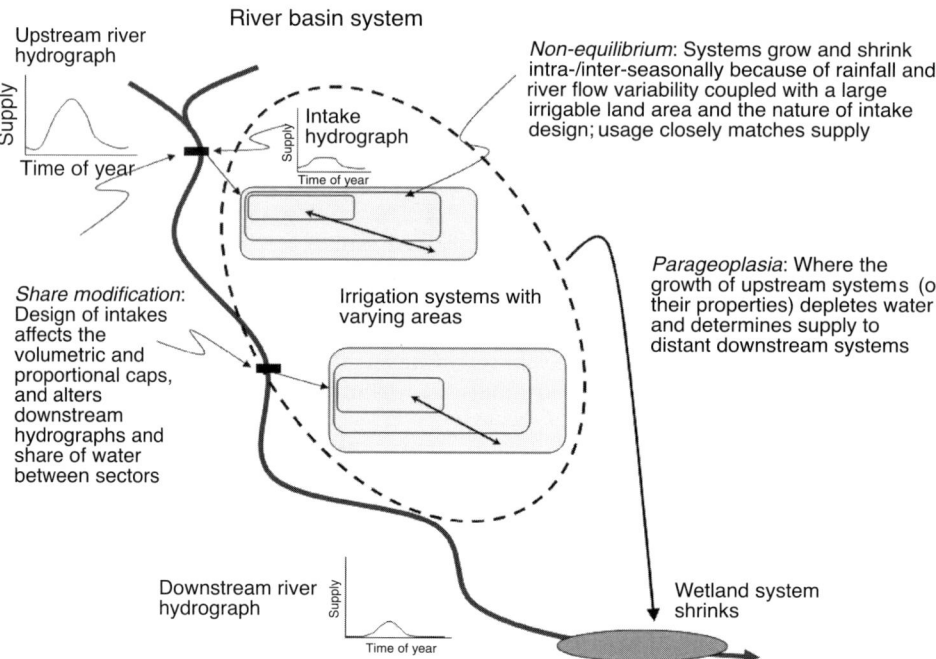

Fig. 8.2. Concepts of basin behaviours resulting from growth of irrigation.

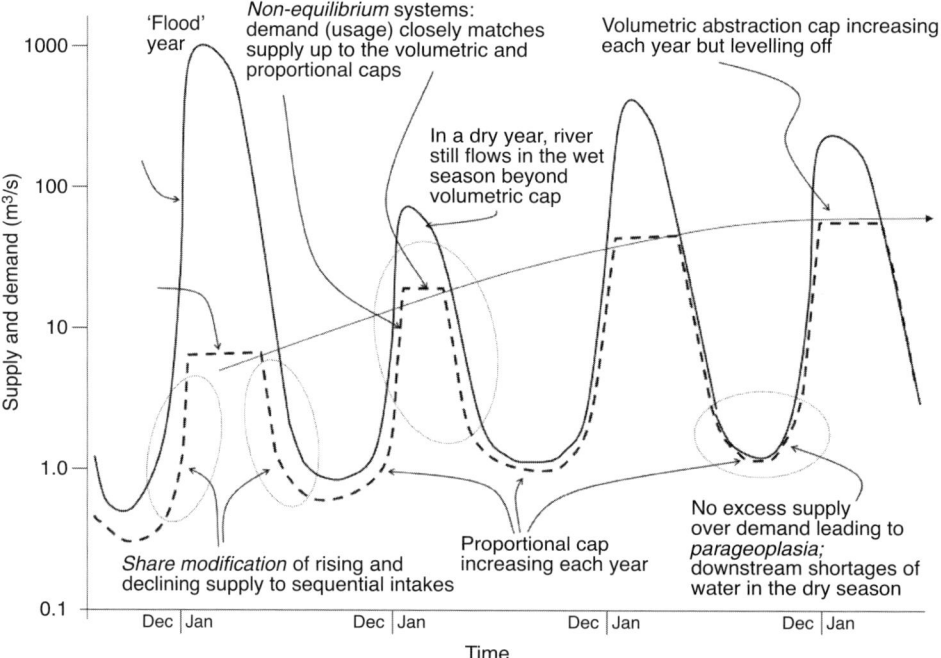

Fig. 8.3. Depiction of basin behaviours via a supply and demand hydrograph.

basin, prompting new behaviours as downstream users react to water shortages. Parageoplasia is captured in Fig. 8.2, where a downstream wetland experiences water shortage. The example in this chapter is of dry-season water shortages in the Ruaha National Park, caused by irrigation abstraction 100 km further upstream. Parageoplasia is defined as depletion or usage causing external symptoms of water shortages in a locality elsewhere in the basin.

Non-equilibrium behaviour

This is observed when demand closely follows and matches supply intra- and inter-seasonally. Figure 8.3 shows this as the demand (dotted) line rising and falling in line with the supply line. This occurs in southern Tanzania because the irrigated area rapidly increases to approximately 40,000 ha in a wet season (with normal rainfall) from about 5000 ha in the dry season (Lankford and Beale, 2007). By contrast, an 'equilibrium' situation might be characterized as one where demand is more restricted, so that an enlarged supply cascades a surplus to downstream users (or, in other words, where demand increases by a maximum of 50–100%, rather than 800% in the non-equilibrium case). Figure 8.3 demonstrates a rising trend of increased wet-season demand over time – notice the volumetric cap increases each year as more intakes are developed or modernized; the figure also shows that the area between the solid and dotted lines diminishes over time, indicating that the proportional abstraction cap increases with time, resulting in less water passing downstream (see Lankford and Mwaruvanda, 2007).

Share modification

This describes purposive or inadvertent changes in shares of water between sectors and/or users in the face of a declining or increasing flow rate resulting from existing or redesigned (new) river flow division infrastructures. Modification of shares is particularly prevalent with conventional designs of irrigation intake infrastructures combined with highly varying flows. On the other hand, proportional designs of river infrastructures help to reproduce the shape of the river flow curve proportionally between the offtaking canal and the downstream section of river.

In summary, these phenomena are realized through the evolving trajectory of the case study basin via three main facets: (i) the growth of irrigation area and demand over time; (ii) the presence of a variable sub-Saharan climate; and (iii) a combined effect of both the choice (intentional or otherwise) and density of infrastructure technology mediating the share of water between sectors.

Study Area and Background

Water resources and location

Tanzania faces perceived (and sometimes real) water scarcity problems at local levels despite the fact that, on average, it has abundant water resources to meet most of its present needs. However, while a third of these resources lie in highland areas, with precipitation in excess of 1000 mm, about one-third of Tanzania is arid or semi-arid, with rainfall below 800 mm. The major river systems constitute the principal surface water resources of the country, with mean annual runoff of about 83 billion m^3 and an estimated groundwater recharge of 3.7 billion m^3. Half of the surface runoff flows into the Indian Ocean from the Pangani, Wami, Ruvu, Rufiji, Ruvuma, Mbwemkuru and Matandu river systems. The remainder drains northward, into Lake Victoria, westward, into Lake Tanganyika, and southward, into Lake Nyasa. Some of the runoff also flows into internal drainage basins with no sea outlets. These include the Lake Rukwa and central Internal Drainage basins.

However, greater demand for water for irrigation and the long dry season (June to October) result in low river flows and seasonal scarcity (World Bank, 1996). As evidenced by the case study in this chapter, this has resulted in conflicts between hydropower and irrigation sectors, between irrigation and livestock sectors, and between upstream and downstream water users within the irrigation sector. Tanzania also lacks the economic resources to harness water and to overcome the extreme temporal and spatial variability in rainfall and surface flow.

The Great Ruaha River catchment (GRRC) is located in south-west Tanzania (Fig. 8.4). It has a catchment area of 83,979 km² and a population of 480,000, according to the 2002 national population census (TNW, 2003). Headwaters rise in mountains to the south, in the Poroto and Kipengere ranges, and drain onto the alluvial Usangu plains. The catchment can be divided into three major agro-ecological zones, which have different characteristics. The upper zone (1400–2500 masl) is semi-humid to humid, highly populated and has high rainfall, deep soils and intensive agricultural production. In this zone, both rainfed and irrigated agriculture is practised all year round. The intermediate middle zone (1160–1400 masl) is characterized by a high concentration of irrigation systems on alluvial fans, and here a limited presence of dry-season irrigated agriculture is an important means of livelihood. Therefore, this is an area of high competitive water demand and hence persistent water conflicts.

The lower zone (1000–1160 masl) is semi-arid with alluvial soils, with a low population density and a high concentration of livestock, particularly cattle. Here, the Great Ruaha River (GRR) and other tributaries pass through seasonally inundated grassland and permanent swamps, which are ecologically significant, supporting a considerable biodiversity, notably its extremely high bird-life diversity (SMUWC, 2001). The GRR discharges from the northern end of the plains at NG'iriama, an outlet of the permanent Ihefu swamp. The catchment area at this point is 21,500 km², and is commonly termed the Usangu basin, synonymous with the upper Great Ruaha River catchment (UGRRC). About 30 km further north, the river passes through the Ruaha National Park, and from there further north-east to Mtera and Kidatu reservoirs. During the dry season, from July to November, the river is the major source of water for much of the wildlife in this park.

As is the case in most of sub-Saharan Africa, the livelihoods of the majority of people in the Great Ruaha River catchment are largely dependent on agriculture. However, the area is characterized by high variability (with an average annual coefficient of variation of 24%), uncertainty, and poor and uneven distribution of rainfall during the crop-growing seasons (SMUWC, 2001; Rajabu et al., 2005). Despite the fact that the rainfall regime is unimodal, with a single rainy season (with a mean annual

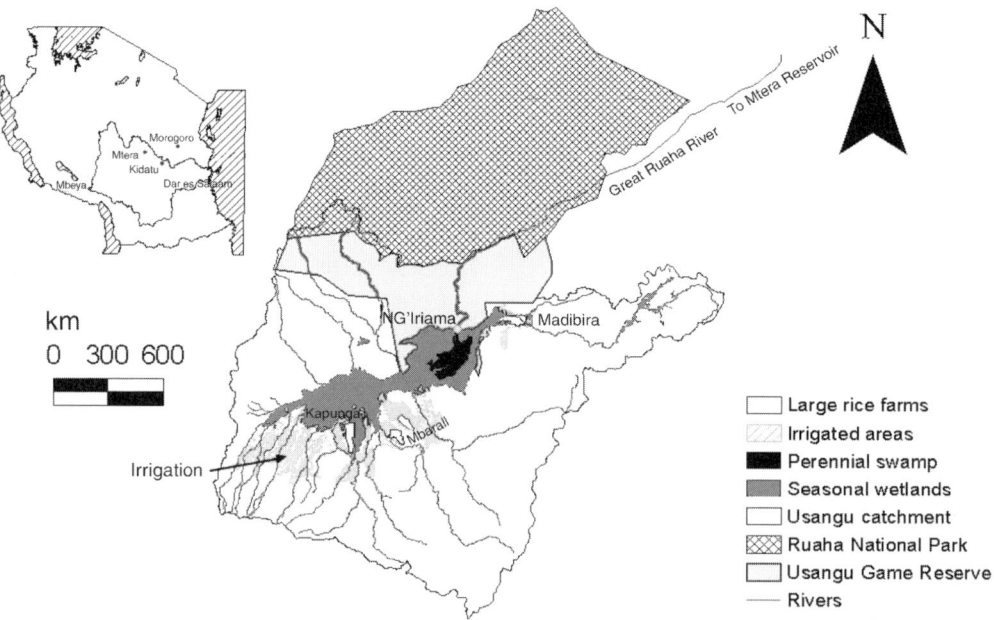

Fig. 8.4. Map of the Usangu basin within the Great Ruaha River basin.

areal rainfall over the UGRRC of 959 mm), the onset and duration of the rains vary from zone to zone and both are unpredictable in timing. Whereas the rainy season for the upper zone (highlands) runs from October to May, the rainy season for the middle and lower (the plains) zones runs between November and April. Of particular consequence for cropping on the plains is the fact that rainfall is between 500 and 700 mm on average, a marginal amount for rainfed maize, and necessitating supplementary irrigation for rice production.

Further analyses of the causes of hydrological changes and background to the area can be read in a number of additional articles (van Koppen et al., 2004; Lankford et al., 2007; McCartney et al., 2008), while additional information on the prevailing political and institutional context can be found in Lankford et al. (2004).

Farming systems and water users

As a strategy to cope with the uncertainty and poor distribution of rainfall during the crop-growing season, the local farming systems in the UGRRC have constructed diversions to abstract water from rivers for supplementary irrigation in order to minimize risks of crop failure. There are three types of irrigation systems, which are:

1. Traditional systems, which comprise village irrigation, based on the diversion of perennial or seasonal flows, used mainly for the production of rice, vegetables and other relatively high-value crops. These are self-sustaining systems, initiated, financed, developed and owned by the farmers themselves, without any external assistance.
2. Improved traditional systems are traditional systems that have received government- or donor-assisted interventions to improve the headworks and water control structures, and, on occasion, farmer training.
3. Modern large-scale schemes that comprise large-scale farms (such as Kapunga, Mbarali and Madibira rice farms) built with the aid of international finance.

In nearly all of these systems, basin irrigated rice (paddy) is grown, to the extent that the Usangu basin contributes about 15% of the rice production in Tanzania and supports the livelihoods of about 30,000 poor households in Usangu (Kadigi et al., 2003).

Below the irrigation systems are the seasonal wetlands of the Usangu plains, containing the permanent wetland of the Ihefu, an area of about 80–120 km^2. The seasonal and permanent wetlands once contained significant numbers of fisherfolk and livestock keepers, but following their forcible removal by government authorities as a result of the formalization (gazetting) of the Usangu Game Reserve, these numbers have been greatly reduced. An examination of the contribution to the local economy is conducted and implications of this intervention are described below.

Further downstream, the total power-generating capacity of the Mtera and Kidatu plants is 284 MW, which is 51% of the total hydropower capacity of Tanzania (TANESCO, 2008). A fuller history of this hydropower development is given below, along with an analysis of the water management of the two dams.

After Mtera, the Kilombero Sugar Company abstracts water from the river for irrigation and cane processing. The company is located in the flat, fertile areas at the base of the Udzungwa mountains in the Msolwa and Lower Ruembe valleys in the Morogoro region of Tanzania. The mean annual rainfall in this humid region is 1347 mm, although moisture deficits are evident from June to December. Thus while crop moisture requirements are generally satisfied by rainfall between the months of January and April, irrigation is required to maximize growth during the remainder of the year and to allow planting operations to take place in the dry months. The sugar company has a year-round water right of 8.5 m^3/s from the Great Ruaha River.

Hence, six main river water users from upstream to downstream can be identified: domestic water users, in the high catchment and plains; irrigators, mainly on the plains; pastoralists and fisherfolk, in the seasonal wetlands and the Ihefu; wildlife and tourists, in the Ruaha National Park; electricity producers, at the Mtera and Kidatu power plants; and sugarcane producers.

For the analytical purposes of this chapter, these water users are divided into two main high user groupings: irrigated users and downstream users, split on the basis of level of abstraction of water into a first group of irrigation systems on the plains (mostly rice growers), and a second group comprising water users downstream of the main irrigation area on the plains (fisherfolk and wildlife in the wetlands, tourists and wildlife in the national park, and power generators). There are domestic and irrigation water users upstream of the plains and irrigators in the mountain watersheds, but these are minor in extent and quantity of water use, given higher rainfall and lower evaporation rates at these altitudes. These users are not shown in Fig. 8.4.

Water accounting[2]

Utilizing the water-accounting methodology of the International Water Management Institute (IWMI) (Molden, 1997; Molden et al., 2001), we have generated a 'finger diagram' of water flows for a normal-to-wet hydrological year in the Usangu basin (Fig. 8.5). It should be recognized that the non-linear behaviour of the catchment, with variable surface areas of irrigation, wetlands and storage by the Mtera hydroelectric dam/reservoir, imply a highly variable model of water flows and partitioning. The finger diagram (Fig. 8.5) should not be interpreted as a static model of water apportionment. The key features are as follows:

1. The calculations represent surface flows only.[3] Catchment precipitation and green water evapotranspiration are not included. With regard to losses in groundwater, studies by the Sustainable Management of the Usangu Wetland and its Catchment (SMUWC) project and observations on the ground show that water losses of about 10% occur when rivers transit the geological fault-line of the East African Rift Valley from the high catchment to the plains. While this water supports perennial flows in the Mkoji subcatchment and some domestic use elsewhere, little of it creates a water table that can be used for substantial irrigation withdrawals or flow augmentation. The Usangu plains are typical African savannah plains rather than flood plains in the Asian sense. Thus groundwater losses are shown as losses from the gross inflow rate.

2. Two types of beneficial depletion occur: non-process (not intended), via evaporation of water from the wetland, and process, via net irrigation demand, and domestic and livestock

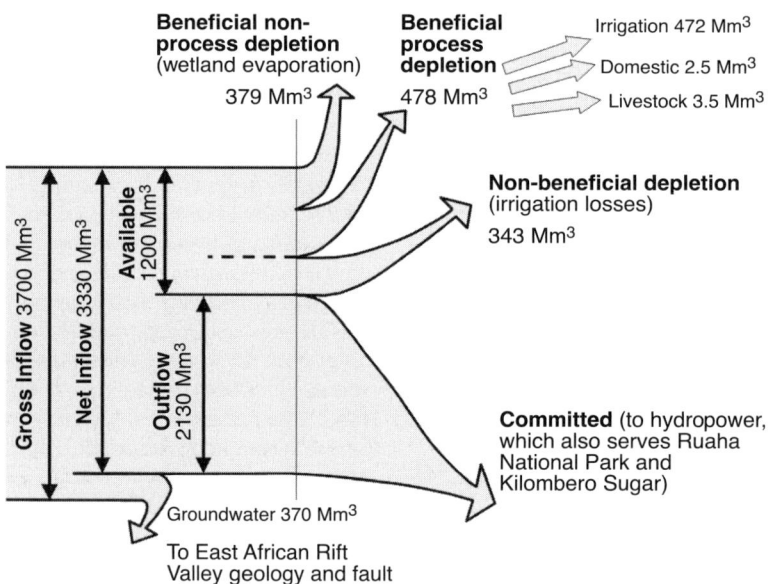

Fig. 8.5. Surface water accounting of the Great Ruaha River basin.

use. Irrigation losses represent the principal non-beneficial depletion (occurring mainly through non-recovered losses to groundwater and unproductive evapotranspiration). Livestock usage relates only to calculations of drinking water – note that green (soil) water is not calculated. These rates are shown in Fig. 8.5.

3. The fourth flow is a committed outflow to provide storage in the Mtera reservoir for evaporation and discharge through the Mtera turbines, which annually have a potential useful power-generating requirement of a flow of 96 m^3/s. This discharge flow and the dam evaporation combine to establish approximately 3800 Mm^3 annually. Hydrological analyses show that 56% of this is contributed by the upper Great Ruaha catchment, approximately 2130 Mm^3. This value very closely corresponds to the determination of the outflow of 2130 Mm^3 at the exit of the Usangu wetland (in other words the surplus water to that utilized in the UGRRC). This demonstrates the analysis by Yawson et al. (2003) that, during an average hydrological year, flows to hydropower storage are sufficient to meet generating needs, despite the common assertion that upstream irrigation is in direct competition with hydropower (Kikula et al., 1996; Mtahiko et al., 2006[4]).

Introduction to policy stakeholders

In addition to the users mentioned in the previous section, throughout this chapter a number of key stakeholder groups are discussed, many of whom have converged and overlapped in influencing policy and providing supporting services to water management in the basin. They are briefly introduced here.

In 1996 (partly as a response to hydropower electricity power cuts during the mid-nineties), perceiving water resources management in Tanzania to be hampered by uncoordinated planning for water use, incomplete policies, inadequate water resources data and inefficient water use, the government of Tanzania, with the assistance of the World Bank (1996), initiated a sustained programme of reform. Tanzania adopted a river basin management approach for water resources management, in which the country was divided into nine river basins for water resources administration. These are Pangani River basin, Wami/Ruvu River basin, Rufiji River basin, Ruvuma River basin, Lake Nyasa basin, the Internal Drainage basin, Lake Rukwa basin, Lake Tanganyika basin and Lake Victoria basin. To manage each of these basins, a basin water office was created. The main activities of the basin water offices include: (i) regulating, monitoring and policing of water use in the basin; (ii) issuing formal water rights; (iii) facilitating and assisting in the formation of water user associations; (iv) billing and collection of water user fees; (v) awareness creation of water users regarding water resources management; and (vi) monitoring and control of water pollution (NORPLAN, 2000; Mutayoba, 2002).

A substantial programme of reform, centred on two pilot basins, the Pangani and the Rufiji, was implemented through the decade from the mid-1990s onwards, through the River Basin Management and Smallholder Irrigation Improvement Project (RBMSIIP), via a loan of US$21 million.[5] The smallholder component of RBMSIIP was deployed principally via the local district council (Mbarali), with significant assistance from the zonal irrigation office, located in Mbeya, and central support from the Ministry of Agriculture in Dar es Salaam.

In the late 1990s, the UK's Department for International Development (DfID) assisted RBMSIIP via a technical assistance project implemented by consultants. The project, SMUWC,[2] determined the cause of the hydrological changes in the GRRC and contributed to the development of water strategies that could be applied in other basins with wetlands in Tanzania. Despite its significant scientific findings, and also incorporating stakeholders, the project was discontinued in 2001, when DfID switched to development assistance via budget support. In recognition of this break, the Knowledge and Research division of DfID (KaR), with the assistance of the IWMI, funded a small project from 2001 to 2005, termed RIPARWIN (Raising Irrigation Productivity and Releasing Water for Intersectoral Needs),[2] and designed to complete some of the studies started by SMUWC.

From 2000 onwards, an increasingly important role has been taken by the World Wildlife Fund (WWF), which has culminated in its ongoing project, the Ruaha Water Programme. In addition, the environmental group 'Friends of Ruaha'[6] has played a number of political advocacy roles in drawing attention to the consequences of water management.

The Mbarali District Council also was a key player. Despite a counter-productive effect on meat revenues, Mbarali district (almost synonymous with the Usangu plains, see Fig. 8.4) was a key advocate of gazetting the Usangu Game Reserve. Furthermore, because of the council's developmental concerns, manifested by support for irrigation, it sought to diminish the conflicts between rice growers and cattle keepers by removing the latter and by siding with the mainstream governmental view that the river should be restored to year-round flow through the construction of improved intakes (also a counter-productive move for reasons explained elsewhere in the chapter).

Historical Trends and Changes in the Basin

As Table 8.2 testifies, the upper Great Ruaha basin has seen many changes over the last 50 years or so, mostly related to population increases associated with greater utilization of natural resources. Associated with this have been major land-use changes. The natural vegetation of the alluvial fans has been largely cleared and replaced with rainfed and irrigated cultivation and grazing areas. Other events listed in Table 8.2 are discussed below and elsewhere in the chapter.

Growth in population, livestock and irrigated area

Between 1950 and 2003, the population in the UGRRC increased from less than 50,000 to approximately 480,000 (TNW, 2003), largely through in-migration from other regions of Tanzania. This growth has also been mirrored in the expansion of the largest urban conurbation, Mbeya, just outside of the catchment in the south-west.

In the plains, most people are farmers, cultivating rainfed and irrigated plots, but a smaller number are pastoralists, who have brought more cattle into the plains. Livestock numbers also increased, although these probably peaked

Table 8.2. Summary of historical events occurring in the upper Great Ruaha catchment.

Period	Events and notes
1935–1967	Pristine condition, pre-El Niño flood event in 1968. Estimated total area of rice reported in 1958 was 3000 ha, at end of 1967 = approx 10,000 ha
1962	Kilombero Sugar Company first factory commissioned
1969–1973	Estimated total area of rice at end of 1973 was approximately 14,000 ha
1970	Kidatu dam constructed (100 MW), with another 100 MW added in 1976
1972	Mbarali rice farm constructed
1974–1985	Post-Mbarali, pre-expansion in rice. Estimated total area of rice at end of 1985 = approximately 25,000 ha
1978	Hazelwood and Livingstone report filed
1980	Mtera dam completed and started to fill
1986–1991	Expansion in rice, pre-construction of Kapunga scheme
1992	Kapunga is constructed; weirs across Chimala river
1992–2000	Post-Kapunga and Chimala river changes, continued expansion of rice, construction of upgraded intakes, introduction of widespread dry-season irrigation, Madibira constructed in 1998. Estimated total area of rice at end of 1999 = approximately 40,000 ha
1996	RBMSIIP project, which was the forerunner to the wider Water Sector Support Project with funding from 2007 to 2012 (both World Bank funding)
1999–2001	SMUWC project (DfID funding)
2001–2005	RIPARWIN project (DfID funding)

in the early 1980s, at around 550,000. In 2000, the number of cattle was estimated at around 300,000 head, with about 85,000 other livestock (SMUWC, 2001). The pastoralists moved into the Usangu catchment in search of pastures, following long periods of drought or competition over resources in their home villages. The areas include central and northern areas of the country, namely Dodoma, Singida, Shinyanga and Arusha regions, although commonly they are known collectively as the Sukuma. The numbers of cattle, goats, donkeys and sheep in the catchment has been a source of scientific debate for the last 10 years. While regional authorities proffered a figure of one million cattle (largely to support arguments that the plains were being degraded by overstocking), various study reports give different levels of the stock in the catchment. The livestock census conducted in 1984 showed for Mbarali district a herd of about 513,600 animals, of which 438,000 were cattle (SMUWC, 2001).

During this 50-year period, the area irrigated in the wet season has increased from approximately 3000 ha to 40,000–44,000 ha (Fig. 8.6), although the area varies significantly from year to year, depending on rainfall. In dry and wet years, the total area can swing from 20,000 ha to 40,000 ha, respectively. It is this growth in area that has led to increased competition and conflict over water, particularly in the dry season, and has led to the emergence of the three behaviours seen and characterized in this chapter.

The bar line in Fig. 8.6 indicates the extent of variability in the area under cultivation from wet to dry years (SMUWC, 2001, adapted).

Environmental changes downstream

Many of the environmental changes in the area were associated with this growth of irrigation; however, most publicly noted has been rapid hydrological change. This is testified by visible changes in the flow of the major river draining the plains. The Great Ruaha River used to be perennial – river flow lasted throughout the dry season. However, since the early 1990s, the discharge through the Ruaha National Park has altered, becoming seasonal, with flows ceasing during part of the dry season. This cessation is explained by water levels in the eastern wetland dropping below the crest of the rock outcrop at NG'iriama (see Fig. 8.4), resulting in the wetland being unable to feed the river downstream. An analysis of flows measured at Msembe Ferry, a gauging station located approximately 80 km downstream of NG'iriama, indicated an increasing frequency and extension of zero flow periods between 1990 and 2004 (Kashaigili et al., 2006) of between 15 and 100 days, depending on rainfall and upstream abstraction, with no discernible upward or downward trend during that time. Coinciding with low flows in the mid-1990s were a series of electricity power cuts from Mtera and Kidatu, fuelling speculation that upstream irrigation was depleting water destined for downstream ecological and economic purposes.

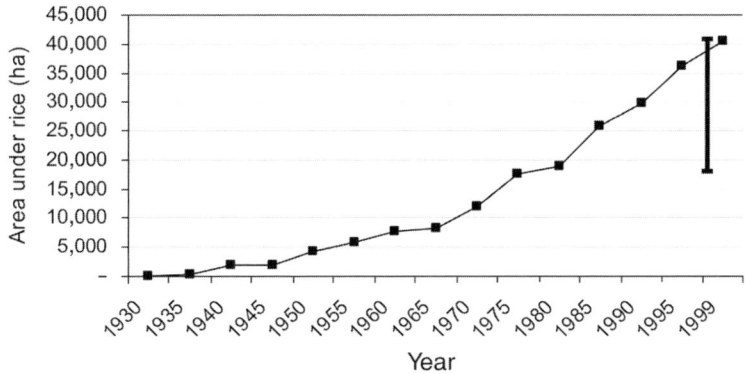

Fig. 8.6. Growth in irrigated area surrounding the Usangu catchment.

Other environmental changes include an encroachment of cultivation into the wetland and a marked decline in wildlife species – most striking of all is the replacement of wildlife herds by cattle. The combination of cultivation and grazing has resulted in a reduction of grass species and a concomitant rise in woody shrub species, which otherwise would have been kept at bay by natural flooding and grassland fires; both suggest a progressive degradation of the alluvial fans and plains. In the highlands, changes have perhaps been less dramatic. However, ever-increasing areas have been, and are still being, converted to cultivation and settlement; erosion on steep slopes is advanced in places; and even where the woodland is relatively intact, it has been exploited for the important timber species.

An analysis of declining dry-season flows and wetland area shows that between 1958 and 2004 the dry-season minimum area decreased significantly, but there was no clear trend in the wet-season maximum area. Overall, the dry-season minimum area was found to have decreased from an average of about 160 km^2 (1958–1973) to approximately 93 km^2 (1986–2004), i.e. a proportional decrease of approximately 40% (Kashaigili et al., 2006). Average dry-season inflow to the Usangu wetland (the Ihefu) between 1986 and 2004 was estimated to be 76 Mm^3, compared with 200 Mm^3 between 1958 and 1973. Although rainfall over these two periods was not exactly the same, this nevertheless indicates a reduction of dry-season flows of approximately 60%, and in some months (e.g. September and October) the reduction was closer to 70% (Kashaigili et al., 2006). However, these data cover the period when the gate closure programme was coming into effect and so slightly underestimate historic water withdrawals. Flow measurements made by the SMUWC project at the end of the dry season in 1999 found that 91% of upland flow was being abstracted and, overall, it was estimated that, on average, 85% was being withdrawn in low-flow months (SMUWC, 2001). More recent studies conducted in 2003 and 2004 in the Mkoji subcatchment, the most heavily utilized for irrigation, continue to show dry-season abstraction levels in excess of 90% on some rivers (Rajabu et al., 2005).

Mtera–Kidatu hydropower

The presence of nearly 50% of Tanzania's electricity generation downstream of the upper Great Ruaha catchment has imposed a particular character to the debates and narratives about water development and management in the basin, and thus we provide here a historical background to the development of hydroelectricity.

In response to growing electricity demand, the decision to construct the Kidatu hydropower station was taken by the government of Tanzania in 1969. The 204 MW Kidatu dam and hydropower plant was the first phase of the Great Ruaha Power Project, funded via loan agreements between the Tanzanian government, the Tanzania Electric Supply Company, the Swedish government and the World Bank, with the Swedish company SWECO as energy consultant. As the demand for electricity further increased, phase two of the Great Ruaha Power Project was considered by constructing a dam at Mtera. The government agreed to the proposal, as the purpose of this reservoir was essentially as an upstream reservoir to ensure there would be sufficient water reserved throughout the year, and especially during the dry season, to supply Kidatu. By December 1980, Mtera dam was completed. Following further consultations, it was proposed that another smaller power station of 80 MW should be built at Mtera, an addition not originally foreseen in the planning of the 1960s. The water stored in the reservoir would generate power before flowing downstream to Kidatu to generate 204 MW of power again. The 80 MW Mtera power station became Phase III of the Great Ruaha Power Project and started operating in 1989.

Mtera has a total storage of about 3600 Mm^3 and a live storage of 3200 Mm^3, when, at the maximum (full) supply, the water level is 698.50 masl. The minimum water supply level allowed for normal power generation is 690.00 masl. Below this level, down to the bottom at 686.00 masl, is a 'dead storage' volume of about 500 Mm^3 of water, which may be used only when there is an emergency such as a national power crisis. Although SWECO's report indicated that the water in 'dead storage' could be used during emergencies, it added that emptying the reservoir below 690.00 masl would have adverse effects on the ecosystems that had developed in and around the dam. The reservoir-operation simulation conducted by SWECO in 1964 illustrates that about 25% of the inflow into the reservoir was lost by evaporation because of the ratio of the very large surface area to the volume.

Irrigation governance narratives

Associated with changes in the basin are narratives regarding irrigation development and governance. There is not enough space here to deal with a wider treatment of the Tanzanian political economy in a post-colonial era, particularly the agrarian impacts of the socialist government of Julius Nyerere arising from villagization and farming collectivization. Instead we concern ourselves with two narratives that pertain to irrigation and basin development: first, agricultural growth and modernization from 1960 to 1990 and then, linked to it, a narrative of efficiency, environmentalism and water reallocation during the period 1995–2005. The former spans the period in which water and land were seen to be abundant, while the latter drew from perceptions regarding a finite supply of water and concerns over power cuts, described in the previous section.

1960s to 1990 – expansion and modernization of irrigated agriculture

The contemporary tension between the two agendas of developmental modernization and environmental protection can be traced to government intentions from 1960 to the 1980s to utilize the water resources of the upper Great Ruaha for irrigation. The key development projects of the formal, state-run irrigation schemes of Mbarali (1972) and Kapunga (1992), plus the concerted efforts to 'improve' traditional intakes, can be traced to the 1978 Hazelwood and Livingstone study of the economic options available to the government of Tanzania in developing the Usangu plains (Hazelwood and Livingstone, 1978a), commissioned by the Commonwealth Fund for Technical Cooperation (CFTC). The request came as a 'pre-feasibility study with the aim of elucidating the nature of development problems of the plains, determining the appropriate pattern of development, assessing the potential for development and identifying projects for detailed feasibility study' (Hazelwood and Livingstone, 1978a: vol. 3). The objective is stated as 'to assess the potential of Usangu for development and for contributing to national economic goals', while it also says 'that its total programme should be seen as a long term plan for the eventual full exploitation of the resources of Usangu' (Hazelwood and Livingstone, 1978a). The ongoing concerns in the 1960s and 1970s with generating economic growth in the region, typified by the study by Hazelwood and Livingstone, were heralded in 1961 by the FAO Rufiji basin study (FAO, 1961) and a US Bureau of Reclamation (USBR, 1967) study offering similar visions of large-scale irrigation development, limited only by water availability and labour, and unencumbered by economic, social or environmental constraints (Palmer-Jones and Lankford, 2005).

Although the formal schemes for Mbarali and Kapunga amount to a total of 6800 ha, there can be no doubt that the Hazelwood and Livingstone work stimulated further developments in the region. Some are directly attributable to this work: for example, prior to 1978, 16 intakes of informal schemes were concrete but since then an estimated 40 intakes have been upgraded by a variety of donors, including the government of China, JICA, the World Bank and FAO. This probably allowed an additional 10,000 ha of rice to be cultivated, and is certainly one major reason for the growth of irrigation from 17,500 ha, recorded in 1978, to nearly 40,000 ha, recorded by SMUWC in 2000. This hectarage makes Usangu one of the single most significant rice-producing areas in Tanzania, contributing 15% of the national total (Kadigi et al., 2003). Other major projects were followed through: the Madibira scheme (3000 ha) was directly supported by Hazelwood and Livingstone and saw its first irrigated planting in 1999/2000. Overall, the development of natural resources has sustained very high population growth in the Mbarali district, with 4–5% annual growth rates.

1990s onwards – irrigation efficiency, environmentalism and allocation

Irrigation efficiency is of significant importance in the discourse on irrigation and river basin management in Tanzania, and since the mid-1990s it has been at the heart of attempts to reallocate water downstream to meet hydropower and wetland water requirements. Raising water-use efficiency was the key rationale for the River Basin Management and Smallholder Irrigation Improvement Project, initiated in the RBMSIIP project funded by the World Bank (World Bank, 1996).

Setting aside the incorrect claims for upstream water originating from powerful interests allied to power generation (as serious though that may be), the economic return on the US$22 million loan to the government of Tanzania was predicated upon the argument that water saved in irrigation through raising efficiency would pass through the turbines at Mtera/Kidatu, generating considerable financial and economic benefits. The single tenet underlying gains in efficiency was that if traditional intakes were improved by the use of a sluice gate, set in concrete headworks, this would give control over abstraction and thus reduce the volume taken into irrigation systems during wet periods. The project also matched intake improvements with 'demand management' through the selling of water rights, as this would regulate upstream demand and send more water downstream.

Interestingly, this discourse was initiated in the 1970s when Hazelwood and Livingstone explored differences between the Mbarali system (perceived to be modern and to have adjustable headworks control) and traditional farmers who employed traditional intakes made of local materials (Lankford, 2004a). Hazelwood and Livingstone (1978b:207) demonstrate prevailing views regarding the waste of water by smallholders:

> The possibility exists of controlling agricultural practices of peasants particularly at the time at which they plant, because an efficient irrigation system requires a considerable degree of water management. It is true that in the area with which we are dealing the limited peasant irrigated cultivation that at present takes place uses irrigation constructions which are largely unplanned and not professionally designed, and for which there is effectively no control or administration of the distribution of water. But this system is very wasteful in its use of water, it is also wasteful of land because cultivable areas are lost through flooding, and it is inequitable in its allocation of water between individual farmers.

The contribution of Hazelwood and Livingstone to this debate should not be underestimated. By publishing figures early on, they affected, perhaps even underwrote, the present-day view that smallholders are less efficient than larger-scale farmers (JICA, 2001; Kalinga et al., 2001). The case study in Usangu provides an example of the errors in scientific understanding of irrigation efficiency. The RBMSIIP was based on the premise that the project could raise efficiency from 15 to 30%, allowing substantial reallocation of water, as the quote below from the appraisal report explains, and that this would be achieved by improving intakes, selling volumetric water rights and training farmers.

> In order to illustrate this effect, the 'savings' in water which result from the improvement of some 7000 ha of traditional irrigated area under the project (this includes both basins) are valued using their capacity to generate electricity in the downstream turbines. An average 'in the field' requirement of 8000 m^3 of water, for one ha of rice production, implies withdrawal of 53,300 m^3 from the river, with an irrigation efficiency of 15 percent. Following improvements in irrigation infrastructure and an increase in irrigation efficiency to 30 percent, the withdrawal requirement from the river drops to 26,700 m^3 per hectare. This releases some 26,700 m^3 for every hectare of improved irrigation, to be used for hydropower generation downstream. For this exercise, the water is valued at US 5 cents per m^3, the valuation for residential electricity use (34 percent of all electricity use, an intermediate point between the two alternate values)
> (World Bank, 1996:42).

Yet closer measurement indicates that effective efficiency was probably in the region of 45–65%, precisely because of reuse of drain water by tail-enders (Machibya, 2003). The erroneous assumptions contained in this quote are that: (i) the efficiency was very low; (ii) the losses were depleted from the basin; (iii) improving intakes would reduce losses; and (iv) savings would automatically move downstream to the hydropower reservoirs. The failure to ground-truth some of these assumptions is evident in that the project went ahead as planned.

The fact that the RBMSIIP programme sought to increase efficiency by upgrading intakes rather than by tackling in-field water management is indicative of the viewpoint of Hazelwood and Livingstone that it is the lack of control at the headworks river intake that reduces efficiency. This understanding fails to recognize that farmers use high flows to cascade water through their system, expanding the cultivation area at tail-end reaches, which in turn places an efficiency emphasis on cascade management rather than what is happening at a single point on the river intake.

Environmental governance stakeholders and impacts

Arguably, the upper Ruaha has become a cause célèbre for a number of individuals and organizations. Foremost has been the interest shown by WWF, an international NGO in the restoration of year-round flows via the establishment of its Ruaha Water Programme. This programme has been working closely with local stakeholders to improve water management, with the aim of returning the river to year-round flow by 2010. It is also thought that WWF successfully obtained high-level support for environmental interventions by the government of Tanzania, manifested by the promise by former President Sumaye (speaking at the Rio +10 preparatory meeting, 6 March 2001, London) to re-establish 'year-round flow' by 2010.

The government of Tanzania, via the Ministry of Tourism and Natural Resources (which also manages the Ruaha National Park), agitated for the gazetting of the Usangu wetland and surrounding plains into a Game Reserve, thereby legitimizing the removal of human inhabitants from the area (Moirana and Nahonyo, 1996). Thus, in March 2006, the government, through the office of the vice president, issued a statement declaring to evict pastoralists and agro-pastoral and smallholder communities from the Usangu catchment and Kilombero valley in Mbarali and Kilombero districts, respectively (PINGOs, 2006). The reasons put forward mainly included, *inter alia*, environmental degradation as a result of overstocking beyond the carrying capacity, land-use conflict between different user groups, and poor agricultural and irrigation techniques. The statement further pinpointed issues of scarcity of water flows in the Ruaha River and subsequent low water levels at the Mtera dam (low hydropower productivity). Omitted from these reasons were the perceived territorial advantages of drawing the wetland and plains into the larger Ruaha National Park and the financial gains to the government via the licensing of game hunting.

In the period from May 2006 to May 2007, large numbers of Sukuma agro-pastoralists and Taturu and Barabaig pastoralists and their livestock were evicted from the Usangu plains in the Mbarali district, Mbeya region (IWGIA, 2008). It is reported that most have now moved to Kilwa and Lindi districts. It is estimated that more than 400 families and 300,000 livestock were involved in this move, and that a large number of livestock died or were lost in the process. The same action was taken against the fisherfolk of the Usangu wetland, including the impounding of bicycles and other belongings. Although some surreptitiously remain, most have returned to their villages and fields, dispersed throughout the Usangu basin.

This action has potentially reversed two opportunities for the management and sharing of environmental services and benefits. The first is that taxes on livestock and meat sales through the Mbarali town livestock market generated an estimated 52% of district council income in 1998 (livestock taxes generated US$0.2 million; SMUWC, 2001). Then, as now, there appears to be no contingency plan in place to suggest how such an income foregone might be compensated for.

Second, the removal of wetland livestock keepers and fisherfolk precludes the establishment of a co-management plan for the Usangu wetland. Such a plan could have allowed local people to stay in the area in return for channelling and directing water flows through the wetland in order to ensure a small dry-season flow at the exit of the wetland. Calculations show that an exit flow of 0.5 m³/s could be generated by a reduction in the dry-season wetland area of approximately 10% (McCartney et al., 2007). A co-management plan would then generate environmental benefits for the district council, the Ruaha National Park and local people. Although this idea has been proposed to local stakeholders since the year 2000, sadly there has been little sign of its uptake.

Summary

Thus, in summary, the upper Great Ruaha has experienced new and changed 'drivers' of water abstraction: increasing area of irrigation in both wet and dry seasons, a rising number of irrigation intakes, and a shift in the design of irrigation intakes from traditional to an 'improved' (but conventionally designed) intake. This has led to a variety of symptoms of problematic water sharing, declining downstream flows and a rise in competition over water. Associated with these trends have been a number of governmental and non-governmental interests in the region, which, among other discourses of natural resource governance, focused on interventions that first helped to drive up water abstraction from rivers for irrigation and, second, attempted to redress the balance of supply between agriculture and downstream needs.

Interactions and Competition

Introduction

In this section, we explain some of the other interactions and conflicts found in the upper Great Ruaha, taking the opportunity also to explore the political construction of upstream scarcity to explain electricity shortages, and to briefly outline the three concepts that appear to be central to understanding how the basin might be managed.

Hydropower claims for upstream water

Here, we explore the 'water scarcity' claims by the representatives and allies of Mtera–Kidatu of overuse of upstream water. A series of analyses demonstrates that despite claims by power-generation authorities, the power cuts experienced from 1992 onwards were largely due to improper dam operation and not to upstream depletion of water – put simply, low water levels at Mtera have recurred almost every year, regardless of the year being dry or wet.

In 1992 and 1994, the Mtera reservoir experienced water shortages for the first time since commissioning and, consequently, TANESCO was forced to impose electricity rationing, with serious consequences for the country's production and economy. Reflecting its unexpected suddenness, there have been controversies over the causes of the low water level. The scantiness of existing data often meant that their interpretation became informed by the partisan interests. It was argued, often via the national press, that the power cuts and water shortages were caused by droughts or by upstream water use and other impacting activities. The activities accused were rice irrigation, deforestation and soil erosion in catchment areas, and valley-bottom agriculture along streams. However, other analyses pointed to the operation of the reservoir, as explained below.

In 2004, the situation became so critical that the Mtera reservoir was operated by utilizing the dead storage. The move was sanctioned by the government, despite advice to the contrary from the Rufiji Basin Water Office (RBWO) and the ministry responsible for water. In fact, the then Minister for Water and Livestock Development, on learning that there were low inflows and very little water in the Mtera reservoir, issued a decree that the power company should not use any more water from Mtera beyond the dead storage level. This announcement by the minister was not heeded. We do not have information regarding why this was the case, but one might assume that the government deemed power generation to be the more expedient decision.

Faraji and Masenza (1992) carried out a hydrological study for the Usangu plains. They compared monthly and annual flow volumes entering during the years 1989–1992 and found that the amounts that went into the reservoir were within the magnitude of the range of the long-term mean. They concluded that, although irrigation had increased over the years, its effects did not show up in the volumes that went into the Mtera reservoir. They suggested the combined management of the two reservoirs was an important dimension, given that, although irrigation was not invoked as a problem during the period 1980–1988, critically, there was no power generation facility at Mtera.

A DANIDA/World Bank study (1995) analysed 30-year annual flows of the Great Ruaha. The results also gave no evidence either of a trend towards decreased runoff from the basin or of any aggravating impact on the droughts in 1965/67, 1975/77, 1981/82 and 1991/92. They were unable to link upstream activities directly with decreasing water levels in Mtera.

Investigations and analyses conducted by SMUWC (2001) revealed that, although there was widespread and significant abstraction of water for irrigation in the Usangu catchment, the critical impact period was in the dry season. However, volumetrically, most of the reservoir recharge occurs during a period of 3–4 months in the rainy season, and thus dry-season flows had always been very small and added little to the total flow. SMUWC argued that the Mtera reservoir receives most of its flow during the peak rainfall months, and power generation is dependent on the storage and management of that flow during the remaining, dry, part of the year. The study also refuted strongly held beliefs (Kikula et al., 1996) that changes in rainfall and, in particular, deforestation were causes of reduced base flows of rivers flowing off the escarpment.

Since the commissioning of the Mtera reservoir, there have been enormous changes in both the demand and supply of electricity in the country, not adequately adapted to by the dam operators. The mismanagement of water in the Mtera–Kidatu system was confirmed by a further study on the system. Yawson et al. (2003; see also Machibya et al., 2003) investigated possible causes for the failure of the Mtera–Kidatu reservoir system within the Rufiji River basin in Tanzania in the early 1990s. Application of the TALSIM model (Froehlich, 2001) to the Mtera–Kidatu system revealed the presence of unaccounted for or unnecessary spillage from the reservoirs. They proposed that the core issue regarding the error-prone management of the Mtera–Kidatu system was that flows generated within the intervening catchment (i.e. the catchment between Mtera and Kidatu) were neglected, while simultaneously pursuing a policy to generate maximum power most of the time. Mtera should only generate power during the dry season, utilizing water being released to Kidatu. They concluded that if these rules (also recommended by the consultants, SWECO (1994)) were followed, then Mtera would not have gone dry in the 1991–1994 period. The validity of this assertion was tested with the TALSIM 2.0 model and an efficiency of 95% was achieved, indicating a very good correlation with the investigative techniques employed in the study.

Parageoplastic behaviour

The salient feature of Usangu's parageoplastic behaviour is that the growth in rice area did not generate symptoms of downstream water shortages during the wet season but it did during the dry season. The total mean annual flow into the Ihefu under natural conditions is estimated to be approximately 3330 Mm^3. Currently, average annual water withdrawals are estimated to be approximately 820–830 Mm^3, just slightly more than the mean annual volume of evapotranspiration from the wetland (790 Mm^3) but less than the net loss (of approximately 390 Mm^3) once rainfall received by the wetland is taken into account. However, both the annual and dry-season volumes abstracted vary considerably from year to year, both in absolute terms and as a proportion of the flow. Hydrological analyses using linear regression confirm a statistically significant decreasing trend in dry-season flows (Fig. 8.7), based on the Student's t-test). While there is a downward trend in total annual flows over the same period, this is not statistically significant. Thus, while the basin witnessed the most visible changes in dry-season flows, the flow volumes during this period represent just a small proportion of the total annual flow (of approximately 6–10%).

The declining wetland area is also associated with the drying of the Great Ruaha River.

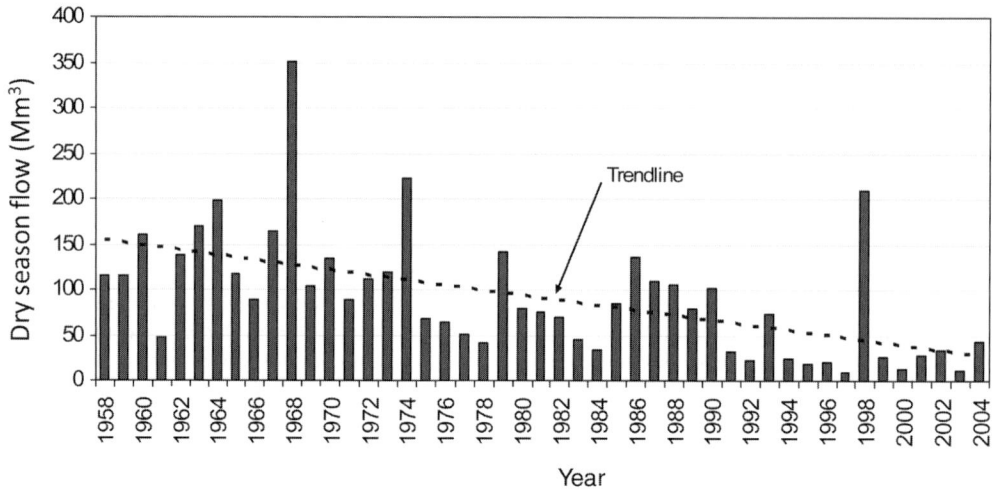

Fig. 8.7. Trend-line in dry-season flows in the Great Ruaha River at Msembe Ferry (1958–2004).

Although systematic surveys have not been recorded, there is widespread agreement that the hydrological change has considerably altered the ecology of the park near the river. Lack of water directly caused the death of hippopotami, fish and freshwater invertebrates, and disrupted the lives of many others that depend on the river for drinking water. The WWF reports that freshwater oyster populations have disappeared from the river, along with the clawless otters that lived on them. It is estimated that for animals that must remain within 1 km of water to survive (e.g. buffalo, waterbuck and many waterbirds), the lack of water has reduced the dry-season habitat by nearly 60% (Coppolillo et al., 2004). The movement of animals outside the park in search of water has led to increasing conflict with local human populations and the death of some animals. Overcrowding of hippopotami in shrinking water pools has led to eutrophication and anoxic water, as a result of which many animals have succumbed to infectious diseases (Mtahiko et al., 2006).

To summarize, the parageoplastic connection between upstream irrigation and downstream shortages in the Ruaha National Park arose from excessive abstraction of water through an increasing number of modernized intakes in the dry season. Although the area of dry-season irrigation was measured by SMUWC at approximately 5000 ha, large amounts of water were abstracted inadvertently through concrete intakes and 'spilled' on to fields that had been cultivated but harvested by that time, leading to unproductive evaporation. The presence of wet-season rice combined with modernized intakes appeared to increase the total length of the season of abstraction. Thus the rice-irrigating season has increased from approximately 150–200 days, observed by Hazel and Livingstone, to 250–350 days, seen in the last decade.

Non-equilibrium behaviour and basin governance

The second basin behaviour revealed by the case study is the inter-annual swing in the rice cultivated area, from approximately 20,000 to 40,000 ha, also mirrored in individual farmed areas, which change from a fraction of a hectare in a dry year to many hectares in a wet year. A second intra-annual fluctuation takes place when the wet-season area cultivated shrinks to approximately 3000–5000 ha during the dry season, seen as a core area made possible by the perennial rivers found on the plains.

Figures 8.2, 8.3 and 8.6 show this dynamic in various ways. The considerable change in cultivated area is forced by three factors: (i) a highly varying river flow; (ii) a large amount of irrigable land on the plains; and (iii) the ability of a large number of intakes to abstract more water when the rivers increase in supply, up to a cap set by the limitations of the intake dimensions. The dynamic is termed 'non-equilibrium' because it establishes an environment that does not lend itself to predictable regulatory water management, thus providing a remarkably different context in which to frame and formulate irrigation planning. This contrast between equilibrium and non-equilibrium thinking is captured in Table 8.3.

Table 8.3 proposes that marked contextual differences exist between equilibrium and non-equilibrium irrigation and water management. The key issue is how the management of the plateau part of the river-basin development curve is theorized (assuming that in the earlier stages of development, supply outstrips demand in both equilibrium and non-equilibrium contexts). For example, in equilibrium basins (or basins deemed to behave within predictable parameters) supply can be raised by adding storage, and demand management is fostered through regulatory and price-based reforms. In non-equilibrium basins, while these measures might apply in theory and be adopted in practice, their intended outcomes of creating further headroom are either limited or unpredictable. Thus, in a basin where the upward potential for unmet demand is so large (e.g. say because of irrigable land), additional storage may not bring intended equitable benefits for all users if the distribution of that additional water is not governed adequately or hard-wired into the infrastructure. The use of normative irrigation planning procedures in widespread use (FAO, 1998) can lead to designs of abstraction headworks that significantly desiccate catchments during the dry season when river flows are negligible (Lankford, 2004b). Furthermore, demand management in a basin where demand already 'crashes' due to a natural supply deficit must also be carefully considered.

Particular dimensions of the River Basin Management Project (the RBM component of RBMSIIP) applied to the non-equilibrium Usangu basin throw light on the ill-considered design of the project. The Rufiji Basin Water Office (RBWO), supported by RBMSIIP, designed a water rights system (see also MWLD, 2002) in order to effect regulatory demand management, which was wholly unsuitable for the basin for a variety of reasons (van Koppen

Table 8.3. Comparing equilibrium and non-equilibrium irrigation and river basin governance.

	Equilibrium	Non-equilibrium
Observation	Irrigation area and demand for water are fixed within limitations	Irrigation area and demand for water vary widely with supply
Inter-annual area of irrigation	Fluctuates <100%	Fluctuates <1000%
Irrigable land	Constrained by planning, soil type, gradients or zoning restrictions	Large area of high potential land available
Climate	Tends to be temperate, tropical oceanic, which reduces water availability	Tends to be semi-arid with a high coefficient of variation of rainfall
Irrigation planning	FAO-type methodology for determining fixed/adjustable peak irrigation demand	Requires a river-centred approach allowing for proportional intakes
Water rights and permits	Defined by quanta (e.g. l/s)	Defined by proportions of river flow (%)
Basin development curve	S-shaped, rising to a stable plateau	S-shaped to high variable supply/demand curve
Supply, demand, share management	Adding storage, applying demand management	Storage and demand management, share modification
River basin governance	Suggests normative forms of regulatory management	Suggests modular and localized models to meet local apportionment

et al., 2004, 2007; Lankford and Mwaruvanda, 2007). The key reason the adopted system was faulty was its choice of a fixed quanta for a water right (e.g. 250 l/s). This specified flow rate implied that the water abstracted into an irrigation system in the Usangu would be measured. Yet, with the exception of the Mbarali intake and occasional record keeping at the Kapunga intake, no intake is monitored in this fashion, principally because there is no evidence for the existence of flow measurement structures.[7] The consequences of this are that farmers do not regulate (throttle back) their abstraction when they exceed their water right, in terms of either discharge or annual volume. For abstraction during the wet season, it should be noted that many intake dimensions do not correspond with the formal entitlement, either in the initial design stage or by further flow calibration (Rajabu and Mahoo, 2008). It should also be stated that the water rights are not calculated systematically using any meaningful algorithm – not least because command areas fluctuate and an excessively high rice water duty of 2.0 l/s/ha is widely employed. Studies by SMUWC (2001) found that the water duty was closer to 1.0 l/s/ha because irrigation is mostly supplemental to the 600 mm or so of annual rainfall. Thus, having paid their water right, there is no mechanism for farmers not to exceed their right. This situation becomes untenable in the dry season, when river flows are a tenth or less of their wet-season flows, leading to officially sanctioned water rights and concrete intake designs that far exceed the actual water available. Indeed, the hydrological conditions in which some water rights might apply accurately in combination with other water rights on a stretch of river to cumulatively add up to an irrigation sector cap (therefore giving rise to a surplus for downstream needs) are statistically quite rare because the river fluctuates markedly above or below the level at which demands were calculated. At most, the system can be employed administratively as a record of intakes, names and owners.

Managing the allocation of water in different contexts also suggests a rethink, given that normative regulation is questionable in a non-equilibrium context. To explain this, a new dimension to water allocation – share modification – is explored in the next section.

Share modification

Modification describes implicit and unintended contemporaneous changes in the share of water between users or sectors as a result of a changing supply being modulated by existing institutional and infrastructural architecture (Lankford, forthcoming). Thus, while 'allocation' applies to longer-term applications of intersectoral sharing, or where an equilibrium climate (e.g. oceanic, temperate) exists, modification of shares of water has greater relevance to non-equilibrium, pulse-driven semi-arid climates. The upper Great Ruaha case study shows that when supply variability is marked, leading to greater amplitude of hydrological events, and abstraction infrastructure is 'fixed', share modification and its management become more important. Here, a variable water supply (where supply increases or decreases over orders of magnitude within relatively short periods of time) 'forces' disproportionate shifts in usage in different sectors, depending on how users differentially abstract an increasing or decreasing rate of supply. This can be seen as a modification of the supply variability upon the proportions of shares to users and intakes.

Share modification is best explained via the case study typical of the Mkoji subcatchment in the Usangu, where an intake of say 250 l/s continues to abstract that fixed amount in the face of a declining river flow supply. Thus, if the flow rate declines from a peak of about 3000 l/s during the wet season down to about 50 l/s during the dry season, the 250 l/s abstraction leads to a concomitant reduction in downstream supply, and eventually to a zero flow. This behaviour contrasts with a proportional abstraction, where the intake takes might be redesigned to abstract a percentage of whatever flow is present, so that the surplus percentage flows downstream. It is the application of many intakes in the Usangu with fixed abstraction design parameters that leads to an uneven allocation of water between upstream irrigation and the downstream wetland during the end of the wet season, which runs into the dry season.

Another interesting example of share modification that influences water distribution between the wetland and the downstream riverine stretch through the Ruaha National

Park arises via the natural rock outcrop that holds back the wetland water, leading to zero flows in the river when the water level drops below the sill level. The SMUWC and the WWF Ruaha projects both considered that installing a weir or a pipe with an adjustable sluice gate would enable more water to be held back in the wetland and also provide some controllability of distribution of environmental flows. This type of infrastructure provides additional levels of proportionality to an otherwise on/off system.

Conclusions

In the last 60 years in the Great Ruaha basin, modernist and progressive narratives regarding water development and conservation have reified into local and external donor initiatives and projects. The period 1950 to the mid-1980s was marked by an expansionist, developmental narrative, resulting in the construction of formal irrigation systems with large engineered headworks to abstract river water. While we might not judge harshly those decisions taken, given the era in which they were formulated, we can be much more critical about a continuing and related set of ideas around regulatory, efficiency and technological improvement approaches to river basin management that have contemporary significance. From the last quarter-century to the current day, we see that ideas of irrigation headworks' construction are still promulgated as a part of an 'efficiency' and volumetric water rights narrative, resulting in an era of contested solutions in attempting to balance allocation between multiple calls on limited water.

An unforeseen complex set of interlinked dynamics has emerged as a result of evolving abstraction and depletion of water in this highly variable river basin. Upstream access to water was further captured by irrigated agriculture, partly led by state interventions such as publicly owned schemes and donor-funded improvement programmes using justifications based on intake upgrading and irrigation efficiency, resulting in inequitable and inefficient allocation across the river basin, and the prompting of new behaviours downstream as downstream users react to non-local, internal and external hydrologic perturbations.

Using three ideas, we have critiqued the efficiency and water management found in the Great Ruaha catchment. In studying the responses of users along these interlinked river sub-basins the authors coined the term 'para-geoplasia' to explore how distant symptoms and behaviours arise from non-local depletion. Simply put, headwork designs that aimed to regulate upstream water abstraction during the wet season led to unforeseen dry-season para-geoplastic impacts some 50–300 km further north in the wetland and the Ruaha National Park.

Using ideas of non-equilibrium water theory, we see that attempts to use fixed volumetric water rights to regulate flows in an environment where flows vary weekly, monthly and seasonally through several orders of magnitude were also misplaced. Instead, proportional water rights and headwork structures should be regarded as a starting point for upstream–downstream water allocation and distribution.

Related to this, water-share modification contrasts further the differences between equilibrium and non-equilibrium environments. Share modification describes the differential uneven apportionment of water to intakes sequenced on a river as a result of the interaction between a declining or increasing flow rate over time and the design parameters of the headworks. A series of proportional intakes would result in a more even distribution of water shares than a series of fixed or regulated orifice intakes, with a percentage of flow designed to pass downstream to the wetland.

How do these ideas relate to river basin development? They underline the high level of interconnectedness between differing subsystems behaving in unforeseen ways in different periods of the hydrological calendar. In particular, theories that underpin water resources development during a growth phase of a river basin (in this case headworks designed using unrealistic water duties supported by standard irrigation design methodologies) might store up problems for governing water during the plateau phase of a river basin's development. Additional signals of wet and dry periods bring a variable supply of water to a basin, which imposes further challenges in the management of demand and allocation. If the plateau phase is not stable or varying within predictable peaks and troughs, but is highly dynamic, then demand management has to be rethought, because the basin is more driven by a non-equilibrium collapse of demand with supply. This, in turn, means that the basin, in the absence of large-scale storage or groundwater, has to welcome expansion of demand during wet years but facilitate a contraction of demand across all users during dry periods. In a maturely developed basin such as the upper Great Ruaha, these effects and behaviours point significantly to proportional water rights and infrastructure as the key departure point for managing surface water flows, combined with domestic provisioning for dry periods.

A key problem is nevertheless the vexed issue of how to cap an upper limit of irrigation abstraction during wet seasons so that water passes downstream for other sectors. While an individual proportional intake can be designed with an upper flow limit, the problem of growth of the number of intakes, seen in the recent past, remains a risk in the future, regardless of the approach to individual intake design. The RBWO is considering an approach which provides a single volumetric water right to a subcatchment (acting as the volumetric cap) so that the user association decides how to share this out among users. With this in place, it will then be necessary to revisit a catchment's intakes to ensure that intra-intake shares are coordinated and that the catchment as a whole provides a downstream proportion during times other than the wet season. A fuller explanation of an approach to volumetric and proportional caps is given in Lankford and Mwaruvanda, 2007. It is not yet clear how this will be fully adopted by Usangu farmers and supported by local government services.

Thinking wider afield and more generically, our ability to select governance theories for future phases of the basin trajectory in different types of basins will be paramount, not least because basin interconnectedness will grow, uncoordinated experimentations with storage and river infrastructure will continue, and hydrometeorological extremes – and transitions between those extremes – may become more commonplace in sub-Saharan Africa.

Notes

1 This word is coined from Greek: 'para' meaning beyond, 'geo' meaning earth or land, and 'plasia' meaning something made or formed. The term is inspired by the concept of 'paraneoplasticity', derived from medical research into cancer, which describes how, in the body, other cancer-related tumours start to occur remotely from the first and main tumour.

2 There is not enough room to describe in detail the productivity analyses of water conducted by the RIPARWIN (Raising Irrigation Productivity and Releasing Water for Insectoral Needs) project, which was funded by DfID (UK Department for International Development) and succeeded another DfID-funded project, SMUWC (Sustainable Management of the Usangu Wetland and its Catchment). It is worth mentioning, however, that productivity is highest for localized livelihoods supported by livestock, brick-making and domestic uses, averaging at around US$1.00/m^3 of depleted water (Kadigi et al., 2008). In addition, the productivity of irrigated rice (US$0.02/m^3 of water abstracted) can be compared with the value of water when it is used to generate and sell electricity – generating about ten times the amount, or US$0.2/m^3.

3 The runoff coefficient for the basin was calculated by SMUWC (see Note 2). It studied three time windows in its hydrological analysis; pre-1974, 1974–1985 and 1986–1998. The runoff coefficient for the first window is 14%, while it is 9% for the second window and 13% for the third window. If the heavy flooding years of 1998 and 1968 are excluded from calculations, then the resulting runoff coefficients are 12, 9 and 10%, respectively, for the three windows.

4 The paper by Mtahiko has a number of errors in it, including citing the SMUWC study for asserting that upstream irrigation resulted in less water for hydropower.

5 Recently, the World Bank (2007) has upgraded its assistance to Tanzania with a US$200 million Water Sector Support Project.

6 See www.friendsofruaha.org

7 Flow can be measured from the properties of the intake flume combined with knowledge of the head difference of water levels, taking the long-crested weir sill height as a datum. In reality, flow-gauging plates are not installed or monitored.

References

Carnell, J., Lawson, J.D., von Lany, H. and Scarrott, R.M.J. (1999) Water supply and demand balances: converting uncertainty to headroom. *Journal of the Chartered Institution of Water and Environmental Management* 13(6), 413–419.

Coppolillo, P.B., Kashaija, L., Moyer, D.C. and Knap, E. (2004) *Technical Report on Water Availability in the Ruaha River and State of Usangu Game Reserve, November 2003*. Wildlife Conservation Society and WWF-Tanzania Program, Tanzania.

DANIDA (Danish International Development Agency)/World Bank (1995) Water resources management in the Great Ruaha basin: a study of demand driven management of land and water resources with local level participation. A report prepared for the Rufiji Basin Water Office, Ministry of Water, Energy and Minerals. DANIDA, Ministry of Foreign Affairs of Denmark, Copenhagen/World Bank, Washington, DC.

Donkor, S.M.K (2003) Development challenges of water resource management in Africa. *African Water Journal* 1–9. UN Water/Africa. Pilot Edition.

FAO (Food and Agriculture Organization of the United Nations) (1961) *The Rufiji Basin, Tanganyika*. FAO report to the government of Tanganyika on preliminary reconnaissance survey of the Rufiji basin. FAO, Rome.

FAO (1998) *Crop Evapotranspiration – Guidelines for Computing Crop Water Requirements*. FAO Irrigation and Drainage Paper 56. FAO, Rome.

Faraji, S.A.S. and Masenza, I.A. (1992) *Hydrological Study for the Usangu Plains with Particular Reference to Flow Entering the Mtera Reservoir and Water Abstractions for Irrigation*. Ministry of Agriculture and Livestock Development, Irrigation Division, Dar es Salaam, Tanzania.

Froehlich, F. (2001) Application of the TALSIM 2.0 model on the Mtera–Kidatu reservoir system in the Rufiji River basin in southern Tanzania. Research undertaken at the University of Dar es Salaam, Tanzania and submitted to the Technical University of Darmstadt, Germany. (www.ihwb.tu-darmstadt.de/ihwb/c-vertiefer/Froehlich/vertiefer-Froehlich.html)

Hazelwood, A. and Livingstone, I. (1978a) *The Development Potential of the Usangu Plains of Tanzania*. Commonwealth Fund for Technical Cooperation, The Commonwealth Secretariat, London.

Hazelwood, A. and Livingstone, I. (1978b) Complementarity and competitiveness of large- and small-scale irrigated farming: a Tanzanian example. *Oxford Bulletin of Economics and Statistics* 40(3), 195–208.

IWGIA (International Work Group for Indigenous Affairs) (2008) Eviction of pastoralists from Usangu plains, Tanzania. www.iwgia.org/graphics/Synkron-Library/Documents/Noticeboard/News/Africa/2008/Memo%20Usangu%20Plains.all.pdf. Accessed 5th July 2008.

JICA (Japan International Cooperation Agency) (2001) *The Study on the National Irrigation Master Plan in the United Republic of Tanzania*. Japan International Cooperation Agency, Ministry of Agriculture and Food Security, Government of Tanzania, Dar es Salaam, Tanzania.

Kadigi, R.M.J., Kashaigili, J.J. and Mdoe, N.S. (2003) The economics of irrigated paddy in Usangu basin in Tanzania: water utilization, productivity, income and livelihood implications. 4th WARFSA/Waternet Symposium. Gaborone Botswana, 15–17 October 2003.

Kalinga, G., Simba, A.H. and Temu, R.J.M. (2001) National Irrigation Development Plan – Update. In: *Proceedings of the 1st National Irrigation Conference, Morogoro, Tanzania, 20th to 22nd March 2001. Theme: 'Irrigated Agriculture for Food Security and Poverty Reduction', Vols I and II*. Ministry of Agriculture and Food Security, Dar es Salaam, Tanzania.

Kashaigili, J.J., McCartney, M.P., Mahoo, H.F., Lankford, B.A., Mbilinyi, B.P., Yawson, D.K. and Tumbo, S.D. (2006) *Use of a Hydrological Model for Environmental Management of the Usangu Wetlands, Tanzania*. IWMI Research Report 104. International Water Management Institute, Colombo, Sri Lanka.

Kikula, I.S., Charnley, S. and Yanda, P. (1996) *Ecological Changes in the Usangu Plains and their Implications on the Down Stream Flow of the Great Ruaha River in Tanzania*. Research Report No. 99. Institute of Resource Assessment, University of Dar es Salaam, Tanzania.

Lankford, B.A. (2004a) Irrigation improvement projects in Tanzania; scale impacts and policy implications. *Water Policy* 6(2), 89–102.

Lankford, B.A. (2004b) Resource-centred thinking in river basins: should we revoke the crop water approach to irrigation planning? *Agricultural Water Management* 68(1), 33–46.

Lankford, B.A. (forthcoming) Responding to water scarcity – beyond the volumetric. In: Mehta, L. (ed.) *The Limits to Scarcity: Contesting the Politics of Allocation*. Earthscan, London.

Lankford, B.A. and Beale, T. (2007) Equilibrium and non-equilibrium theories of sustainable water resources management: dynamic river basin and irrigation behaviour in Tanzania. *Global Environmental Change* 17(2), 168–180.

Lankford, B.A and Mwaruvanda, W. (2007) A legal–infrastructural framework for catchment apportionment. In: van Koppen, B., Giordano, M. and Butterworth, J. (eds), *Community-based Water Law and Water Resource Management Reform in Developing Countries*. Comprehensive Assessment of Water Management in Agriculture Series, CAB International, Wallingford, UK, pp. 228–247.

Lankford, B.A., van Koppen, B., Franks, T. and Mahoo, H. (2004) Entrenched views or insufficient science? Contested causes and solutions of water allocation; insights from the Great Ruaha River basin, Tanzania. *Agricultural Water Management* 69(2), 135–153.

Lankford, B.A., Merrey, D., Cour, J. and Hepworth, N. (2007) *From Integrated to Expedient: an Adaptive Framework for River Basin Management in Developing Countries*. Research Report 110. International Water Management Institute. Colombo, Sri Lanka.

Machibya, M. (2003) Challenging established concepts of irrigation efficiency in a water scarce river basin: a case study of the Usangu basin, Tanzania. PhD thesis, University of East Anglia, Norwich.

Machibya, M., Lankford, B. and Mahoo, H. (2003) Real or imagined water competition? The case of rice irrigation in the Usangu basin and Mtera/Kidatu hydropower. In: *Proceedings of the Tanzania Hydro-Africa Conference, 17–19 November 2003, at Arusha International Conference Centre, Arusha, Tanzania*. International Centre for Hydropower, Norway.

McCartney, M.P., Lankford, B.A. and Mahoo, H.F. (2007) *Agricultural Water Management in a Water Stressed Catchment: Lessons from the RIPARWIN Project*. Research Report 116. International Water Management Institute. Colombo, Sri Lanka.

McCartney, M., Kashaigili, J.J., Lankford, B.A. and Mahoo, H.F. (2008) Hydrological modelling to assist water management in the Usangu wetlands, Tanzania. *International Journal of River Basin Management* 6(1), 51–61.

Milly, P.C.D., Betancourt, J., Falkenmark, M., Hirsch, R.M., Kundzewicz, Z.W., Lettenmaier, D.P. and Stouffer, R.J. (2008) Stationarity is dead: whither water management? Policy Forum. *Science* 319(5863), 573–574.

Moirana, L. and Nahonyo, C.L. (1996) Why the Usangu plains should be an environmentally protected area (Usangu Game Reserve). Internal Report. Ruaha National Park, Tanzania National Parks, August 1996.

Molden, D. (1997) *Accounting for Water Use and Productivity*. SWIM Paper 1. International Irrigation Management Institute (now International Water Management Institute), Colombo, Sri Lanka.

Molden, D., Sakthivadivel, R. and Habib, Z. (2001) *Basin-level Use and Productivity of Water: Examples from South Asia*. Research Report 49. International Water Management Institute, Colombo, Sri Lanka.

Molden, D., Sakthivadivel, E., Samad, M. and Burton, M. (2005) Phases of river basin development: the need for adaptive institutions. In: Svendsen, M. (ed.) *Irrigation and River Basin Management: Options for Governance and Institutions*. CAB International, Wallingford, UK, pp. 19–29.

Molle, F. (2003) *Development Trajectories of River Basins: a Conceptual Framework*. Research Report 72. International Water Management Institute, Colombo, Sri Lanka.

Mtahiko, M., Gereta, E., Kajuni, A., Chiombola, E., Ng'umbi, G., Coppolillo, P. and Wolanski, E. (2006) Towards an ecohydrology-based restoration of the Usangu wetlands and the Great Ruaha River, Tanzania. *Wetlands Ecology and Management* 14(6), 489–503.

Mutayoba, W. (2002) Management of water resources in Tanzania through basin management. In: *Proceedings of 3rd WATERNET/WARFSA Symposium: Water Demand Management for Sustainable Development 30–31 October 2002*. WaterNet, Dar es Salaam, Tanzania.

MWLD (Ministry of Water and Livestock Development) (2002) *National Water Policy (NAWAPO)*. Ministry of Water and Livestock Development, P.O. Box 456, Dodoma, Tanzania.

NORPLAN (Norwegian Consulting Engineers and Planners) (2000) *Review of Water Resources Management Policy, Legislation and Institutional Framework. River Basin Modelling*. Norwegian Consulting Engineers and Planners. Ministry of Water, Tanzania.

Palmer-Jones, R. and Lankford, B.A. (2005) Agricultural development: reflections on the study of the irrigation potential of the Usangu plains of Tanzania. In: Tribe, M., Thoburn, J. and Palmer-Jones, R. (eds) *Development Economics and Social Justice. Essays in Honour of Ian Livingstone*. Ashgate. Aldershot, UK, pp. 141–158.

PINGOs (Pastoralists Indigenous Non-Governmental Organisation's Forum) (2006) Resource mis-management and the misery of pastoralists in Usangu basin. A draft baseline report of the fact finding mission on proposed eviction of livestock keepers submitted to members of Pastoral Civil Society Organisations, Arusha, Tanzania, April, 2006. http://pingosforum.org/Docs/usangu_research_report.pdf (accessed 5th July 2008).

Rajabu, K.R.M. and Mahoo, H.F. (2008) Challenges of optimal implementation of formal water rights systems for irrigation in the Great Ruaha River catchment in Tanzania, *Agricultural Water Management* 95(9), 1067–1078.

Rajabu, K.R.M., Mahoo, H.F., Sally H. and Mashauri, D.A. (2005) Water abstraction and use patterns and their implications on downstream river flows: a case study of Mkoji sub-catchment in Tanzania. In: Lankford, B. and Mahoo, H.F. (eds) *Proceedings of the East Africa Integrated River Basin Management Conference*, 7–9 March 2005, ICE Conference Hall, Sokoine University of Agriculture, Morogoro, Tanzania, pp. 233–245.

SMUWC (Sustainable Management of the Usangu Wetland and its Catchment) (2001) *Final Project Reports*. Directorate of Water Resources, Ministry of Water, Government of Tanzania, Dar es Salaam, Tanzania.

Sullivan, S. and Rohde, R. (2002) On non-equilibrium in arid and semi-arid grazing systems. *Journal of Biogeography* 29(12), 1595–1618.

SWECO (1994) *Follow up Study of the Environment at Mtera Reservoir and Region – Hydrology Draft Report*. SWECO AB, Stockholm, Sweden.

TANESCO (Tanzania Electric Supply Company Limited) (2008) Tanzania Electric Supply Company Limited (TANESCO) web site 'Generation'. http://www.tanesco.com/ (accessed 6 April 2008).

TNW (Tanzania National Website) (2003) The 2002 population and housing census. Tanzania National Website. http://www.tanzania.go.tz/census/districts (accessed 24th July 2005).

USBR (United States Bureau of Reclamation) (1967) *Rufiji River Basin*. United States Bureau of Reclamation, Washington, DC.

van Koppen, B., Sokile, C., Hatibu, N., Lankford, B.A., Mahoo, H. and Yanda, P. (2004) *Formal Water Rights in Tanzania: Deepening the Dichotomy?* Working Paper 71. International Water Management Institute, Colombo, Sri Lanka.

van Koppen, B., Sokile, C., Lankford, B.A., Mahoo, H., Hatibu, N. and Yanda P. (2007) Water rights and water fees in Tanzania. In: Molle, F. and Berkoff, J (eds) *Irrigation Water Pricing: the Gap between Theory and Practice*. International Water Management Institute/CAB International, Wallingford, UK, pp. 143–164.

Weatherhead, E.K. (2007) *The 2005 Irrigation Survey for England and Wales*. Report for Department for Environment, Food and Rural Affairs (DEFRA). Cranfield University, Cranfield, UK.

World Bank (1996) *River Basin Management and Smallholder Irrigation Improvement Project (RBMSIIP)*. Staff Appraisal Report, Washington, DC.

World Bank (2007) *Water Sector Support Project*. Project Appraisal Document, Washington, DC.

Yawson, D.K., Kashaigili, J.J., Mtalo, F.W. and Kachroo, R.K. (2003) Modelling the Mtera–Kidatu reservoir system to improve integrated water resources management. In: *Hydro Africa 2003*, CD Proceedings, 17–19 November, Arusha, Tanzania.

9 Buying Respite: Esfahan and the Zayandeh Rud River Basin, Iran

François Molle,[1]* Iran Ghazi[2]** and Hammond Murray-Rust[3]***

[1]*Institut de Recherche pour le Développement, Montpellier, France;*
[2]*Esfahan, Iran;* [3]*ARD Inc., Burlington, Vermont, USA ;*
e-mails: *francois.molle@ird.fr; **iranghazi@yahoo.com;
***hmurray-rust@ardinc.com

Introduction

Oases pose a particular challenge to water resources development: they are tightly dependent upon the sources of water that they are able to access and strongly constrained in their growth by the utter scarcity that comes with aridity. Some of the oases – think of Marrakesh, Samarkand or Baghdad – are located in desert or semi-desert areas but are supplied by a river that starts its course in rainier, and often distant, regions. For such large cities, the time eventually comes when expansion of both the city and its surrounding fields and orchards, which thrive on the association of sun, water and dry air, encounters the limits established by nature.

Esfahan, in central Iran, is one such city. The story of Esfahan, with its rich and long history, and of its lifeblood, the Zayandeh Rud River, vividly illustrates the challenges faced by societies in situations of water scarcity. In the past, user communities have developed robust institutions to share springs, *qanats* (human-made underground galleries that drain aquifers), intermittent streams, or river flows. Yet, basin closure – a state where all resources are fully committed and where water only reaches the terminus of the basin in exceptional years – coupled with the expansion of state power, characterized by the reshaping of waterscapes by large-scale interventions, has made local systems dependent on decisions taken at other scales. Competition for resources and basin closure generate both increased hydrological interconnectedness between users and entanglement of governance and legal management regimes.

This chapter first describes the physical and human setting of the Zayandeh Rud, then reviews ancient and recent water resources development in the basin, and finally reflects on the hydrological, social and institutional consequences of basin closure. The Zayandeh Rud basin provides a vivid account of an oasis buying respite by implementing successive water imports from neighbouring basins. It also offers a textbook illustration of both the process of continuing river basin overbuilding and its consequences.

Physical and Human Context

The Zayandeh Rud basin covers 41,500 km^2 in the centre of Iran (Fig. 9.1). The river rises in the bleak and craggy Zagros mountains (northwest of the basin), which reach over 4500 m, traverses the foothills in a narrow and steep valley, and then bursts forth onto the plains at an altitude of some 1800 m. However, the splendour of the river is short lived: reduced towards the east by natural seepage losses, evaporation and more recent extractions for irrigation, and urban and domestic uses, the river eventually dies out in the Gavkhuni lake, a vast expanse of white salt that forms the bottom end of the basin, lying at an altitude of over 1200 m. In this naturally confined (or endoreic) basin, the flows reaching the lake are now much reduced compared with natural conditions, and there are extended periods when no water flows in the tail reach of the river (Fig. 9.1).

The total length of the river is some 350 km, but it is the central 150 km of the flood plain to the east and west of Esfahan that provides the basis for intensive agriculture and large settlements. Along this strip soils are deep and fertile, predominately silts and clay loams, and slopes are gentle, ideal for the irrigated agriculture built up over many centuries. The river indeed forms an oasis in the desert (Murray-Rust and Droogers, 2004).

The climatic conditions in the mountains are markedly different, as shown by data from Kuhrang, which lies just to the west of the Zayandeh Rud basin (Fig. 9.1). Situated at an elevation of almost 2300 m, precipitation averages 1500 mm, much of it in the form of snow, and snow remains on the ground throughout winter, only melting when temperatures warm up from April onwards (Murray-Rust and Droogers, 2004). In contrast, the city of Esfahan only receives 130 mm of rainfall each year, on average (Fig. 9.2).

The primary source of water in the basin is, thus, the upper catchment of the Zayandeh Rud. Lateral tributaries joining the river in the plains are mostly non-perennial, have little regional importance and do not reach into the main part of the basin, except during winter months and rare flash floods, although subsurface runoff accrues to the main stream. Runoff generated in the upper basin is strategically

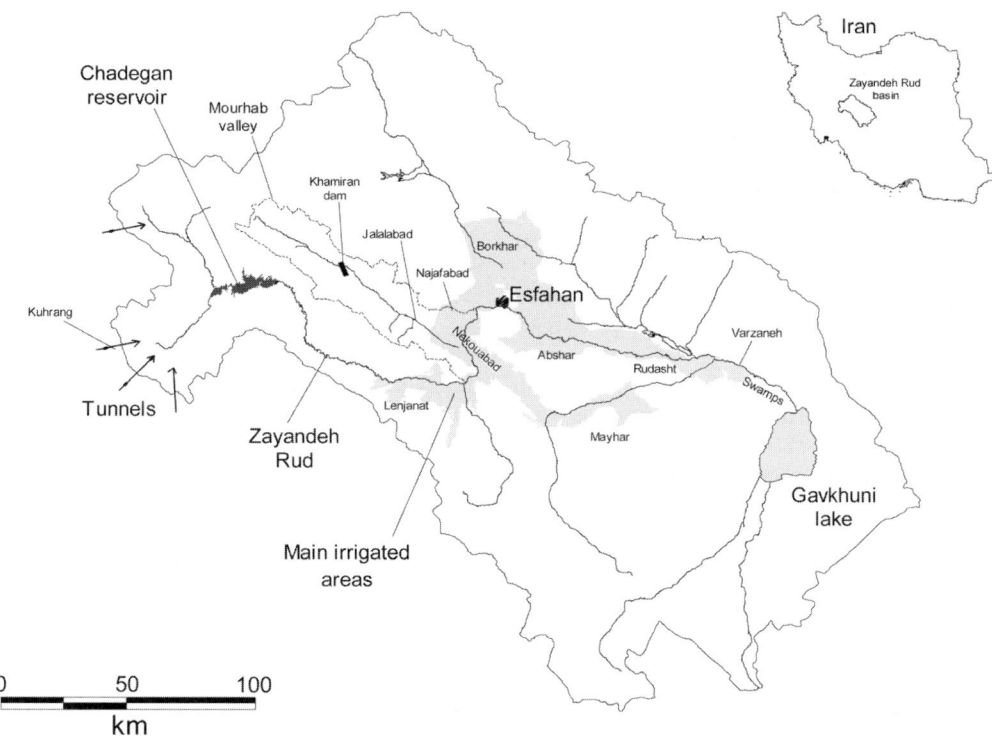

Fig. 9.1. The Zayandeh Rud basin.

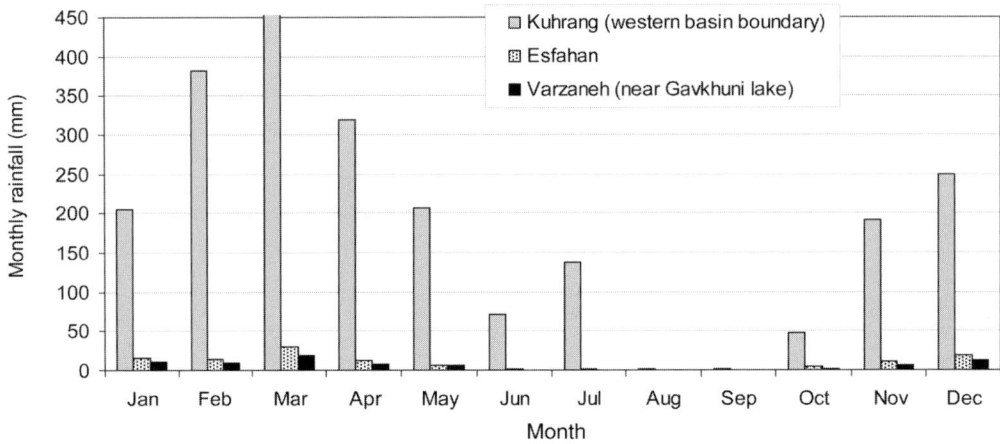

Fig. 9.2. Average monthly rainfall: Kuhrang, Esfahan and Varzaneh (1988–1999).

stored in the Chadegan reservoir, constructed just above the point where the Zayandeh Rud enters the flatter parts of the basin (Fig. 9.1). From September until February, inflows only average between 50 and 75 Mm³ per month (20–30 m³/s), reflecting both the dry conditions of summer and then the cold conditions dominated by accumulation of snow in the upper parts of the basin. From March onwards snowmelt increases and discharges normally peak in April or May, with average flows of 125–150 m³/s. In June and July, the discharge slowly declines to the low-flow conditions. The peak flows from April to June provide the basis for widespread downstream irrigation using simple diversion structures.

The Zayandeh Rud basin has seen a dramatic population increase in the past 45 years. According to the 1956 census, the population in the basin was some 420,000, while in 2000 the total population was estimated at 2.3 million. This is an annual growth rate of 5.9%. Figure 9.3 shows population growth in the basin and in Esfahan since 1956, projected to 2020 with a 2% annual growth rate from 1996 onwards. Growth has not been uniform. The fastest growth was between 1956 and 1986, averaging close to 7% a year, but in the past 15 years it has slowed down to 2–2.5% a year. Initially, Esfahan city grew faster than the rest of the basin, but this is no longer the case: The growth rate of Esfahan is close to 2%, while outside the city it has risen to 2.5–3% a year.

Early Water Use in the Zayandeh Rud River Basin

Although water use around Esfahan is as old as the city itself and although there are records of water management dating back to the 3rd century BC, when Ardeshir of Babak (the founder of the Sassanid dynasty) sent an engineer to fix the 'disorders [that] appeared in the regulation of the Zayandeh Rud waters' (Hossaini Abari, 2006), historical documents on water use are scarce. Rusteh (1889), for example, who wrote in the early 10th century, mentioned that water use was unrestricted up to the district of Alandjan, while the distribution to the downstream districts of Djay, Marbin, Alandjan, Baraan, Rud and Rudasht was organized following 'rules established by Ardeshir Ibn Babak'. Hawqal, four decades later, also reported that the sharing of the Zayandeh Rud water was 'calculated so that no water would be lost'.

The earliest-known detailed regulation of the Zayandeh Rud was unearthed by Lambton (1938). Riparian rights in the 16th century are described in detail in a *tumar* (an edict) attributed to Sheikh Bahai, which specifies the water apportioned each month to each *boluk* (district) and village. The river was managed by a *mirab* (water master) elected by 33 *boluk* (representatives), who selected six assistants, who, in turn, appointed *maadi salars*, heads of each *maadi* (main run-of-river diversion canal) that branched off the river. According to Lambton (1953), the introduction of the edict states that:

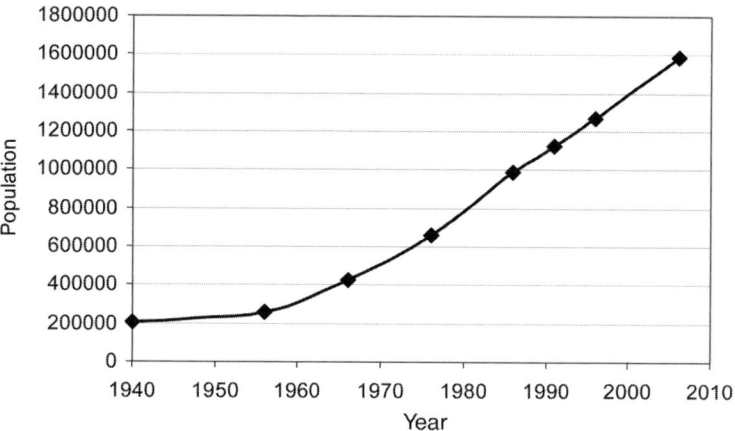

Fig. 9.3. Population growth in the Zayandeh Rud basin, 1956–2020.

(...) the competent authorities of the State should appoint a few persons of the reliable and aged men to establish, under the signatures of the exalted and honourable mostawfis and the confirmation of the kadkhodas and rish-safids of the boluks which share the water of the Zayandeh Rud, honestly and to the best of their knowledge, the shares and lot of each village and hamlet in each boluk, according to its capacity and need, and to enter in the registers under guarantee, so that regulation (of the waters) should be put into execution.

Water was divided into 55 primary shares, which were further subdivided 'into 276 secondary shares associated with the major irrigation canals or *maadi* and into 5105 tertiary shares at the village level' (Hossaini Abari, 2000). Managers were paid by users, in due proportion to the amount of water received, and were dispensed with if their services were judged to be unsatisfactory (Hossaini Abari, 2006). Where there was no *maadi*, water could be lifted from the river or from drains using animal-driven Persian wells (Murray-Rust and Droogers, 2004). The application of the *tumar* was discontinued by invasions and some rulers, but was renewed in 1927, when about 500 rights-holders met to demand the reinstatement of the rules. With some modifications in the 1930s, these were enforced until the early 1970s (Pirpiran, 2007).

In lateral valleys, such as the Mourhab valley, which rejoins the Zayandeh Rud's left bank west of Esfahan (see Fig. 9.1), the use of surface water was also socially controlled. In the 1960s, the water of the Mourhab River was allocated according to rules that villagers also trace back to Sheikh Bahai. The rules determine which village can divert which proportion of the river flow during which period, and they were equally enforced by a powerful *mirab*.[1]

The village of Jalalabad, located in the lower part of the Mourhab valley (see Fig. 9.1), provides a very good picture of water rights and management at the village level (Molle et al., 2004). The main sources of supply to the village until the 1960s were two *qanats*, in addition to whatever surface water could be diverted from the Mourhab River according to the rules. Land was apportioned among the six main lineages of the village in the beginning of the last century, and up to the present, *qanat* water rights have been defined at the plot level, in terms of minutes of use per 6-day turn. These rights can be reallocated among plots, temporarily lent, ceased or leased, or permanently sold and transferred. No one in the village is aware of the full details of the system. This striking lack of centralized control goes together with a strict adherence to the established rights and schedules. Spooner (1974b) posits that this can be partly ascribed to the fact that since 'any disturbance of the temporal distribution systems affects all shareholders adversely, the normal premium on social order is increased'. Out of equity, each lineage was given plots both at the beginning and at the end of the canal system. Maintenance of the ditches was undertaken collectively and that of the *qanat* was entrusted to specialized workers; these workers, as well as the water masters, were paid by the users themselves, a system still in use.

More generally, *qanats* were considered as the private property of those who had invested in their excavation. Owners could be individuals, groups of families, or wealthy merchants, as in the case of Najafabad city, which used its wealth to tap the water of 17 *qanats* distant from the city by as far as 100 km and collected by a canal that follows the Mourhab valley and, even today, irrigates the lush gardens of the city. Rules have defined protected areas to prevent conflicts between *qanat* users (Foltz, 2002). Areas like those of Borkhar, north of Esfahan, were well known for their high density of *qanats* (see Fig. 9.4).

Ancient water-use systems thus involved village ditch managers, system overseers and valley *mirabs* (in both the main and the lateral valleys), who were all nominated and paid by the users in their jurisdiction, with well-accepted and well-enforced rules for sharing the resource. The cultivation area and irrigation doses were attuned to the available river flow and to the discharge of the *qanats*, which served as 'phreatic barometers' (Lightfoot, 2003), their flow varying in line with the level of the aquifers. Likewise, gardens formed the core of the irrigated area but were not overextended so that they could stand water shortages. In case of excess surface water, short-cycle crops were cultivated on adjacent lands; this was the way to deal with the variability of the resource. As far as one can judge from available evidence, the system appears to have been strongly based on local governance and quite resilient. Hydrological interconnectedness was not critical because the density of *qanats* was regulated, and lateral valleys would contribute both surface flow to the Zayandeh Rud in excess years and a subsurface flow at least during a large part of the year.

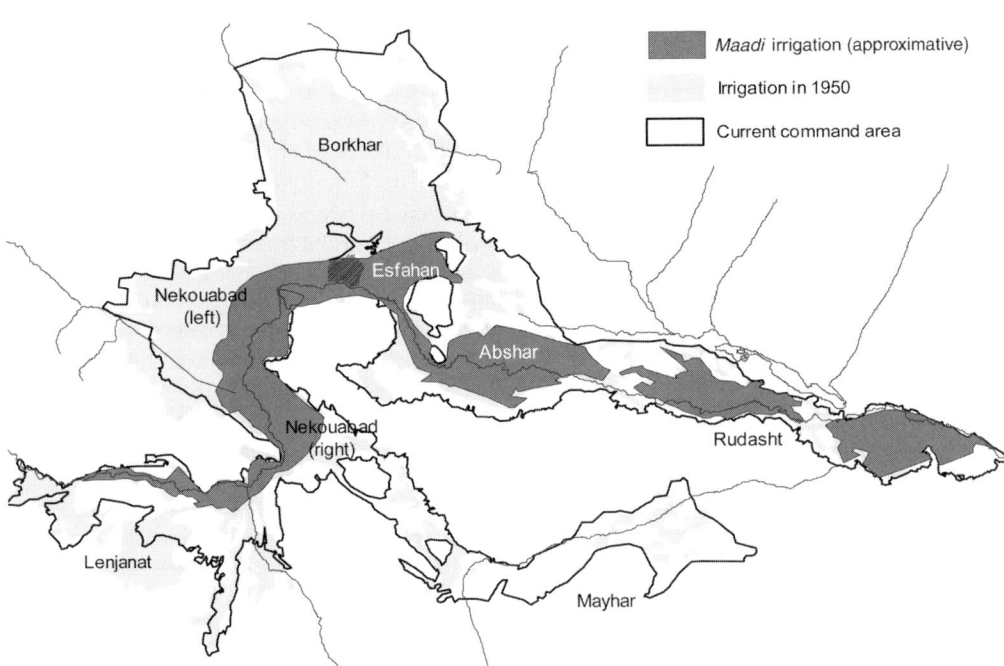

Fig. 9.4. Ancient and current irrigation areas in the main plain.

Recent Water Resources Development in the Basin

Large-scale state interventions

Agricultural and urban development in the Zayandeh Rud basin has always been constrained by water availability. But the history of the basin's water development is not (yet) a story of limits. It shows that demand – largely generated by expansion of irrigation schemes – always exceeded supply, despite the successive increases in available water brought by reservoirs and interbasin transfers. 'New' water was, each time, committed outright.

The basin resources were first augmented in 1953, when a first interbasin tunnel diverted water from the Kuhrang River to the Zayandeh Rud basin, adding 340 Mm^3/year to a natural runoff of about 900 Mm^3 (Abrishamchi and Tajrishy, 2002). In 1970, the completion of the 1500 Mm^3 capacity Chadegan reservoir (see Fig. 9.1) allowed the regulation of the water regime. With these two works, water supply and storage in the basin dramatically increased. This date also almost coincides with the nationalization of water resources in 1968 (and the establishment of regional water authorities, subordinate to the Ministry of Energy) and signals the new power acquired by the state to control the lifeblood of the region and to design the expansion of the irrigation area in the valley, where an area of 76,000 ha provided with modern hydraulic infrastructure was established. Yet, in many cases, these modern schemes were superimposed on the ancient network of *maadi* and *qanats*, and the gains were thus limited, although double-cropping became possible in most of the valley (Fig. 9.4). The *maadi* system and its attendant social organization and local knowledge were thus overridden and replaced by a state agency in charge of operation and maintenance. The intakes of most *maadi* were obstructed and instead the river was barred at two points (Nekouabad and Abshar) by major regulators that distributed water to new, large main canals, one on each bank of the river. Likewise, overseers and heads of *maadi* were replaced by state-appointed technicians.

With the opening of a second interbasin tunnel from the Kuhrang River in 1986, another 250 Mm^3 was made available annually.[2] This spurred the rehabilitation of the old Rudasht scheme, at the tail-end of the valley, and the extension of the irrigated area by some 40,000 ha (Borkhar and Mayhar schemes). Part of these two districts was already irrigated with groundwater, but overexploitation had generated problems of declining water quality, which new surface water was first supposed to mitigate; whatever fresh water was available in excess would be used to expand cultivation.

The increased available supply, in addition to being committed to new irrigation areas, also met the growing needs of Esfahan (with its population now totalling 1.6 million, and a growth rate that reached 5% in some years) and of neighbouring industries. The industrial sector now needs over 100 Mm^3 annually.

In 2009, an additional 260 Mm^3 will be made available through the third Kuhrang tunnel, together with 200 Mm^3 diverted from the Dez River upper catchment (the Lenjan tunnel). This will more than double the natural annual runoff (see Fig. 9.1). Another tunnel, the Behesh Abad tunnel, is under study. It would bring 700 Mm^3 downstream of the Chadegan dam but would require a very costly 75-km-long tunnel (Abrishamchi and Tajrishy, 2002; Morid, 2003).

The evolution of surface water supply and use is shown in Fig. 9.5. Inflow into the valley (measured at Pol-e-Kaleh station) is completely diverted and consumed, except in wet years, when part of it reaches the Gavkhuni lake (flow at Varzaneh). The additional inflow to be brought by the two new tunnels is likely to be fully allocated and consumed as soon as it is made available. At best, within a few years, they will help to replenish aquifers if farmers can use more surface water instead of groundwater.

There is no significant year-to-year carry-over storage in the Chadegan reservoir because almost all of the flood water entering the reservoir is released prior to the next flood season. This maximizes the production from irrigated agriculture (at the expense of security in supply), and part of the variability in supply is handled by resorting to groundwater. This buffering role of aquifers was critical in the 1999–2001 drought (see later) (Molle et al., 2008). Yet this role is gradually weakened by the decline of the aquifers, and they will not be

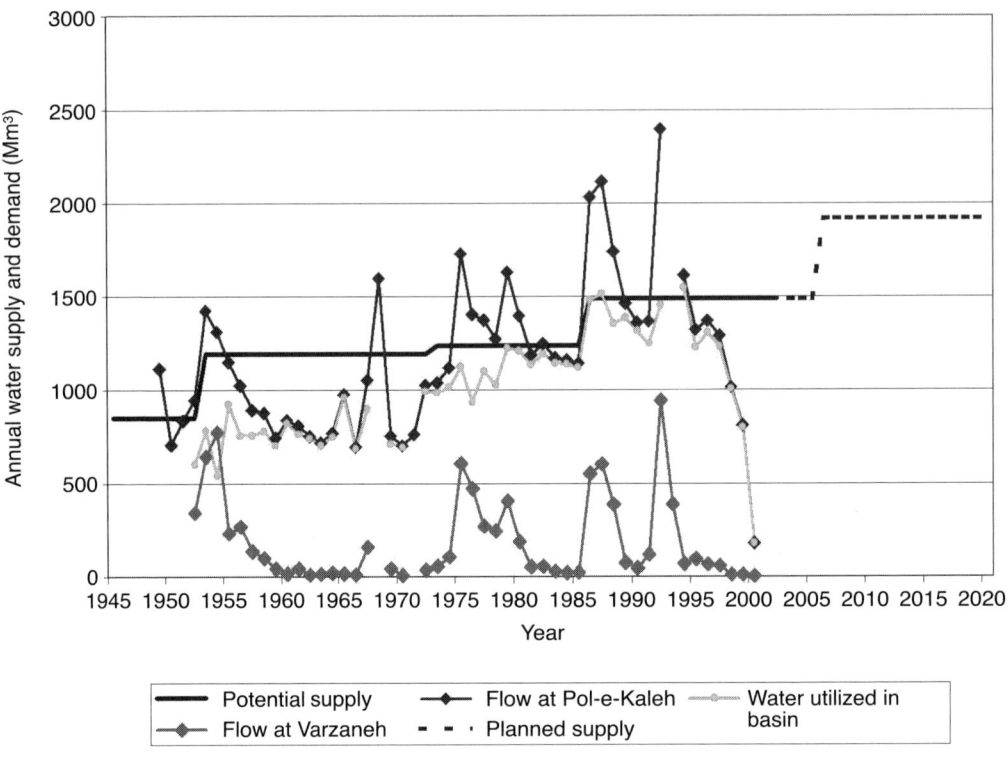

Fig. 9.5. Evolution of supply and use of surface water in the Zayandeh Rud basin (Murray-Rust and Droogers, 2004).

able to compensate for dwindling surface water in the long run.

State investments and regulation did not remain confined to the main valley: they also expanded into lateral valleys, such as the Hana and Mourhab valleys. In the latter, for example, in the late 1980s the Ministry of Jihad undertook the construction of the Khamiran dam, with the objective of increasing storage and local water use (Molle et al., 2004). The dam was completed in 1992 and has a capacity of 6.8 Mm³. Instead of the natural system of aquifer recharge through the stream, which had prevailed for centuries, the dam is now supplying water to downstream villages through a lined canal approximately 40 km long. To increase the value and usefulness of the Khamiran dam and extend the benefits of the Chadegan reservoir to other valleys, a plan was drawn up to pump water from the reservoir over the mountain ridge into the Khamiran dam. In 1991, the Karvan pump station was constructed for that purpose, but it faced severe technical problems and its operation was discontinued after some 3 years (Newson and Ghazi, 1995).

Local water resource development

Notwithstanding these state-initiated projects, villagers at the local level have also been actively looking for ways to respond to population growth by increasing supply from aquifers, through *qanats* or wells. The growing intervention of the state after 1968 came together with a modernist ethos that considered traditional village irrigation as primitive, backward and inefficient (McLachlan, 1988; Ehlers and Saidi, 1989). Modernization required technology and modern water-lifting devices, and the development of pumps and wells was seen as very advantageous compared with *qanats*, because the fluctuating discharge of the latter was considered as hindering agriculture. This considerably boosted the expansion of wells, which started in the late 1950s. While in the 1950s the contribution of tube-wells was negligible and existing *qanats* were serving 1.2 million ha of irrigated land in the whole of Iran, by the mid-1970s wells were already providing 8 billion m^3 against 9 billion m^3 by *qanats* (McLachlan, 1988).

The post-revolution period was marked by the continuing development of shallow wells. This was part of a policy emphasizing self-reliance and the development of production, coupled with a strong stance in favour of population growth (which reached a rate of 3.8% in the 1980s). This development seems to have been based on inadequate hydrological analyses, and villagers got into the business of well-digging despite reservations and awareness that *qanats* might be impacted. In Jalalabad, for example, the wells did bring a substantial increase in water supply. Jalalabad received an authorization to sink eight wells around the village, and these were used to expand the garden area. In addition, villagers obtained a permit to dig 15 wells within the existing orchards, as a way to boost the available water per hectare of garden. As a result, however, the discharge of one of the two *qanats* used by the village soon started to dwindle and eventually dried up. The impact of the development of wells on the discharge of the *qanats* confirmed local knowledge about the interconnectedness of the different water sources.

Studies conducted by the Esfahan Water Authority (EWA) in 2000 revealed that several aquifers were being overexploited, especially in some of the irrigated areas (Morid, 2003). Presently about 21,200 tube wells, 1726 *qanats* and 1613 springs exploit a total of 3619 Mm^3 of groundwater annually. This is more than twice the surface water diversions, which (although both sources are partly interdependent) gives an idea of the importance of groundwater in the Zayandeh Rud basin.

Socio-hydrological Interconnectedness

Despite the periodic transfer of additional water from neighbouring basins, these changes in water resources development and use point to a constant overcommitment of resources. The increase in the abstraction capacity, notably because of the overdevelopment of irrigated areas, created a very tight river basin system, where some water paths disappeared or were reversed and where users in the basin are increasingly interdependent. What is stored, conserved or depleted at one point dictates what is available at another point further downstream; externalities travel across the basin in a way that is blurred by the irregularity and partial invisibility of the hydrological cycle (Molle, 2003). This section illustrates several social/spatial competitions and allocation conflicts which result from this growing interconnectedness.

Upstream versus downstream

In the absence of clear and enforced water rights, upstream areas are in an advantageous position. In closed basins, new upstream abstraction merely shifts the benefits of water use from downstream to upstream areas. A typical example of such a shift in the Zayandeh Rud basin is occurring between the Chadegan reservoir and Lenjanat (the beginning of the main plain; see Fig. 9.1). Traditionally, irrigation was restricted to the narrow valley bottom (to areas which, altogether, might nevertheless amount to 40,000 ha) and occurred through gravity. Numerous private, large diesel pumps now abstract water to supply 10,000 ha of nut and almond orchards located on the plateau, 150 m above the valley floor (Murray-Rust and Droogers, 2004). These orchards, often irrigated with drippers, may be in the order of 10,000 ha and are rapidly expanding. One bank of the river belongs to the Chaharmahal-va-Bakhtiari province, which – in the absence of interprovincial allocation agreements – is supporting this development, based on the perception that the river is also 'theirs'.

Other upstream capture of resources is apparent in the unbalanced share of water delivered to the different irrigation schemes (see Fig. 9.4). The Nekouabad schemes receive, on average, 39% of the total irrigation supply, although they only make up 18% of the irrigated area. Expansion of irrigation facilities to the Borkhar and Mayhar schemes has also reduced the amount of water flowing downstream. Increasing water scarcity (and resulting soil salinization) in the downstream area can be contrasted with its affluent past: strikingly, in the 10th century, Hawqal (1889) reported that the [tail-end] districts of Rudasht and Baraan constituted 'an important region in which ten mosques can be found. Harvests are abundant and *all the supply of Esfahan comes from it*' (emphasis added). Benefits from water use have clearly been shifted upstream.

Wells versus *qanats*

A prime example of reallocation is, of course, that of wells depleting local aquifers. Development of wells is tantamount, at least partially, to a reallocation of water from *qanat* (sometimes spring) owners to the well owners, and – oftentimes – from collective to individual use and management. These owners may or may not be the same persons, but those with the financial capacity to drill wells tend to get the upper hand. The development of wells eventually reduces groundwater flows to downstream areas. Jalalabad's farmers, in the Mourhab valley, understand that groundwater is not a static resource and that the issue is 'pumping groundwater before it flows downstream', as one of them expressed.

The history of the destruction of *qanats* by wells, in Iran and elsewhere, is documented by several studies (e.g. Ehlers and Saidi, 1989; see also Lightfoot, 1996 and Mustafa and Usman Qazi, 2007, for examples from Morocco and Baluchistan, respectively). It is likely, however, that in some areas the potential of groundwater was higher than what the *qanats* were extracting, but insufficient control of their number and location eventually led to competition with the *qanats*. The *qanats* of the Borkhar area, for example, a flourishing cultivated area north of Esfahan, were destroyed by the spread of deep wells sunk to irrigate summer crops and orchards (Lambton, 1969).

Qanat discharges are determined by the height of the water table, which determines the length of the water-bearing section (Beaumont, 1989). Wells, in contrast, ensure a more or less constant discharge, irrespective of the depth of the water table (at least in a certain range and in the short term). They are not only less sensitive to variations in the groundwater stocks but may also abstract more water out of the aquifer than what comes in as recharge. The 'mining' of aquifers had little short-term impact but proved to be unsustainable after a few years, especially when the 1999–2001 drought occurred.

Lateral plains versus the main plain

Depletion of groundwater in both the main and lateral valleys has inverted the total net underground flow to the Zayandeh Rud. In the Mourhab valley, for example, the cumulative impact of the Khamiran dam and the wells and the *qanats* on the groundwater flow to the Zayandeh Rud itself has been dramatic, although partly invisible, since water was 'retained' in the valley. Likewise, Gieske and Miranzadeh (2003) have estimated that approximately 250 Mm^3 out of an annual yield of 275 Mm^3 of lateral groundwater flow to the Lenjanat alluvial fan aquifer is now tapped. These examples show how base-flow water formerly used by agriculture downstream in the main valley was reallocated almost 'invisibly' to provide benefits to upstream farmers.

Further down the valley it is, in all likelihood, the river which now recharges the valley aquifers, an aspect which is often overlooked (Morid, 2003). By drawing down the water table, well users (including the city that sank deep wells to irrigate large 'green belts' of trees planted 'for the environment') not only tap underground flows that used to contribute to the base flow of the river but also 'drag' water from the river bed to lateral aquifers, to the detriment of irrigation downstream of Esfahan.

City versus agriculture

As in many regions of the world, the combination of water scarcity and urban sprawl results in water being reallocated out of agriculture to the domestic and industrial sectors. In the Zayandeh Rud basin such reallocation is left to the discretion of the Ministry of Power, which controls the allocation of the Chadegan dam water and accommodates demands and requests from MPs or other political constituencies (Ghazi, 2003). For example, factories generally have no problem in getting supply from irrigation canals since their demand is allegedly limited and the Ministry can sell water to them at a much higher price. The interests of construction and landscaping companies notoriously involved in kickback practices are also more easily catered for (Foltz, 2002).

That priority in allocation is given to non-agricultural uses was well illustrated in 2001, when, at the peak of the drought, diversions to agriculture were reduced to zero during the whole season and cultivators were left solely with their groundwater resources, despite water releases from the dam still amounting to 39% of yearly average values (Molle et al., 2008). Power asymmetries were made patent when business owners (and angry residents alike) in the city asked for water to be released from the dam, claiming that national coverage of the crisis in the basin (children playing soccer in the river bed) was detrimental to the flow of tourists which normally converged to the city. As the attractiveness of Esfahan is strongly related to the spell of its gardens and bridges, water was released to the Zayandeh Rud (literally the 'life-giving river') to restore their magic and save the tourist season.

Greater Esfahan, with its population of 1.6 million and its current annual growth rate at 2.3%, receives an increasing share of water, estimated at 250 Mm^3/year. In the 1970s, the Zayandeh Rud basin was the focus of specific government policies to increase industrial production outside Tehran. Esfahan was seen as a prime location, particularly as the Chadegan reservoir had just been completed and it was assumed that water supplies would be readily available. Between 1975 and 1977 four major industries were developed (defence industries, Mobarekh steel mill, Esfahan oil refinery and Sepahan cement factory), with a total annual demand of 60 Mm^3. A polyacrylic factory was added in 1980, with a demand of an additional 5 Mm^3. The war with Iraq halted industrial development, but from 1988 to 1991 more industrial enterprises were established, with a total demand of 39 Mm^3. Total industrial demand is therefore at least 104 Mm^3 (Murray-Rust and Droogers, 2004).

But water is also committed to cities located in much drier areas (Yazd, Rasfanjan, Kashan) and outside the basin. Yazd receives 90 Mm^3 annually through a pipeline, and diversions of 42 Mm^3 to Kashan and Sahr Kurd will soon start (Abrishamchi and Tajrishy, 2002; Morid, 2003). While these cities are more distant from the Zagros 'water-tower' and their situation is somewhat worse, these transfers are also political decisions, which are probably not unrelated to the fact that Yazd and Rafsanjan are home to former Presidents Khatami and Rafsanjani.

Human use versus nature

Abstraction of all the water available in the river has been the rule since the mid-1960s, when the basin closed and the flow to the Gavkhuni swamp and lake was limited to flood periods and excess years (see Fig. 9.5). As a result, the Gavkhuni swamps, an important wetland for migratory birds and registered as a Ramsar site, became degraded. Salinity of soil and water in Rudasht – the tail-end agricultural area – is on the rise; yields are the lowest in the valley, and some plots are now left uncultivated (Morid, 2003; Murray-Rust and Droogers, 2004).

More generally, reduced diversions to irrigation also means that percolation and leaching of salts have been reduced, while the groundwater used as a substitute is also often of poor quality. Soil management becomes a central issue as more soils are threatened by salinization and by becoming sodic.

With insufficient discharges in the river, river health has also been impacted, and the values of biological oxygen demand from Esfahan downstream are classified as 'bad' (i.e. higher than 10) and reach 23 (Pourmoghaddas, 2006).

Groundwater exploitation versus next generations

Following the construction of the Chadegan reservoir, it appeared that water table levels have risen in many areas, not least in Rudasht, at the tail-end. However, data over the past 10 years indicate that groundwater levels are dropping in all parts of the irrigated areas of the basin, and in some areas they are dropping dramatically. In Najafabad, just west of Esfahan, fruit trees planted 10–15 years ago based on groundwater irrigation are dying due to rapidly declining groundwater, resulting in older wells drying up due to the drilling of larger, deeper wells for urban and industrial water supplies.

While wells have spread in areas formerly exploited through the use of *qanats*, they have also developed in irrigation schemes. Within the irrigation systems, the decline of aquifers has been more or less constant in the past 6 years. In Nekouabad left and right banks, average decline has been 2.5 m/year and 1.5 m/year, respectively, almost certainly exacerbated by domestic and industrial installation of wells. In Abshar it has declined by some 0.4–0.6 m/year, in Borkhar by 0.8 m/year, and even in Rudasht, where water quality is poor, groundwater tables have dropped by 0.25 m/year. This suggests that somewhere around 250–600 mm/year are being pumped for agriculture and are not being recharged (Murray-Rust and Droogers, 2004).

Aquifers definitely have a crucial buffering role in compensating for deficient surface water supply in dry years. A fascinating measure of their importance was provided in 2001, when no water was delivered to irrigated areas but the cropping area was still at 60% of its value in a normal year (Molle *et al.*, 2008). This role, however, can only be sustained if aquifers are replenished; it is hard to imagine, at the moment, why and how this could occur. In addition, it is also unclear to what extent the overdraft of the aquifer can continue without incurring changes in the water's salt content.

Main Issues and Responses to Basin Closure

Allocation mechanisms and basin governance

The problems of competition highlighted above signal a situation in which water is constantly reallocated through the decisions of both local actors (e.g. spread of wells) and the state (e.g. construction of irrigated schemes, export of water, etc.), with negative consequences in terms of equity and environmental sustainability, and externalities concentrating on downstream rural users, the environment and the next generations. Overallocation (due to an abstraction capacity far above available resources) and reallocation (whether implicit or explicit, intended or not) are due to both the lack of control/monitoring of who gets what and when, and the absence of a system of entitlement or rights.

The Civil Code, following Islamic Law, gives priority to established owners of land over newcomers, and upstream over downstream users of water (Ghazi, 2003). Prior appropriation rights were protected by a clause stipulating that the use of water by newcomers should not impact on the interests of existing users. However, McLachlan (1988) reports that:

> the legal frameworks from Islamic Law and the Civil Code that surrounded water use were powerfully supplemented by customary practices ('*urf*') ... These local regulations governed to a large degree the access to, and use of, water in irrigation within what was a complex organization of supply in an uncertain physical environment.

The need to protect springs, wells and *qanats* was addressed by defining a *harim*, or an area with extraction around these sources prohibited (Foltz, 2002). While these socially controlled modes of water exploitation were efficient at the scale of communities, they were eroded by the lack of control and hydrological criteria regarding the drilling of wells.

The nationalization of water resources was introduced in 1967 as the tenth point of the Shah's 'White Revolution', and regional boards were established to assess and control water use and to charge for its consumption. The 1968 Water Law was intended generally to end the traditional system of water rights, based primarily on the riparian doctrine, and replace it with a system of rights based on water-use permits for the purposes of beneficial and reasonable use of these resources (Beaumont, 1974). The state thus gained wide power of control and taxation of private/communal ownership. In several instances, the state took over the management of minor schemes and abolished customary rights, with mixed results (Lambton, 1969; Ghazi, 2003), but this seems to have happened on a case-by-case basis.

In the valley itself, with the superimposition of concrete canals over the network of ancient *maadi* in the early 1970s, the state largely overrode the riparian rights enshrined in Sheikh Bahai's regulation. Yet the administration could not fully erase these rights, and a study of water allocation within schemes has shown that ad hoc distinctions were made between canals built in former *maadi* areas and those in newly reclaimed areas (Hoogesteger, 2005). In the Mourhab valley, traditional rights on the river water were equally eroded. The redistribution of water in the Mourhab valley after the construction of the Khamiran dam was a non-transparent process with no direct participation of the population concerned.[3] Some villages that had developed quite lately and had no right to water were allocated part of the water coming from the dam. In contrast, other former rights-holders, like Jalalabad, lost the benefit of the river.

The examples given above make it clear that some sort of basin-level coordination body is needed to analyse hydrological data, establish transparent allocation schemes (through a system of entitlements or otherwise), discuss priorities and development plans, and integrate representatives from the different socio-economic sectors. Such participation is, however, unlikely to be very effective under present circumstances, since representation of the civil society is still weak (Namazi, 2000). The state is likely to retain full control of the decision-making power of such a vital resource. Establishing a sound water regime at the basin level is thus a monumental task, which needs governance patterns that are yet to emerge.

Limited scope for (real) efficiency gains

In a basin with hardly any water reaching its terminus, water can only be 'saved' by limiting unproductive evaporation. There are not so many opportunities to achieve such a reduction. Conventional conservation efforts impact water pathways and merely reallocate water: canal lining in Jalalabad 'saves' water, which can then be spread over a larger area, increasing not only local production but also water depletion, to the detriment of downstream users who were tapping subsurface flows. The canal that collects *qanat* water for Najafabad city has also been lined to offset declining supply, thus increasing the flow to Najafabad but, at the same time, decreasing groundwater recharge in the Mourhab valley.

Little is known about the efficiency of large-scale irrigation in the valley. In Iran, as elsewhere, gravity irrigation is stigmatized as a process wasteful of precious resources and micro-irrigation is held as a natural solution to this state of affairs. In the particular setting of the valley, however, it is dubious that much improvement can be brought about: there already exists extremely efficient recycling of 'losses' at the plot level (pumping of groundwater), at the scheme level (pumping from drains) and at the valley level (the return flow from one scheme – 30% of gross diversion values on average – is part of the supply to the following one).

Micro-irrigation is believed to reduce unproductive soil evaporation, but even this benefit is unclear and has been found by some researchers to be sometimes illusory (Burt et al., 2001).[4] In any case, there are also a number of constraints to the adoption of micro-irrigation. First, not all crops (e.g. rice or lucerne) are suitable for such a technique; second, the investment cost is very high and can never be offset by whatever saving in the water bill (Perry, 2001); and third, such investments only make sense for high-value crops for which security of supply is essential (as such, they are more likely to be adopted where groundwater is abundant and used).

Efficiency gains have also been sought in improvements of scheme management. A few years ago the government contracted out the operation and maintenance of irrigation systems to parastatal enterprises, cleverly referred to as the *mirab*: as in many other countries, the ideology of efficiency that favours private rather than state operators has allowed former staff from state agencies to form their own companies and to perform the same service but with some private benefit to themselves (although workers who moved along from one structure to the other lost their former state privileges and saw their working hours increase markedly; see Hoogesteger, 2005). Possible efficiency gains are undocumented but the administration claims that costs have been cut by 15–20%.

Groundwater control

The control of groundwater use has been problematic, even though the drilling of new wells is checked by the local farmers themselves (who do not want to see more local abstraction) and by a control of the activities of drilling companies. The right to access groundwater is officially regulated by the granting of permits by state authorities. Permits have been administered centrally, with limited knowledge of local hydrology, transparency and control by interested populations. This has opened the way for bribery and for powerful people to obtain well permits thanks to their political clout.

Control of groundwater abstraction is an intractable problem worldwide. As supply in public schemes becomes deficient, farmers resort to wells as a compensation. It would be politically very hard for the state to parallel its failure to deliver reliable water by a crackdown on self-funded private wells; indeed, the administration acknowledges that illegal drilling of wells is a pervasive problem (Hoogesteger, 2005). Overcommitment of resources and the resulting decline of supply to agriculture are likely to reinforce the shift to groundwater and the dropping of water tables.

Water quality, wastewater and health

With reduced flows and recurring shortages, and pollution from both agriculture and industries, the health of the Zayandeh Rud River has been affected. The solute content of the irrigation return flow into the aquifers and the river, combined with urban and industrial effluents, is much higher than that of the water flowing in the river. The mixing leads to progressively increasing levels of salinity (measured as EC, electrical conductivity) and total dissolved solids (TDS) along the Zayandeh Rud.

Pourmoghaddas's (2006) study of water quality in the Zayandeh Rud between 1989 and 1999 (not including drought years) shows that the average value of EC is around 250 mS/m before the river enters the plain, rising to 700 mS/m after receiving industrial effluents and to 1200 mS/m in Esfahan, increasing to 4500 mS/m as the river receives return flow from the Abshar irrigation scheme, and peaking at 19,600 mS/m in the terminal reach of the river. The pattern is similar for non-agricultural pollution. The concentration of the major cations and anions follows the same increasing trend as one goes downstream. The concentration of heavy metals (Pb, Ni, Cd) increases tenfold as the river passes through Esfahan, to levels of 0.1 mg/l for Pb, 0.07 mg/l for Ni and 0.02 mg/l for Cd (four times WHO's standards) (Vahid, 1996). A sharp decrease in dissolved oxygen (DO) is observed at the Pole Chom station, where effluent of the wastewater treatment plant discharges into the river.

A hydrochemical analysis of groundwater from boreholes along the Zayandeh Rud River reveals the same pattern, which is not surprising as the aquifers are recharged both by the river water and by return flow and leakage from the irrigation schemes. A detailed hydrochemical study of a small subcatchment (Lenjanat) along the Zayandeh Rud upstream of Esfahan over a 10-year period has shown that the groundwater composition is subject to long-term trends (Gieske et al., 2000). In some parts of the aquifer, salts are being slowly flushed out, whereas in other parts concentrations are rising. It appears that the groundwater composition is slowly changing in response to expanding or variable cultivation practices. Other studies on shallow wells (1.5–9.5 m) also showed that pollution has been transferred from the river to aquifers (Pourmoghaddas, 2006).

Such levels of pollution may create public-health hazards, as during the 1999–2001 drought, when the treatment station of Esfahan could not handle the quality of the incoming water, resulting in serious health problems in the city. The effluents of Esfahan are also increasingly reused by agriculture, but the health impacts are not well known at the moment. Tourist and urban development around the Chadegan dam not only extracts water from the lake but also pollutes it in return, impacting the quality of water at its source. In sum, degraded water quality results in various health and environmental impacts, which tend to get worse both in the long run and in times of shortage.

Vulnerability to droughts

When basin water resources are overcommitted and fully depleted there is no more slack in the system and all the hydrologic variability in supply is passed on to users. Since urban uses receive priority, agriculture (not to mention the environment) has to cope with a supply that basically varies each year and bears the brunt of climatic variability. The 1999–2001 drought has put this fact in sharp relief (Molle et al., 2008). The third year was very critical, since diversions of surface water decreased down to 39% of average values, with the irrigation share at only 3% of its pre-drought average. Yet, contrary to this dramatic drop in supply, cropping areas were curtailed by 39% only, although there was a degree of shift to crops with lower water requirements and average yields were slightly affected (by 12%).

Farmers have responded to the drought and to pervasive water scarcity in the past 20 years in different ways, as illustrated by a study of farmers' coping strategies in the Abshar irrigation system (Hoogesteger, 2005). At the outlet level, some user groups defined priority rules (e.g. priority to smallholdings) to allocate limited water; in others, some farmers ceded their share to others and left their land fallow; elsewhere, farmers joined together to drill collective wells. At the individual level, farmers' responses included: increased use of groundwater by drilling or deepening of wells; use of untreated effluents from Esfahan; a shift to less-sensitive crops, such as fodder maize; migration to other regions unaffected by the drought to rent land; and lease or sale of land (Molle et al., 2008). Despite this adaptive capacity, recurring shortages tend to affect the weakest farmers and to drag them out of business in a context of high unemployment.

Reopening the basin?

The history of the Zayandeh Rud basin has shown repeated resorts to water import as a means of solving the recurring and marked imbalances between supply and demand. At first sight this would appear to merely result from population growth (Esfahan sheltered refugees from western provinces during the war with Iraq, when its population grew at an annual rate of close to 7%), industrial development and the needs of agriculture. This latter sector, although subject to irregular supply, still totals 66% of water diversions in an average year and there are serious questions about the reasons for continuing investment in irrigation infrastructure.

It seems somewhat contradictory that while large-scale irrigation systems established 30 years ago (the Nekouabad and Abshar schemes), let alone the traditional systems that go back hundreds of years, are struggling to get sufficient water, new irrigation developments continue apace in the basin. Many of the reasons 'why enough is never enough' (see Molle, 2008, for an examination of the societal drivers of basin overbuilding) possibly apply to the present case. The financial and political benefits accruing to a set of decision makers and entrepreneurs may have played a role in the extension of canals to Borkhar and Mayhar areas (Foltz, 2002). At a minimum, the design hypotheses and justifications for such works, in a context where water is increasingly exported to large cities in neighbouring basins, are likely to have been dubious.[5] While in the current situation of high unemployment agriculture remains a sector which cannot be neglected, it is also not clear what categories of farmers benefit most from these investments.

All in all, it may well be that this benefit will be very limited, since supply is likely to be limited and intermittent. A perverse consequence of such overdevelopment of irrigation infrastructure, however, is that it 'mechanically' generates water scarcity, exposes 'beneficiaries' to the precariousness of uncertain supply, and creates the political conditions for justifying further development. With this logic at work, further and highly costly imports of water are expected to be effected. It can be equally feared that the next abundance of water will be absorbed by waiting fields in the Borkhar and Mayhar areas, and perhaps in new areas, which will be planned to raise the design economic benefits of the new transfer.

While the basin is buying respite at a high cost[6] (although this cost is largely shifted to the national level), one may wonder what the limits of such a process are. It is already apparent that 'donor basins' are complaining about the diversions and that these are only made possible because of the overriding decision-making power of the central government. During the drought, for example, people in the lower Dez basin (of which the Kuhrang is a tributary) suffered shortages and severe health problems. Diversions also take their toll on hydropower generation, since the Kuhrang feeds into the Karun and its four dams (the first hydropower complex in the country). These externalities imposed on donor basins should certainly be considered in order to get a clearer picture of the full costs of these transfers.

Conclusions

No doubt sprawling urban oases with growth dynamics that largely lie beyond the question of water availability are faced with critical challenges. In the Zayandeh Rud basin, increase in population, decline in farm size and agricultural income, environmental degradation and growing sectoral competition for water appear to be at loggerheads with the finite and circumscribed nature of the water generated in the Zagros mountains. Yet, while oasis culture is characterized by frugality and attention to nature's limits, the Zayandeh Rud basin seems to have developed without a sense of limits. Esfahan and its surroundings have been planned to become major urban and industrial poles during both the Shah and the post-revolution periods; irrigation infrastructure has been repeatedly overdeveloped, leading to suboptimal cropping intensities and forcing farmers to complement canal supply with groundwater. At each step of the Zayandeh Rud basin development, these contradictions were – albeit briefly – dissolved by the construction of a dam or by an interbasin transfer which 'reopened' the basin. Despite these interbasin transfers, which double the availability of surface water in the basin (in 2009), and a total use of groundwater estimated at 3500 Mm^3/year (i.e. 72% of all water use), only less than 2% of the natural flow of the river reaches the Gavkhuni marshes (Management and Planning Organization, 2002). Considering the overdraft of aquifers signalled by dropping water tables (on average, 2.5 m/year), water use in the basin exceeds renewable resources. By all definitions, the Zayandeh Rud basin is closed.

In such arid areas where land is abundant, any possible excess of water will be readily absorbed by waiting fields or expanding cultivated land if no regulation control is exercised; likewise, unchecked drilling of wells will also tend to exhaust aquifers and, in places, cancel the historical investments and rights vested in the qanats. Imperative demands from neighbouring desert cities with even less available supply also contribute to sucking up whatever additional water is made available. The basin has thus been buying respite by ever-increasing capital investments in tunnels, but this logic now collides with the financial costs of the works required and the externalities generated on donor basins.

The spatial pattern of water resources development induced a gradual shift of benefits upstream: the Gavkhuni Ramsar site and the lush gardens of Rudasht of bygone days are the obvious victims of that shift of water use to upstream urban areas, almond tree orchards and tourist resorts around the lake. The study provides instructive and graphic examples of how water gets redistributed between surface water and groundwater, upstream and downstream, the lateral and the main valleys, wells and qanats, between villages, and between rural and urban users. All human interventions induce hydrological changes that travel across scales and time, and across levels of social and political control. This interconnectedness across scales has critical implications for societies, since it links macro-level management and decision making to local processes.

The absence of clear allocation rules or water rights means that interventions, re-appropriation and redistribution, with their impacts across scales and social groups, are a sizeable reality. The three main losers of this lack of overall control over resources use in the Zayandeh Rud are, not surprisingly, those most commonly affected in closing basins: the downstream users, the next generations and the environment, in decreasing order of bargaining power. The environment bears the brunt of the reduction of flows at a time when more water is generally needed to dilute pollution and to leach the salt. The next generations are affected by the gradual and continued depletion of groundwater resources. Agriculture, as the residual user, has to deal with a supply that basically varies each year. There is no slack in the system and the only buffering capacity or flexibility is provided by declining aquifers.

A consequence of the closure of the basin that cannot be overemphasized is the logical impossibility of overall water conservation, except where unproductive evapotranspiration can be reduced. Local conservation measures are possible but they necessarily have third-party impacts. Therefore, while such local measures may have benefits for the users involved, they are – just like additional abstraction or diversions – eventually tantamount to a mere reallocation of water within the basin. Shifting the benefit of water may be desirable or not, but it is rarely explicit and raises questions on equity, water rights and third-party impacts.

The complexity of social and hydrologic macro–micro interactions makes the state incapable of reordering the basin water regime by its sole action or by legislation. Constructing a sound and sustainable water regime is contingent upon enabling multi-level governance patterns, which allow interest groups to negotiate arrangements that bring more certainty, social value and equity to the sharing of water. This does not mean that the power of centralized management agencies should be eliminated. Rather, the nested nature of hydrologic scales and the overriding importance of dam management and bulk allocation call for forms of co-management (Sneddon, 2002), with management power and responsibility 'shared cross-scale, among a hierarchy of management institutions, to match the cross-scale nature of management issues' (Folke et al., 2007).

In the Zayandeh Rud basin, the challenge could be to re-establish the earlier stakeholder-controlled allocation (when *mirabs* were elected). An ancient source quoted by Spooner (1974a) stresses that the *mirab* 'must prevent the powerful from trespassing on the weak with regard to the shares of water', and referee water disputes 'with the confirmation and approval' of the local leaders. According to Hossaini Abari (2006), 'the management of the Zayandeh Rud was entirely in the hands of local people; the system was democratic and the government or state governors rarely had a direct role', while Ghazi (2003) underlines the strict enforcement of the rules. Whereas this management seems to embody what would nowadays qualify as subsidiarity and 'stakeholder empowerment', it must now be carried out in a much more complex physical and social setting than in the past, demanding both an increasing knowledge of the basin hydrology and expanded arenas of representation and negotiations.

Notes

1 The valley probably remained relatively underpopulated since the invasion and the destruction wrought by the Afghans (circa 1725). Around 1900, Zélé Sultan, the governor of Esfahan, tried to revitalize the valley by bringing people in from other regions (such as Yazd province). It is thus doubtful that water-sharing rules were established in the 16th century, but this shows the mythical role acquired by Sheikh Bahai in the celebration of past water wisdom in the area (Pirpiran, 2007).

2 There are large discrepancies in the average volumes transferred, according to source: Murray-Rust and Droogers (2004) refer to 250 Mm^3, and Abrishamchi and Tajrishy (2002) to 160 Mm^3. Morid (2003) reports that tunnels Kuhrang 1 and 2 (together?) divert 300–400 Mm^3 of water per year.

3 This change did not remain unchallenged. Villagers organized themselves and demonstrated against this change in Tiran and other places. These demonstrations ended up with some fatalities, but to no avail. The dam had a dramatic impact on the hydrology of the Mourhab valley. It was probably based on the common – yet radically wrong in the present context – idea that surface storage is beneficial because it may regulate water that would otherwise flow downstream unused. But springs and *qanats* feed on the huge natural water storage provided by the alluvial aquifer of the valley. This natural reservoir has overwhelming advantages over a dam: (i) it incurs no loss by evaporation; (ii) it is distributed all along the valley, allowing access to almost all villages; (iii) this distribution is free and requires no intervention; and (iv) water use was quite finely attuned to the available resource. In addition, the remaining flows, if any, were not lost, as often perceived, but used further downstream in the main valley.

4 A smaller fraction of the soil surface is saturated after irrigation, thus reducing soil evaporation losses, but more frequent irrigation increases the

average humidity content of the top layers; the two effects cancel each other.
5 Satellite images confirm that the Borkhar and Mayhar areas only have interspersed cultivation and are therefore irrigated far under their design levels.
6 While the Kuhrang 1 and Kuhrang 2 tunnels are 2.8 km long each, the Kuhrang 3 and Lanjan tunnels are 23 km and 15 km long, respectively. The Behesh Abad tunnel, under study, would be 75 km long (Abrishamchi and Tajrishy, 2002). This gives a measure of the corresponding increase in costs solely for the drilling of tunnels.

References

Abrishamchi, A. and Tajrishy, M. (2002) Interbasin water transfers in Iran. In: National Academy of Sciences (ed.) *Water Conservation, Reuse, and Recycling: Proceedings of an Iranian–American Workshop*. National Academy of Sciences, Washington, DC, pp. 252–274.

Beaumont, P. (1974) Water resource development in Iran. *Geographical Journal* 140(3), 418–419.

Beaumont, P. (1989) The qanat: a means of water provision from groundwater sources. In: Beaumont, P., Borine, K. and McLachlan, K. (eds) *Qanat, Kariz, and Khattara*. School of Oriental and African Studies, London, pp. 13–31.

Burt, C.M., Howes, D.J. and Mutziger, A. (2001) *Evaporation Estimates for Irrigated Agriculture in California*. ITRC paper P01-002. Irrigation Training and Research Center, Fresno, California.

Ehlers, E. and Saidi, A. (1989) Qanats and pumped wells: the case of Assad'abad, Hamadan. In: Beaumont, P., Borine, K. and McLachlan, K. (eds) *Qanat, Kariz, and Khattara*. School of Oriental and African Studies, London, pp. 89–122.

Folke, C., Pritchard, L., Berkes, F., Colding, J. and Svedin, U. (2007) The problem of fit between ecosystems and institutions: ten years later. *Ecology and Society* 12(1), 30. [Online] URL: http://www.ecologyandsociety.org/vol12/iss1/art30/

Foltz, R.C. (2002) Iran's water crisis: cultural, political, and ethical dimensions. *Journal of Agricultural and Environmental Ethics* 15(4), 357–380.

Ghazi, I. (2003) Legislative and government intervention in the Zayandeh Rud basin, Iran. Paper presented at the workshop on the 'Comparative river basin study, Comprehensive Assessment of Water Management in Agriculture', Embilipitiya, Sri Lanka, October 2003.

Gieske, A. and Miranzadeh, M. (2003) *Groundwater Resources Modeling of the Lenjanat Aquifer System*. Research Report 15. Iranian Agricultural Engineering Research Institute, Karaj, Iran/Esafahan Agricultural Research Center, Esafan, Iran/International Water Management Institute, Colombo, Sri Lanka.

Gieske, A., Miranzadeh, M. and Mamanpoush, M. (2000) *Groundwater Chemistry of the Lenjanat District, Esfahan Province, Iran*. Research Report No 4. Iranian Agricultural Engineering Research Institute, Karaj, Iran/Esfahan Agricultural Research Center, Esafan, Iran/International Water Management Institute, Colombo, Sri Lanka.

Hawqal, Ibn (1889) Surat al-'ard. In: de Goeje, M. (ed.) *Bibliotheca Geographorum Arabicorum II*. Leyde [Translated into French by Kramers, J.H. and Wiet, G., as *Configuration de la terre, 2 vols*, Paris and Beirut, 1964].

Hoogesteger, J.D. (2005) Making do with what we have: understanding drought management strategies and their effects in the Zayandeh Rud Basin, Iran. MSc thesis, Wageningen University, Wageningen, The Netherlands.

Hossaini Abari, S.H. (2000) *Zayandeh Rud from its Source to its Mouth*. Golha, Esfahan, Iran (in Farsi).

Hossaini Abari, S.H. (2006) *Ancient Water Management in the Zayandeh Rud River basin, Iran*. Working Paper (Draft report). International Water Management Institute, Colombo, Sri Lanka.

Lambton, A.K.S. (1938) The regulation of the waters of the Zayandeh Rud. *Bulletin of the School of Oriental Studies* 9(1), 663–676.

Lambton, A.K.S. (1953) *Landlord and Peasant in Persia*. Oxford University Press, Oxford.

Lambton, A.K.S. (1969) *The Persian Land Reform 1962–1966*. Clarendon Press, Oxford.

Lightfoot, D. (1996) Moroccan khettara: traditional irrigation and progressive desiccation. *Geoforum* 21(2), 261–273.

Lightfoot, D. (2003) Traditional wells as phreatic barometers: a view from qanats and tube wells in developing arid lands. Paper presented to the UCOWR Conference: Water security in the 21st Century. Washington, DC..

Management and Planning Organization (2002) *Studies of Land Use Projects of Esfahan Province*. Management and Planning Organization, Esfahan, Iran.

McLachlan, K. (1988) *The Neglected Garden: the Politics and Ecology of Agriculture in Iran*. L.B. Tauris & Co. Publishers, London.

Molle, F. (2003) *Development Trajectories of River Basins: a Conceptual Framework*. Research Report No. 72. International Water Management Institute, Colombo, Sri Lanka.

Molle, F. (2008) Why enough is never enough: the societal determinants of river basin closure. *International Journal of Water Resource Development* 24(2), 247–256.

Molle, F., Mamanpoush, A. and Miranzadeh, M. (2004) *Robbing Yadullah's Water to Irrigate Saeid's Garden: Hydrology and Water Rights in a Village of Central Iran*. IWMI Research Report No 80, International Water Management Institute, Colombo, Sri Lanka.

Molle, F., Hoogesteger, L. and Mamanpoush, A. (2008) Macro- and micro-level impacts of droughts: the case of the Zayandeh Rud river basin, Iran. *Irrigation and Drainage* 57(2), 219–227.

Morid, S. (2003) *Adaptation to Climate Change to Enhance Food Security and Environmental Quality: Zayandeh Rud Basin, Iran*. ADAPT Project, Final Report. Tabiat Modares University, Tehran, Iran.

Murray-Rust, H. and Droogers, P. (eds) (2004) *Water for the Future: Linking Irrigation and Water Allocation in the Zayandeh Rud Basin, Iran*. International Water Management Institute, Colombo, Sri Lanka.

Mustafa, D. and Usman Qazi, M. (2007) Transition from karez to tubewell irrigation: development, modernization, and social capital in Balochistan, Pakistan. *World Development* 35(10), 1796–1813.

Namazi, N.B. (2000) *Non-governmental Organizations in the Islamic Republic of Iran: a Situation Analysis*. United Nations Development Programme, Tehran, Iran.

Newson, M.D. and Ghazi, I. (1995) River basin management and planning in the Zayandeh Rud basin. *Esfahan University Research Bulletin* 6(1–2), 40–54.

Perry C.J. (2001) *Charging for Irrigation Water: the Issues and Options, with a Case Study from Iran*. Research Report 52. International Water Management Institute, Colombo, Sri Lanka.

Pirpiran, M. (2007) History of water distribution in the Zayandehrood river and Sheikh Bahai's scroll. Paper presented at the International History Seminar on Irrigation and Drainage, Tehran, Iran, 2–5 May.

Pourmoghaddas, H. (2006) Water quality and health issues in the Zayandeh Rud basin. Research report presented to the International Water Management Institute. Draft.

Rusteh, Ibn (1889) *K. al-A'laq al-nafisa*. In: de Goeje, M. (ed.) *Bibliotheca Geographorum Arabicorum, VII*. Leiden, The Netherlands [translated into French by G. Wiet, Les atours précieux, Cairo, 1955].

Sneddon, C. (2002) Water conflicts and river basins: the contradictions of comanagement and scale in northeast Thailand. *Society & Natural Resources* 15(8), 725–742.

Spooner, B. (1974a) City and river in Iran: urbanization and irrigation on the Iranian plateau. *Iranian Studies* 7(3–4), 681–713. The Society for Iranian Studies, Boston, Massachusetts.

Spooner, B. (1974b) Irrigation and society: two cases from the Iranian plateau. In: Downing, T.E. and McGuire, G. (eds) *Irrigation's Impact on Society*. University of Arizona Press, Tucson, Arizona, pp. 43–57.

Vahid, D.M. (1996) A study of toxic trace elements Cd, Pb, Ni in agricultural products using effluent of wastewater treatment plant. Masters' thesis, School of Public Health, Esfahan University of Medical Sciences, Esfahan, Iran.

10 Rural Dynamics and New Challenges in the Indian Water Sector: the Trajectory of the Krishna Basin, South India

Jean-Philippe Venot

International Water Management Institute, Accra, Ghana; e-mail: j.venot@cgiar.org

Introduction

In many river basins, water use for urban, industrial and agricultural growth is approaching, and sometimes even exceeding, the availability of renewable water resources. The Krishna River basin in South India is a good example: it has witnessed intense water development since India gained independence in 1947, resulting in overcommitment of water and river basin closure.

A generally accepted definition of a closed river basin is a basin where most or all available water is committed (Molden, 1997; Falkenmark and Molden, 2008) and river discharge falls short of meeting environmental functions (flushing out sediments, diluting polluted water, controlling salinity intrusion, sustaining estuarine and coastal ecosystems; Molle *et al.*, 2007). The process of basin closure intensifies the interconnectedness of ecosystems and water users across the basin. When river basins close, supply development projects and demand management reforms eventually tend to result in a regional or sectoral redistribution of water, along existing economic, political and social forces (Molle *et al.*, 2007). Early warnings of such an evolution are emerging in the Krishna basin. During the recent 3-year drought (2001–2004), surface water resources were almost entirely committed to human consumptive uses, groundwater was overabstracted and the discharge to the ocean was almost nil. The absence of any basin-wide strategy for water management has led to an uncoordinated expansion of surface water infrastructure and groundwater abstraction.

As the Krishna basin closes, recurring water conflicts suggest that there is not enough water for all current users and the environment: while more water is diverted than ever before, the security of supply to all existing users naturally declines – fuelling a feeling of scarcity and leading inevitably to conflicts over access and allocation.

This chapter attempts to unpack the forces that drove the closure of the Krishna basin. The first section presents the main features of the Krishna basin. The second section recounts the history of water development in the basin. The third section provides a water accounting method to quantify past and current water uses in the Krishna basin. The fourth section describes the main policy interventions that have affected the basin over the last 50 years, and the fifth section identifies some ways forward to slow down the process of river basin closure. The final section offers some conclusions.

Human and Physical Setting of the Krishna River Basin

The Krishna River basin is the fifth largest river system in India. The Krishna River originates in the Western Ghats, drains the dry areas of the Deccan plateau, and forms a delta before discharging into the Bay of Bengal. The main stem of the Krishna River has two major tributaries, the Bhima River in the north and the Tungabhadra River from the south (Fig. 10.1).

The Krishna basin drains an area of 258,514 km^2 in three states (Andhra Pradesh, Karnataka, Maharashtra). Most of the basin lies on crystalline and basaltic rocks associated with hard rock aquifers with low groundwater potential. The Krishna River basin is subject to both the south-west and the north-east monsoons; the average rainfall in the basin is 840 mm, of which approximately 90% occurs during the monsoon from May to October. The climate of the Krishna basin is predominantly semi-arid, with potential evaporation (1457 mm a year, on average) exceeding rainfall in all but 3 months of the year, during the peak of the monsoon. Irrigation is needed for agricultural development (see Biggs et al., 2007, for details). In 2007, the basin's population was 73 million (estimates based on GoI, 2001, assuming a growth of 1.5% per annum), with 48 million in rural areas. The rural population is highest in the Krishna delta and the central west of the basin, and lowest in the centre and south-west. The main city is Hyderabad, the capital of Andhra Pradesh, which accommodates a population of 7 million.

Water Resources Development and Rural Changes in the Krishna Basin

Originally, water in the Krishna basin was managed through small-scale and locally

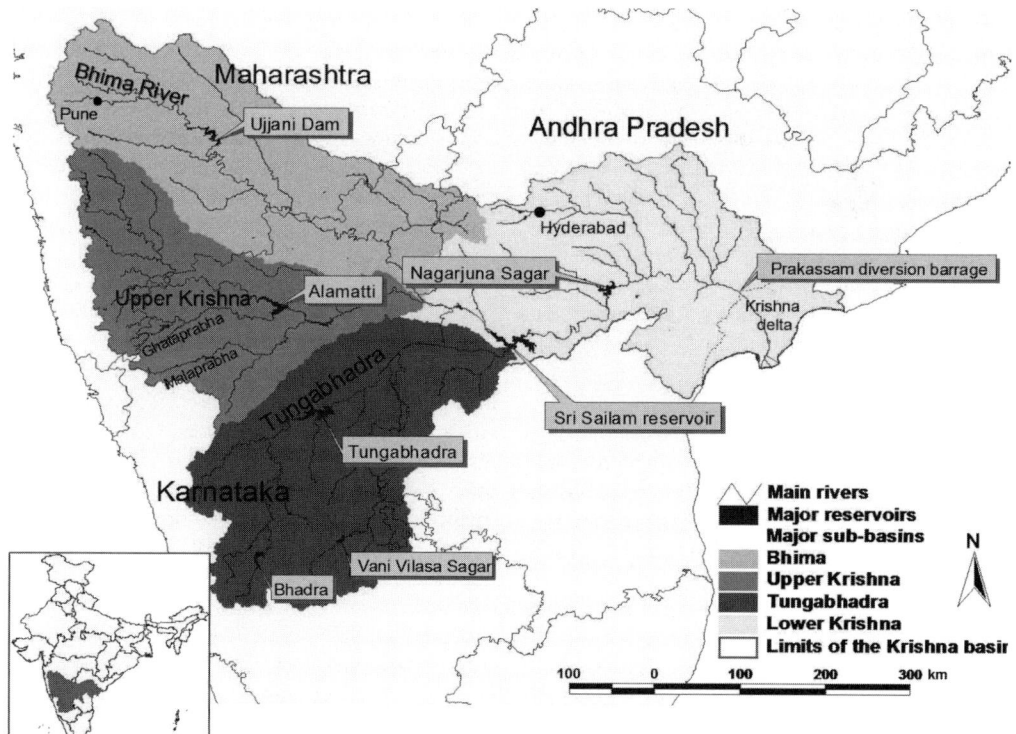

Fig. 10.1. The Krishna River basin, South India.

managed structures (tanks) fed with water diverted from small streams (Wallach, 1985). The first large-scale water diversions took place in the Krishna delta in the 1850s. Between the 1850s and 1947, efforts to promote irrigation focused on the dry areas of the Deccan plateau, and on providing protection against the droughts and famines which regularly struck the region. During that period (1850–1947), no large-scale expansion of agriculture occurred in the lower Krishna basin, where irrigation continued to be practised based on local tanks (Venot et al., 2007). The pace of irrigation development accelerated with the modernization of the Krishna delta project (1954–1957), which irrigates 540,000 ha, and the construction of several multi-purpose reservoirs (irrigation and hydropower production) in the 1970s and 1980s of which the major ones were: Nagarjuna Sagar (1967) and Sri Sailam (1983) in Andhra Pradesh; Bhadra (1953), Malaprabha (1973), Ghataprabha (1977) and Alamatti (1990) in Karnataka; and Koyna (1964) and Ujjani (1981) in Maharashtra (Fig. 10.1).

At the end of the 1980s and in the early 1990s, the pace of large-scale infrastructural development levelled off (see Fig.10.2), and attention was directed towards improving the management and the performance of existing irrigation systems. With the liberalization of the economy in the early 1990s, the strong state support to agricultural development slowed down (Suri, 2006). However, local private or community initiatives (tanks, contour ditches, check dams) continued to be heavily promoted all over South Asia (Barker and Molle, 2005), and the Krishna basin was no exception. Simultaneously, scattered irrigated plots multiplied due to the availability of private pumps, shallow tube-wells and subsidized electricity; Shah et al. (2003) describe this process for South Asia).[1] The groundwater situation has raised much less public concern than disappearing river flows but raises equally important issues in terms of management.

The total storage capacity in the medium and major reservoirs of the Krishna basin multiplied eightfold, to reach about 54 billion m^3 in the early years of the 21st century, i.e. 95% of the pre-1965 river runoff. Further, minor

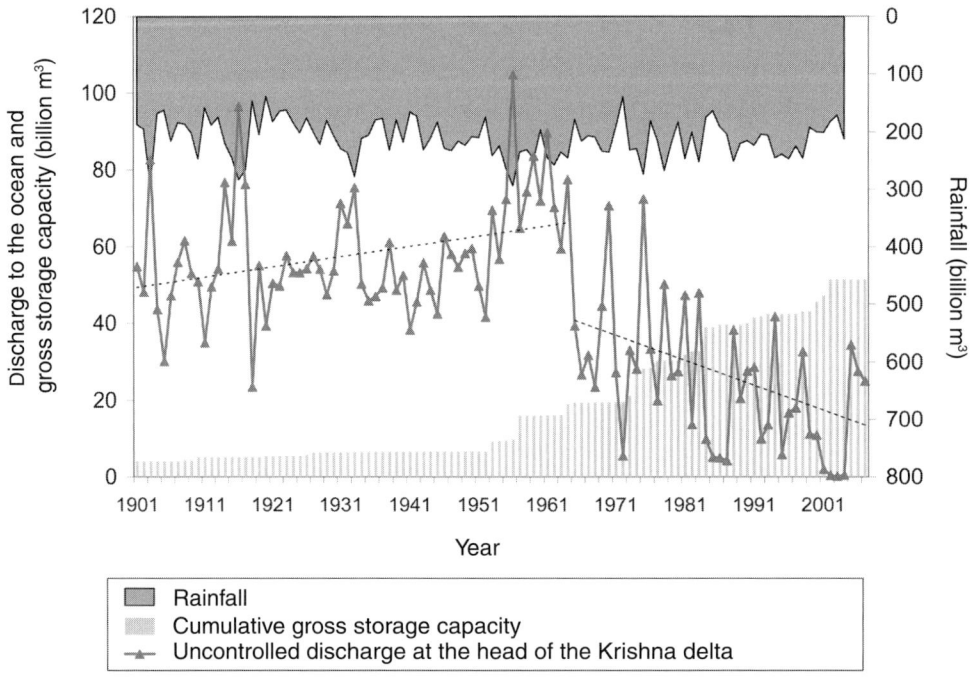

Fig. 10.2. The Krishna basin is closing: a declining discharge (Venot et al., 2007).

surface irrigation projects and groundwater irrigation have also boomed. Although their total storage capacity is not well known, minor irrigation projects are likely to significantly affect the basin water balance: the minor irrigation census of 2001 estimated that about 175,000 minor irrigation structures could irrigate an area of about 1 million ha (Mha) in the basin.

As a result of this infrastructural development, the net irrigated area in the Krishna basin increased more than twofold during 1955–2000, from about 2.2 to 4.8 Mha, and the average cropping intensity rose from 107 to 118%.[2] Cultivating during the dry season became more common as irrigation expanded. The cropping pattern of the basin dramatically changed as rainfed coarse grains (sorghum and millet) were progressively replaced by rice and cash crops (pulses, oilseeds, chillies, cotton). In the early years of the 21st century, about 50% of the irrigated area was irrigated with groundwater, against 36% in 1955–1965: the Krishna basin is in transition, with groundwater becoming one of the main sources of water supply for the farmers. Box 10.1 further describes the agrarian transformations that have affected the rural landscape of the Krishna basin over the last 60 years. In a context of basin closure, this shift towards more local water control is not neutral: it affects existing patterns of water use and spatially reallocates water from downstream to upstream regions.

The discharge from the Krishna River to the ocean gradually decreased from the 1960s, providing an indication of river basin closure (Figs 10.2 and 10.3). Before 1960, the river discharge to the ocean averaged 57 billion m^3/year (i.e. a rainfall:runoff coefficient of 0.29; Biggs et al., 2007). Since 1965, it has steadily decreased, falling to 10.8 billion m^3/year in 2000, and to almost nil in 2004 (0.4 billion m^3/year).[3] The high discharges observed in 2005–2007 (29 billion m^3/year, on average) illustrate that the Krishna River basin is in transition: droughts intensify the interconnectedness of water users and lead to water shortages downstream. As this might be a harbinger of the future, defining management interventions for sustainable water use at the basin level, especially during low-flow years, is increasingly needed. This requires an identification of the spatial and historical dynamics of water use and an understanding of the drivers of the closure of the Krishna basin.

Assessing Water Availability and Use in the Krishna Basin

Methodology

The water accounting method presented here uses the water balance categories proposed by Molden (1997) and is presented diagrammatically in Fig. 10.4. The water balance is based on estimates of water depletion and follows the principle of mass balance, where total input equals the total of outflows and change in storage.

The inflow to the basin includes the mean annual rainfall P, measured from the CRU (Climatic Research Unit, University of East Anglia, UK) data set (CRU, 2007) and district statistical handbooks (GoM, 2005a; GoAP, 2006; GoK, 2006), and water transfers entering the Krishna basin (imports, T_{in}), estimated from government statistics and data from water supply projects.

This inflow is partly depleted through evapotranspiration of vegetation (Dpl);[4] domestic, industrial and livestock processes (U);[5] water transfers out of the Krishna basin (exports, T_{out}); the net change in water storage in newly built reservoirs (ΔV)[6] and in aquifers (ΔGW). ΔGW is estimated as the aquifer recharge (R)[7] minus the groundwater demand by groundwater-irrigated areas (Gw_i),[8] minus the groundwater base flow (Bf, estimated as the difference between R and Gw_i), minus the net groundwater stock depletion (δs), calculated to close the water balance.[9]

The remainder is the discharge to the ocean Q_{out}, measured at the head of the delta, at the Prakassam barrage, located downstream of the diversions to the canals of the Krishna delta. In what follows, the term 'net inflow' designates the sum of mean annual precipitation and imports to the Krishna basin, minus the net groundwater stock depletion (P + T_{in} – δs). Average annual estimates for periods of 5–10 years are used. Although inter-annual variability is important in terms of management, this chapter focuses on long-term trends revealed by average balances.

Box 10.1. The Indian agrarian economy: past trends and current challenges.

Gigantic dams (the 'modern temples of India' according to Jawaharlal Nehru), shifting cropping patterns, the multiplication of small private pumps, and expanding irrigated lands are the most visible signs of the transformations that have affected Indian agriculture since the early 1950s. A technological 'triptych' (irrigation; high-yielding varieties; and widespread use of fertilizers, and, to a lesser extent, pesticides), price support policies (for both production and inputs), and institutional reforms (access to credit, land reforms, etc.) have been the drivers of a 'Green Revolution' (1964 onwards) (Landy, 2008), when the annual growth of the agricultural gross domestic product (AGDP) averaged 2.44% (GoI, 2007a) and yields tripled for food grains (rice, wheat) and increased fourfold for crops such as cotton, oilseeds and sugarcane. But beyond the apparent success story of a booming Indian agriculture, many challenges remain: the growth of the agriculture sector has slowed down from the mid-1980s onwards; the overall productivity of Indian agriculture is low compared with those of China, Vietnam and Thailand; access to food is highly unequal; food security is far from being a reality; and more than 250 million Indians are still malnourished (IFPRI, 2005; Landy, 2008).

The agrarian structure has undergone significant changes in the last 50 years. Post-independence land reforms aimed at limiting land concentration and consolidating landholdings to create a class of owner–operators more likely to invest in irrigation and modern agricultural techniques (see, for example, Upadhya, 1988, describing the emergence of 'farmer–capitalists' in the Krishna delta). Results have been uneven according to the states. The abolition of intermediaries (*Zamindars*) was relatively successful, but tenancy reforms, land redistribution and land ceilings met with less success and most of large owners managed to keep their land within the family: in Andhra Pradesh, for example, only 4% of the total sown area was redistributed (2% at the all-India level) and 12.5% of unappropriated government land was assigned to landless poor (GoAP, 2007; see Mearns, 1999, and Deshpandhe, 2007, for further information on land reforms in India). The average farm size in all three states of the Krishna basin is under 1.8 ha (which is more than the all-India average of 1.3 ha: in the rainfed Deccan plateau, landholdings are relatively larger but far less productive than in irrigated areas; GoI, 2007b). Marginal and small farmers (less than 2 ha) account for 83, 73 and 75% of all farmers in Andhra Pradesh, Maharashtra and Karnataka (without accounting for concealed tenancy), respectively, while indebtedness concerns 82, 55 and 62% of the households in these three states, respectively (GoI, 2007b).

At the macro level, the contribution of agriculture to the Indian economy has decreased from 61% (in 1950–1951) to 20% (in 2003–2004) of the total GDP of the country. In the Krishna basin, the importance of agriculture is highest in Andhra Pradesh (22% of the state GDP and 62% of the total workforce) and lowest in Maharashtra (11% of the state GDP and 55% of the total workforce). From 2002–2005, the growth of the Indian AGDP was as low as 0.89.[10] While other economic sectors are booming (with an overall economic growth of 8% and more during recent years), social disparities between rural and urban areas and between the agriculture and other economic sectors increase: distress in rural areas becomes pervasive due to the lack of non-farm employment. Agricultural growth cannot sustain a quantitatively stable rural population anymore and India is facing an agrarian crisis, recently epitomized by highly publicized farmers' suicides: high indebtedness (for annual agricultural expenses, often from the informal market), and dependency on volatile cash crop cultivation (oilseeds, cotton, chillies) are seen as some of the main reasons behind farmers' distress.

A landscape and waterscape dominated by rural changes

Figure 10.5 maps the evolution of the Krishna basin water balance since the mid-1950s. Four main regions have been delineated: the Bhima sub-basin in the north-west (located in Maharashtra, Karnataka and Andhra Pradesh); the upper Krishna sub-basin (Maharashtra and Karnataka); the Tungabhadra sub-basin (Karnataka and Andhra Pradesh) and the lower Krishna Basin (Andhra Pradesh) (Fig. 10.1). The main trend is a dramatic increase in water depletion by irrigation, from 17.1 billion m^3/year during 1955–1965 to 44.3 billion m^3/year during 1990–2000. This means a 19% rise in total depletion over the period 1955–2000. The total depletion amounted to 181 billion m^3/year during 1990–2000, i.e. 88% of the net inflow to the Krishna basin. Consequently, the discharge to the ocean dramatically decreased and amounted only to 10% of the net

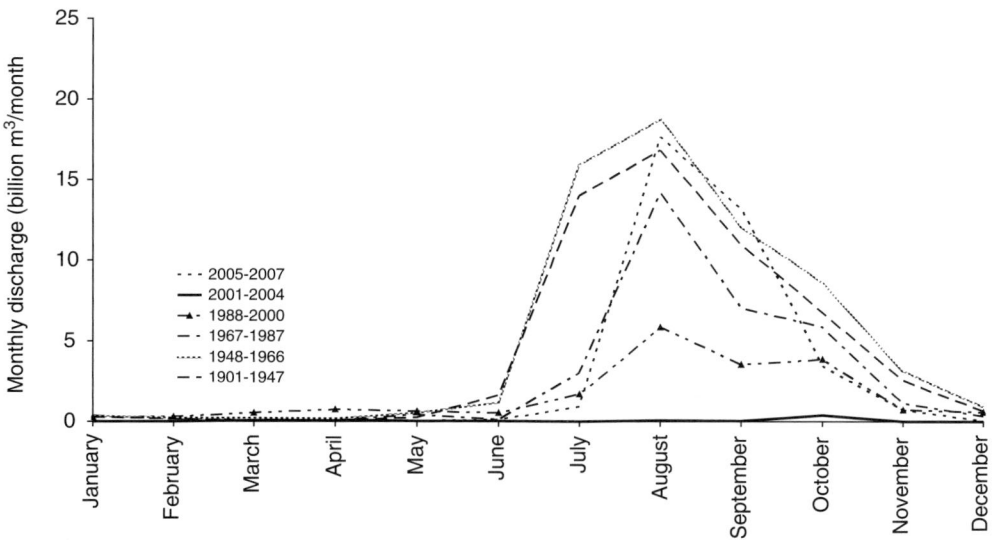

Fig. 10.3. Average monthly discharge to the ocean for different periods of time.

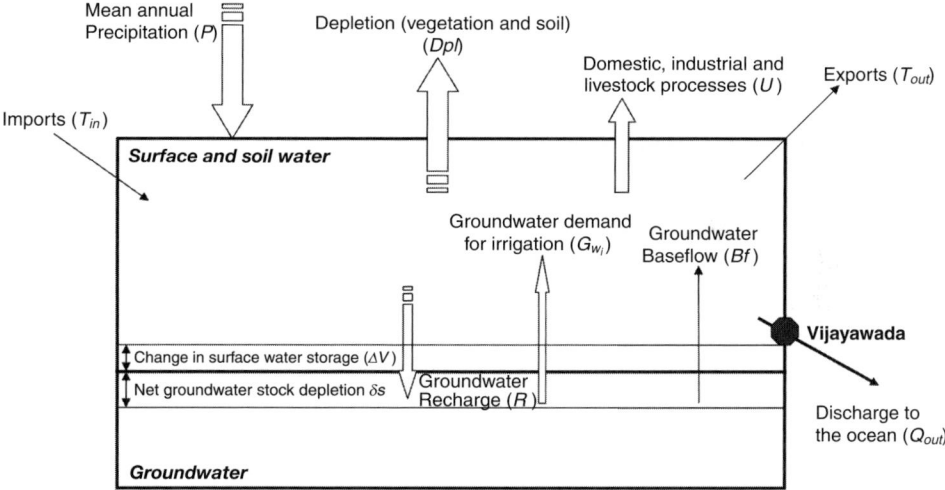

Fig. 10.4. Water flows and uses in the Krishna basin.

inflow during 1990–2000 (2% of the net inflow was exported to other basins).

Soil moisture and prospects for increased basin efficiency

The high level of water depletion as early as 1955–1965 (69% of the net inflow) highlights the importance of rainfed agriculture and natural vegetation in the water balance of the Krishna basin. Rainfed agriculture has always been the main user of water because of its large areal coverage in the dry areas of the Deccan plateau. Depletion in rainfed agriculture has increased slightly since the late 1950s, revealing widespread supplemental irrigation of formerly rainfed crops. Supplemental irrigation takes place through groundwater abstraction or diversion of small streams in secondary upstream basins and is rarely reported in governmental statistics. These findings are consistent with those of Biggs et al. (2007),

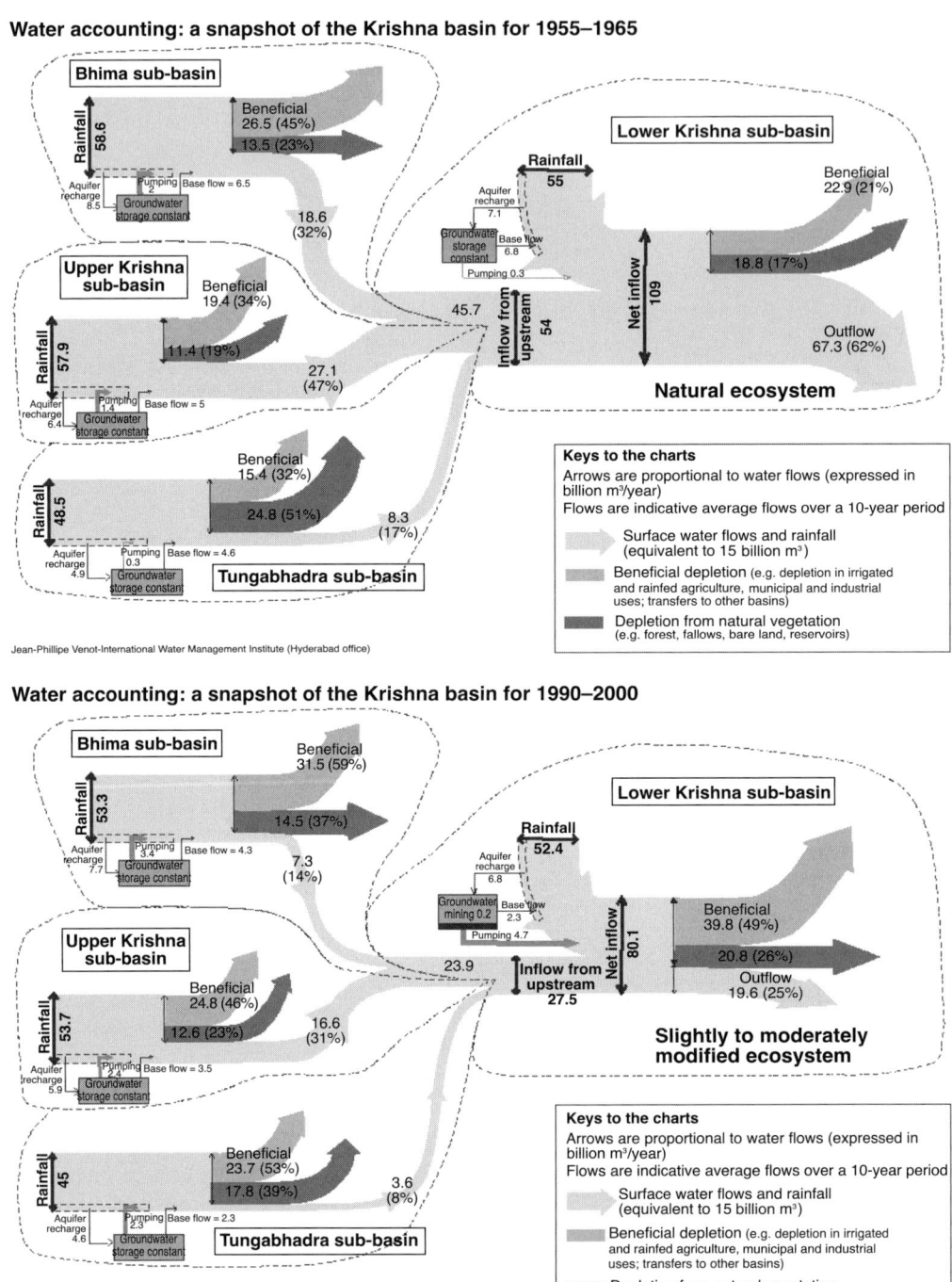

Fig. 10.5. Water accounting of the Krishna basin: an evolution for 1955–2000.

who pointed to the widespread nature of small irrigated patches in rainfed areas. In 1990–2000, depletion from rainfed agriculture and natural vegetation together accounted for 54% of the total rainfall in the Krishna basin. These values illustrate that sustainable and equitable water management – rainfed agriculture is the main livelihood for the poorest communities – can only be achieved through an increase in the productivity of agriculture in semi-arid, rainfed areas. Small-scale supplemental irrigation from rainfall is promising, but has to be cautiously planned and downstream impacts carefully assessed (see below).

*Development of surface water:
a state-wise approach*

The planning and development of irrigation projects in the three states that share the Krishna water have always led to acute conflicts, highlighting the need for formal interstate allocation rules, because no state has ever considered the potential third-party impacts of its own development. Major interstate disagreements led to the setting up of the Krishna Water Disputes Tribunal (KWDT) in 1969, and to the agreement, in 1976, on formal allocation procedures, which apportioned the 75%-dependable flow of the Krishna River (58.2 billion m^3/year, the value exceeded in 75% of the years)[11] as follows: 15.8; 19.8 and 22.6 billion m^3/year to Maharashtra, Karnataka and Andhra Pradesh, respectively.[12] Andhra Pradesh is also entitled to use any *surplus water*, with the caveat that it shall not acquire any formal right to it (GoI–KWDT, 1976).[13] This formal process of water apportionment did not slow down the pace of infrastructural and irrigation development. Between 1955 and 2000, depletion in surface irrigation projects increased from 11.2 to 28.4 billion m^3/year (irrigation is at its highest in the lower and upper Krishna basin). The KWDT award expired in May 2000; a new tribunal was constituted in 2004 and is expected to reach a decision for allocating water between the three states around 2010. It is crucial that this new tribunal acknowledges and quantifies surface water/groundwater interactions, and accounts for small-scale surface water use (minor irrigation; rainwater harvesting and watershed programmes) and groundwater abstraction, which have both skyrocketed during the last five decades and have impacted the availability of surface water downstream (see below).

*Uncoordinated groundwater abstraction and
small-scale irrigation*

All over India, one of the most striking features of irrigation development during the past five decades has been the rapid growth in the use of groundwater (Vaidyanathan, 1999). This trend was supported by the government through: (i) rural development projects that targeted rural areas earlier neglected by the 'Green Revolution' (due to the relatively poor conditions that prevailed there for agriculture); and (ii) policies subsidizing electricity for agricultural uses. According to remote-sensing and census data, today groundwater and minor irrigated areas cover more land than medium and major irrigation projects (Biggs et al., 2007). This raises many management issues. Although the nature and extent of surface water/groundwater interactions are not well known, the water balance presented in this chapter highlights that increasing groundwater abstraction (from 5.9 to 18.1 billion m^3/year between 1955 and 2000) led to decreasing base flows (minus 8.7 billion m^3 between 1955 and 2000)[14] and aquifer overdraft (minus 0.25 billion m^3/year). Scattered surface-water and groundwater irrigation developments in upper secondary catchments significantly reduced surface water flows and reliability to downstream water users. In a context of basin closure, this shift towards more local water control is tantamount to a re-appropriation of water and might raise tensions, as supporting minor or major irrigation has become highly political, because of different social and economic implications (Dhawan, 2006).

**The emergence of new large-scale
water users**

Hydroelectricity generation

The increase in electricity needs has led to the completion of several hydropower projects.

Major interbasin transfers take place in the Western Ghats of Maharashtra (~3.5 billion m^3 are transferred each year to the western coast because of a much higher head). Hydroelectricity generation is a major concern for the government of Maharashtra, which is contemplating increasing the capacity of these transfers. These plans are strongly opposed by downstream states because they would reduce water availability down the river system. In other parts of the basin, hydropower projects do not deplete water. They have impacted the hydrological regime of the river but reservoirs have minimized impacts on downstream agricultural uses (Venot et al., 2007). Given the increasing importance of hydropower generation (and possible related conflicts), there are plans to pump the water discharged to produce hydroelectricity back into the reservoirs for further reuse (for example, for this purpose, a 'tail pond' downstream of the Nagarjuna Sagar reservoir is under construction). Impacts of the growing need for electricity on existing water uses and the environment need to be further studied.

Domestic and industrial uses

Industrialization and urbanization are fast developing in the Krishna basin (van Rooijen et al., 2008). The demand for domestic and industrial water keeps growing, notably around the megalopolis of Hyderabad (with a population of 7 million) and around Pune (3 million), which are increasingly supplied from distant sources, by shifting water out of agriculture.

At the basin level, domestic and industrial water uses have trebled during the last 50 years but still represent less than 1% of all depleted water in the Krishna basin (and 3% when compared with depletion by irrigation). These percentages highlight that intersectoral reallocation of water from agriculture to more productive uses is unlikely to shape the future waterscape of the Krishna basin in average years. However, urban and industrial uses will receive priority in case of drought (Molle and Berkoff, 2006), when intra- and inter-basin transfers could impact users in rural areas and sharpen local conflicts (see Celio, 2008, for a case study of Hyderabad).

The environment: a water user in its own right?

Water and infrastructural development to meet growing human consumptive uses has resulted in significant degradation of various ecosystems. Although the impacts of reduced flows on ecosystems are not well quantified, there is well-documented evidence of downstream environmental degradation in the lower Krishna basin, manifesting itself by soil and groundwater salinization, increasing pollution, disappearing mangroves and wetland (the Kolleru Lake) desiccation (Venot et al., 2008). With increasing evidence of the adverse impacts of water and land degradation on people's livelihoods, environmental concerns have started to gain strength, and the notion of environmental flows is establishing itself and challenges the very notion of 'surplus water' that is commonly called upon to justify new infrastructure (Molle et al., 2007).

According to a simple desktop assessment method proposed by Smakhtin and Anputhas (2006) to quantify environmental water needs in data-scarce river basins of developing countries, preserving the ecosystems of the Krishna basin in their current status would require an environmental flow allocation of about 6.5–14.2 billion m^3/year. Water resources commitment would then reach 94–98% (the discharge to the ocean averaged 19 billion m^3/year in 1990–2000), showing that resources would be fully committed under average conditions.

Implementing environmental flows is a highly sensitive question and presents a great challenge to current water users as: (i) the volumes at stake are large; (ii) the science of environmental flows is relatively new; (iii) there are recurrent questions on how to assess the environmental status of river basins and how environmental degradation relates to altered flow regimes; and (iv) preserving the environment is often perceived as anti-poor and anti-development, especially in the developing world. In an era of economic liberalization, which pushes Indian decision makers to seek overall growth, the key question is to understand and quantify the benefits that letting a given volume of water free to flow to the ocean can yield to the society as a whole, while keeping in mind that using this water elsewhere in the basin for other purposes also has some costs and benefits.

Finally, taking up the challenge of environmental preservation not only requires maintaining a given flow to the ocean but also the implementation of other policies from the local level (regulating farmers' practices and controlling the current mode of natural resources extraction) to regional (the creation of integrated management zones, which would be defined on the basis of agroecological features) and state (rural development policies) levels (Venot et al., 2008). Putting environmental issues on the Indian agenda of water resources policies and reforms is a challenging task and requires a shift in the governance structure of the sector to allow the poorest communities, who often depend on fragile ecosystems for their livelihoods, to voice their demands.

Transferring water

In addition to the water transfers from the Godavari basin to Hyderabad, implemented in the 1990s, several projects withdraw water from the lower Krishna basin and transfer it south-east to irrigate some dry areas of Andhra Pradesh (Fig. 10.1) and supply the water-scarce megalopolis of Chennai in Tamil Nadu. These projects have performed well below expectations, even at times of abundant water availability, and their full implementation would further increase the pressure on the Krishna water (Venot et al., 2007).

Foreshadowing the future: the drought of 2001–2004

Between 2001 and 2004, rainfall was 12% below the long-term average in the Krishna basin. Such droughts of 3 years or more had a return period of 1 in 15 years over the last century (CRU, 2007) and are likely to herald the future average water availability in the lower Krishna basin, given continued upstream development of irrigation infrastructure. The lower Krishna basin, which bears the brunt of any intervention upstream, was the region most affected by the drought: its net inflow fell dramatically to 57.2 billion m^3/year and total surface water availability (18.4 billion m^3/year, including local runoff) was close to the allocation of the KWDT (18.6 billion m^3/year for the lower Krishna basin).[15] During 2001–2004, almost no water reached the ocean, as the basin consumed or stored 99.5% of its net inflow.[16] Any further development of water use in the basin will impinge on existing uses, and the consequent reallocation is likely to exacerbate competition and conflicts. This happened even though the water depleted in surface irrigation projects always remained (for all three states and for the past 5 decades), within the limits of the KWDT award. This observation calls for the design and effective implementation of new allocations, within the framework of the present KWDT, which needs to reassess the dependability of river flows and account for a booming groundwater economy, to provide a platform to manage access and use of water during years of low flow. In the following section, we highlight why the KWDT has not been instrumental in limiting infrastructural development in the Krishna basin.

Past Policy Interventions and River Basin Closure

The literature on river basin development identifies three main ways through which societies address their water resources problems: supply augmentation, conservation strategies and water allocation (Molden et al., 2005). These three types of answers are often linked to three phases of river basin development, e.g. development, utilization and reallocation, respectively. However, Molle (2003) highlighted that different types of societal responses may occur concomitantly, at different stages of basin development, and that their relative importance is context specific. The following paragraphs describe how and why different interventions have been combined (both spatially and temporally) by different users in the Krishna basin since the late 1950s and how this has resulted in the current overexploitation of water resources.

Supply augmentation

Green revolution or rural development?

Societies typically resort to supply augmentation projects to meet their growing water demand. The dynamics of irrigation in the Krishna basin reveal the tension between two approaches to rural development in India, regardless of the stage of development of the river basin. Since 1947, two main clusters of rural development policies have been implemented in India, concomitantly or otherwise (Landy, 2008). These two types of policies have translated into two different modes of access to, and use of, water in different regions of the Krishna basin. Broadly, the first group of policies aims at 'efficiency in development' and concentrates financial and institutional investments on those social groups and areas that offer the highest potential for development. They are the technologies of the Green Revolution, adopted in medium and large irrigation projects, and, more recently, in attempts at integrating agriculture into agribusiness chains.[17] The second group aims at 'equity in development' and advocates rural development programmes through strong state planning and public investments in remote areas. They are watershed and tank rehabilitation programmes, and minor irrigation projects in upper secondary catchments (Landy, 2008). This need to balance economic efficiency and equity in rural development has been a major driver of the spatial distribution of water use in the Krishna basin over the last 50 years.

Interbasin transfers

Interbasin transfers, often costly, generally occur when the pressure on water is at its highest, in order to reopen closed or closing river basins. In the Krishna basin, water was transferred *out* of the basin as early as the mid-1970s because of much needed hydroelectricity, despite the costs of such projects, and at a time when water was considered to be abundant. The first transfers *into* the Krishna basin (from the neighbouring Godavari basin) date back to the early 1990s, at a time when basin water availability was still high and Hyderabad was 'thirsty'. This highlights that assessing overall water availability is not enough to explain river basin development: sub-basins, regions or cities might be 'closed' in an open river basin. Finally, the creation of the National Water Development Agency (NWDA) in 1982 paved the way for the National River Linking Project (Venot et al., 2007)[18] and the construction of the Polavaram–Vijayawada link between the Godavari and the lower Krishna river basins (see below).

Groundwater access

Despite evidence of aquifer depletion, free access to groundwater remains the rule: a draft groundwater bill has been contemplated for several decades but has not been implemented due to its high transaction (political) costs (Shah, 2007). Low-cost electricity and preferential loans have been the major drivers of significant groundwater abstraction since the mid-1980s.

Institutional arrangements and conservation strategies

Institutional arrangements and conservation strategies are generally associated with the 'utilization phase' of river basin development, when the focus is on improving efficiency in the context of recurring shortages. Public interventions in the water sector are generally designed at the federal or state level and focus on measures as diverse as modernization and rehabilitation of existing water supply projects, technical on-farm improvements, rainwater harvesting, participatory irrigation management (PIM), water-pricing policies, private-sector participation, water legislation and policies, institutional support, etc.

Many such reforms have been recently implemented in the Krishna basin and are clearly driven by growing evidence of an Indian water crisis. But other typical demand management options have long been relevant and dealt with in the Indian water sector long before water availability was seen as a constraining factor in the Krishna basin. Therefore, pressure on water resources is not the only motive behind the adoption of demand management: low return on investments, physical deterioration of low-performing infrastructure, poverty and inequality within irrigation systems, international donors and environmental issues have also triggered irrigation reforms in the Krishna basin.

Institutional initiatives at the federal level

In 1987, the Indian federal government enacted its first Water Policy. In 2002, a revised National Water Policy was finalized, which incorporated the principles of integrated water resources management at the basin level (GoI, 2002). These institutional moves are characteristic of countries whose river basins are under growing stress, but Shah (2007) highlights that 'nothing in the way India's water sector functions has changed as a result of these [policies].' Following the National Water Policy of 2002, several states have issued their own water policies (Karnataka in 2002; Maharashtra in 2003; Andhra Pradesh has not yet enacted any such policy), but these generally fail to shape relevant strategies (Mohile, 2007) and supply augmentation options remain prevalent.

In contrast, federal initiatives to enhance rural development have had significant impacts on irrigation development. The creation of the National Bank for Agriculture and Rural Development (NABARD) in 1982 to facilitate farmers' access to credit is one of these institutional interventions: NABARD has been one of the main drivers of the development of groundwater abstraction in upper secondary catchments (through credit for well drilling). Today, it is also used as an indirect lever to restrict institutional credit for private investments in areas where groundwater is overexploited (Shah, 2007).

Federal involvement in irrigation programmes

Two main types of all-India irrigation programmes (in line with the two different approaches to rural development described above) are presented here, and are often associated with the idea of demand management (conservation, increased efficiency), although their implementation in India has led to large-scale irrigation development.

First, the command area development (CAD) programmes were initiated as early as 1974 to enhance irrigation efficiency and improve agricultural production and productivity in surface-irrigation projects. These programmes aimed at bridging the gap between the actual irrigated area and the existing irrigation potential (Hashim Ali, 1982). Box 10.2 describes how the South India approach to irrigation development is at the origin of this perceived underutilization and how local politics led to overbuilding in the Krishna basin.

The CAD programme represented the largest single investment of the World Bank in India, the biggest bank client at the time (Wade, 1976). The programme envisaged a comprehensive approach to water management but mainly focused on interventions 'below the outlet', and did not acknowledge that the institutional and managerial set-up of the main canal system was decisive in shaping water

---

Box 10.2. Protective irrigation, local politics and overcommitment of water.

Large surface-irrigation projects in South India have been built along *protective* lines, i.e. they aimed at spreading available water resources thinly over a large area and to a large number of farmers (supplementary irrigation is implied; see Mollinga, 2003, on protective irrigation). Denying the natural trend towards intensification (as population increases and landholdings shrink), protective irrigation is hardly viable in the long run. But despite the acknowledgement of these problems, this model of irrigation development remains central in South Indian irrigation policies: it provides convenient legitimacy for the state for infrastructural development (on the basis of equity and poverty-alleviation principles), and may even have cemented a social justification, called upon by local politicians, for overbuilding. The continued existence of protective irrigation (and promises of large irrigation projects, regardless of water availability and often presenting doubtful figures of 'potential irrigated area') lies in the populist character of Indian politics: irrigation projects are major means of securing the support of rural constituencies (Ramamurthy, 1995; Venot *et al.*, 2007).

availability and farmers' practices in large gravity-fed irrigation systems (Wade and Chambers, 1980; see Chambers, 1988 for further descriptions of the CAD programmes). The CAD programmes (along with other modernization projects such as the National Water Management Program) did receive further impetus by the mid-1980s.[19] They were relatively successful in increasing the actual irrigated area and in reducing conveyance losses (thus increasing water use); but the lack of: (i) proper operation and maintenance (O&M) of the main network; (ii) coordination between institutions in charge of intervening 'above' and 'below' the outlet (the creation of the ICAD (Irrigation & Command Area Development Department aimed at enhancing these two points); and (iii) any involvement of the farmers in the design and implementation of the interventions considered were identified as crucial issues that needed to be addressed (IRDAS, 1996) and would trigger Participatory Irrigation Reforms in the late 1990s (see below).

Other federal programmes focus on rural development in upper secondary catchments through integrated watershed development and rainwater-harvesting projects.[20] Recently, the focus has been on users' participation and aquifer recharge (Vaidyanathan, 1999), but irrigation development is implicitly targeted: Aubriot (2006) highlights the lack of any demand management component in these projects and shows that artificial recharge programmes encourage farmers to deepen their wells and further deplete the aquifers.

Participatory irrigation management (PIM)

Since the late 1990s, PIM measures have flourished in the Krishna basin. In 1997, Andhra Pradesh was the first Indian state to embrace the PIM rhetoric on a large scale, and water user associations (WUAs) were created in the entire state, with mixed results (Mollinga et al., 2004; Nikku, 2006). In 2005, Maharashtra also initiated a state-wide PIM programme, and in Karnataka PIM is being promoted through large-scale pilot projects. Gulati et al. (2005) identify states' fiscal deficits (partly due to high subsidies for irrigation) and the physical deterioration of low-performing irrigation systems as the main motives behind these reforms, not the looming water crisis. International funding agencies and domestic debates on 'underutilization' of irrigation potential were pivotal in driving PIM in the Krishna basin.

Water pricing: supply augmentation or demand management?

Water pricing and the profitability of irrigation systems were major issues in British India, with projects being identified as *productive* or *protective*, depending on their rate of return (Mollinga, 2003). Bolding et al., 1995, also show that the objective of a volumetric water pricing shaped the way water supply infrastructures of large irrigation projects were designed. At that time, bulk allocation and crop-wise differential rates were identified as possible options to enhance the financial viability of irrigation projects through an increased water use. Such pricing policies were, for example, the rule in the Nira irrigation project of the upper Bhima sub-basin (Attwood, 1992).

In independent India, water charges were increased several times: significant increases in the mid-1970s and mid-1980s were part of the CAD programme; the most recent increase, in 2001, was part of the PIM package. The call for PIM also pushed Maharashtra to return to the system of bulk allocation (along with rotational water supply) to manage water demand (GoM, 2005b). There are similar recommendations in Andhra Pradesh and Karnataka. Recent increases in water charges have improved the cost recovery ratio of most irrigation projects in the Krishna basin – this was their main objective, not water savings – but the financial viability of state-managed irrigation remains one of the main problems of the Indian water sector (Gulati et al., 2005).

Autonomous irrigation agencies: motives and expectations

The involvement of the private sector in irrigation management was envisaged during the British period to ease the financial burden that irrigation development represented to the Crown. The experiment failed (Atchi Reddy, 1990). The creation of financially autonomous irrigation agencies is often referred to as a renewed attempt to bring the private sector back into irrigation management, as local governments face serious financial difficulties, with broad consequences for rural services. Experiences in the three states sharing the Krishna water yielded mixed results. In Andhra Pradesh, the Andhra Pradesh Water Resources Development Corporation was set up in 1997 but has not undertaken any significant activity (Madhav, 2007). In Karnataka and Maharashtra, financially autonomous irrigation agencies were set up in the 1990s to raise financial resources from the market to build irrigation structures, notably in the Krishna basin (Gulati et al., 2005). They were set up to overcome reduced budget allocations (and uncertainties around external funding),[21] and in response to the KWDT award, to make sure ongoing projects would be completed (and water used) by 2000, when the award was to be renegotiated (Gulati et al., 2005); these measures were mainly about increasing supply. Gulati et al. (2005) studied the achievements of, and the problems faced by, the agency entrusted with the task of developing the Upper Krishna Project (Alamatti/Narayanapur dams) in Karnataka. They highlighted that the agency had been successful in raising funds (thanks to the support of the government of Karnataka) and in completing construction work, thus contributing to river basin overbuilding. However, its overall financial situation is not good (the agency is not financially autonomous: it does not generate its own income, which should have come from better cost recovery, and depends on state support); there has been little improvement in irrigation system management and the agency functions as a government line department.

Clearly, financially autonomous irrigation agencies have failed to live up to researchers' expectations of demand management; but the motive behind the creation of such agencies has always been supply augmentation. This shows the permeability of socially constructed, scale-dependent and interrelated categories such as 'supply augmentation' and 'demand management'.

While these corporations have mainly been about 'developing' large surface irrigation projects, the recent Maharashtra Water Resources Regulatory Authority Act of 2005 makes them de facto 'river basin agencies', whose work is to be coordinated by a public body: the Water Resources Regulatory Authority (GoM, 2005b). The law highlights the need for integrated resources management, but river basin agencies are mainly entrusted with the task of further developing water resources within the limits of the state of Maharashtra. Such calls for basin-wide water management have a long history in India, but much has yet to be done to implement the idea (Vaidyanathan, 1999).

Conservation measures

With evidence of a growing water crisis and low returns from irrigated agriculture, several on-farm 'water-saving' technologies are being promoted in India. Two cases in point are: (i) the development of drip irrigation (for profitable fruit and vegetable cultivation, notably in Maharashtra), although it remains marginal at the all-India level (less than 1% of the total irrigated area); and (ii) a new way of cultivating rice, known as the system of rice intensification (SRI), which demands less water and fewer inputs than common paddy cultivation. More research on the viability of these techniques and the scope for water saving (to be reallocated to other users) is needed.

Challenges to demand management in India

Aubriot (2006) and Shah (2007) underline that Indian agricultural and food policies are often at loggerheads with the objectives of water demand management: the public procurement system ensures minimal prices for water-consuming crops such as sugarcane and paddy (wheat in northern India), which together represent 55% of all irrigated crops in the Krishna basin (Biggs et al., 2007). Rationalizing agricultural policies on the basis of water availability is needed, and requires a better understanding of the hydrology of the main Indian river basins and of the spatial dimension to equity and efficiency in water use.

Water allocation

Water allocation policies are generally associated with late phases of river basin development and 'mature' (closing or closed) basins, where the pressure on water resources and the likelihood of conflicts are at their highest. But conflicts about the sharing of the Krishna's water have brewed since the late 19th century (D'Souza, 2006), leading to the setting up of the KWDT as early as 1969. Formal water allocation was established in 1976, in an early development stage of the basin, while harnessing water was still the motto of the three states. This sharing agreement has been instrumental in the overbuilding of the basin.

Questioning the notion of 'surplus water'

The KWDT allocated the 75%-dependable flow of the Krishna River and explicitly recognized that all water in excess of this flow had to be considered as 'surplus water' (and free to be used by the downstream most riparian state: Andhra Pradesh). But as environmental concerns take centre stage, it has become clear that the notion of 'surplus water' involves a value judgment (Mohile, 2007). Smakhtin *et al.* (2007) also question the notion of surplus water from a hydrological point of view. As river basins close, policies, rules and infrastructures based on this notion will become increasingly counter-productive (Wester *et al.*, 2005): they create new demands for water already committed and are likely to sharpen conflicts.

Water rights and provisional allocation

There are no clear water-allocation guidelines in India, and negotiated agreements, such as the KWDT award, accommodate different water rights regimes, depending on the principles that each party (here the three states) wants to be considered in the allocation process (Mohile, 2007). The KWDT first endorsed the *riparian rights* of the three states, and within the states, bureaucratic decisions of water allocation remained the norm. Second, it explicitly recognized *prior appropriation rights* by protecting existing uses – at the project and state levels – and, third, it sanctioned the rights of the states to further develop water resources by considering planned future uses, generally on the basis of ongoing projects (Venot, 2008b). The KWDT award of 1976 was to be revised in 2000, and the states sharing the Krishna water engaged in massive development of their hydraulic infrastructure[22] (with serious economic and fiscal damage)[23] to lay claim on water resources and ensure they would be holding a prevailing position when the KWDT award was renegotiated (Gulati *et al.*, 2005).[24] Interstate competition unquestionably resulted in overcommitment of water. But the politics of water also play a major role at other levels (see Mollinga, 2001, for a conceptual presentation), such as the regional level in a single state (see Box 10.3), in shaping water availability and use in large river basins. The fact that state governments are allocated funds from the planning commission proportional to the number of medium and major ongoing irrigation projects has also driven infrastructural development (Sengupta, 2005).[25]

Box 10.3. Intrastate politics and overcommitment of water: the case of Andhra Pradesh.

Politics in Andhra Pradesh have strongly influenced water use in the lower Krishna basin. Since India gained independence, access to irrigation facilities and state funds for irrigation have been contested by the various regions of the state (e.g. Rayalaseema, Telangana and coastal Andhra), which are unevenly developed. Government-funded canal irrigation is concentrated in coastal Andhra, while Rayalaseema and Telangana mainly rely on groundwater for irrigation (Ratna Reddy, 2006). Despite reduced investments in the agriculture and irrigation sectors from the mid-1980s onwards, the government of Andhra Pradesh has always tried to balance rural development across the three regions of the state to attenuate inequities rooted in the natural, historical and political context of Andhra Pradesh, characterized by the dominant influence of the coastal region and of its entrepreneurs (Venot *et al.*, 2007). This has led to promises of new surface irrigation projects both within and outside the lower Krishna basin, planned on the expectation of 'surplus' water from the Krishna River. Although these projects comply with the KWDT award, they are based on a notion of 'surplus water', which needs to be questioned (see above). Preventing regional tensions and state implosion – under the pressure of independent claims from all three regions of the state – have been major drivers of infrastructural development in the lower Krishna basin (Venot *et al.*, 2007); political and socio-economic concerns for poor regions have also promoted the overbuilding of the basin, as observed in many other cases (Molle, 2008).

Technical limits of the KWDT award

Technical limits explain why the KWDT did not prevent the closure of the Krishna basin. First, by allocating fixed volumes of surface water on the basis of the 75%-dependable flow, it did not offer an adapted platform to manage low-flow years. Second, it neglected the relationships between surface water and groundwater systems and endorsed *capture rights*, as it mentioned that the three states 'will be free to make use of underground water within their respective territories in the Krishna river basin [and that] use of underground water shall not be reckoned as use of the water of the river Krishna' (GoI–KWDT, 1976). But groundwater abstraction has skyrocketed over the last decades and has contributed to streamflow depletion, notably in upstream river valleys, where shallow alluvial aquifers and river systems are highly connected (see above; Hanumantha Rao, 2006). The 75%-reliability of the KWDT allocations is now jeopardized: irrigation development is based on an overestimation of available surface water (Biggs *et al.*, 2007), also highlights that river runoff has decreased for all probability levels due to irrigation development) and security of supply for existing users is at risk. Understanding the interactions between groundwater and surface water, especially with regard to dry periods, is critical in defining allocation rules that would consider both surface water and groundwater resources and cap their respective uses.[26] Third, more accurate estimation of return flows in irrigated areas is needed (Biggs *et al.*, 2007): this is a challenging task, as actual return flows vary with supply, due to farmers largely resorting to drain-water, especially during low-flow years. Fourth, while the KWDT mentioned that 'beneficial use shall include uses [...] for domestic, municipal, irrigation, industrial, production of power, navigation, aquaculture, wildlife protection and recreation purposes' (GoI–KWDT, 1976), it did not mention the relative shares allocated for these consumptive and non-consumptive uses. As domestic and industrial demands increase and potentially conflict with other uses, there is a need to formally quantify the water entitlement for cities and industries (this is likely to be done at the state level, but the current KWDT tribunal could provide guidelines) that are likely to be supplied as they grow (Molle and Berkoff, 2006). Environmental needs have to be recognized and formally quantified as well (at the state and basin levels), and water-quality issues need to be considered so that productive use of wastewater is maintained with minimum hazard to public health.

Local adjustments for water allocation

Local adjustments and the adaptive capacity of farmers are major 'buffers' in the face of low-flow years (see Box 10.4), but they remain largely overlooked. Further, community management and local institutions for water allocation and conflict resolution have all existed

---

Box 10.4. Farmers' coping strategies: the case of the Nagarjuna Sagar project.

Nagarjuna Sagar is a large irrigation project (900,000 ha) located in the lower Krishna basin (Andhra Pradesh). Inflow to the reservoir has been dramatically curtailed due to upstream water development. During the dry years of 2002 and 2003, canal water availability was 20% of the long-term average. Consequently, in 2002, the total irrigated area decreased by more than 85%, showing that most farmers did not engage in cultivation or lost their crops due to lack of water. In 2003, the cropped area was 77% of the area irrigated during years with normal water availability. This illustrates the resilience of irrigated farming systems and the 'learning' abilities of farmers, who engaged in early rainfed cultivation with the monsoonal rains rather than waiting for canal water. Farmers resorted to diverse strategies, such as: (i) increasing irrigation efficiency through better land preparation and better O&M of field canals and drains; (ii) pumping groundwater (the number of wells dramatically increased in the region); (iii) changing cropping patterns, such as leaving part of the land fallow and planting dry crops such as millet and sorghum for self-subsistence; (iv) seeking work outside the agriculture sector; and (v) selling their livestock – a primary source of revenue and workforce for the poorest households. There is also anecdotal evidence of tampering with the irrigation system and of the collusion of farmers with politicians and employees of the Irrigation Department (Venot, 2008c).

from the earliest stages of river basin development in South India (see Mosse, 2003, on tank management). However, the 73rd Constitutional Amendment of 1993 (and the related Panchayati Raj Act) constitutes the first national attempt to devolve powers to local bodies: the panchayats are notably made responsible for local management of water-related services and water bodies. In the field of irrigation, this means the O&M of minor irrigation projects. While much hope has been vested in the decentralization process for increasing the efficiency of water use, it is clear that the panchayat raj system will be unable to address some of the core problems of the water sector, which need to be tackled at a regional level (Mollinga, 2005). Two of the three states that share the Krishna basin provide striking examples of the limits to the ongoing decentralization process in India: in Karnataka, there is a trend towards political re-centralization (after a period of strong state support to panchayats), and in Andhra Pradesh, state poverty-reduction programmes and the creation of WUAs are thought to have diluted the power and autonomy of the panchayats, as a non-elected bureaucracy has bypassed locally elected institutions (Johnson, 2003; Mooij, 2003).

New Challenges and Ways Forward: for a New Governance Structure

Supply augmentation projects remain the most common option resorted to in order to face increasing water demand. The project of interlinking the Indian rivers would, for example, result in an 11.6 billion m³/year increase in water availability in the lower Krishna basin, a fifth of which would be 'reserved' for the Krishna delta to support agriculture, counterbalancing the observed decline in discharge of the Krishna River and limiting environmental degradation (NWDA, 2007). This could alleviate the situation of crisis, which is likely to recur in the near future, but attendant plans to expand irrigation with this transferred water are likely to defeat this objective. The construction of a diversion scheme from the Godavari basin began in 2006 and demonstrates the commitment of the state of Andhra Pradesh to this project. A transfer from the Alamatti dam to the Pennar River basin to the south-east is also contemplated, but plans to develop irrigation en route might result in water being used up before it crosses the boundaries of the Krishna basin. Finally, the state of Maharashtra is planning transfers from the upper Krishna to the upper Bhima sub-basins, to meet irrigation water demand in the latter.

However, as the water sector faces new challenges, and water management organizations are expected to focus not only on agricultural growth and increasing food production but also on broader social (enhancing equity, reducing poverty) and environmental goals, there is a need for institutional reforms to trigger a change in the governance structure of natural resources. But current water management organizations are often structured to address past challenges, and are unable to internalize these new priorities (Merrey et al., 2007).

Mollinga (2005) identifies the resilience of civil-engineering-dominated water bureaucracies as one of the main obstacles to change in the Indian water sector. The Indian water resources governance structure and policy process remains one of centralization and hierarchy, inherited from the post-1947 planned development approach and the structure of the government administration. Much hope was vested in devolution and decentralization processes, and in PIM programmes (see above), but these experiments did not bring a socio-political revolution in rural governance (Jayal et al., 2006). The Irrigation Department is reluctant to devolve its power to WUAs (Mollinga et al., 2004). Broader decentralization policies do not challenge the prevailing dual system between federal and state levels[27] and seem to be hardly taken up by local populations (Johnson, 2003; Mollinga, 2005). Radical changes in the balance of power in favour of water users and major restructuring of hydro-bureaucracies have yet to happen (Merrey et al., 2007), and are unlikely to take place through further institutional development. A shift in the governance structure would require that the state concentrate on its regulatory role because of its responsibility for providing public goods and for ensuring equity and sustainability. The main challenges are to improve effectiveness of the state itself and to find the right balance between state action and other institutional actors (Merrey et al., 2007). This requires creating political room for mobilization and empowerment of disenfranchised stakeholders (Molle et al., 2007).

Other challenges to water resources management in India are the organizational and policy fragmentation (Mollinga, 2005), and the highly informal and dispersed character of the water economy, with millions of farmers securing private access to groundwater (Shah, 2007). Water and irrigation management reforms are not linked to broader agricultural or rural development policies (which are the remit of other governmental departments), and have generally focused on single organizations (this happens in most countries; Merrey et al., 2007). The dynamics of agricultural water use in the Krishna basin and in India as a whole (notably the increasing groundwater abstraction) are not reflected in institutional and financial investments, which are mainly targeted at medium and large surface-irrigation projects (although free electricity could be seen as an 'investment' for minor irrigation). While such investments are needed, it is crucial to adopt a more integrated approach to land and water resources management (Mollinga, 2005). There are examples of such approaches in Indian history, notably during the 1980s, when Prime Minister Rajiv Gandhi implemented intersectoral development programmes whose coordination was taken up at the district level (Shah, 2007).

As demand for water by different user groups is increasing, river basins are closing, and interconnectedness between users intensifies, leading to decreasing security of supply to all existing users; allocation of water takes centre stage in integrated approaches to water resources management. However, existing institutions in India are unlikely to offer a space for effective negotiation at relevant scales (Mollinga, 2005); past experiences of basin water allocation (e.g. by the KWDT) did not prevent basin closure and lie at the root of basin overbuilding (see above). These allocation mechanisms – centred on surface blue water – do not allow internalization of the multi-level drivers of river basin closure.

The current KWDT tribunal is expected to reach a decision by 2010. At present, this tribunal mainly involves decision makers and bureaucrats from the three states and needs to be made more responsive to the demands of local communities by involving local users in the negotiation process for both social and environmental benefits (refer to Iyer, 2003, and Janakarajan, 2007, for the prospects and problems related to involving farmers in the resolution of water-sharing conflicts). The coming KWDT allocation rules should: (i) be defined at the basin level; (ii) be based on a comprehensive and transparent understanding of the hydrology; (iii) internalize the variability of water availability – and give special attention to allocation during dry years; (iv) recognize the interactions between surface water, groundwater and green (soil) water (soil moisture is crucial for rainfed agriculture and many livelihoods); (v) estimate long-term reliable supplies in any part of the basin in light of actual and projected use; (vi) recognize customary rights, local strategies and local adjustments (Venot, 2008a); and (vi) be implemented and monitored with a more transparent and effective data collection. The KWDT should also provide mechanisms for stopping any project in contradiction to the award of the tribunal. As allocation is likely to take place mainly on economic (increasing water productivity) and political grounds, primary attention is to be given to equity and environmental principles through a reserve for both productive water for the poor and the environment (environmental flows). As attention to environmental conservation is often perceived as anti-poor (Merrey et al., 2007), there is a need for an economic and social valuation of the goods and services provided by ecosystems (see Pearce et al., 2006); this is one of the most promising ways of making decision makers commit to the objective of environmental preservation in the process of water allocation.

Finally, sectoral water resources policies alone cannot slow down river basin closure. There is a need to resort to integrated rural development policies that would ensure the rural population alternatives within (diversification, organic farming, integration in agribusiness chains, etc.) or outside the agriculture sector.

Conclusion

Overcommitment of water resources and signs of basin closure are apparent during dry periods in the Krishna basin. Overexploitation comes along with changing patterns of access to water and declining security of supply to all existing users, as the basin is almost fully allocated and demand greater than supply. This inevitably leads to conflicts of access and allocation. As early as the mid-1970s, through the KWDT, the government of India legislated for allocation of water to the three states of Andhra Pradesh, Karnataka and Maharashtra. The KWDT award did not offer a means to manage crises and low-flow years. On the contrary, it encouraged the three states to invest massively in the development of their hydraulic infrastructures, often with dramatic economic and fiscal consequences. In parallel with supply augmentation, the three states also implemented water demand management options, but almost all these measures had as a prime objective the bridging of the gap between actual and potential irrigated area, and resulted in a further commitment of water resources.

Unpacking the drivers of river basin closure requires going beyond a linear trajectory of river basin development, where strategies of supply augmentation, demand management and water allocation are successively called upon. In the Krishna basin, all past interventions combined to lead to the closure of the basin. Indian debates on the 'underutilization' of the irrigation potential and the nature of rural development policies (oscillating between equity and growth) lie at the root of the overexploitation of water resources in the Krishna basin.

River basin closure means that, in hydrologic terms, overall basin efficiency is close to its maximum. The scope for effective water savings has become very limited: supply augmentation as well as demand management options eventually lead to reallocation of water. Current institutions are not adapted to face these new challenges and deal with basin water allocation. A formal and clear apportionment of water between different user groups is also difficult because of the complexity and limited knowledge of the basin hydrology.

What is needed is a new democratic governance system, allowing for the demands of local communities to be voiced through a platform of negotiations where state, civil society and economic spheres could interact to allocate basin water. This is a challenging task, which goes beyond reforming the water sector and requires a broader change in the nature of the state and democracy in India (Mollinga, 2005).

Acknowledgments

This research was supported by a grant from the Australian Centre for International Agricultural Research (ACIAR) and contributes to the PhD thesis of the author. The author thanks Trent Biggs, Sylvain Massuel, François Molle, Madar Samad and Hugh Turral for thoughtful comments in the early stages of this research. Thanks are also due to the team at the Hyderabad office of the International Water Management Institute.

Notes

1 According to the minor irrigation censuses of 1994 and 2001, the number of shallow tube-wells in the Krishna basin increased from 35,000 to 515,000 between 1987 and 2001, while the number of deep tube-wells increased from 14,000 to 82,000 during the same period. In 2001, 1.1 million dug wells were registered (515,000 in 1987).
2 Estimates are based on district-wise land use data (presented in GoAP, 2006; GoK, 2006; GoM, 2005a; EPW, 2005), and on data available online, with a subscription, at www.indiaagristat.com
3 From 1990–2001, the rainfall:runoff coefficient was only 0.07 (Biggs et al., 2007).
4 Depletion from irrigated fields and reservoirs is calculated from climatic data and the Penman–Monteith equation (Allen et al., 1998); depletion in rainfed ecosystems is estimated using the methods of Ahmad et al. (2006), Bouwer et al. (2007) and Immerzeel et al. (2007). To compute basin depletion, land cover is estimated from land-use statistics at the district level.
5 Domestic and industrial uses are computed according to van Rooijen et al. (2008), assuming a water-use efficiency of 70% in both sectors. Livestock consumption is computed according to

Peden *et al.* (2007) and livestock statistics at the district level.

6 The live storage capacity of existing reservoirs is used and replenished every year and inter-annual variation of storage in these reservoirs is considered negligible.

7 That is, the rain that infiltrates to the groundwater, calculated as a constant fraction of precipitation, depending on the sub-basin considered, on the basis of estimates of the National Water Development Agency (NWDA).

8 Estimated as 70% of all depletion in groundwater-irrigated areas.

9 Groundwater outflows to the sea from the Krishna delta are ignored.

10 Overall agricultural growth remained positive due to improved productivity in livestock- and fish-production activities, and in cultivation of high-value fruits and vegetables, but the crisis affecting the crop sector (and marginal family farmers) is even more acute than these macro-indicators suggest.

11 The KWDT uses the term 'flow' to designate the 'naturalized runoff' (i.e. the observed discharge to the ocean + an estimate of basin water uses at the time of the tribunal). Interestingly, the first formal allocation rules, defined in 1951, apportioned a much lower flow of 48.5 billion m^3/year (Venot *et al.*, 2007), which is exceeded in more than 90% of the years, if data from GoI-KWDT (1976) are considered.

12 These allocations include 'local runoff': the KWDT did not only share water from the main-stem Krishna but also runoff from sub-basins (e.g. the Andhra Pradesh allocation includes water from sub-basins that are entirely located in Andhra Pradesh, such as the Musi, Palleru and Muneru sub-basins: only 12.8 billion m^3/year of the 22.6 billion m^3/year allocated to Andhra Pradesh has to come from upstream).

This volumetric apportionment of water is known as 'Scheme A', and constitutes the default scheme to be implemented. Proportional apportionment of water had been contemplated in what was called 'Scheme B': allocations would then depend on water availability, and either scarcity, or surplus water, would be proportionally shared by the three states. Scheme B was supported by upstream states and opposed by Andhra Pradesh, and was thus never implemented (Sajjan, 2005).

13 If upstream states take their 'share' (as mentioned by the KWDT), there is a high probability (0.25) that the Andhra Pradesh allocation will be jeopardized and downstream uses constrained: in these conditions, it is crucial for Andhra Pradesh to discuss provisions during years of low flow.

14 Declining base flows means that the 75%-dependable flow of the Krishna River, as estimated by the KWDT, is exceeded less than one year in two. This is consistent with results from Biggs *et al.* (2007).

15 Naturalized runoff from 2001 to 2004 can be estimated at 52.8 billion m^3/year (e.g. 9% lower than the 75%-dependable flow, as estimated by the KWDT; inflow from upstream was about 10 billion m^3/year (i.e. 2.8 billion m^3 lower than the amount needed for Andhra Pradesh to receive its full KWDT allocation).

Even though water availability was close to the allocation defined by the KWDT, the drought had dramatic consequences on downstream irrigated agriculture (see Box 10.4); this is because diversions to downstream irrigation projects (Nagarjuna Sagar, the Krishna delta) had been – for most of their history – much higher than the volumes 'protected' by the KWDT (Andhra Pradesh diverts significant volumes of 'surplus water'; Venot *et al.*, 2007). Farmers have been used to high – but highly unreliable – supplies and have been dramatically affected by years of low flows: overcommitment of water can 'artificially' create a situation of scarcity.

16 The deficit in the discharge to the ocean was sharpened by the increased storage capacity (3 billion m^3) of the Alamatti reservoir, whose level had been raised in the early 2000s.

17 Growth in well-endowed areas is supposed to trickle down to other regions and social groups.

18 This long-mooted project would consist of inter-linking the Indian rivers through a 'national water grid', and in transferring water from 'water-abundant' basins to water-scarce basins, such as the Krishna and the Cauvery, further south. After being given a strong impetus in the late 1990s, it is not clear if the project has achieved consensus among national politicians and decision makers (Venot *et al.*, 2007): indeed, this 'all-India' mega-project seems to have been put off, and only a few links are being contemplated or constructed, but without further references to an all-India grid.

19 In Andhra Pradesh, the work of the Commission for Irrigation Utilization (Hashim Ali, 1982) was key in highlighting the need for such technical improvement programmes.

20 The National Watershed Development Project (early 1980s) and the National Watershed Development Project for Rain-fed Areas are two examples.

21 The World Bank, for example, has withdrawn more than once from the controversial Alamatti project (in Karnataka) and returned to it each time (D'Souza, 2006).

22 Intrastate political motives (to obtain the support of particular social groups or regions) have also

23 largely driven the construction of irrigation projects (see Box 10.3 and Venot et al., 2007).
23 Briscoe and Malik (2007) evaluate that 18% of Maharashtra's fiscal deficit is due to the construction of dams whose primary purpose was to lay claims for water from the Krishna.
24 Notably, upstream states were worried that their allocation would be revised downward if they could not justify water use equal to, or higher than, their 1976 allocation.
25 In 1996, the federal government also initiated the Accelerated Irrigation Benefit Scheme (a loan system) for the states to complete irrigation projects that were already at an advanced stage of construction.
26 Shah et al. (2003) highlight that capping groundwater abstraction is one of the most challenging tasks in the Indian water economy.
27 For example, financial decentralization is not part of the reforms.

References

Ahmad, M.D., Biggs, T.W., Turral, H. and Scott, C.A. (2006) Application of SEBAL approach and MODIS time-series to map vegetation water use patterns in the data scarce Krishna river basin of India. *Water Science and Technology* 50(10), 83–90.

Allen, R.G., Pereira, L.S., Raes, D. and Smith, M. (1998) *Crop Evapo-transpiration – Guidelines for Computing Crop Water Requirements.* Irrigation and Drainage Paper 56. Food and Agriculture Organization of the United Nations, Rome.

Atchi Reddy, M. (1990) Travails of an irrigation canal company in South India 1857–1882. *Economic and Political Weekly* 25(12), 619–628.

Attwood, D.W. (1992) *Raising Cane: the Political Economy of Sugar in Western India.* Westview Press, Boulder, Colorado.

Aubriot, O. (2006) Baisse des nappes d'eau souterraine en Inde du sud: Forte demande sociale et absence de gestion de la ressource. *Géocarrefour* 81(1), 83–90.

Barker, R. and Molle, F. (2005) *Evolution of Irrigation in South and Southeast Asia.* Research Report 5. Comprehensive Assessment of Water Management in Agriculture Secretariat, Colombo, Sri Lanka.

Biggs, T.W., Gaur, A., Scott, C.A., Thenkabail, P., Rao, P.G., Krishna, G.M., Acharya, S. and Turral, H. (2007) *Closing of the Krishna Basin: Irrigation, Streamflow Depletion and Macroscale Hydrology.* Research Report 111. International Water Management Institute, Colombo, Sri Lanka.

Bolding, A., Mollinga, P.P. and van Stratten, K. (1995) Modules for modernisation: colonial irrigation in India and the technological dimension of agrarian change. *The Journal of Development Studies* 31(6), 805–844.

Bouwer, L., Biggs, T. and Aerts, J. (2007) Estimates of spatial variation in evaporation using satellite-derived surface temperature and a water balance model. *Hydrological Processes* 22(5), 670–682.

Briscoe, J. and Malik, R.P.S. (2007) *Handbook of Water Resources in India: Development, Management, and Strategies.* World Bank, Washington, DC/Oxford University Press, New Delhi.

Celio, M. (2008) Transferring water from agriculture to growing cities: the Hyderabad case (South India) and its implications for intersectoral water reallocation research and policies. PhD thesis, Water Engineering and Development Centre, Loughborough University, UK.

Chambers, R. (1988) *Managing Canal Irrigation: Practical Analysis from South Asia.* Oxford University Press, New Delhi.

CRU (Climatic Research Unit, University of East Anglia) (2007) Monthly precipitation data. http://www.cru.uea.ac.uk (accessed 11 May 2007).

Deshpandhe, R.S. (2007) *Emerging Issues in Land Policies.* INRM Policy Brief No.16. Asian Development Bank, India Resident Mission, New Delhi. Available online at http://www.adb.org/Documents/Reports/Consultant/TAR-IND-4066/Agriculture/deshpande.pdf (accessed 20 August 2008).

Dhawan, B.D. (2006) India's irrigation sector: myths and realities. In: Mujumdar, N.A. and Kapila, U. (eds) *Indian Agriculture in the New Millennium: Changing Perceptions and Development Policy*, Vol. 2. Academic Foundation, New Delhi, pp. 39–52.

D'Souza, R. (2006) *Interstate Disputes over Krishna Waters – Law, Science and Imperialism.* Orient Longman, Hyderabad, India.

EPW (*Economic and Political Weekly*) (2005) *Statistical Abstract of Maharashtra.* Economic and Political Weekly, Mumbai, India.

Falkenmark, M. and Molden, D. (eds) (2008) Special issue: closed basins – highlighting a blind spot. *Water Resources Development* 24(2).

GoAP (Government of Andhra Pradesh) (2006) *Statistical abstract of Andhra Pradesh (1955–56 to 2004–2005)*. Ministry of Agriculture, Government of Andhra Pradesh, Hyderabad, India.

GoAP (2007) Andhra Pradesh Human Development Report: Chapter 6: Agriculture. Available online at www.aponline.gov.in/Apportal/HumanDevelopmentReport2007/APHDR_2007_Chapter6.pdf (accessed 20 August 2008).

GoI (Government of India) (2001) *All-India Census*. Government of India, New Delhi.

GoI (2002) *National Water Policy*. Government of India, New Delhi.

GoI (2007a) *Eleventh Five Year Plan: Chapter 1: Agriculture*. Central Planning Commission, Government of India, New Delhi.

GoI (2007b) *Agricultural Statistics at a Glance 2006–2007*. Government of India, New Delhi.

GoI–KWDT (Government of India – Krishna Water Disputes Tribunal) (1976) *The Further Report of the Krishna Water Disputes Tribunal with the Decision*. Government of India, New Delhi.

GoK (Government of Karnataka) (2006) *Statistical Abstract of Karnataka 2000–2001*. Government of Karnataka, Bangalore, India.

GoM (Government of Maharashtra) (2005a) *Statistical Abstract Maharashtra State 1997–98*. Government of Maharashtra, Mumbai, India.

GoM (2005b) *Maharashtra Water Resources Regulatory Authority Act of 2005*. Government of Maharashtra, Mumbai, India.

Gulati, A., Meinzen-Dick, R. and Raju, K.V. (2005) *Institutional Reforms in Indian Irrigation*. Sage, New Delhi.

Hanumantha Rao, C.H. (2006) Integrated water resources development in Andhra Pradesh – Problems faced in 'Jalayagnam' and solutions. Available at http://www.indianfarmers.org/History.htm (accessed 31 August 2007).

Hashim Ali, S. (1982) *Report of the Commission for Irrigation Utilisation*. Government of Andhra Pradesh, Hyderabad, India.

IFPRI (International Food and Policy Research Institute) (2005) Indian agriculture and rural development strategic issues and reform options: a strategy paper prepared by IFPRI's senior management team for consideration by the policymakers of the government of India. Available online at http://www.ifpri.org/pubs/ib/ib35.pdf (accessed 18 May 2008).

Immerzeel, W.W., Gaur, A. and Droogers, P. (2007) *Remote Sensing and Hydrological Modelling of the Upper Bhima Catchment*. Research Paper 3. Future Water, Wageningen, The Netherlands.

IRDAS (Institute of Resources Development and Social Management) (1996) *Evaluation of Command Area Development Program Nagarjuna Sagar Project – Andhra Pradesh*. Institute of Resources Development and Social Management, Hyderabad, India.

Iyer, R. (2003) Cauvery dispute: a dialogue between farmers. *Economic and Political Weekly* 38(24), 2350–2352.

Janakarajan, S. (2007) Rethinking on policy: can stakeholders' dialogue contribute to conflict resolution, social learning and sustainable use of water? Paper presented at the seminar Water, Politics and Development, 12 August 2007, World Water Week, Stockholm.

Jayal, N.G., Prakash, A. and Sharma, P.K. (2006) *Local Governance in India. Decentralization and Beyond*. Oxford University Press, New Delhi.

Johnson, C. (2003) *Decentralisation in India: Poverty, Politics and Panchayati Raj*. Working Paper 199. Overseas Development Institute, London.

Landy, F. (2008) *Feeding India. The Spatial Parameters of Food Grain Policies*. Manohar, New Delhi.

Madhav, R. (2007) Irrigation reforms in Andhra Pradesh: whither the trajectory of legal changes. International Environmental Law Research Center Working Paper. Available at http://www.ierlc.org/content/w0704.pdf (accessed 14 May 2008).

Mearns, R. (1999) *Access to Land in Rural India: Policy Issues and Options*. Policy Research Working Paper Series No. 2123. World Bank, Washington, DC.

Merrey, D.J., Meinzen-Dick R., Mollinga, P.P. and Karar, E. (2007) Policy and institutional reform: the art of the possible. In: Molden, D. (ed.) *Water for Food, Water for Life: a Comprehensive Assessment of Water Management in Agriculture*. Earthscan, London; International Water Management Institute, Colombo, Sri Lanka, pp. 193–231.

Mohile, A.D. (2007) Government policies and programmes. In: Briscoe, J. and Malik, R.P.S. (eds) *Handbook of Water Resources in India – Development, Management, and Strategies*. World Bank, Oxford University Press, New Delhi, pp. 10–46.

Molden, D. (1997) *Accounting for Water Use and Productivity*. SWIM Paper 1. International Irrigation Management Institute, Colombo, Sri Lanka.

Molden, D., Sakthivadivel, R., Samad, M. and Burton, M. (2005) Phases of river basin development: the need for adaptive institutions. In: Svendsen, M. (ed.) *Irrigation and River Basin Management: Options for Governance and Institutions*. CAB International, Wallingford, UK; International Water Management Institute, Colombo, Sri Lanka, pp. 19–29.

Molle, F. (2003) *Development Trajectories of River Basins – a Conceptual Framework*. Research Report 72. International Water Management Institute, Colombo, Sri Lanka.

Molle, F. (2008) Why enough is never enough: the societal determinants of river basin closure. *International Journal of Water Resources Development* 24(2), 217–226.

Molle, F. and Berkoff, J. (2006) *Cities versus Agriculture: Revisiting Intersectoral Water Transfers, Potential Gains and Conflicts*. Research Report 10. Comprehensive Assessment Secretariat, Colombo, Sri Lanka.

Molle, F., Wester, P. and Hirsh, P. (2007) River basin development and management. In: Molden, D. (ed.) *Water for Food, Water for Life: a Comprehensive Assessment of Water Management in Agriculture*. Earthscan, London; International Water Management Institute, Colombo, Sri Lanka, pp. 585–625.

Mollinga, P.P. (2001) Water and politics: levels, rational choice and South Indian canal irrigation. *Futures* 33(8), 733–752.

Mollinga, P.P. (2003) *On the Waterfront: Water Distribution, Technology and Agrarian Change in a South Indian Canal Irrigation System*. Orient Longman, New Delhi.

Mollinga, P.P. (2005) *The Water Resources Policy Process in India: Centralisation, Polarisation and New Demands on Governance*. ZEF Research Report 7. Center for Development Research, Bonn, Germany.

Mollinga, P.P., Doraiswamy, R. and Engbersen, K. (2004) The implementation of participatory irrigation management in Andhra Pradesh, India. *International Journal of Water* 1(3/4), 360–379.

Mooij, J. (2003) *Smart Governance? Politics in the Policy Process in Andhra Pradesh, India*. Working Paper 228. Overseas Development Institute, London.

Mosse, D. (2003) *The Rule of Water: Statecraft, Ecology, and Collective Action in South India*. Oxford University Press, New Delhi.

Nikku, B.R. (2006) The politics of policy – Participatory irrigation management in Andhra Pradesh. PhD thesis, Wageningen University, Wageningen, The Netherlands.

NWDA (National Water Development Agency) (2007) http://nwda.gov.in/indexmain.asp?linkid=84&langid=1 (accessed 5 February 2007).

Pearce, D., Atkinson, G. and Mourato, S. (2006) *Cost–Benefit Analysis and the Environment. Recent Developments*. OECD Publishing, Paris.

Peden, D., Tadesse, G. and Misra, A. K. (2007) Water and livestock for human development. In: Molden, D. (ed.) *Water for Food, Water for Life: a Comprehensive Assessment of Water Management in Agriculture*. Earthscan, London; International Water Management Institute, Colombo, Sri Lanka, pp. 485–514.

Ramamurthy, P. (1995) The political economy of canal irrigation in South India. PhD thesis, Graduate School of Syracuse University, Syracuse, New York.

Ratna Reddy, V. (2006) 'Jalayagnam' and bridging regional disparities – Irrigation development and distribution in Andhra Pradesh. *Economic and Political Weekly* 41(4344), 4613–4620.

Sajjan, S. (2005) Availability of water in Krishna basin and scope for scheme-B irrigation projects. Fifth International R&D conference, 15–18 February, Bangalore, India.

Sengupta, N. (2005) Institutions against change. In: Shivakoti, G.P., Vermillion, D.J., Lam, W.-F., Ostrom, E., Pradhan, U. and Yoder, R. (eds) *Asian Irrigation in Transition: Responding to Challenges*. Sage, New Delhi, pp. 99–123.

Shah, T. (2007) Institutional and policy reforms. In: Briscoe, J. and Malik, R.P.S. (eds) *Handbook of Water Resources in India – Development, Management, and Strategies*. World Bank, Washington, DC/Oxford University Press, New Delhi, pp. 306–326.

Shah, T., Deb Roy, A., Qureshi, A.S. and Wang, Z. (2003) Sustaining Asia's groundwater boom: an overview of issues and evidence. *Natural Resources Forum* 27(2), 130–141.

Smakhtin, V. and Anputhas, M. (2006) *An Assessment of Environmental Flow Requirements of Indian River Basins*. Research Report 107. International Water Management Institute, Colombo, Sri Lanka.

Smakhtin, V., Gamage, N. and Bharati, L. (2007) *Hydrological and Environmental Issues of Interbasin Water Transfers in India: a Case of Krishna River Basin*. Research Report 120. International Water Management Institute, Colombo, Sri Lanka.

Suri, K.C. (2006) Political economy of agrarian distress. *Economic and Political Weekly* 41(16), 1523–1529.

Upadhya, C.B. (1988) The farmer–capitalists of coastal Andhra Pradesh. *Economic and Political Weekly* 23(27/28), 1376–1382, 1433–1442.

Vaidyanathan, A. (1999) *India's Water Resources: Contemporary Issues on Irrigation*. Oxford University Press, New Delhi.

van Rooijen, D.J., Turral, H. and Biggs, T.W. (2008) Urban and industrial water use in the Krishna basin. *Irrigation and Drainage* 57, 1–23.

Venot, J.P. (2008a) Why and where are the Krishna waters disappearing? Sustainability at risk in a South Indian river basin. *Economic and Political Weekly* 43(8), 15–17.

Venot, J.P. (2008b) Shifting access and rights to irrigation water in a context of growing scarcity: the case of the Krishna basin, South India. Paper presented at the international conference on 'Spatial Justice', 12–14 March 2008, Nanterre, France.

Venot, J.P. (2008c) Water scarcity and farmers' adjustments in South India: the Nagarjuna Sagar irrigation project. Paper and poster prepared for the 8th IFSA Symposium on 'Empowerment of rural actors: a renewal of farming systems perspectives'. 6–10 July, Clermont-Ferrand, France.

Venot, J.P., Turral, H., Samad, M. and Molle, F. (2007) *Shifting Waterscapes: Explaining Basin Closure in the Lower Krishna Basin, South India.* Research Report 121. International Water Management Institute, Colombo, Sri Lanka.

Venot, J.P., Sharma, B.R. and Rao, K.V.G.K. (2008) Krishna basin development: intervention to limit downstream environmental degradation. *Journal of Environment and Development* 17(3), 269–291.

Wade, R. (1976) How not to redistribute with growth – The case of India's command area development program. *Pacific Viewpoint* 14–17, 95–104.

Wade, R. and Chambers, R. (1980) Managing the main system: canal irrigation's blind spot. *Economic and Political Weekly* 15(39), A107–A112.

Wallach, B. (1985) British irrigation works in India's Krishna basin. *Journal of Historical Geography* 11(2), 155–173.

Wester, F., Shah, T. and Merrey, D.J. (2005) Providing irrigation services in water-scarce basins: representation and support. In: Svendsen, M. (ed.) *Irrigation and River Basin Management: Options for Governance and Institutions.* CAB International, Wallingford, UK, International Water Management Institute, Colombo, Sri Lanka, pp. 231–246.

11 Pumped Out: Basin Closure and Farmer Adaptations in the Bhavani Basin in Southern India

Mats Lannerstad[1]* and David Molden[2]*

[1]*Linköping University, Stockholm, Sweden;*
[2]*International Water Management Institute, Colombo, Sri Lanka;*
e-mails: *mats@lannerstad.com; **d.molden@cgiar.org

Introduction

This chapter describes and analyses water resource development in the Bhavani basin, a sub-basin of the Cauvery basin, in southern India. The Bhavani basin is almost entirely located within the semi-arid state of Tamil Nadu. This is an area where, for centuries, agriculture has had to cope with erratic rainfall. Up to the middle of the 20th century, the vagaries of the monsoons resulted in feasts or famines. While irrigation development offered some stability in the second half of the 20th century, a rapidly growing population amplified the difficulty of matching food production with human needs. Water is now a key limiting factor to agricultural growth, and it is imperative to have different strategies to increase the availability and productivity of water resources (Baliga, 1966; Mohanakrishnan, 2001).

The Cauvery basin is the most important surface water source in Tamil Nadu. In the 2nd century AD the Grand weir was constructed across the Cauvery River. It serves 350,000 ha in the delta and was the first major, and is still the largest, command area in the basin. During the 20th century, development of irrigation infrastructure in the Cauvery basin increased the (gross) total irrigated area from about 600,000 ha to 1.9 million ha and brought the entire basin to closure (GoI, 2005). The Bhavani basin was essentially brought to closure in the middle of the 1950s, and since then only a fraction of the natural outflow reaches the Cauvery River (Lannerstad, 2008).

A primary response to the lack of additional surface water for irrigation has been a marked increase in groundwater use. In the 1980s, the Tamil Nadu farmers were given free electricity, irrespective of the quantity consumed, and subsidies allowed groundwater use to flourish (Kannan, 2004; Shah, 2007). This chapter demonstrates spatially and statistically how the initiative during the last decades has moved from the state, which earlier constructed the large-scale projects in the basin, towards individual farmers pumping water from many different sources, and thus depleting aquifers and perturbing allocation schemes for surface water.

The analysis is based upon statistics, water-use data, interviews, reports and spatial analyses from maps, GIS and remote sensing. The water dynamics in the Bhavani basin provide examples of a complex web of interconnections and redistribution – not only upstream–downstream, but also downstream–upstream. This chapter describes how intensification of agricultural water use continues after basin closure. It has a special focus on the expanding use of pumps to extract water from aquifers, rivers and canals to enable further intensification in spite of an apparent constraint on overall water resource availability.

Agricultural Water Use in the Bhavani Basin

Agriculture has been the dominant water and land user in the Bhavani basin for hundreds of years. However, the way water is used for agriculture has changed significantly and is expected to change further and rapidly.

The Bhavani basin

The Bhavani basin is the fourth largest (6500 km^2) sub-basin in the Cauvery basin (81,000 km^2) (Fig. 11.1). The western part (Western Ghats) is hilly terrain of 300–2400 masl. The northern (eastern Ghats) side of the basin is dominated by rugged, discontinuous hills, with an elevation of 300–1000 masl. The Bhavani valley, the south-eastern part, is flat terrain (NWDA, 1993). In the upper Bhavani basin (4100 km^2), the average annual precipitation is 1600 mm and the potential yearly evapotranspiration is about 800 mm. In the lower Bhavani basin (2400 km^2), the conditions are the opposite, with an annual rainfall of around 700 mm and a potential evapotranspiration of 1600 mm (von Lengerke, 1977).

Forest reserves and plantations dominate the upper and the northern part of the lower Bhavani basin. Tea and coffee plantations, vegetables (carrots, potatoes and cabbages) and spices (cardamom and ginger) characterize the cultivated areas in the Nilgiris district areas. In the parts of the basin falling within the Erode and Coimbatore districts, there are irrigated lands with canals and groundwater, and rainfed croplands, often with supplemental groundwater irrigation. Cultivated crops are paddy, groundnuts, sorghum (for fodder), pulses, sugarcane, coconuts, sesame, turmeric and bananas (SCR, 2005).

The population in the Bhavani basin has increased by about 200% during the last 50 years, reaching around 2.5 million (GoI, 2005). More than 50% of the workforce in the Nilgiris district is involved in livestock, forestry, fishing, hunting, plantations or orchards, and 14% are employed in the agriculture sector as cultivators or agricultural labourers. In the Erode district, almost 55% work in agriculture. Industrial development is now increasing in the basin and is fuelled by the rapid development of Coimbatore, the industrial and second biggest city in Tamil Nadu, and Tiruppur, the textile centre of southern India, which are both located in the neighbouring Noyyal basin (Census of India, 1991a, b).

Historical development of gravity irrigation

The major part of the cultivated areas in the Bhavani basin are located in the Coimbatore and Erode districts, which are described as an area 'of exceptional dryness' with 'not less than two-thirds of the seasons' as 'unfavourable'. The years 1804–05, 1806, 1808, 1812, 1813, 1823, 1831, 1832, 1834, 1836, 1861, 1866, 1876–1878, 1891–92, 1892–93, 1894–95, 1904–05 and 1905–06 all had serious water scarcity, often leading to 'scarcity, desolation and disease' or 'famine, sickness and death.' In 1808, failure of both monsoons caused a famine 'that carried off half the population', while 'The Great Famine' in 1876–1878 was described as 'more disastrous in effect than any of its predecessors' (Madras Presidency, 1902; Baliga, 1966).

The great annual and seasonal rainfall variability and the hot climate pose an agricultural challenge in Tamil Nadu. Most parts of the state rely upon the unpredictable and erratic north-eastern monsoon (October–December), which is characterized by cyclones and short and heavy downpours (Mohanakrishnan, 2001). The people in the Erode district are fortunate as the upper part of the Bhavani basin receives ample rainfall during the south-west monsoon (June–September), which contributes to a perennial flow in the Bhavani River. To utilize this flow, two important weirs were constructed by local rulers during historic times. In the 13th century, the Kalingarayan weir (serving 4800 ha) was constructed across the Bhavani River just above the confluence point with the Cauvery

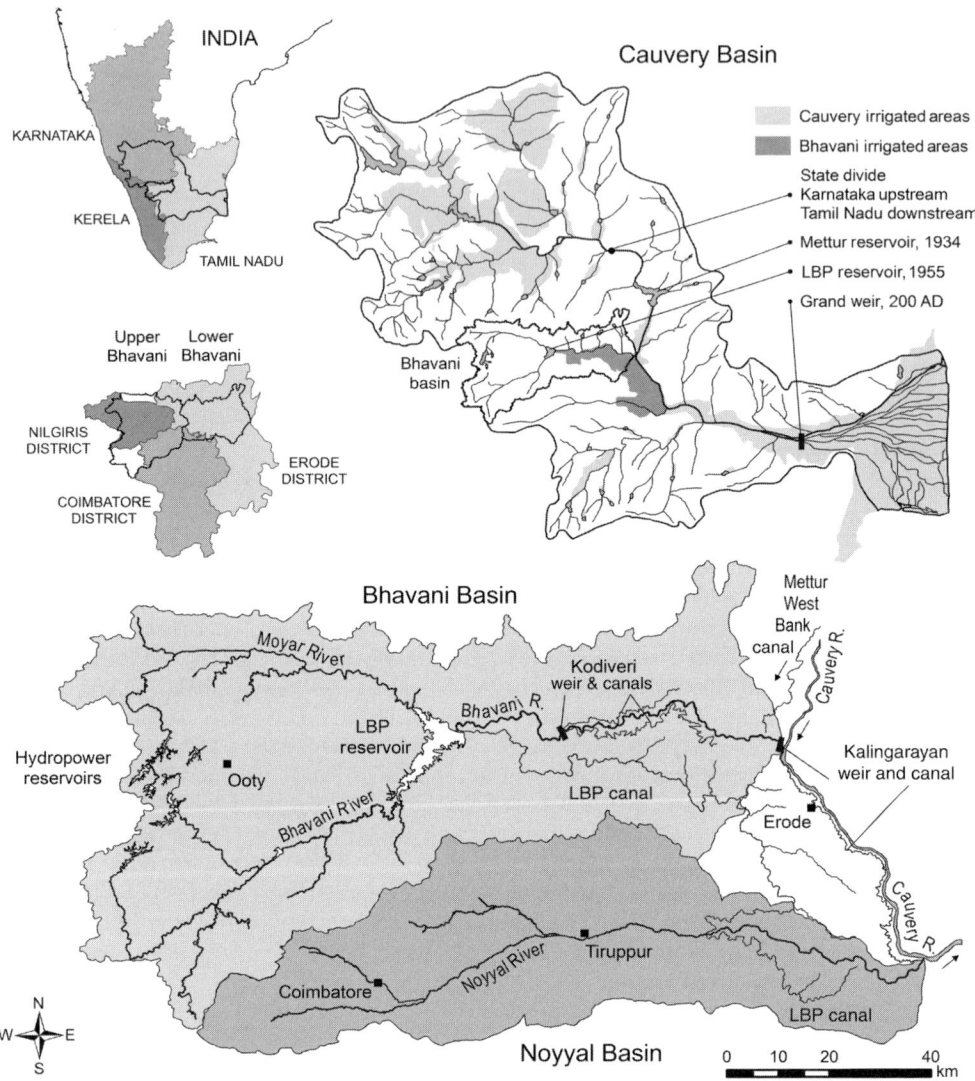

Fig. 11.1. Bhavani basin from an administrative, hydrological and irrigation perspective (Source irrigated areas Cauvery basin: Government of India, 2007, not to scale).

River. In the 17th century, the Kodiveri weir, which serves the Arakkankottai (serving 2000 ha) and Thadapalli (serving 7200 ha) canals, was constructed 50 km further upstream. Together with Kanniyampalayam weir (serving 160 ha) there were three weirs across the Bhavani River in 1940. At that time water was already diverted from the Kallar, Coonoor and Gandaipallam tributaries into a number of small canals, such as the Nellitturai and Maruthavalli canals (Fig. 11.2). In total, about 1100 ha were irrigated from these minor canals (Baliga, 1966; MIDS, 1998).

Compared with many other districts in Tamil Nadu, tanks play only a minor role in the Coimbatore and Erode districts. In 1903, long before the expansion of canals and groundwater use, only about 5% of the irrigated area was under tank irrigation (SCR, 1903). The total tank-irrigated area in Bhavani basin is

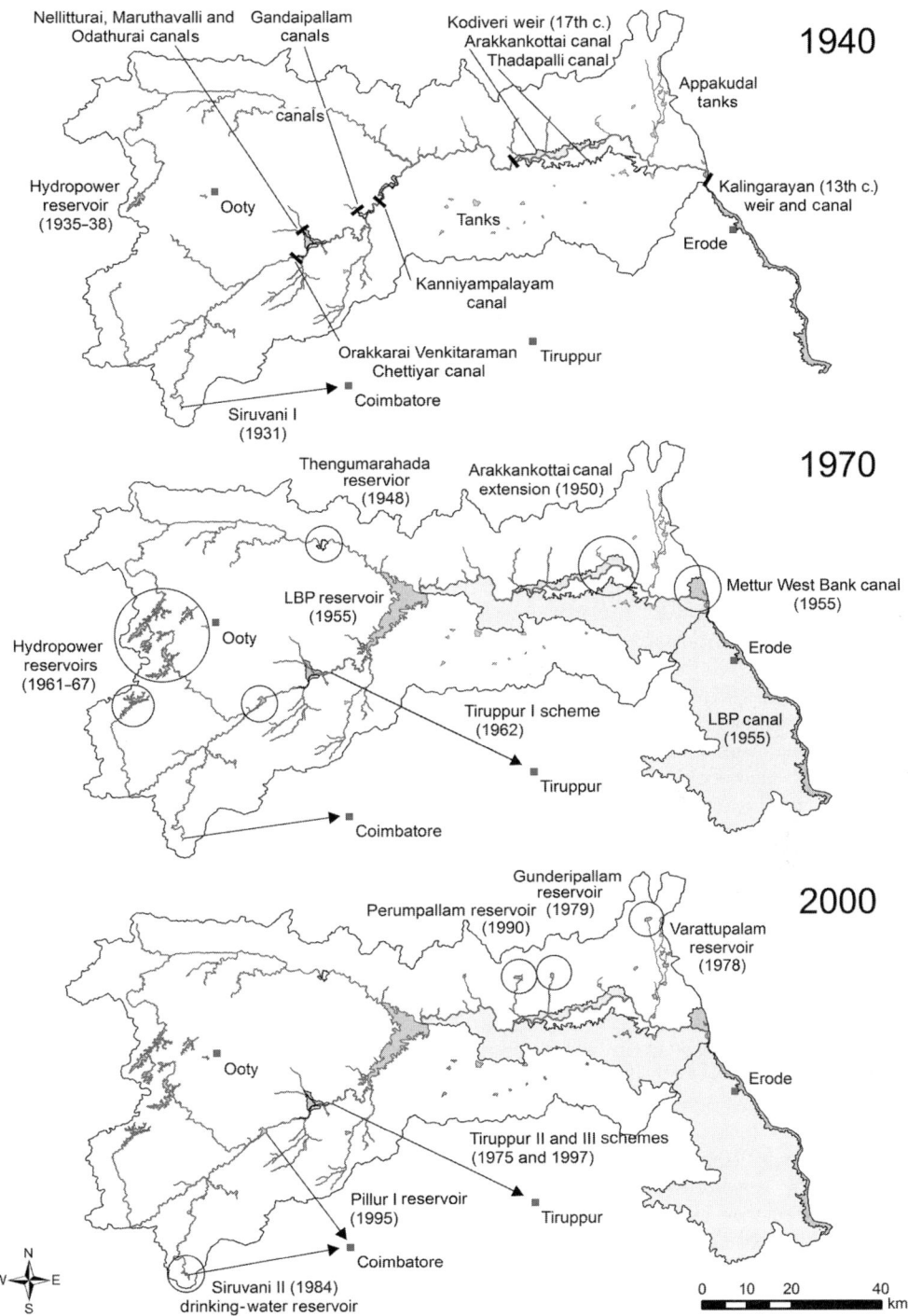

Fig. 11.2. Surface water resource development in Bhavani basin by 1940, 1970 and 2000.

currently about 2000 ha (MIDS, 1998). Most of the tanks are fed by local runoff and a few by canals. The Appakudal tank cascade, for example, is fed by various 'jungle streams' from the Bargur Hills (Baliga, 1966).

During the Second World War, the food problem became acute on a national scale in British India. To quickly increase food production, the Grow More Food (GMF) campaign was launched. Within this programme, the 'Minor Irrigation Programme', focused on irrigation works that could be rapidly implemented and did not demand large funds. The programme aimed at both private works (such as wells, tanks and water-lifting devices) and public measures (such as channels, embankments, tube-wells, public tanks, etc.) (GoI, 1952). Within the GMF campaign, the Public Works Department (PWD) started several projects in the Bhavani basin. The Arakkankottai canal was extended to include another 800 ha in 1950. From Mettur reservoir, one canal was constructed on each side of the Cauvery River. The Mettur West Bank canal from 1955 (Baliga, 1966) irrigates an area of about 800 ha in the Bhavani basin (NWDA, 1993) (Figs 11.1 and 11.2). The Thengumarahada Co-operative Farming Society is a small-scale example of how the food shortage inspired individuals, sanctioned by local authorities, to develop a new area. Since 1948 about 200 ha have been irrigated with water sourced from a weir across the Kukkulthorai River, a small tributary to the Moyar River (Seetharaman, Nilgiris district, Tamil Nadu, India, 2007, personal communication).

The major canal-irrigated cropland expansion in the Bhavani basin after 1940 was the development of the Lower Bhavani Project (LBP), which is based upon plans mainly designed during the last decades of British rule, and was sanctioned by the newly independent government of India and the government of Madras state in 1947. In times of national food-shortage emergency, the LBP was of national importance and was built as a 'Post-War Development Scheme'. The project was completed in 1955 and included a reservoir with a storage capacity of 930 Mm^3 and an irrigated area that straddles the Bhavani basin boundary (with about 31,500 ha falling within and 52,500 ha outside) (see Fig. 11.2 and Lannerstad, 2008).

With the completion of the LBP, the possibilities for larger canal projects were exhausted. Since 1970, only three minor irrigation reservoirs, in the north-eastern part of the basin, have been constructed by the PWD. Their total storage capacity is 10 Mm^3 and the total designed command area is about 2000 ha (TWAD, 2000). In the upper Bhavani, about 200 ha are irrigated in Karnataka in the northwest and about 700 ha in Kerala in the southwestern corner (NWDA, 1993).

Between 1940 and 2000, primarily in the period 1950–1955, the net canal-irrigated area within the Bhavani basin (not considering cropping intensity) increased fourfold, from about 12,000 ha to about 48,000 ha. The total net command area designed to rely upon surface water generated within the Bhavani basin increased from about 17,000 ha to 105,000 ha, a fivefold enlargement. The annual demand for surface water generated within the Bhavani basin for gravity irrigation increased from about 600 Mm^3 per year to more than 2000 Mm^3 per year.

Urban water demands and hydropower storage

Several non-agricultural, large-scale, surface water, resource development projects can also be found in the basin (Fig. 11.2). About 35% of the hydropower generation in Tamil Nadu, 640 MW, is produced in the basin (NWDA, 1993). The development of hydropower reservoirs started in 1938 and multiplied during the 1960s, when the Tamil Nadu Electricity Board (TNEB) increased the current storage capacity to 500 Mm^3, with 26 dams and weirs (TNEB, Coimbatore, 2005–2006, personal communication). The hydropower storage represents more than half the storage of the LBP reservoir, and the timing of the releases impacts the water availability for downstream irrigation farmers. Evaporation from the reservoirs reduces the flow from the upper Bhavani basin.

The first out-of-basin transfer to Coimbatore city was completed in 1931, with 4 Mm^3 diverted annually from the Siruvani River (Saravanan and Appasamy, 1999); this transfer was increased to 37 Mm^3 in 1984. In 1995, a pipeline from the lowest hydropower reservoir, behind the Pillur dam, was completed and transfers about 48 Mm^3 every year. To meet the needs of Tiruppur city, an additional volume of 19 Mm^3 is withdrawn each year from the Bhavani River through three schemes. Every year, as a result, 104 Mm^3 are transferred to meet urban and rural needs in the Noyyal basin. Annual municipal withdrawals within the Bhavani basin total about 23 Mm^3 above the LBP reservoir and 34 Mm^3 below, all mainly developed after 1970. Total industrial withdrawals amount only to about 20 Mm^3 per year (TWAD, 2000; TWAD, Coimbatore and Nilgiris Circle, 2005–2006, personal communication). Municipal and industrial abstractions together amount to almost 10% of the annual demand of the major canals in the basin.

Groundwater irrigation

Historically, wells were the major source for irrigation in the Coimbatore and Erode districts and were described as 'the heart and life of the district' at the end of the 19th century. At the beginning of the 20th century, these open wells were mostly situated in 'little valleys and hollows', and most of them were found on tank-fed lands (Baliga, 1966). According to statistics, the groundwater-irrigated area amounted to about 20% of the cultivated area at the beginning of the 20th century and around 25% in 2000 (Krishnaswami Ayyar, 1933; SCR, 2005).

Hard rock underlies the lower parts of the basin. Groundwater is found in the porous and granular weathered mantle and in the joints, fissures and fractures in the shallow depths underlying the weathered zone. There is also groundwater in the narrow, deep-seated fracture zones in the fresh crystalline rock. The aquifers can thus be differentiated into shallow aquifers at a depth of 10–30 m and deep aquifers down to a depth of about 200 m (TWAD, 2000).

The traditional well in the Bhavani basin is an open, rectangular well of about 7 × 7 m with a depth of 15–40 m. The statistics show that open wells dominate and only a fraction are bore wells (7%) (SCR, 2005). Today, a bore well is, however, often the preferred option for a new well. In the 1970s, bore wells used to be about 80 m deep. After the year 2000, bore wells down to 250 m depth became common (TWAD, 2000) (for more details see 'Intensification of Agricultural Water Use').

Changes in Intensification of Land Use and Water Use

A description of canal-irrigated areas by scale and location gives a static perspective. The real agricultural land and water uses are more dynamic. Different kinds of data, including remote sensing, can be used in different ways to understand what has happened over time.

Erode district cropland statistics

One way to understand the development in the Bhavani basin is to analyse the agricultural statistics for the Erode district (Krishnaswami Ayyar, 1933; SCR, 2005). The entire command area of all three major canals taking off from the Bhavani River falls within the district, i.e. the LBP, Kodiveri and Kalingarayan canals.

A comparison of data for 1926 with figures for the period 1980–2005 shows the same net sown area, about 300,000 ha. The canal- and tank-irrigated area increased from 18,000 ha to 90,000–100,000 ha, following canal expansion. The groundwater-irrigated area increased from about 60,000 ha to slightly less than 80,000 ha, a 30% increase.

Areas irrigated with water pumped directly from rivers and canals should come under the category of 'other sources', but this category totalled 1700 ha and 1300 ha in 1926 and 2005, respectively, which clearly shows that the statistics fail to capture this water-use development. The statistics also fall short of differentiating conjunctive water use, e.g. considerable areas within the 84,000 ha LBP command area. Out of the entire gross irrigated area of 178,000 ha in the Erode district only 800 ha are registered as an area where wells are used for supplementary irrigation.

In 1925/26, the 'area cropped more than once' was about 10%. According to statistics for Erode, the overall cropping intensity in 2004/05 was still about 110%, with about

121% for canal-irrigated areas and 118% for well-irrigated areas. This does not reflect today's reality. Technical advances during the last decades, with electricity, pumps, bore wells, long-distance pipes and different irrigation combinations point to a quite different situation.

Remote-sensing analysis of cropland areas

Remote sensing offers an alternative option to reveal the extent and, above all, the spatial location of cultivated areas in the basin. Three Landsat satellite images (27 February 1973, 9 November 1999, 3 March 2001) enabled a comparison of land-use change over time and between seasons. During the north-east monsoonal period, the maximum crop extent in the basin, both rainfed and irrigated, can be captured. During the dry 'summer' months, crop areas indicate irrigation, as rainfall is normally too meagre to sustain rainfed crops.

During the dry season in 1973, the total crop area was 55,000 ha, and for the same season in 2001 the area was 95,000 ha, an increase of about 70% (Table 11.1). In non-command areas in the lower Bhavani basin (OL or 'Other Areas Lower'; see Fig. 11.3 and Table 11.1), the crop area increased from 22,000 ha to 44,000 ha between 1973 and 2001. This points to at least a 100% increase in water-lifting irrigated areas, surface water and groundwater, in this area. In the LBP command, the crop area increased by almost 75%, indicating increased water lifting from aquifers, canals or the Bhavani River in this area.

The cropping seasons in the part of the Nilgiris district included in the Bhavani basin (N) are different. Analysis of image data shows that one or two crops are grown on around 6500 ha, while 2500 ha are classified as fallow (indicating crop cultivation during other parts of the year). Altogether, a total cultivated area of 9000 ha can be found in this part of the basin. There has been a very strong trend towards increasing vegetable cultivation in this area, through both the number of crops per year and the expansion of cultivation, replacing traditional rainfed crops (SCR, different years).

Table 11.1. Crop areas in the Bhavani basin in summer 1973 and 2001, and north-east monsoon 1999, based on analyses of satellite images (Thenkabail et al., 2005).

Season Rainfall situation Date Year			Summer Dry 27 Feb 1973	Summer Dry 3 Mar 2001	NE monsoon Wet 9 Nov 1999
Total area (ha)	Upper Bhavani basin	Code	414,900	414,900	414,900
Fallow area	Nilgiris district	N	4,300	4,300	4,400
Crop area	Nilgiris district	N	300	4,900	4,700
Crop area	Other areas upper	OU	15,600	17,600	21,600
Crop area	Total upper		15,900	22,400	26,300
Total area (ha)	Lower Bhavani basin		243,400	243,400	243,400
Crop area	Arakkankottai command	A	900	1,600	2,800
Crop area	Thadapalli command	T	2,300	3,600	7,100
Crop area	LBP command	LBP	13,500	23,500	31,400
Crop area	Other areas lower	OL	21,800	44,100	67,600
Crop area	Total lower		38,700	72,800	108,900
Total area (ha)	Entire Bhavani basin		658,300	658,300	658,300
Crop area	Entire Bhavani basin		54,600	95,200	135,200
Crop area	Basin without Nilgiris		54,300	90,400	130,500

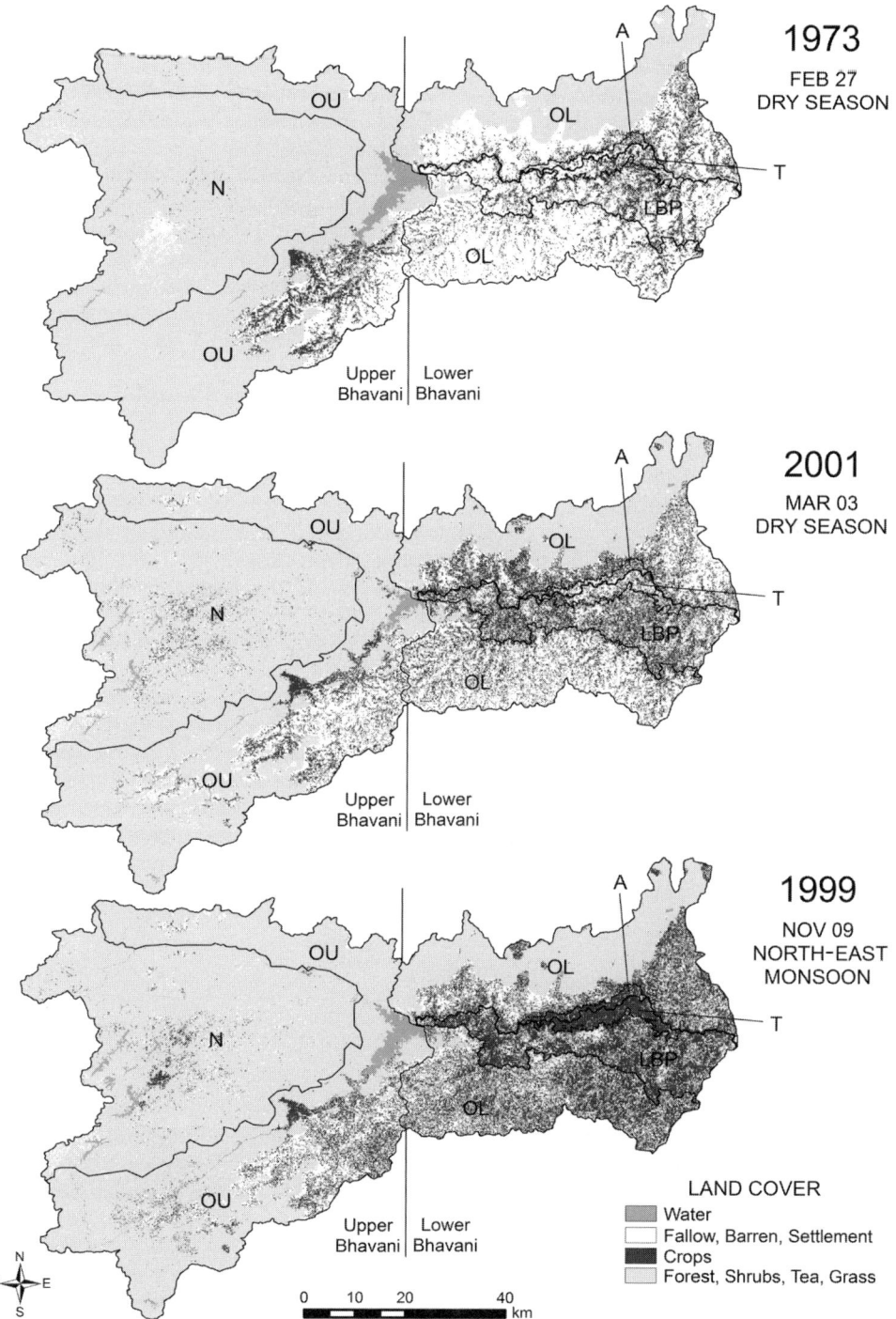

Fig. 11.3. Land cover in the Bhavani basin during the dry season in 1973 and 2001, and during the north-east monsoon in 1999, based on satellite images.

When excluding the high-altitude Nilgiris district and analysing the remaining part of the basin, there was a net crop area of almost 159,000 ha around the year 2000. With about 59,000 ha with double cropping, the cropping intensity is 137%.

Intensification of agricultural water use

After the completion of the LBP reservoir in 1955, the basin essentially closed, in that all average available surface water resources were put to use by agriculture, with little remaining for the environment and the Cauvery River. In spite of this, and increasing in- and out-of-basin drinking-water diversions, agricultural systems continue to intensify, with groundwater and surface water pumping playing a key role.

Pump-based Irrigation

In 1930, each well on the plains of the Bhavani basin was used to irrigate about 1.5 ha, on average (Krishnaswami Ayyar, 1933). The number of wells has increased substantially during the 20th century. Figure 11.4 shows how the total number of wells in the Erode district has increased threefold. The mode of lifting is now fully mechanical and bullocks have been replaced by electricity and diesel engines. Palanisami (1984) showed that, at the end of the 1970s, the average extraction from a well with diesel or electrical pumps was four times greater than from bullock bailing, indicating up to a 12-fold increase in groundwater abstraction capacity over the last century. Today, most farmers use 3–10 hp pump sets (Kannan, 2004).

In Tamil Nadu, a fierce farmers' movement protested against the cost of electricity during the 1970s and 1980s. A state-wide general strike in 1972 was met by a strong police response, with 15 farmers killed in the confrontation. The development spurred the formation of The Tamil Nadu Agriculturalist's Association, which reached across the entire state. Farmers refused to pay their electricity bills and the Electricity Board responded by trying to disconnect the power lines. More violence followed, and 13 people were killed in 1978–1979. In 1980–1982, 'Operation Disconnection' was intensified and electricity prices were raised. In 1982, the farmers launched a political party, 'The Indian Farmers' and Toilers' Party', with the aim of seeking legitimacy and ensuring protection from state government repression. Although the farmers' party earned very few votes in the elections that followed, the protests eventually resulted in a political decision in which the ruling party, as an act of appeasement and a final move to undermine the farmers' movement, decided to accept the farmers' demands (Lindberg, 1999). From September 1984 onwards, electrical power was supplied free to 'small farmers',

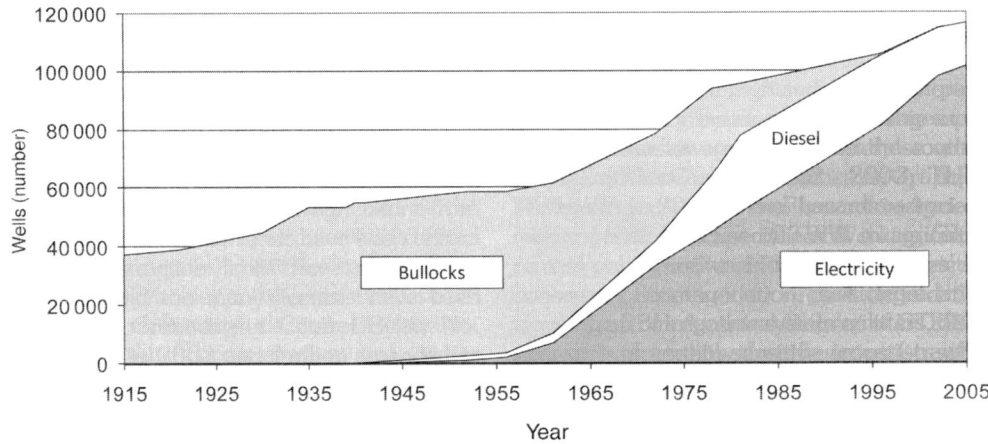

Fig. 11.4. Wells in the Erode district 1915–2005, number and mode of lifting (Lannerstad 2008).

irrespective of the quantity of electricity consumed, a policy extended to all farmers in 1992 (Kannan, 2004; TNEB, different years).

The free electricity policy has, predictably, driven a demand for electricity connections. Since the end of the 1980s, the TNEB has limited the increase to about 40,000 new connections per year in the whole state (Fig. 11.5). During the end of the 1990s, the increase in the Erode district was 1500–2400 connections per year (PWD, 2002). From 1 million pump sets in Tamil Nadu by 1984, the total is approaching 2 million in 2008. From the 1970s up to 1984, the total electricity consumption followed the same trend as the number of connections and stayed at around 2300 kWh per electric pump set and year. In 2005, the consumption per pump set had increased to more than 5000 kWh. This is more than a 100% increase in energy consumed per electricity connection over less than 20 years. It is important to observe that there is no difference in the statistics between pumping from difference sources. Pumping of surface water from rivers, streams and canals is included in the statistics for all electricity connections and the total electricity consumption by agriculture, often assumed to refer only to groundwater.

Remote-sensing statistics showing a crop area increase from 22,000 to 44,000 ha from 1973 to 2001 during the dry season for the non-command areas (Table 11.1) confirm the increased amount of water lifting. A survey in 2000 of irrigation wells in the lower Bhavani south of the LBP command area showed that there were large areas with crops such as sugarcane, turmeric (*Curcuma longa*), coconuts and bananas (TWAD, 2000) irrigated with groundwater. Groundwater levels in the surveyed area have fallen during the last 20 years. Initially many farmers resorted to drilling vertical as well as horizontal bores inside the open wells to augment the yield. This only lasted for a few years. Instead, the farmers now have bore wells of 100 m to as much as 250 m depth, with submersible pumps or air compressors (TWAD, 2000).

Canal releases and lifting from aquifers, rivers and canals

Canal irrigation within the Bhavani basin has increased in area. By the end of the 1950s and the beginning of the 1960s, it had also increased in both intensity and number of seasons.

The LBP command area was designed for one season with irrigated dry crops, such as cotton and groundnuts, and paddy cultivation was limited to zones affected by seepage from the unlined LBP canal. Having learned from the historic command areas, the LBP farmers wanted to cultivate paddy, and violated cropping regulations. The PWD engineers tried to stop this trend, but the farmers complained to

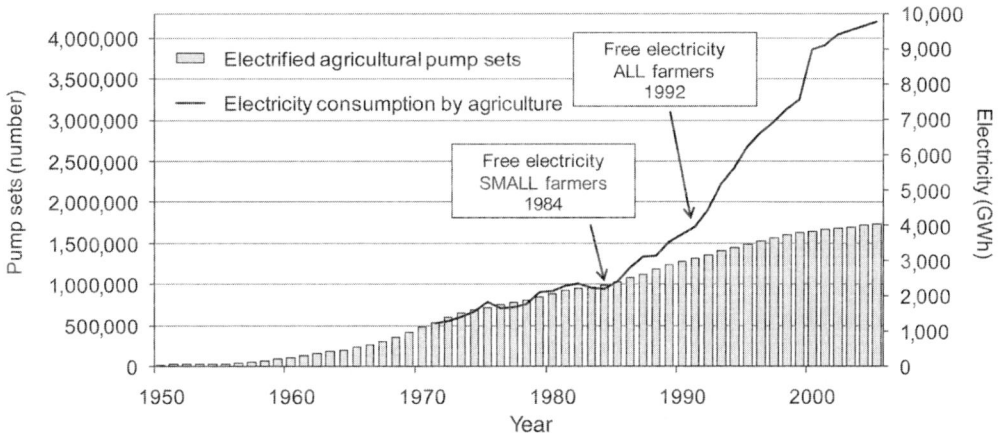

Fig. 11.5. Electrified agricultural pump sets and electricity consumption by agriculture in Tamil Nadu, 1950 to 2005 (TNEB, different years; TNEB, 2008).

the state government. The government, facing urgent food-shortage problems, made it clear 'that they were very anxious that the farmers should have no cause of complaint'. The engineers had to accept keeping the water flowing in the canal, and the area under paddy increased from the intended 4000 ha to 27,000 ha. The canal, designed for irrigated 'dry crops', can only convey water for a paddy crop on about half the command area and the tail-end farmers suffered.

After public meetings between the farmers and the government authorities, a system with two seasons with paddy crops was tried. Such a system demands about twice the average annual water available for the LBP and had to be abandoned. In 1964, the LBP system was eventually changed to a yearly alternating system, where in the first year half of the command area is supplied during two seasons: one (monsoonal) season with wet crops (680 Mm^3), mainly paddy, and one season with irrigated 'dry crops' (340 Mm^3), such as groundnuts. The other half of the command area gets no supply at all during this year and roles are reversed during the second year. This system increased the water demand by 60%, from the originally estimated average water availability of 650 Mm^3/year for the LBP to 1020 Mm^3/year. Higher supply from the dam means that there are almost no carry-over stocks and, with a larger designed demand relative to average availability, the frequency of seasons without planned canal supply will inevitably be higher.

The farmers have adapted by developing groundwater resources. During supply seasons conjunctive use is common, and during non-supply seasons groundwater is the main irrigation water resource. Several farmers have also turned to external surface water resources, and lift water into the command area from the Bhavani River or the Thadapalli or Kalingarayan canals. Some farmers pump water from the Kalingarayan canal more than 7 km into the LBP area (PWD, Erode, Erode district, Tamil Nadu, India, 2004–2007, personal communication).

Out of the 31,500 ha LBP command area located inside the Bhavani basin only half, or about 16,000 ha, receives canal supply during each of the two seasons. Remote-sensing statistics show an area of 31,400 ha with crops during the north-east monsoon in 1999. Dry-season figures from 1973 and 2001 show that the cultivated area has increased by 75% to as much as 23,500 ha. This development points to the importance of increased lifting of groundwater and surface water in the LBP command area.

At the end of the 1950s, the command area served by the Kodiveri weir was, as part of the GMF campaign, shifted from a single paddy crop area to a double-crop area with 10 months of continuous canal supply (GoM, 1958, 1963). While the canal is closed (15 February–15 April), only minor water quantities are released for 'standing crops' (PWD, 1984; PWD records). The satellite image from 2001 showed perennial crops on 50% of the Kodiveri command area during canal closure. This is a 60–80% increase compared with the figures from 1973. Farmers in the area confirmed this by describing the development of 20-m-deep open wells since the mid-1990s. Farmers with access to groundwater are able to bridge the 2-month gap in canal supply and cultivate perennial crops such as sugarcane, which were not present at all in the basin in 1926 (Fig. 11.6).

This 10-month supply has inspired farmers outside the area to lift water from canals to expand cultivation out of the command area. North of the Arakkankottai canal, most farmers within about 150 m of the canal have a pipe connection between the canal and an open well. Water is pumped up to 5 km away. A farmer cultivating sugarcane and bananas exemplifies this development: originally coarse grains were irrigated through bullock bailing from an open well from the 1920s. In the middle of the 1970s, the well was connected to the Arakkankottai canal by an underground pipe, and the farmer was one of the first in the village to install a diesel pump. After 2 years, electricity replaced diesel, and today two 10 hp pumps are used.

The Kalingarayan canal has continuous supply during 10.5 months of the year, with two paddy crops a year since as early as the end of the 19th century (Madras Presidency, 1902), and up to three crops nowadays. The ample canal supply also spurred water lifting to irrigate lands on the elevated west side at the beginning of the 20th century. The area was

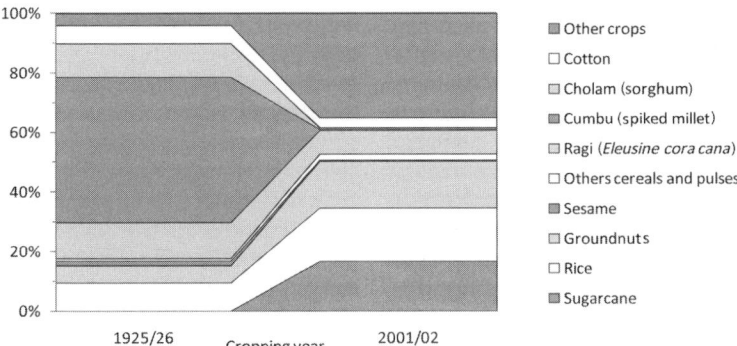

Fig. 11.6. Crop statistics for Sathymangalam and Gobichettipalayam taluks for 1925/26 and 2001/02. (Source: Krishnaswami Ayyar, 1933; Agricultural Department, Agriculture Directorate, Chepauk, Chennai, 2006, personal communication.)

originally limited by the chief engineer to about 200 ha, with 'dry crops' irrigated through bullock bailing. In a stepwise process starting in the 1940s and 1950s, diesel and electricity pumps for wet-crop cultivation have taken over (Saravanan, 2001; PWD, n.d.). Local and national food shortages from 1940 until the end of the 1960s, with policies like the GMF campaign, were the main reason behind letting water lifting increase against regulations. The number of unauthorized pumps (some with a capacity of 20 hp) has gradually mushroomed to 1000–2000, and the total lift-irrigated area is estimated at more than 7000 ha (PWD, Erode, Erode district, Tamil Nadu, India, 2004–2007, personal communication).

During the drought years, 2002–2004, the diversions into the Kodiveri and Kalingarayan canals were cancelled or much less than normal. Immediate water scarcity along the canals intensified well development; interviews show that Kodiveri farmers have drilled bore wells of about 100–200 m depth, and that the Kalingarayan farmers have mainly increased the number of open wells. This will make it possible for the farmers to cope with drought years, diversify their cropping patterns and also intensify water use further during normal years.

Lifting from rivers and streams

The best source of surface water for lift irrigation in the basin is the Bhavani River itself. Just as for canals, the free electricity, and diesel pumps and pipe technology make it possible to pump water several kilometres away.

Each of two interviewed farmers (among several others) pumping water from the Bhavani River cultivates an area of about 8 ha. One farmer cultivates banana and coconut trees and has one piece of land next to the river and another piece of land 2.5 km away. He has an electricity connection that was formally granted for an open well. The well does not yield any water and was set up only to get the electricity connection endorsed, to make it possible for the farmer to use the free power for river pumping. Recently he has drilled a new 'fake' bore well to get a second connection. The other farmer cultivates sugarcane and bananas on his land about 300 m from the river. Some 10 years ago he started to irrigate additional areas 1 km away. Both examples show how entrepreneurial farmers with access to river water (thanks to pump and pipe technology) can expand their cultivation to additional land acquired further away.

With the construction of the LBP reservoir, the historic Kanniyampalayam command area was submerged. As can be seen on the 2001 satellite image (compare Fig. 11.2, 2000, and Fig. 11.3, 2001), farmers cultivate parts of the reservoir bed during the dry season, when the reservoir is almost empty. Water is pumped from the Bhavani and Moyar rivers. This practice is another example of how farmers effectively utilize available land and water resources.

Another illustration is the Thengumarahada Farming Society, located next to the Moyar River (see Figs 11.2 and 11.3), where, since the 1980s, electric pumps have allowed farmers to permanently lift water directly from the perennial Moyar River to their canal system.

The PWD engineers, researchers and many farmers describe how farmers purchase '1 cent of land' (1/100 acre or 40 m^2) close to the river or a canal. On the land they dig an open well and through an unauthorized underground pipe obtain access to the canal or the river. With electricity or diesel pumps, water is then often pumped far away. Many persons refer to distances of more than 5 km. Farmers often install a more powerful pump than stipulated when they get a new electricity connection approved and can thus pump more water and expand the irrigated area. The three-phase electricity for water pumps is limited to 12 h/day. Some farmers overcome this constraint by installing a 'condenser' and use the two-phase electricity to pump water 24 h/day anyway. The same also takes place in other places, such as Gujarat (Shah, 2007).

According to information from the PWD, in the middle of the 1990s, the number of authorized pumps along the Bhavani River was about 900 upstream of the LBP reservoir (sometimes financed by a group of farmers) and around 600 between the reservoir and the confluence with the Cauvery River. The total irrigated area was then estimated to be almost 1100 ha (MIDS, 1998), increasing to 7000 ha by 2000 (TWAD, 2000), and it must have increased further since then. The PWD tries to bring several cases to court, and the exact data on the number and capacity of unauthorized river pump sets is sensitive information and is not made public.

In the Nilgiris district, the increased cropping intensity is based upon hose and sprinkle irrigation from streams or shallow dug wells initiated in the 1970s. With pipes, areas uphill and at longer distances can be irrigated from the streams (Lannerstad, 2008). Statistics show that the cultivated lands for vegetable crops have more or less stayed around 10,000 ha during the last century (SCR, different years), but the crop intensity and water use have increased, with a shift from one or two to two or three crops per year.

Altered crop choices

With the expansion of the areas irrigated through gravity and water lifting, and shifts in consumer preferences, cropping patterns have changed. Almost all cultivated areas (gross) of about 100,000 ha in the Sathymangalam and Gobichettipalayam subdistricts (*taluks*) fall within the Bhavani basin and include rainfed lands, water-lifting irrigated areas, Arakkankottai and Thadapalli command areas, and some of the LBP command areas. The statistics for 1925/26 are most probably valid up until the end of the 1940s. A comparison thus shows dramatic changes during the 50 years till 2000 (Fig. 11.6). Water-intensive crops, such as paddy and sugarcane, have increased from 9 to 35%, with perennial sugarcane cultivation increasing from 127 to 16,000 ha and paddy from 8000 to 18,000 ha.

The subdistrict-level statistics available for the crop year 2001–2002 unfortunately do not specify the 'other crops' category. District-level data indicate that this group includes crops such as turmeric, bananas, tobacco, maize, flowers, spices, coconut trees, garden produce and sorghum (for fodder), found on 20% of the cultivated area. This means that the area cultivated with less water-demanding coarse grains, such as sorghum, spiked millet and ragi (*Eleusine coracana*) has decreased by 60%, while sorghum, which was earlier one of the staple food crops, is now instead used for fodder. This trend means higher water requirement per hectare. The water demand for increased cropping intensity and higher yields is, however, to some degree, compensated by new short-duration crop varieties.

The general trends in the Bhavani basin are valid for the whole of Tamil Nadu. Statistics comparing cropping data for 5 years around the years 1955 and 2000 show that areas with coarse cereals have gone down by 50–75%, while areas with sugarcane have increased by 800%. Yields have increased dramatically: rice from 1.3 to 3.5 t/ha, spiked millet from 0.4 to 1.4 t/ha, and groundnut from 1.0 to 1.8 t/ha. The figures for Tamil Nadu also show that the yield for irrigated sorghum, spiked millet or ragi is about twice the rainfed yield (SCR, 1958, 2002).

Hydrological, Livelihood and Environmental Implications

Cropping pattern statistics, remote-sensing figures, canal irrigation expansion and water lifting from aquifers, streams, rivers and canals all point to an intensified use of water. This trend has a number of implications, which are apparent in the Bhavani basin.

Falling and rising groundwater levels

Groundwater observation wells within the Bhavani basin have been continuously monitored every month since 1971. The groundwater levels range from less than 10 m during the post-monsoonal period to as deep as 50 m during the hot summer months. The deepest annual water level is observed during May, which marks the end of hot weather months and the beginning of the south-west monsoon (TWAD, 2000). Groundwater level changes based on values in May over 30 years are visualized in Fig. 11.7.

The light and dark grey dots indicate a drop in groundwater levels of 6 to 14 m and are all found on the plains outside the command areas. A black cross shows that many of these observation wells have been dry during a month or more. A study of irrigation wells in the lower Bhavani south of the LBP command area in 2000 showed that groundwater levels in the surveyed area have fallen during the 1980s and 1990s, and left most of the open irrigation wells dry during the summer months or the entire year (TWAD, 2000). Within the LBP command areas, the groundwater level is stable or slightly falling. Even if groundwater lifting and conjunctive use are common, seepage from the fields and canals appears to almost recharge the aquifers. Normally, the water table is shallow in canal- and tank-irrigated regions (PWD, 2002)

The grey squares indicating rising groundwater levels of 6–10 m are located just north of the Arakkankottai command area or the Bhavani River. Most of the white squares (rising groundwater levels of 2–6 m) are also found on non-command areas next to the Bhavani River or next to the LBP canal. One plausible explanation for this pattern must be the increased pumping from the Bhavani River and the canals. Over the years, the return flows have locally raised the groundwater level by several metres. Free electricity thus not only resulted in aquifer overexploitation but also in locally rising groundwater levels.

Government groundwater authorities estimate that the total groundwater draft within

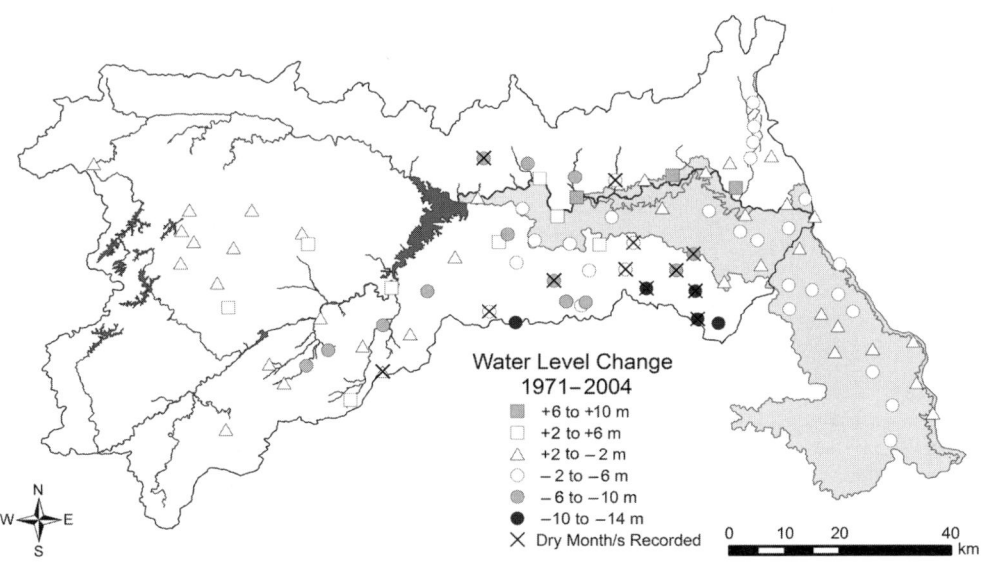

Fig. 11.7. Observation well water level changes 1971–2004 (Data source: PWD records).

the Bhavani basin has reached about 350 Mm^3 per year, with an annual overexploitation of about 30 Mm^3 (PWD 2002, 2003).

Changed surface water dynamics

The inflow into the LBP reservoir reflects changes in the upper part of the Bhavani basin. Despite a high variability in flow at the reservoir site (Fig. 11.8), it is possible to discern a trend of falling inflow of about 500 Mm^3 during the past few decades, a reduction of about 25% of the inflow. This is due to several factors: the 104 Mm^3 of drinking water transferred out of the basin to Coimbatore and Tiruppur cities; the evaporation of about 10 Mm^3 from domestic water use (assuming a 50% return flow) within the upper part of the basin; the evaporation of about 40 Mm^3 from the hydropower reservoirs since the 1960s (NWDA, 1993); the general trend of conversion of grazing land and natural forest towards tea and tree plantations (SCR, different years), with higher transpiration (Wilk, 2000) and thus increased consumptive water use and reduced runoff from these rainfed areas; and water lifting from streams, rivers and aquifers. River pumping affects the inflow to the LBP reservoir, especially during the dry season (PWD, Ooty, Nilgiris district, Tamil Nadu, India, 2004–2007, personal communication).

The LBP farmers have the lowest priority of water supply in the Bhavani basin (PWD, 1984). The water quantity released to the LBP farmers is decided by the inflow to the LBP reservoir, the water to be shared and the demands of the downstream Kodiveri and Kalingarayan canals. The records from the PWD show that as soon as the inflow to the LBP reservoir is less than 1500 Mm^3 per year the LBP farmers lose one, two, or both seasons of canal releases. In addition, in some years, during episodes with high water demand in the delta, water is released to the Cauvery delta farmers, thus reducing the water available within the Bhavani basin.

Return flow from the LBP and Kodiveri command areas increases the flow in the Bhavani River to be diverted into the Kodiveri and Kalingarayan canals. When the intensity of water use increases in the command areas (notably through pumping and recycling), less water is drained to the river. Pumping along the Bhavani River also reduces the water quantity to be diverted into the historic canals.

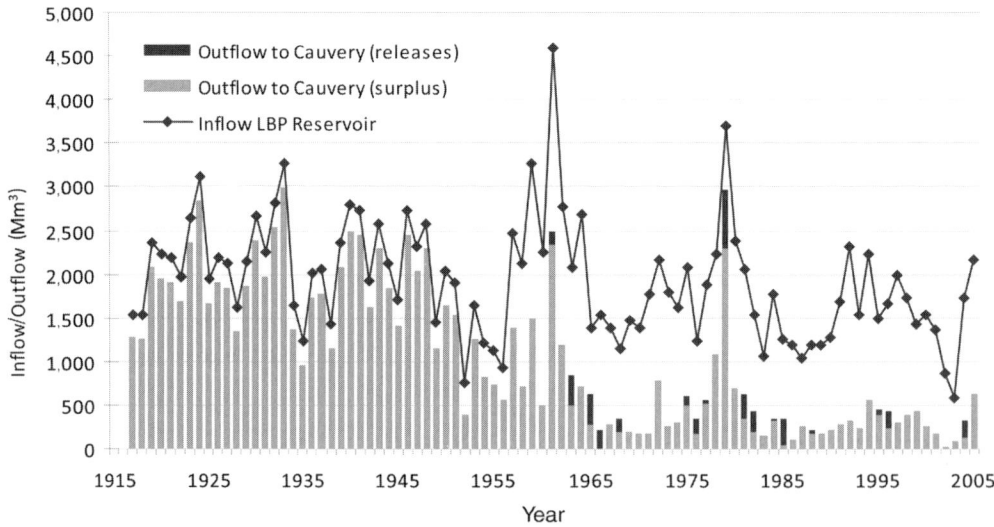

Fig. 11.8. Inflow to the LBP reservoir and basin outflow to the Cauvery River. Actual inflows 1955 onwards, actual outflows 1975 onwards, and other values estimated from the Kodiveri weir flow measurements (Source: NWDA, 1993; GoM, 1965; PWD records).

Water lifting from the Kodiveri and Kalingarayan canals reduces the availability of water along the canals and more water has to be released from the reservoir to ensure water to all canal farmers. This, in turn, leaves less water in the LBP reservoir to be supplied to the LBP command area, prompting LBP farmers to further intensify their use of water and to tap groundwater.

Analysis of return flows and local runoff at the basin level at the end of the 1990s illustrates the importance of the return flows and shows that while annual overall releases from the LBP reservoir totalled about 1650 Mm^3, diversions into the four main canals amounted to 2000 Mm^3. Yearly data records show that during years with full supply to the LBP canal, the releases from the reservoir only amount to 75% of the water actually diverted into the Kodiveri canals. During years with limited or cancelled supply into the LBP canal, releases from the reservoir have to compensate for losses along the Bhavani River and correspond to about 115% of actual diversions. Releases from the LBP reservoir for the Kalingarayan canal normally equal 50% of actual diversions (PWD records).

All these fluctuating flow paths mean that water is constantly spatially reallocated and that, consequently, conflicts arise. The conflicts over canal-lifting and water-scarcity problems started along the Kalingarayan canal at the end of the 1940s (Saravanan, 2001). Over the years, the authorities have made several attempts to gain control over the number of pumps and the water quantity withdrawn, in order to reduce tail-end problems, but they have ended up diverting more water into the canals to compensate for these withdrawals. Diversions into the Kalingarayan canal have increased from 310 to 380–400 Mm^3 per year (PWD, n.d.). Likewise, PWD records show that the annual diversions into the Kodiveri canals have increased by about 30%, to 600 Mm^3 per year.

Overall, the average impact of surface water pumping on water demands downstream of the LBP reservoir amounts to almost 310 Mm^3 per year, with almost 250 Mm^3 falling within the basin boundaries. This includes reservoir releases of 90 Mm^3 to compensate for river pumping and increased diversions of up to 150 Mm^3 into the Kodiveri canals, and around 70 Mm^3 into the Kalingarayan canal.

Basin closure

An analysis of the water situation at the basin level shows how, since the construction of the LBP reservoir, the Bhavani basin is a closing basin (Lannerstad, 2008). The discharge over the Kalingarayan weir to the Cauvery River can be of two types: releases from the LBP reservoir, destined to the Cauvery delta farmers, and basin surplus outflows.

The annual average unintended discharge over the Kalingarayan weir during the last 25 years has been about 240 Mm^3, or 10–15% of the outflow present before the completion of the LBP (Fig. 11.8). There is no storage below the LBP reservoir to capture the intensive downpours during the north-east monsoon and the return flows from the LBP and Kodiveri command areas during the annual 6-week closure of the Kalingarayan canal. The outflow to the Cauvery River can thus only fall close to zero when local runoff and return flows are so small that all can be diverted into the Kalingarayan canal. The small outflows during the past few decades indicate that the consumptive water use within Bhavani basin has reached a maximum level and cannot increase further.

Livelihood repercussions

Farmers lifting water from the Bhavani River or pumping from wells generally point to electricity subsidies as a 'good subsidy'. It is aimed directly at the individual farmers and there is no middleman who can take a share. The resulting falling groundwater, however, distresses many marginal and small farmers, who find it difficult to invest in drilling bore wells deeper (TWAD, 2000). Small farmers turn to buying water from large farmers who can afford to deepen their wells, and local water markets emerge (GoTN, 2002). The original objective behind the free electricity policy to support small farmers has thus partly failed.

Falling groundwater levels negatively affect domestic water supply on the plains of the Bhavani basin. Dug wells and shallow bore wells equipped with hand pumps dry up, especially during dry summer months. To secure water supply, the Tamil Nadu Water Supply and Drainage Board (TWAD) has drilled a large number of deep bore wells of 150–275 m depth (TWAD, 2000). The falling water tables and failing wells force many women to spend 3–4 h/day fetching water from far-off places, including wells in the fields. Many local communities (the village panchayats) have to pump for 10–12 h to fill up the rural water supply overhead cisterns and find it difficult to bear the electricity costs, often amounting to US$350–600 (Rs 15,000–25,000) per month. The water situation drives many people to migrate and settle in nearby towns, thus speeding up urbanization (TWAD, 2000).

To increase the water availability in these areas, a groundwater recharge project was planned in 2000. The project, not yet implemented, aims to withdraw a quantity of about 40 Mm^3 during the monsoonal months from the Bhavani River above the LBP reservoir to fill up 48 tanks and 213 ponds (inside and outside the Bhavani basin) for groundwater recharge to increase water supply for rural habitations (TWAD, 2000). Such diversions will further decrease the water availability for downstream users.

Environmental consequences

The water resources development initiated in the 13th century has turned the Bhavani basin into a complex human-regulated system, where all normal flows are controlled. The natural seasonality, important for many organisms, with annual runoff peaks during June–August and October–December along the Bhavani River, has been replaced by an almost steady flow during the entire year. The weirs and reservoirs across the Bhavani River and the tributaries effectively stop the natural migration upstream of the river system by, for example, some fish species, and retain the silt from the upper Bhavani, whose earlier fertilizing effect was highly appreciated by the delta farmers (GoM, 1965).

A number of small-scale textile, bleaching and paper industries upstream of the LBP reservoir and the four major municipalities in the basin discharge their effluents and sewage water without any treatment. The strong annual river flow, however, dilutes the pollution, and the surface water of the Bhavani River generally meets water-quality standards. There is, in contrast, a major pollution problem along the Kalingarayan canal, where 26 tannery and 32 textile-processing units discharge untreated effluents into the canal. This especially affects the tail-end farmers. The annual canal closure period leads to accumulation of effluents, and farmers have to wait until pollution is flushed out to the Cauvery River before using the water (Appasamy et al., 2005).

There are generally no problems with groundwater quality in the basin, apart from increased nitrate levels in areas with intensive agriculture. In areas where industries discharge the effluents on land, and in some places where the polluted water is even used for irrigation, the groundwater is locally contaminated and drinking water schemes have to be provided by the industries or the municipality (PWD, 1999; Appasamy et al., 2005).

Future Challenges

The competition for water resources in the Cauvery basin and its sub-basins is increasing and the dynamics of water use are rapidly changing. The two following sections consider the future viewed from different perspectives and scales.

The interstate Cauvery basin perspective

There have been discussions over how to share the Cauvery River flow for centuries between the two major states in the basin, Karnataka (earlier Mysore) and Tamil Nadu (earlier Madras). In the beginning of the 20th century, the British Madras Presidency decided to construct the Mettur reservoir (Fig. 11.1) across the Cauvery River. The reservoir should protect the delta farmers by moderating the floods and droughts, following the monsoonal climatic variability, and increase the irrigated area by more than 100,000 ha. However, the design also started another 'Madras–Mysore Dispute' over water. The project was postponed and not started until after the interstate agreement in 1924. In 1934, Tamil Nadu's largest reservoir, with a capacity of 2650 Mm^3, was completed (Barber, 1940; GoM, 1965).

The irrigated areas in the Cauvery basin have increased considerably since the 1924 agreement. Tamil Nadu has increased the irrigated areas (including a second crop) from about 620,000 ha to 850,000 ha, about a 60% increase. Karnataka has about doubled the area, from 430,000 to 850,000 ha (GoI, 2007). The water demand has increased proportionately.

In January 2007, the Indian National Court of Arbitration delivered 'The Report of the Cauvery Water Disputes Tribunal with the Decision' to resolve the last Cauvery dispute, which had been going on between Karnataka and Tamil Nadu states since the 1970s. The decision settles the amount of water that each of the Cauvery basin states – Tami Nadu, Karnataka, Kerala and Pondicherry – can utilize. Three parts of the agreement might affect the Bhavani basin (GoI, 2007).

First, Kerala state has not developed much of the runoff generated within its Cauvery basin areas. The state, according to the decision, has the right to withdraw an additional 170 Mm^3 per year from the upper Bhavani basin.

Second, among the post-1924-agreement water developments, only those approved by earlier interstate agreements were considered when establishing the new shares for the different states. The second crop, along the Kodiveri canals for example, is not recognized by the court. About half the diversions for these canals are thus, according to the water-use account, supposed to be used for irrigation and cities in other parts of Tamil Nadu's portion of the Cauvery basin.

Third, the Cauvery tribunal decision states that there should be an environmental flow of 25.5 m^3/s from February until June below the Mettur reservoir, about 280 Mm^3/year, to maintain the freshwater–seawater interface in the Cauvery estuary to protect the mangrove forest. It is Tamil Nadu that controls the releases from the Mettur reservoir and thus has to ensure the environmental flow. In case of shortage, water might have to be released from the LBP reservoir, the only other major storage in Tamil Nadu after the Mettur reservoir.

The Tamil Nadu and Bhavani basin perspective

The recent Cauvery tribunal decision places the water use in the Bhavani basin within a larger Cauvery basin and in an interstate context. There are, however, already within Tamil Nadu increasing demands for the runoff generated in the Bhavani basin.

The rapidly growing Coimbatore and Tiruppur cities along the ephemeral Noyyal River depend upon drinking water from the Bhavani River above the LBP reservoir. Coimbatore city already faces scarcity during normal conditions and rationing during drought periods. A second pipeline is planned from the last hydropower reservoir and will increase abstractions by 46 Mm^3 to 150 Mm^3/year (TWAD, Coimbatore and Nilgiris Circle, 2005–2006, personal communication).

The Lower Bhavani Project was designed as a 'surplus project', intended only to impound and use water quantities in excess of the water rights of the Cauvery delta farmers and the farmers under the historic Kodiveri and Kalingarayan canals. Up to now, there have only been (limited) releases for the Cauvery delta farmers a few times during the last 20 years (Fig. 11.8). With increased competition for water in the closed Cauvery basin, there is a risk that more water will be requested from the LBP reservoir.

Within the Bhavani basin, farmers compete for the same water resources. The LBP farmers often express a desire to renegotiate the water rights in the basin. They want the releases from the LBP reservoir to be more evenly distributed, with less for the Kodiveri and Kalingarayan farmers and more for the LBP command area. A change in allocations is a political decision. The pumping from aquifers, canals, rivers and streams occurs at an individual level and will probably continue to increase unless, somehow, it can be regulated. One option for the government is to take control over the electricity, with metering and electricity charges. Another option is to control the use of different water sources, i.e. aquifers and surface water sources, and make sure that unauthorized actions are limited or closed down.

It is highly likely that less water will be available within the Bhavani basin in the future, as more water will need to be released for downstream uses and environmental flows. Upstream, the state of Kerala is likely to exercise its right to abstract more water. Less water for agriculture in the Bhavani basin would probably stimulate further pumping, but could also substantially change the way agriculture is practised within the basin. Competition will increase both within the Bhavani basin and from outside actors.

Discussion

The Bhavani basin is located in a historically famine-prone area, with meagre and unreliable rainfall. Already during British rule different 'improvements and extensions of irrigation' were considered, and some carried out (Madras Presidency, 1902). The LBP had been under consideration for decades but was not sanctioned, since the investment would not meet the British requirements for economic return. This criterion was disregarded by the Indian National Government after independence. 'Minor', 'medium' and 'major' irrigation projects were launched on a broad scale. The projects implemented at the end of the 1940s and during the 1950s in the Bhavani basin are all examples of these ambitions (Mohanakrishnan, 2001).

Food shortages on a national level in the 1950s were met through large food-aid imports, mainly from the USA, under the Public Law 480 (PL 480). Still as late as 1965–1967, India has witnessed serious drought and near-famine conditions. In 1976, food production self-sufficiency targets were met for the first time (del Ninno et al., 2005). The food context, together with a constant population increase, is one explanation behind the goal, renewed in each Tamil Nadu state 5-year plan, of bringing more and more areas under irrigation (Mohanakrishnan, 2001); for example, the three small irrigation reservoirs constructed in the Bhavani basin from 1978 to 1990.

Today, India is a food-exporting nation and agriculture only accounted for 11% of the Tamil Nadu 2004/05 net state domestic product. With the rapid societal change taking place in India, agriculture is today regarded by some as 'the parking place of the poor'. Yet, 47% of the 28 million workforce in Tamil Nadu are classified as either 'cultivators' or 'agricultural labourers' (GoTN, 2005). Water plays an imperative role in the daily survival of many small and marginal farmers, and the large numbers of farmers still constitute an important political power in the Tamil Nadu democracy; but water management itself is fraught with several difficulties.

Water-use complexity

Changing societal demands and drivers add complexity to the status of water use, making it even more difficult to manage it sustainably and equitably. A farmer with pumps can utilize up to five different water sources: rain, canal, drain, river or groundwater. While it is convenient to categorize water use by water source, in fact the situation is much more complex, because farmers typically use more than one source of water on the same field. This has made the statistics fuzzier. It is clear that in Tamil Nadu the traditional division of cultivated land into the three categories – 'dry', 'garden' and 'wet', assuming rainfed, well-irrigated and canal-irrigated lands – is not valid anymore.

With an increased ability to withdraw water from different sources, individual farmers are ready to tap water whenever and wherever water happens to be available, as recharged groundwater, rainfall-generated runoff or canal and river flows, following allocation decisions by the irrigation authorities.

Groundwater complexity

Protective well-irrigation has been practised in India for at least a thousand years. When seasonal supplemental irrigation of 1.5 ha with coarse grains is altered to 1.5 ha with sugarcane, there is a risk that a 'race to the bottom' has started. Aquifer depletion is prevalent on the plains outside the command areas, where only bore wells of more than 200 m depth offer a reliable water source. One problem of how to deal with this unsustainable use is that the information provided by the authorities is inadequate to address the situation. First, the exploitation situation is analysed and presented according to the administrative boundaries and not according to hydrological realities as shown in Fig. 11.7. Second, the 40–50-m-deep groundwater observation wells from the 1970s cannot monitor water use from deep aquifers extending as far as 200 m.

Clearly, there are water-use conflicts between users of the same aquifer – farmers with shallow wells versus farmers with deep wells, and farmers versus drinking water needs. The link between groundwater and surface water in the Bhavani basin is not clear, and it is difficult to say whether groundwater lifting competes with farmers depending on surface water.

A dilemma of groundwater use is that, on the one hand, it has provided food security and economic growth and has been extremely important in allowing farmers to cope with water stress, but, on the other hand, it is not sustainable in the long run, with some farmers dependent on groundwater going out of business. In spite of this, existing policies, such as free electricity for agriculture, will encourage farmers to use more groundwater in the future. The reality of groundwater needs to be brought to the forefront of water policy and not to be hidden, as it appears to be today. Forward-looking policies need to anticipate that this resource will not be able to sustain indefinitely the farming sector as we know it today.

Allocations and local perceptions

As agricultural water-use intensification progresses, the different users become aware of the water-use linkages in the Bhavani basin. During interviews, command area farmers criticize the pumping from the Bhavani River. Kalingarayan farmers claim that lifting has reduced the inflow to the LBP reservoir. Some Kodiveri farmers believe the unauthorized pumping from the Bhavani River downstream of the LBP reservoir amounts to 340 Mm^3/year. They think that the return flow from the Kodiveri scheme is enough for Kalingarayan and that all releases from the LBP reservoir for the Kalingarayan canal are, in reality, aimed at meeting the river pumping demand. The LBP farmers also talk about river pumping decreasing the inflow to the LBP reservoir and always point out that they get much less water per hectare compared with historic canals, demanding that water allocation in the Bhavani basin be reassessed.

The Kalingarayan command area farmers have little reason to feel concerned about river pumping, since water rights secure a 10.5-month supply. The Kodiveri farmers have some reason to fear an increased competition. During scarcity conditions, the second crop can, according to regulations, be cancelled (GoM, 1963). This happened during some years in the 1980s. But it is the LBP farmers who should be most worried about the increasing use of the Bhavani water: with the weakest water rights, they are last in line for water and bear the brunt of the hydrological variability (Lannerstad, 2008).

This seasonal and yearly variability in water availability in the Bhavani basin masks the trend over time. The outflow from the Bhavani basin also shows that there is no surplus and no 'wasted water' leaving the basin. Famers and water users are becoming more aware of this increased interconnectedness, and people are more likely to question new water development within the closed basin.

Canal and river pumping

The decision to permanently increase water allocation for the LBP command area by 60% in 1964 was the last major intensification of surface water use in the Bhavani basin. Since then, individual investments have increased the area irrigated with surface water pumped from the rivers (7000 ha), the Kodiveri canals (6000 ha), the Kalingarayan canal (7000 ha), and streams (6500 ha in the Nilgiris district).

Abstraction of surface water has resulted in less water reaching the LBP reservoir; more water has thus to be released into the Bhavani River to compensate for losses along the river and for water pumped from the historic canals. From time to time, canal pumping results in water scarcity and elicits complaints from tail-end farmers (TWAD, 2000; Saravanan, 2001). The farmers along Kalingarayan describe how they secured a court decision during the drought year 2002/03 to disconnect the electrical pumps used to irrigate non-command areas, to make sure water reached the tail-end farmers. The lifting of surface water thus increases the competition for decreasing water resources in the basin and in many ways disturbs the functioning of the entire water allocation system in the Bhavani basin.

The individual lifting initiatives can also be regarded as an efficient way to use available water resources. A comparison of water supply in gravity-irrigated command areas shows that the LBP farmers get 2100 mm, the Kodiveri farmers 6000 mm and the Kalingarayan farmers 8300 mm per year. However, when the diversions into the historic canals are divided over both gravity and estimated lift-irrigated areas, the yearly average supply is reduced to about 3300 mm/year (two seasons), which appears more reasonable. The majority of the farmers pumping water from the Kalingarayan and Thadapalli canals belong to the LBP command area, and these farmers thus contribute to rebalancing the unequal water supply given to the historic canals and also increase the efficiency of water use in the basin.

The fact that individual actions collectively add up to a detrimental situation for the basin as a whole poses a further dilemma for the irrigation authorities and policy makers. During pre-closure, these individual actions to withdraw more water may be justifiable, but after closure these diversions incur costs on other water uses. Overcoming this dilemma will require tighter control on individual actions and shared management of the basin.

Basin closure implications across multiple scales

Water resource development in the Bhavani basin and a context of basin closure clearly impact actors and water-use sectors at different scales. Within the lower Bhavani basin, the allocations for the command areas are perturbed by individual farmers acting at field level. Inside the Bhavani basin, there is a rivalry over water resources between the historic command areas and the LBP command area, the lower and upper parts of the basin, and the Bhavani and the Noyyal basins, both closed sub-basins in the Cauvery basin. Finally, competition is likely to get more serious between the Bhavani basin and the Cauvery delta.

The delta command area is many times larger than the LBP command area, and the political power of the many delta farmers is likely to be decisive in the future. As the extreme drought of 2003/04 showed, water can be released for the Cauvery delta, while no water is given to the command areas in the Bhavani basin.

Both the economic and political powers of the urban sector drive the drinking water diversions to Coimbatore and Tiruppur. According to existing plans, the drinking water siphoned off to the Noyyal basin will increase by almost 50%. Total drinking water abstractions in the Bhavani basin will, as a result, increase to more than 10% of the annual designed diversions into the four major Bhavani basin canals. It is often claimed that drinking water diversions are very small compared with irrigation demands and are therefore negligible. In a basin with different water rights, such as the Bhavani, the reduction in water availability for agriculture is not evenly spread. It is the dry-season releases for the LBP farmers that are reduced first. The total future drinking water abstractions of 210 Mm^3 equal more than 60% of the dry-season supply (340 Mm^3) and are therefore not as insignificant as many may think.

Basin closure at different scales leads to competition for the same water. When sub-basin closures multiply over a larger basin, water managers face a dilemma whereby water resources that were earlier thought to be sufficient for a given sub-basin may be requested or claimed from outside. So even if closure within a sub-basin is successfully handled it can be disrupted by demands from other parts of the larger basins that are in even worse conditions.

Electricity subsidies and energy consumption

Free electricity is an important factor fuelling the increasing number of wells and the water quantities being pumped from aquifers, canals, streams and rivers in the Bhavani basin. In 2005, the total electricity consumption assigned to agriculture in Tamil Nadu was almost 10 GWh, equivalent to 24% of the total electricity consumption in the state (GoTN,

2005). Subsidized agricultural electricity use has created a number of negative externalities. The farmers do not understand the cost of their water pumping, and at the same time groundwater levels fall. The government is supposed to transfer funds from the state budget to cover the costs for the subsidized agricultural use. The TNEB nevertheless, in an attempt to reduce losses, has chosen to raise the tariff rates for non-subsidized sectors and has thus increased the cost of production in the industrial sector. The TNEB must also invest in additional capacity to keep up with a growing demand (GoTN, 2002).

The electricity consumption by agriculture is generally unmetered, and it is consequently difficult to know where in the system the energy use takes place. Free power for agriculture and the higher price charged for other sectors have led to unauthorized consumption of electricity by both farmers and other sectors (GoTN, 2002). Many times, the Indian State Electrical Boards have used the agricultural consumption as a scapegoat to cover up both transmission and delivery losses (Shah, 2007). The more than doubled consumption per agricultural connection in Tamil Nadu during the last decades thus has three explanations: increased water pumping following free electricity, increased power consumption to lift from deeper and deeper levels because of aquifer depletion, and losses and illegal consumption in other segments that are wrongly assigned to agriculture. Researchers in Tamil Nadu have, without considering the losses, estimated that about 30% of the increased power was explained by additional pumping and around 70% by falling groundwater levels (Shah, 2007).

No author discussing the electricity subsidies for agriculture relates this to pumping of surface water from rivers or canals. As shown in this study, the subsidies have encouraged farmers in the Bhavani basin to considerably increase water lifting from rivers and canals. Increased pumping thus not only results in falling groundwater levels but also directly affects surface water flows in rivers and canals and impinges on the entire water use in the basin. Individual or collective pumping from surface water sources is attractive because the source of water is often perennial and energy consumption is lower than for groundwater. The electricity subsidies must thus not only be discussed in relation to groundwater, especially in areas with available surface water sources.

The electricity subsidy has stimulated a difficult dilemma. On the one hand, it has been very effective in increasing groundwater irrigation, and some river/canal pumping, and in alleviating rural poverty, but, on the other hand, with falling groundwater levels, higher energy consumed and impact on allocation schemes, sustainability is at risk. The dilemma is that any benefit given by the society to the individual citizen is very difficult to withdraw, even if totally unsustainable. However, ultimately there will probably be no choice other than changing the existing electricity subsidies.

Conclusions

Through a 'triangulation approach' – using different kinds of data and information – this chapter has shown an increased water-use complexity and interconnectedness throughout the Bhavani basin. Water lifting from aquifers, streams, rivers and canals has affected the water situation and has played an important role in a continuing intensification of agricultural water use in the closed Bhavani basin. While large-scale, state-driven irrigation development peaked in the middle of the 1950s, development was furthered by individual and private investments in pumps, pushing the system further and further into a more 'watertight' situation.

A number of factors have led to water-use intensification after basin closure:

- More and more farmers withdraw surface water from rivers, streams and canals. As a result less water is reaching the LBP reservoir and more water has to be released down the Bhavani River to compensate for abstraction along canals and the river.
- Cropping patterns in the Bhavani basin clearly show a trend towards more water-intensive crops such as sugarcane. In both command and non-command areas, these crops depend on the pumping of surface water and groundwater.
- Since the introduction of free electricity for

agriculture in Tamil Nadu in 1984, the number of connections has increased by 50% and the annual electricity consumption per pump set has doubled. This development should be discussed in relation to groundwater use, but must also be analysed in a context of pumping of surface water from rivers and canals.

Increased cropping intensity, mostly fuelled by water pumping, which in turn was propelled by free electricity, has had initial positive but later serious negative impacts:

- Tail-end farmers along the canals experience water scarcity.
- Small farmers cannot keep up investing in deeper wells and become dependent on buying water from larger farmers.
- With the lowest water right, the LBP command area risks suffering more seasons without water supply.
- Individuals and municipalities relying on groundwater as their drinking water source face many problems with falling water levels.

While there is a reasonably good data-collection system, official statistics do not reflect reality. For example, remote sensing reveals a total cropping intensity in the basin of about 140%, with a 100% increase in cultivation in non-command areas during the dry season from 1970 to 2000. This is evidence of the importance of water lifting and differs from government statistical data.

In the future, more water will be requested for use outside of the Bhavani basin. Drinking water diversions for the cities in the neighbouring Noyyal basin will increase. Kerala might utilize its right to withdraw considerable quantities of water from the upper Bhavani. There is also a risk of an increased frequency of water to be released from the LBP reservoir for the Cauvery delta farmers, or for the environment, as the closing of the Cauvery basin progresses.

There is increased interconnectivity within the Bhavani basin, and in fact within the entire Cauvery basin. The concerted impact from the many individual actions has resulted in a redistribution of the water use and a pressure on prevailing allocation rules. The development of the Bhavani basin illustrates that individual actions taking place after basin closure cannot be ignored by policy makers but should be thought of and recognized before undesirable water-use patterns have established.

The study holds important implications for policy. The main implication is that, after basin closure, means must be found to align individual actions with the objectives of society. This will require a better understanding of what drives individual water use and of the hydrology of the basin, but many problems have built up over time and are difficult to resolve. These include the built-up dependency on electricity for pumping and the increasing pressure on agriculture from other sectors. Eventually, it may be that policies may have to gradually shift people away from agriculture to ease water scarcity.

Acknowledgements

The authors would like to thank Palaniappan Gomathinayagam for sharing his knowledge and insights, and for invaluable support during fieldwork; Chennimalai Mayilswami for support with data collection; Dheeravath Venkateswarlu for remote-sensing analysis; and François Molle, Flip Wester, Jean-Philippe Venot and Malin Falkenmark for editing and valuable review comments.

This research was supported by Formas, the Swedish Research Council for Environment, Agricultural Sciences and Spatial Planning, and contributes to the Comprehensive Assessment of Water Management in Agriculture, which itself was supported by a grant from the governments of the Netherlands and Switzerland to the Comprehensive Assessment.

References

Appasamy, P., Shanmugam, K.R., Nelliyat, P. and Mukherjee, S. (2005) *Environmental Issues in the Bhavani River Basin*. Project Report. Madras. Madras School of Economics, Chennai, Tamil Nadu.

Baliga, B.S. (1966) *Madras District Gazetteers – Coimbatore*. The Director of Stationery and Printing, Government of Madras, Madras, Madras State (Tamil Nadu).

Barber, C.G. (1940) *History of the Cauvery–Mettur Project*. The Superintendent Government Press, Madras, Madras Presidency (Tamil Nadu).

Census of India (1991a) *District Census Handbook, Nilgiri*. Indian Administrative Service, Director of Census Operations, Madras, Tamil Nadu.

Census of India (1991b) *District Census Handbook, Periyar*. Indian Administrative Service, Director of Census Operations, Madras, Tamil Nadu.

del Ninno, C., Dorosh, P.A. and Subbarao, K. (2005) Food aid and food security in the short and long run: country experience from Asia and sub-Saharan Africa. SP Discussion Paper No. 0538. Social Safety Nets Primer Series. World Bank Institute, www.worldbank.org/safetynets (accessed 1 April 2008).

GoI (Government of India) (1952) *Report of the Grow More Food Enquiry Committee*. Ministry of Food and Agriculture, Government of India, New Delhi.

GoI (2005) *Census Info India 2001*, Version 2.0 (CD-ROM). Office of the Registrar General, New Delhi.

GoI (2007) *The Report of the Cauvery Water Disputes Tribunal with the Decision*, Volume I–V. Government of India, New Delhi. www.wrmin.nic.in/index3.asp?sslid=393&subsublinkid=376&langid=1 (accessed 15 May 2007).

GoM (Government of Madras) (1958) *GO MS No. 1827, 1958.07.01*. Government of Madras, Madras, Madras State (Tamil Nadu).

GoM (1963) *GO MS No. 2274, 1963.08.30*. Government of Madras, Madras, Madras State (Tamil Nadu).

GoM (1965) *History of the Lower Bhavani Project, Volume I – Head Works*. The Director of Stationary and Printing, Government of Madras, Madras, Madras State (Tamil Nadu).

GoTN (Government of Tamil Nadu) (2002) *Tenth Five Year Plan (2002–2007)*. State Planning Commission, Chennai, Tamil Nadu. www.tn.gov.in/spc/tenthplan/default.htm (accessed 10 January 2007).

GoTN (2005) *Statistical Hand Book of Tamil Nadu 2005*. Special Commissioner and Director, Department of Economics and Statistics, Government of Tamil Nadu, Chennai, Tamil Nadu.

Kannan, R. (2004) *Perambulation Notes of the Commissioner of Agriculture 2002–2004*. The Special Commissioner and Director of Agriculture, Department of Agriculture, Government of Tamil Nadu, Chennai, Tamil Nadu.

Krishnaswami Ayyar, K.N. (1933) *Statistical Appendix and Supplemental to the Revised District Manual (1898) for Coimbatore District, Madras District Gazetteers*, Volume II. Superintendent, Government Press, Madras, Madras Presidency (Tamil Nadu).

Lannerstad, M. (2008) Planned and unplanned water use in a closed South Indian basin. *International Journal of Water Resources Development* 24(2), 289–304.

Lindberg, S. (1999) When the wells ran dry: the tragedy of collective action among farmers in South India. In: Toft Madsen, S. (ed.) *State, Society and the Environment in South Asia*. Curzon Press, Richmond, UK, pp. 266–296.

Madras Presidency (1902) *Preliminary Report on the Investigation Of Protective Irrigation Works in the Madras Presidency*. Madras, Madras Presidency (Tamil Nadu).

MIDS (Madras Institute of Development Studies) (1998) *Natural Resource Accounting for Water Resources in Bhavani Basin – Phase II*. Interim Report. Madras Institute of Development Studies, Madras, Tamil Nadu.

Mohanakrishnan, A. (2001) *History of Irrigation Development in Tamil Nadu*. Indian National Committee on Irrigation and Drainage, INCID (Constituted by Ministry of Water Resources, Government of India), New Delhi.

NWDA (National Water Development Agency) (1993) *Technical Study No. WB 51, Water Balance Study of Bhavani Sub-basin of Cauvery Basin, (Index No. 63)*. National Water Development Agency, Society under Ministry of Water Resources, Government of India, New Delhi.

Palanisami, K. (1984) *Irrigation Water Management: the Determinants of Canal Water Distribution in India – a Micro Analysis*. Agricole Publishing Academy, New Delhi.

PWD (Public Works Department) (n.d.) *Kalingarayan Anicut System*. Public Works Department, Tamil Nadu, Executive Engineer Erode, Erode Division, Erode, Tamil Nadu.

PWD (1984) *Compendium of Rules of Regulation, Part 1, Rules for Water Regulation.* Office of the Chief Engineer (Irrigation), Public Works Department, Madras, Tamil Nadu.
PWD (1999) *Environmental Action Plan of Bhavani River Basin.* Water Resources Organisation, Chennai, Tamil Nadu.
PWD (2002) *Groundwater Perspectives Profile of Erode District Tamil Nadu.* State Ground and Surface Water Resource Data Centre. Water Resource Organisation, Public Works Department, Government of Tamil Nadu, Tharamani, Chennai, Tamil Nadu.
PWD (2003) *Groundwater Perspectives Profile of Coimbatore District Tamil Nadu.* State Ground and Surface Water Resource Data Centre. Water Resource Organisation, Public Works Department, Government of Tamil Nadu, Tharamani, Chennai, Tamil Nadu.
Saravanan, V. (2001) Technological transformation and water conflicts in the Bhavani River basin of Tamil Nadu, 1930–1970. *Environment and History* 7(3), 289–334.
Saravanan, V. and Appasamy, P. (1999) Historic perspectives on conflicts over domestic and industrial supply in Bhavani and Noyyal basins, Tamil Nadu. In: Moench, M., Caspari, E. and Dixit, A. (eds) *Rethinking the Mosaic: Investigations into Local Water Management.* Nepal Water Conservation Foundation and the Institute for Social and Environmental Transition, Boulder, Colorado, pp. 161–190.
SCR (Season and Crop Report) (different years) *Season and Crop Report.* Annual publication since end of 19th century. Department of Agriculture/Department of Economics and Statistics, Madras/Chennai, Madras Presidency/Tamil Nadu.
SCR (1903) *Season and Crop Report of the Madras Presidency for the Agricultural Year 1902–03.* Board of Revenue, Madras, Madras Presidency (Tamil Nadu).
SCR (1958) *Season and Crop Report of the Madras State for the Agricultural Year 1956–57.* Department of Statistics, Government of Madras, Madras, Tamil Nadu.
SCR (2002) *Season and Crop Report 2000–2001.* Department of Economics and Statistics, Chennai, Tamil Nadu.
SCR (2005) *Season and Crop Report 2004–2005.* Department of Economics and Statistics, Chennai, Tamil Nadu.
Shah, T. (2007) The groundwater economy of South Asia: an assessment of size, significance and socio-ecological impacts. In: Giordano, M. and Villholth, K.G. (eds) *The Agricultural Groundwater Revolution, Opportunities and Threats to Development.* CAB International, Wallingford, UK; International Water Management Institute, Colombo, Sri Lanka, pp. 7–36.
Thenkabail, P.S., Schull, M. and Turral, H. (2005) Ganges and Indus river basin land use/land cover (LULC) and irrigated area mapping using continuous streams of MODIS data. *Remote Sensing of Environment* 95(3), 317–341.
TNEB (Tamil Nadu Electricity Board) (different years) *Tamil Nadu Electricity Board – Statistics at a Glance* (annual publication). Tamil Nadu Electricity Board, Planning Wing of Tamil Nadu Electricity Board, Chennai, Tamil Nadu.
TNEB (2008) Statistics. www.tneb.in (accessed 11 June 2008).
TWAD (Tamil Nadu Water Supply and Drainage Board) (2000) *Project: TWAD–Bhavani River Surplus Flood Water Diversion (Internal Report).* Tamil Nadu Water Supply and Drainage Board, Chennai, Tamil Nadu.
von Lengerke, H.J. (1977) *The Nilgiris: Weather and Climate of a Mountain Area in South India.* Franz Steiner Verlag, Wiesbaden, Germany.
Wilk, J. (2000) Do forests have an impact on water availability? – Assessing the effects of heterogeneous land use on stream flow in two monsoonal river basins. PhD thesis, Department of Water and Environmental Studies, Linköping University, Linköping, Sweden.

12 Much Ado about the Murray: the Drama of Restraining Water Use

Hugh Turral,[1]* Daniel Connell[2]** and Jennifer McKay[3]***

[1]North Carlton, Melbourne, Australia; [2]The Australian National University, Canberra, Australia; [3]University of South Australia, Adelaide, South Australia; e-mails: *hugh.turral@gmail.com; **daniel.connell@anu.edu.au; ***jennifer.mckay@unisa.edu.au

Introduction

The Murray–Darling basin (MDB) in Australia has recently received considerable international exposure and is frequently commended as a working example of interstate cooperation and management of shared basin water resources (World Bank, 2005). This position is usually accompanied by the caveat that the economy and agriculture of Australia have very different structures, strengths and vulnerabilities from those in developing countries. The basin has been extensively developed for agriculture, and water resources are widely thought to be overdeveloped, at the cost of aquatic ecosystems (Cullen et al., 2000). The demand for water for industry and urban settlements has been very limited in comparison with the volumes used in agriculture, stock-rearing and irrigation. The major debate on the allocation of water resources in the basin now centres on reallocation to mitigate the negative environmental impacts of agricultural uses (NWI, 2005b). Recently, the decision to transfer 75 Mm^3 annually from the inland irrigation districts to the rapidly expanding coastal city of Melbourne has added a new dimension to this equation, although other urban transfers are not expected from the basin and the major factor governing the implementation of the transfer is its cost ($A1 billion). This is again in strong contrast to basins in many developing countries, where transfer of water from agriculture to higher-value uses is dominated by rapidly rising urban and industrial demand, with little conscious consideration of environmental water allocation. Nevertheless, most of the irrigation infrastructure in the basin is quite similar to that found in other semi-arid and arid regions, and owes much to the work of engineers who had earlier developed irrigation systems in northern India and in the western USA in the late 1800s (Hallows and Thompson, 1995).

Within Australia itself, the perceptions of the success of the institutional arrangements for water management in the Murray–Darling Basin are changing (Connell, 2007). The emerging view at the federal government level has been that the institutional arrangements for interstate cooperation were flawed, having been based on a voluntary and unanimous agreement. On 25 January 2007, the then Prime Minister John Howard announced that the Commonwealth Government would invest $A10 billion to reform rural water management and take over control of the Murray–Darling basin from the states. Later in the year, the Commonwealth Parliament passed the Water Act, 2007. This is the third attempt since federation in 1901 to design a comprehensive management framework. The previous efforts were in the early decades of the 20th century and in the 1980s. During both these earlier periods, a strong coordinating structure was initially proposed. Ultimately, in both cases, the efforts of the reformers were frustrated by the strength of established interests and concern for state 'rights'. As a result, environmental conditions and the security of water as an economic resource in the MDB have continued to decline.

In November 2007, a Labour government was elected and subsequently, to the surprise of many onlookers, it, in fact, increased the financial allocation to the new arrangements for the MDB to $A12.9 billion and has succeeded in bringing the recalcitrant state of Victoria into agreement, with an updated version of the 'Howard Plan' that is more respectful of state positions and contributions, and pays more attention to water allocation for the environment (Water for the Future). The strength of past arrangements has been that innovative solutions have been required in order to gain unanimous acceptance from the states. This, in turn, has meant that the process has been noisy and slow (especially to those closely involved), whereas from the outside, it has a logic and strength that has put Australia in the forefront of water management. There are many reasons to doubt the efficacy of a centralized approach, especially given experience overseas, and the lack of federal experience and connection with details on the ground. A new chapter will have to be written some years down the track, which will distil the lessons of this latest change in institutional direction in water management in Australia.

The primary focus of this chapter concerns the institutional arrangements developed for integration of management across the states lying within the MDB, set against the evolution of water use and its environmental consequences. It leads up to the recent major change in water resources management in Australia through the ceding of state responsibility to the federal government. The implications of this most recent move are especially pertinent to developing-country water-policy analysts and policy makers. The writing provides some biophysical, economic and technical background on the pillars of Australian water management, including a well-evolved and, by world standards, sophisticated water-allocation and accounting system. This chapter tries to cover the range of perspectives on water management in the MDB, and to link these to broader issues of natural resources management in Australia.

The chapter concludes with a discussion of the current challenges being faced, most notably in environmental water allocation and in adjusting to the significant projected impacts of climatic change. The emerging lessons from Australia are broadly instructive, since its arid climate is already very variable and is expected to experience considerable change, with significant reductions in annual runoff expected in the existing irrigated areas.

Australia – a federation of states and territories

Although the smallest continent on the globe, Australia is a vast and sparsely populated country, with a population of just over 20 million, mostly living on the coastal margins of the 7.79 million km^2 land mass. Climates range from temperate in Tasmania, through Mediterranean on the south-east and south-west coasts, semi-arid in the near-coastal interior and hyper-arid in the centre, to lush tropical in the north.

The Commonwealth of Australia unites six states and two territories under a three-tiered government system, of national, state (and territory) and local jurisdictions. From federation through to 2007, water has been the responsibility of state and territory governments, and each has evolved its own water law and regulation. Australia was federated in 1901, with section 100 of the federal constitution providing that the power over water was to remain with the states. The former colonies viewed water as a key stumbling block to federation and hence extracted a prohibition clause in the new constitution. Section 100 of the treaty, added at the states' insistence, stated: 'The Commonwealth shall not, by any law or regulation of trade or commerce, abridge the rights of the State or of the residents therein to the reasonable use of waters of rivers from conservation or irrigation.'

The contest was really between commonwealth power over water for navigation and the states' desire to use the water for irrigation. As a consequence of Section 100, the states have, for much of the last century, created their own laws, policies and organizations, sometimes without reference to their neighbours. However, the commonwealth has intervened in state water management through section 96 of the constitution, which gives it power to grant financial assistance to the states, contingent on specified conditions being met, and it has been used to tackle basin-wide flood and salinity problems in the MDB.

The national government has progressively asserted its perspectives through financial leverage in a number of ways, and has thus carved a role in shaping national water policy (Bjornland and McKay, 2002). Matters of national significance that concern the states and the federal government are overseen by the Council of Australian Governments (COAG), which deals with specific issues through specialized ministerial councils. A Natural Resources Management Ministerial Council was created in 2001 (replacing the earlier committees, ARMCANZ and ANZECC), and has subsequently had a major impact on water reform.

Agriculture and irrigation in Australia

Historically, agriculture has been a major industry in Australia, with an estimated total farm area of 463 Mha, or roughly 60% of the total land surface area (ABS, 1998). Much of this area is used for pastoralism, with only 4.6% of the total sown to crops and 4.9% to intensive pastures. Up to the late 1950s, agriculture employed a population of up to 450,000 and accounted for 80% of the nation's export earnings. By 2006/7, farm populations had declined to 308,000 and the agriculture share of national GDP had fallen to less than 3%. Nevertheless, the sector still accounts for a significant portion of total exports, around 12%, including goods and services, and 17–18% of all merchandise (ABARE, 2007).

Since European settlement, there has been a strong interest in irrigation, with the Chaffey Brothers establishing the first irrigation system in northern Victoria before the end of the 19th century. This, as with their earlier ventures in California, was not completely successful, and was eventually taken over by the state, although it helped seed a broader momentum in irrigation development (Turral, 1998). The state governments, partly influenced by irrigation development in northern India, began large-scale developments around the turn of the century, peaking in the 1970s in New South Wales (NSW). Soldier settlement, a form of compensation for returning servicemen, was a major plank of agricultural policy between the two world wars and after 1945. The expansion of irrigation relied heavily on the creation of interannual dam storage, with the completion of the Hume dam on the Murray in 1929, and the creation of the Snowy Mountains Hydroelectricity Scheme from 1949 to 1974, which includes interbasin transfers to the Murray–Darling system from a network of 16 dams and seven power stations, linked by 275 km of underground tunnels.

In 1998, the total irrigated area of Australia was about 2.4 Mha, of which about 80% lies in the Murray–Darling basin. Approximately 70% of all water abstracted in Australia is used for irrigation in the MDB, predominantly from surface sources (Table 12.1). Groundwater accounts for about 22% of national water use, on average, but is limited by salinity in shallow groundwater and the depth of pumping for fresh, non-saline water.

In recent years, there has been a growing recognition of the hydrological linkages between land use and salinity in the landscape. Much of Australia is underlain by ancient marine deposits with high levels of natural salinity. The imbalance between precipitation and evaporation has also fostered a gradual concentration of salt in soils from rainwater, and additional saline areas have been formed by wind-blown deposits of such soils. Although irrigation generates salinity due to rising water tables, there are large areas of Australia at risk from dryland salinity, where replacement of deep-rooted vegetation by annual crops has resulted in a gradual but inexorable increase in net recharge to groundwater, contributing to rising water tables and potential hazard to 20 Mha (CSIRO, 2001).

The Murray–Darling basin (MDB)

The Darling (2740 km), Murray (2530 km) and Murrumbidgee (1690 km) are Australia's three

Table 12.1. Allocation and use of surface water and groundwater in states that fall in the MDB (Mm^3), 1995–1996.

State	Surface water allocation	Surface water use	Groundwater use	Total water use	Difference between surface allocation and use	Difference between allocation and use (%)
NSW	10,252	6,139	197	6,336	4,113	40.1
VIC	5,589	3,662	95	3,757	1,927	34.5
QLD	702	574	–	574	128	18.2
SA	296	246	70	316	50	16.9
ACT	63	63	–	63	0	0.0
Total	16,902	10,684	361	11,045	6,218	21.9

Source: MDBC web site.

longest rivers. The Murray–Darling basin region (Fig. 12.1) covers more than 1,000,000 km^2 (14%) of Australia, unevenly spread over the five jurisdictions of Queensland (QLD), New South Wales (NSW), Australian Capital Territory (ACT), Victoria (VIC) and South Australia (SA). The estimated population living in the basin was 1,956,765 in the last census, which corresponds to around 10% of the total Australian population.

The Murray and the Darling are essentially two river systems, with only 16% of the Murray's mean annual flow contributed by the Darling. Population in the basin is increasingly sparse to the north and west, on the edge of the 'outback'. Bourke, an important town in the history and mythology of Australia, has a population of 3000, and Wilcannia, the third largest inland port in Australia in the 1890s, now has a population of just five. Larger country towns such as Broken Hill have populations in excess of 20,000, and those closer to the Murray are substantially larger, with Albury-Wodonga heading towards 100,000.

Three of the five polities in the MDB (NSW, Victoria and South Australia) have developed the greater part of available water resources. The allocation and abstraction of surface water and groundwater in the MDB states for the period 1995–1996 are summarized in Table 12.1; it can be seen that groundwater use accounted for less than 4% of abstractions overall. Although groundwater use increased

Table 12.2. Interstate water shares.

Flows and shares	(billion m^3/%)
Mean annual flow (Bm^3)	13.2
Mean annual diversion (Bm^3)	10.8
Minimum flow to South Australia (Bm^3)	1.8
Share to New South Wales (%)	57.4
Share to Victoria (%)	34.3
Share to South Australia (%)	5.4
Share to Queensland (%)	2.3
Share to Australian Capital Territory (%)	0.6

in terms of volume and proportion through the drought period after 1999, it is still only a small component of allocation. Total abstraction in ACT and Queensland is very limited.

In general, rainfall on the coast is high and falls rapidly towards the interior, following a decreasing trend from east to west (600–200 mm per annum) in NSW and Victoria, with an increase again to the west in the lower basin. Precipitation in the upper catchments of the headwaters of the Murray may reach 900 mm, accounting for the disproportionate contribution to runoff above Albury (33% of mean annual flow in the basin). NSW has the largest surface water runoff and also abstracts the highest volumes of water (Tables 12.1 and 12.2). Victoria's water is supplied from the main river and from the state's own internal storage dams.

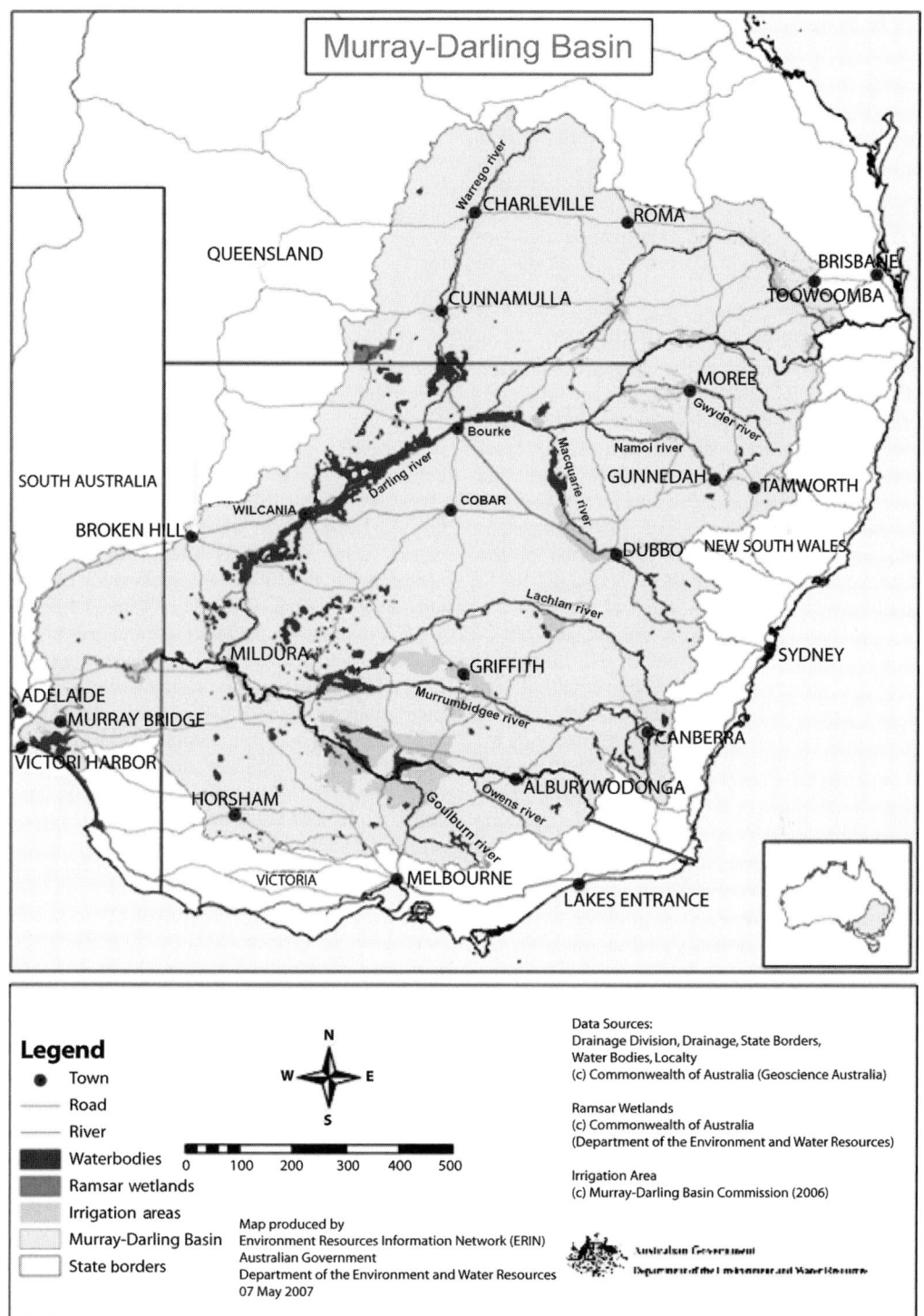

Fig. 12.1. The Murray–Darling basin, south-east Australia (Source: MDBC).

Australian hydrology is among the most variable in the world (McMahon and Mein, 1986) and droughts occur irregularly but may last for several years. Since 2000, there have been 5 years of continuous and unprecedented droughts across different parts of the basin. This high climatic variability has prompted a high level of river regulation of the Murray–Darling basin, which has a dam storage capacity of approximately 18 billion m^3 along the main stem of the river network, with two dams having capacities greater than 3 billion m^3 (the Hume and Dartmouth). Total inflows to major streams and storages in the basin are estimated at around 24–25 billion m^3 on average, although much of this does not reach the main river; for example, the Lachlan River loses most of its tail flows in the Great Cumbung swamp. Large areas of the interior wetland can be seen on the map in Fig. 12.1, especially along the middle and lower reaches of the Darling, where evaporation accounts for a significant proportion of 'internal runoff'.

The average naturalized annual flow out of the Murray mouth is 13 billion m^3, but actual outflows to the sea have been minimal for years (less than 200 Mm^3/month slightly upstream at Euston on long-term averages for 8 months, peaking at around 600 Mm^3/month in September). Irrigation use (around 11.5 billion m^3/year) accounts for 95% of diversions in the basin, from both the main river and other sources (MDBMC, 1995). The water shares, mean annual flow and minimum flows to South Australia, the most downstream riparian state, are shown in Table 12.2.

The shares are based on flow analyses undertaken for the Murray and the Darling at the beginning of the century, whereas the abstractions shown in Table 12.1 account for flows that are sourced from all rivers and groundwater lying in the MDB. The average runoff coefficient for the basin is a low 0.16, with a higher value of 0.2 for the lower-rainfall northern sub-basins and 0.10 for the southern sub-basins.

The average irrigated area in the basin is estimated to be just over 1.47 Mha, but can vary substantially through dry and wet periods. There are over 14,500 irrigated properties producing crops or pasture. The mean size of irrigated properties is larger in the basin than elsewhere, with 70% of the irrigated area farmed by 47% of the total number of irrigators. Cape (1997) estimates that irrigation accounts for 25–30% of the gross value of farm output, or about $A7.2 billion (Table 12.3).

The nature of irrigated agriculture in the basin reflects the security of water supplies. Large surface-irrigation systems were developed in all states, with two covering more than 500,000 ha each (the Goulburn–Murray and Murrumbidgee irrigation areas). In Victoria, there is a high degree of internal storage from catchments that feed into the River Murray, and strong links between different parts of the system (DSE, 2004). This results in a relatively

Table 12.3. The value of irrigation enterprises in MDB (nominal year 2005).

Sector	Area (ha)	Number of farms	Percentage of national area	Value of production (million $A)
Pasture and grains	862,155	8,584	79.8	2,450
Fruit	38,856	2,732	64.5	1,027
Vines	30,492	2,819	69.0	813
Vegetables	23,511	1,106	25.4	1,119
Rice	150,000	–	–	310
Cotton	490,000	–	–	1,128
Total	1,595,014	–	–	–
Horticulture as % of basin irrigated area	5.8	–	–	–

Source: Adapted from MDBC and River Murray web sites (www.rivermurray.com).

secure water supply throughout the year, enabling widespread development of intensive pasture for dairy production as well as horticulture.

By contrast, in NSW, the rivers are long, with little storage downstream of dams located in the upper catchments. Water supplies are more variable, and farming enterprises tend to be larger and more mixed than in neighbouring Victoria, with farmers adapting the balance of irrigated and rainfed production each year. The northern valleys of the Namoi and Gwyder rivers in NSW have a vibrant cotton industry, in part based on the overuse of groundwater and the contentious storage of flood water, and a tough process of renegotiating water use continues to generate controversy (Turral and Fullagar, 2006).

Despite relatively high natural flow variability and salinity in the river water, horticulture and dairy production are the main enterprises in South Australia. Recent trends in water pricing and water trading have tended to see water move from more extensive irrigated pasture to dairy and horticulture, with a rapid expansion of viticulture in the last 10 years (ABARE, 2007).

Regulation of river flows has also had a major impact on the magnitude of high and low flows in the system, with a 'reversal' of flow patterns from winter to summer in the upper reaches of many tributaries, as well as in parts of the main stem of the river. This reversal (more water flowing in summer than in winter) is due to the stocking of dams in winter and the subsequent release of irrigation supplies in the summer (Maheshwari and McMahon, 1995). Flow reversal and reduction in the upper tributaries of the Darling and in the lower Murray have caused great concern to ecologists and other scientists. It has had a profound impact on the in- and near-stream flora and fauna of the river. There are 30,000 wetlands in the basin, including large tracts of riparian forest, and 15 are listed under the Ramsar Convention. Many wetlands, especially in the west of the basin, are naturally saline. A recent audit found that about 10% of bird species and 20% of mammal species are endangered, while 20 mammal species have become extinct.

Concern about the impacts of reduced water quantity, quality and timing of flows has propelled many of the recent changes in water resources management in the basin, with particular emphasis on developing means of reallocating water to the environment.

An Institutional and Political History of the MDB

The story of the efforts to integrate basin water management across multiple state jurisdictions unfolds in three distinct eras. The first saw attempts to coordinate the development and use of water resources, followed in phase two by a gradual appreciation of the need for integrated and sustainable management. The third phase arises out of the perceived failure of a consensual model between the states themselves, with the emergence of a more directive and controlling federal involvement. Each phase is addressed in turn.

Attempt 1 – integrating development

The 1915 River Murray Waters Agreement

After many years of negotiation, a formal organizational structure designed to coordinate management of a limited number of issues by the three state governments in the Murray catchment in the southern section of the MDB was established in 1915 by the River Murray Waters Agreement (RMWA). Neither the commonwealth nor the three state governments (NSW, Victoria and South Australia) wanted the newly born federal government to take a leading role. This agreement was incorporated into identical parallel legislation, which was passed by each of the three state parliaments (Commonwealth Parliament, 1917). The new arrangement had three main components. First, there was a programme of engineering works planned as an integrated whole, with building and operations to be the responsibility of the state within which they were constructed. Construction costs were to be shared near equally (later made equal) by the four governments, and operation and maintenance costs were to be the responsibility of the three states within their jurisdictions. Second were the water-sharing rules, which underpinned Australia's future water allocation process (see Box 12.1). After providing a defined

Box 12.1. Water entitlements in the Murray–Darling basin.

From the early 20th century, robust institutions have evolved in the MDB to underpin the development of irrigation. For example, in 1909, during the first phase of publicly sponsored development, water entitlements were introduced by the state of Victoria, through the administrative issue of water licences on application by potential water users. Water meters were introduced in 1910 to ensure improved water sharing, accounting and charging, and, for more than eight decades, Victoria applied the same concept of a water entitlement, allocated to a specific area of land and with no right to transfer.

Water entitlements in the MDB have two distinct characteristics compared with many international allocation systems (particularly those in the south-western USA). First, most water 'entitlements' in the MDB are defined as a nominal (maximum) volume of water that can be abstracted under a licence. The actual amount allocated in any year (or 'allocation') is determined from the water available (the proportional appropriation doctrine) after the reservation of high-value uses (urban water supply, stock and domestic supply, environmental reserve and permanent plantings). Allocation announcements are given as percentages of entitlement and updated regularly through the water year for each subsystem or valley, and therefore risks are shared equally. This differs from the prior appropriation doctrine, where the entitlement is specified in absolute terms and risks are allocated according to seniority (see Chapter 6, this volume). Second, there are no beneficial use obligations for entitlement-holders in the MDB. The Water Act, 1912 (NSW) contains powers dating back to revisions to the Act in the 1930s that enable the cancellation of inactive water licences, but this provision was not applied successfully. Throughout the reform process, the policy of upholding unused entitlements has been hotly debated, but has been upheld, primarily since entitlement-holders have continuously paid charges to water management authorities to retain their entitlement.

Water entitlements are specified, measured and charged volumetrically. Although the principal technology to measure surface flows, the Dethridge Wheel, was developed in the early 1900s, it only saw widespread application from the 1960s onwards in Victoria and NSW, with volumetric conversion (re-specification of area-based water allocation in volumetric terms) in the major irrigation areas taking place in the 1970s and 1980s in NSW. Surprisingly, the process of volumetric conversion in South Australia only started in 2005, although, as with bulk allocation processes in the other states, the process also includes environmental allocation (SA Government 2005; NWI–South Australian Implementation Plan (NWI, 2005a)).

minimum monthly flow to South Australia (which would vary from month to month, depending on the time of year), NSW and Victoria were to share the flow at Albury equally and have exclusive right to the water in their own tributaries. As recommended by the 1902 Interstate Royal Commission, a proportional share arrangement between the three states was agreed for times of drought. Third, a commission of four members, one from each government and chaired by the commonwealth representative, and supported by a small full-time secretariat, was established to oversee implementation of the works programme and the water-sharing arrangements. In comparison with cross-border river management schemes elsewhere in the world, the division of water by a proportional share approach (Table 12.3) and the creation of a small full-time secretariat, which could later be expanded incrementally, were notable innovations.

Sandford Clark, water law expert and long-term commentator on matters related to the MDB, has argued there is evidence that the RMWA and the River Murray Commission (RMC) that it created were originally intended to be part of a more comprehensive institutional structure than was ultimately the case. Clark (1983) argues there is strong evidence that the RMWA and the RMC were designed to operate in combination with the Interstate Commission, a body intended by the designers of the constitution to be a key part of a federal decision-making system. The legislation establishing the Interstate Commission had wide-ranging clauses describing the scope of its powers to deal with river issues. These plans were frustrated, however, by a High Court decision in 1915, which effectively stripped the Interstate Commission of most of its powers.

An age of water resource development

It was perhaps partly in response to these setbacks that the premiers of the state governments agreed to a number of important changes to the RMWA when they met in May and July 1920. Instead of being the coordinator of three independent state construction bodies, it was agreed that the commission would create a single construction authority to build the dams, locks and other structures that were part of the programme of joint works. Even more intriguing, they agreed to change the voting system for the commission so that a three out of four majority would be sufficient, rather than the unanimous approval previously required for all major decisions. It was this last proposition, however, that resulted in the required legislation being rejected by the NSW parliament, thereby aborting all the amendments approved by the premiers (Eaton, 1946).

Continuing to work with the decentralized organizational model originally approved in 1915, the RMC implemented a major works programme through the 1920s and 1930s. Its main components were:

- A storage on the River Murray upstream of Albury, at Lake Hume (3.6 billion m^3).
- An enlarged Lake Victoria, a natural lake just off the main channel of the River Murray in NSW near the South Australian border, to supply South Australia.
- A diversion channel near Yarrawonga to take water into the flat lands of south-western NSW.
- A series of weirs along the Murray, Murrumbidgee and Darling rivers to support navigation.
- Barrages to separate the lower lakes from the Coorong and the mouth of the River Murray (Eaton, 1946).

The focus during this time was on the management of water quantity rather than on protecting water quality. The aim was to build new communities in the dry inland regions of Australia and to promote economic development. This approach was at its strongest in the decades immediately after the Second World War, pushed by no-nonsense aggressive state boosters such as Henry Bolte, premier of Victoria, and Thomas Playford, the long-serving premier of South Australia (Blazey, 1972). For the South Australian government, the main obstacle to development was its shortage of water. Concerns about water scarcity had dominated the state's history ever since Charles Sturt's expedition down the River Murray in 1830 and his subsequent report recommending the foundation of the colony. Since then, water shortage had severely constrained agricultural settlement and the expansion of Adelaide, the state capital. Access to water for irrigation along the river corridor and the piping of water to Adelaide and other towns from the Murray, undertaken in the 1950s, were seen as solutions to these constraints.

In the late 1940s and 1950s, a growing network of pipelines began to distribute water around the state. Adelaide came online in 1954. Eventually, the city would draw an average of 40% of its water from the Murray, a dependence that has increased up to 90% in times of drought (Hammerton, 1986). This connection between the River Murray and the state capital, where most of the population lived, greatly increased the significance of the River Murray for South Australia. With the benefits came greater dependence and risks. Drought was an obvious danger, but more insidious was the lack of political, institutional and legal protection against the likelihood that the upstream states would continue increasing their extractions. For the upper-catchment states, the main factors limiting the growth of their diversions were the volume of flow that came into the system and the need to provide South Australia's annual entitlement under the RMWA, then 1.5 billion m^3 and later 1.85 billion m^3 (Table 12.2). Extra water only flowed over the border from either the Murray or the Darling because the upper states had not increased their consumption to the level that they were entitled to under the RMWA. In the coming decades, insecurity created by this situation caused South Australia to wage a long campaign to restrain the water-development ambitions of its upstream neighbours.

Attempt 2 – managing overallocation and rediscovering the environment

Salinity problems and institutional responses

Much of the rethinking of the aims and methods of water management in the MDB during the

1970s and 1980s was the product of growing awareness of salinity problems. Management options available to the RMC to control salinity were very limited. Speaking to a workshop conducted at Khancoban late in 1984, Don Blackmore, then Deputy Chief Executive of the River Murray Commission, identified 14 factors influencing the level of salinity in the Murray. Of these, the commission controlled only one, dilution flows (Blackmore, Canberra, personal communication). The need for closer interstate cooperation was made clearer by the early results of a series of major projects undertaken in the following years, which revealed that the rivers draining the MDB overlay a number of naturally saline regional groundwater systems. The research also showed that these basins were filling up rapidly and changing the balance of the hydrogeological system of the basin; in essence, the groundwater systems under the southern MDB are easy to fill but difficult to empty (Williams and Goss, 2002). The high saline water table is the main cause of secondary salinization of soils.

The growth in understanding of the basin's salinity problems in the 1970s and 1980s was based on increasing research and hard-won experience (Fig. 12.2). This new knowledge paved the way for what came to be called integrated catchment management (ICM[1]), a combination of holistic thinking about the biophysical environment and recognition of the need for community involvement and empowerment. A recent survey of ICM throughout

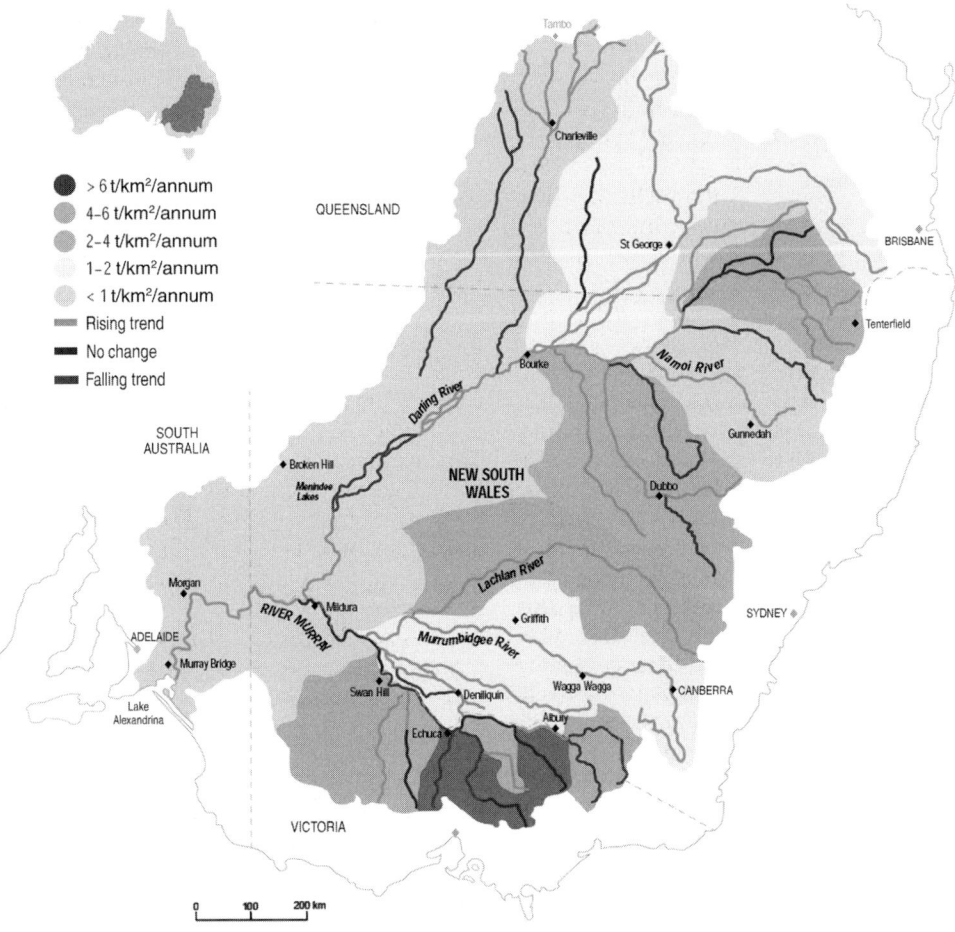

Fig. 12.2. Salinity trends in the land and rivers of the MDB, 1975–1995 (MDBMC, 1987, 1988, 1995).

Australia put forward a useful generic description of the elements that would characterize a mature ICM system (Bellamy et al., 2002). The authors suggest that it would be flexible and adapted to the variability and diversity of the area being managed. Within such a system, communities would be thoroughly involved and aware of the significance of their place in a broader regional context. An ICM system would also be supported by legislation and regulation designed to empower and assist rather than to dominate and unnecessarily restrict Programmes would be well resourced and power devolved down to the appropriate level. A wide range of options for community involvement would be available and participation encouraged by providing opportunities to exercise judgement and discretion. Above all, the people and groups involved would be encouraged to be cooperative and positive in their relationships with one another (Bellamy et al., 2002).

ICM philosophies had a strong influence on the thinking behind the revised organizational framework for the MDB, which was put in place in stages between 1985 and 1988. Known rather clumsily as the MDB Initiative, the new arrangements were incorporated in a revised MDB Agreement, which, for the first time, included Queensland and Australian Capital Territory, although not as fully committed signatories. The key elements of the structure that resulted from the debates of the mid-1980s were the Murray–Darling Basin Ministerial Council (MDBMC), the Community Advisory Committee to the Ministerial Council, and the Murray–Darling Basin Commission (MDBC). The Ministerial Council has two or three ministers from each government. The Community Advisory Committee is made up of selected representatives from the major regions and organizations, such as the Australian Conservation Foundation, the Murray–Darling Association and the National Farmers' Federation. The commission has two representatives from each jurisdiction, usually chief executives from the environment or water agencies, who report to the ministers on the Ministerial Council. All three bodies were assisted by the MDBC office, which had inherited the RMC staff and then expanded it through the 1990s in response to the growing list of issues that came within the widening ambit of basin-wide management.

In parallel with this higher-level institutional development and adoption of ICM principles, there have been innovative, local-scale initiatives aimed at solving local problems in an integrated fashion. Land Care, a community-based land management programme, has mushroomed in Australia to number more than 6000 Land Care groups, in rural, urban and coastal areas. The idea of Land Care emerged spontaneously in Western Australia (with community action on dryland salinity) and in northern Victoria (with Salinity Action Groups in irrigated areas) during the 1980s. A remarkable accommodation between the National Farmers' Federation and the Australian Conservation Foundation oversaw the formalization of Land Care as a regional and national strategy, with increasing levels of federal funding going into establishing and training Land Care groups. The principles and practice of community funding and community direction of public funding (co-financing) have underwritten Land Care. Activities have included the adoption of better practices; for example, salinity mitigation measures, tree planting and buffer-strip management. There are also interesting school-based educational programmes, designed to target adults via their children, and community-based awareness-raising, such as 'Water Table Watch' in Victoria. Essentially, it has been a parallel, and sometimes disconnected, activity to the states' and commonwealth's pursuit of improved water management and environmental sustainability in the MDB. They certainly cross paths in many ways, but Land Care is much more truly integrated at farm and community level. The National Land and Water Audit (1998–2000) was conducted in part because of the realization that there was no well-established baseline from which to measure the impact of 15 or so years and billions of dollars of private and public investment (National Land and Water Audit, 2001). A formal assessment of Land Care is not available, and deriving one presents a formidable challenge in integrating agricultural, environmental and community benefits. Unsurprisingly, the evidence on the success and merits of Land Care activities varies from positive endorsement, through doubt on its economic merits, to disillusion with progress on the ground.

The main environmental concerns in the basin related to salt and altered flow regimes

(Fig. 12.2). Degradation in rivers and changes in habitat arose from salinity, low flows and changed flow patterns. Rising water tables, due to groundwater recharge from irrigation and land clearance for dryland farming, resulted in high salinity, which restricted yields and crop choice. Disposal of salt emerged as the key challenge, with added political weight arising from its deleterious effects on the water supply infrastructure for the city of Adelaide, at the tail of the river. A key strategy has therefore been to prevent salt movement into the river, through interception, diversion and on-farm recycling.

Algal blooms have been a historic feature of Australian rivers since European settlement, but their severity and occurrence are exacerbated by changed (low) flow regimes and the additional influence of phosphorus and nitrogen fertilizers in agricultural runoff.

From water development to water management

In part, the water reforms beginning in the late 1980s were also the product of changing ideas about how public institutions should be organized and operated. There was a widespread feeling that decision making could no longer be left to small groups of engineers who had spent their careers dealing mainly with water resources infrastructure. Under the new arrangements the basin's river system was to be managed to conserve biodiversity and improve sustainability as well as production. The state and commonwealth governments seconded to the MDBC teams of ministers and senior public servants drawn from the agencies who dealt with these often-conflicting responsibilities for production and the environment. This brought the environment and agriculture into the institutional fold along with water management (although other potential contenders, such as tourism, recreation, aboriginal affairs and local government remained outside). In the lead-up to meetings of the Ministerial Council and Commission, each jurisdiction was expected to develop a consolidated position on the various issues to be discussed. These changes were incorporated into new legislation and passed as identical acts in each of the parliaments of the MDB in 1992–1993 (Commonwealth Parliament, 1993).

However, most of the activities incorporated into the new agreement were advisory or discretionary in nature, and needed the enthusiastic cooperation of all the governments and agencies involved before they could be implemented in any significant way. This applied particularly to activities outside the River Murray corridor. In addition, the long-established unanimity principle still applied to all decision-making processes, giving the power of veto to any jurisdiction that wanted an item excluded from the agenda or which was dissatisfied with any decision made. Despite these limitations, however, the early years of the MDB Initiative were a time of considerable achievement.

Re-evaluating natural resources and their management

Soon after the MDBMC was formed in 1985, it commissioned a series of studies to provide the knowledge and outline a new approach to implementation that would support a substantial expansion of interjurisdictional activities. Brought together as the Murray–Darling Basin Environmental Resources Study, the project summarized existing information, identified knowledge gaps, documented the locations of environmental resources that required special protection, recommended actions needed to protect these resources, and nominated further investigations. It also specified the requirements needed for a basin-wide monitoring programme, given that lack of quantitative data was a 'major constraint' on effective policy and management. After noting that 'integrated catchment management with strong community involvement will need to be a fundamental strategy', the study proposed comprehensive action to deal with issues related to agricultural land resources, climatic change, vegetation, groundwater, flora and fauna, aquatic and riverine environments, water quality, water allocation, water-use efficiency, riverine regions, cultural heritage, tourism and recreation (MDBMC, 1987).

The resources study was the precursor of the Natural Resources Management Strategy (NRMS) adopted by the Ministerial Council in August 1990 (MDBMC, 1989). The NRMS was to:

- Prevent further degradation.
- Restore degraded resources.
- Promote sustainable user practices.
- Ensure appropriate resource use planning and management.
- Ensure a long-term viable economic future for basin dependants.
- Minimize adverse effects of resource use.
- Ensure community and government co-operation.
- Ensure self-maintaining populations of native species.
- Preserve cultural heritage.
- Conserve recreational values (MDBMC, 1990).

But the implementation programmes that would have been required to achieve these goals were never prepared. The water-quality policy was typical of a number of policies developed within the NRMS framework. The NRMS outlined a comprehensive view of the problems of the MDB and provided an overarching justification for many projects, both specific and general. What did not happen, despite strong statements in the preparatory work for the required strategy, was the development of a programme of activities that could be seen as a comprehensive response on a scale that matched the extent and dimensions of the problems that had been identified. In the following years, there were Herculean efforts to overcome this gap, but attempts to devise middle-level plans for the range of issues of concern were continually frustrated. Instead, the result was an ad hoc list of projects justified in a general way as contributing to a vaguely defined 'improved sustainability'.

The one issue that did result in focused action under the new arrangements was salinity management, particularly with regard to the manifestations of the problem that were directly relevant to irrigation. The first schedule added to the new Murray–Darling Basin Agreement in 1988 was the Salinity and Drainage Strategy (S&D Strategy), which brought together plans that had been developing independently in Victoria, South Australia and NSW for more than a decade (Commonwealth Parliament, 1993) and had little relevance to the emerging Land Care movement. Once negotiations between the states finally got underway, agreement on the broad outline of the strategy was reached fairly quickly. As well as a number of management changes to reduce evaporation from storages, the new strategy allowed some additional saline drainage from new irrigation projects to flow to the river in the upper catchment states – Victoria and NSW. In return, these states and the commonwealth invested in groundwater interception works in the middle and lower reaches of the Murray River, where the greatest salinity reduction benefits would be obtained.[2] Over time, the aim was to produce a significant drop in net average salinity levels in the Murray, as measured at Morgan in South Australia (Fig. 12.3), and manage flows so as to avoid the short but severe spikes in salinity levels that periodically caused considerable damage in the lower reaches of the Murray (MDBC, 1999).

Central to the S&D Strategy was a register maintained by the MDBC to record the negative salinity impacts of new irrigation developments and the positive impacts of the compensating remedial projects. The currency developed to measure positive and negative impacts was known as electrical conductivity units (ECs), a measure of electrical conductivity in water that indicates its salt content. For the S&D register, the key measurement was the EC reduction or increase at Morgan caused by the activity in question. The 10-year aim of the S&D Strategy was to achieve a net average reduction in salinity of 80 EC at Morgan and an average salinity of less than 800 EC at least 95% of the time.

The strategy was a major success, but its planners were well aware that, rather than solving the problem, they were only buying time against a rising long-term trend, driven by the expansion of areas affected by dryland salinity in the wider catchment (Turral, 1998). Reversing that trend will require large-scale changes to the way in which the wider catchment is managed, and despite much debate and planning that remains an elusive goal (Williams et al., 2002).

Although the MDB Initiative was not able to mount a comprehensive response to the many issues that now need to be managed in the MDB, its programme was still wide-ranging. For example, working on the assumption that sustainable management cannot be based solely on economic considerations, the MDBC

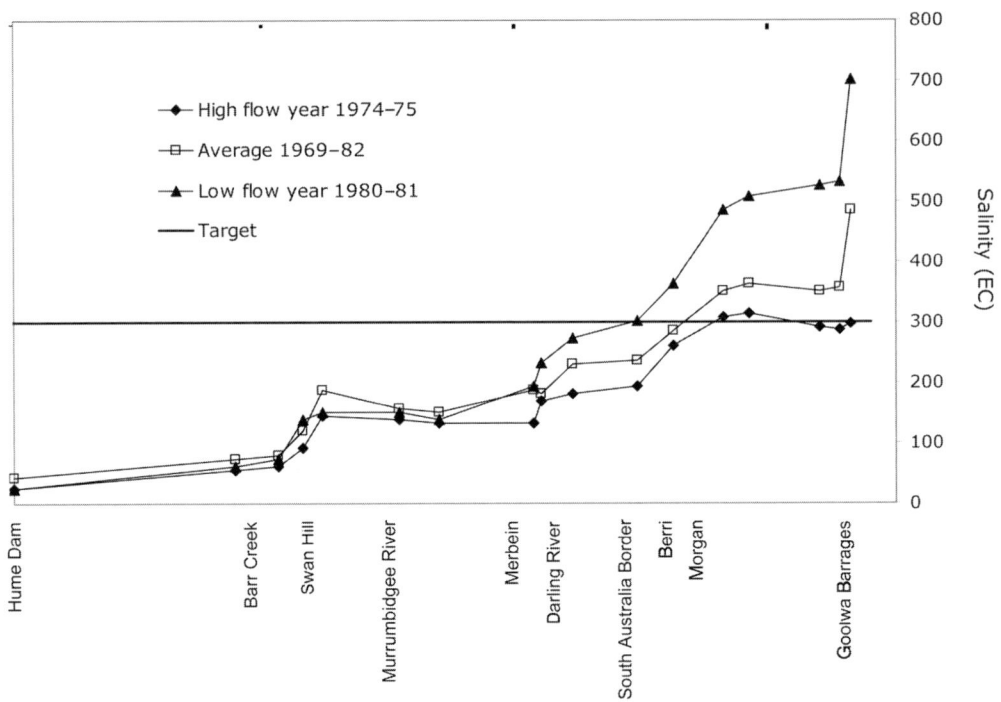

Fig. 12.3. Salinity trends along the River Murray in different flow conditions: upstream is on the left of the chart (adapted from MDBC, 1987).

has been funding programmes such as Special Forever in hundreds of primary schools across the region. This programme uses artistic and literary activities to encourage children to investigate the relationship of themselves and their communities to local streams and catchments. The intention is to foster cultural values that will support more sustainable practices.

Another significant influence on public policy in the MDB in the 1990s and 2000s was the controversy over the future of Lake Victoria, a major water storage body on the River Murray in south-west NSW near the South Australian border (Connell, 2002). The lowering of the lake in 1994 to allow repairs revealed a large number of Aboriginal grave sites, reflecting many thousands of years of occupation. After years of protracted negotiations with the Aboriginal community connected with the lake, and more than $A4 million for conservation work to protect the grave sites and cultural material, a new operational plan was agreed in 2002. The previous plan had focused on supplying water to South Australia and mitigating some of the salinity impacts of the river-management regime then in place. In contrast, the new plan takes account of a much wider range of issues, particularly indigenous and environmental matters (MDBC, 2002). The Lake Victoria project also made the Ministerial Council and Commission aware that many other parts of the riverine system in the basin required similar consideration of indigenous interests, in that rivers and their banks had been prime sites for human habitation and burials for many thousands of years.

Supply, demand and the environment

The current total basin diversion and storage capacity (Fig. 12.4) considerably exceeds actual use (Fig. 12.5). It also illustrates the predominance of run-of-river diversions at early stages of development (during the inter-war years), followed by rapid development of storage capacity, which continued through to the late 1980s. One of the last diversions built (Barren Box, completed in 1988) was developed to

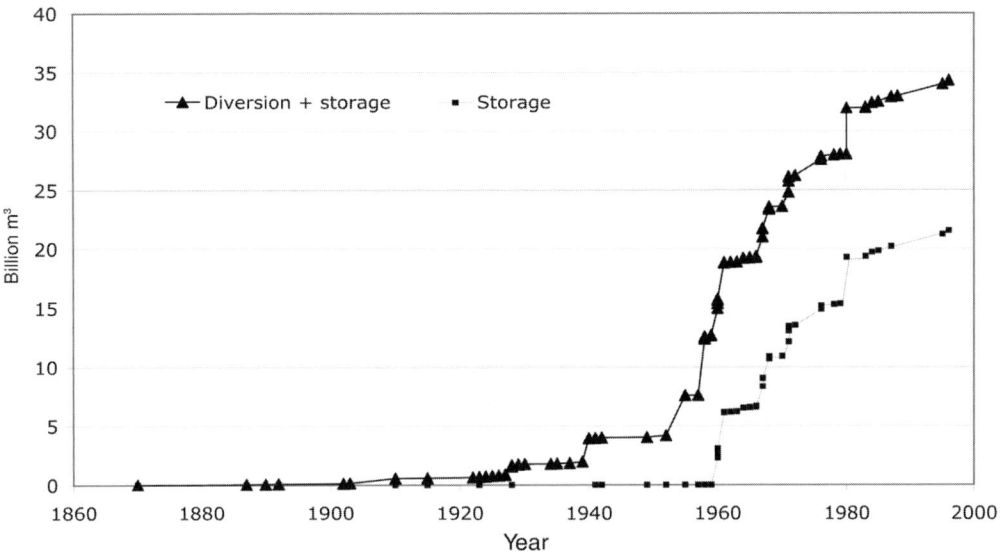

Fig. 12.4. Evolution of storage and diversion capacity in the MDB (adapted and modified from Haismann, 2004).

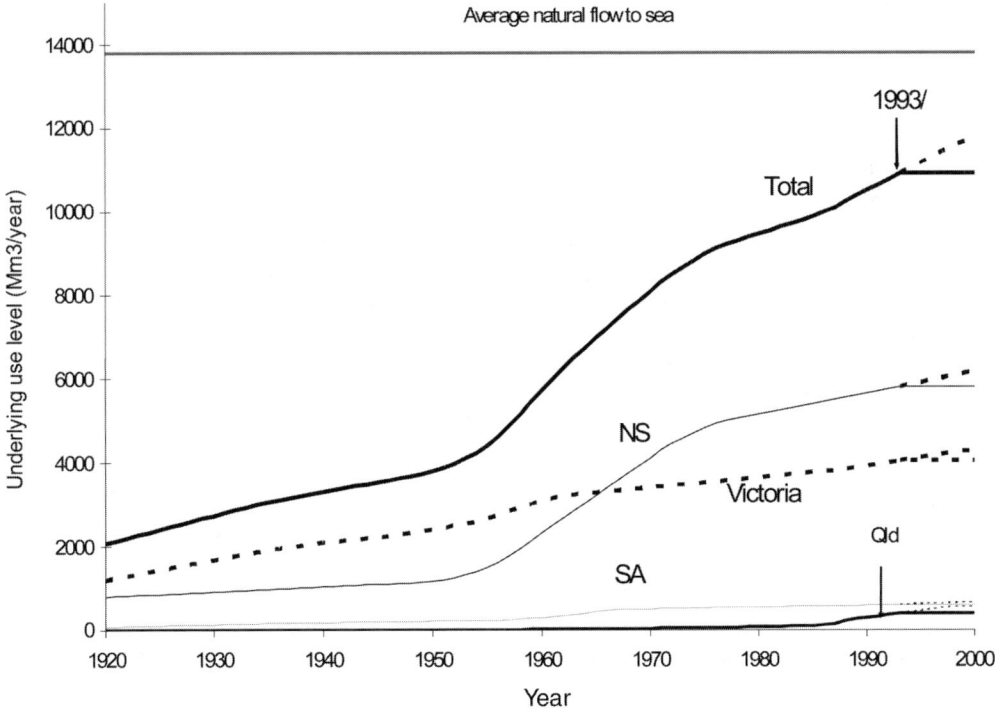

Fig. 12.5. Historical water diversions and projections without the cap (dotted lines) in the Murray–Darling basin (Source: MDBC, 1996).

manage environmental flows to natural but endangered swamps. Although there are different ways of accounting for total storage, the approximate ratio of major storage capacity to mean annual flow of the river system is about 2.3:1, although there are figures as high as 2.8:1 quoted in the literature. The MDB is highly regulated, and storage is primarily intended to improve inter-annual supply security. The pursuit of this goal has come at an increasingly evident and acknowledged cost to river and wetland health.

The institutional changes of the 1980s were largely driven by concerns about the increasing salinization of the streams in the MDB, but that was not the only water-quality issue that had to be managed. By the early 1990s, it had become clear that riverine conditions in the MDB were still deteriorating. A spectacular algal bloom in the summer of 1991/92, which extended along more than 1000 km of the Darling River, gave the issue international prominence. Water use had been growing in the basin throughout the 1970s and 1980s, with continued development, mostly in NSW (Fig. 12.5). Fearing that this would lead to overabstraction, no new licences were issued after 1986, but existing unused licences were not rescinded. In NSW, it is common for licence-holders, particularly stock farms, to keep water rights in reserve for drought periods (known as 'dozers') or not use them at all ('sleepers'). As time went on, more of the sleeper and dozer volume was activated, through property transfers and enterprise diversification, and, more recently, through water trading.

As the proportion of flow allocated to irrigation expanded, it had negative impacts on environmental conditions and reduced the capacity to meet demands in dry years. To halt the ongoing expansion, the Ministerial Council commissioned an audit of water use in the MDB, which was delivered in June 1995 (MDBMC, 1995). Up to that point, the detailed accounting for water use had been undertaken by each state and then reported to the MDBC. Since the MDBC was constituted with the representation of all states, and limited by the requirement of unanimous decisions, it had no executive authority over entitlements, even though actual volumetric allocations were effectively well monitored. The situation was perceived to be further complicated by the differences in the specification of entitlements between states and between regions within the states. Although these differences often have reasonable local justification, they confuse the bigger picture. Two major concerns underwrote the Ministerial Council audit: (i) that total diversions, if unchecked, would consume all available streamflow, despite the cap on licences; and (ii) that the large volume of unused licences in NSW ('sleeper' and 'dozer' licences) could be activated to propel further increases in total diversions.

The Water Audit found that, under 1994 levels of development, median annual flows from the basin at the River Murray mouth were only 28% of what they would have been under natural conditions, and that the percentage of years in which the lower reaches of the River Murray experienced drought had increased from 5% to over 60% (MDBMC, 1996). Furthermore, diversions had grown 8% since 1988, when the S&D Strategy was introduced, and were estimated to have the potential to increase by an additional 15% in the future. This would have severe environmental impacts and so reduce the security and reliability of supplies to existing entitlement holders.

According to the Water Audit, the real constraints on water use were the inadequacies in the infrastructure used to physically distribute water and economic decisions about the profitability of potential activities, rather than the MDB's water management systems. In most years, the total licensed volume was greater than the water available (MDBC, 1995), although actual allocations had to match available supply (see Box 12.1). Water apportionment in the MDB had evolved in response to the imperative to encourage water use to justify the investment in dams and infrastructure by governments. Except in times of drought, controlling diversions had not been a special priority, since there was already a clear allocation procedure that reflected water availability and inter-annual variability. As a result, diversions had tripled in the previous 50 years and most small-to-medium floods were now captured by the storages, thereby severely weakening the linkages between flood plains and stream channels.

The seasonal pattern of flow had also been substantially modified in many parts of the MDB, with much of the late winter/spring flow captured for release in the summer and early autumn, a time of year when flow had previously been at its lowest. Both changes were having substantial impacts on water quality and biodiversity in the riverine corridor. In the northern valleys of the Darling River tributaries, the continued development of irrigation through flood water and runoff harvesting had a severe effect on the flood plains (Kingsford, 2000)

Controlling water abstraction: the cap

In response to the Water Audit, the Ministerial Council introduced an immediate temporary cap on further expansion at 1993/94 levels of development, the irrigation season upon which the Water Audit was based (MDBMC, 2000).

In July 1997, the cap was made permanent (MDBMC, 1999). When it was introduced, the cap was described as the first step in a process that needed extensive development before it could achieve acceptable levels of environmental sustainability and resource security. At a time when the pressure for continued uncontrolled growth was strong, the introduction of the cap was a determined attempt to call a halt, as a precursor to a reassessment and potential winding back of development pressure. The original intent was that the cap would be refined over time, but that has not happened. In addition, despite many official statements of intent, the cap was never extended to groundwater or to much of the northern section of the MDB, as it is a voluntary mechanism, to which Queensland, as a minor and later-developing user, did not subscribe. The cap and the policy statements that accompanied its introduction showed that the Ministerial Council was aware of the large-scale changes that were needed. Its failure to complete implementation of the cap by extending it to include groundwater and the northern part of the MDB, however, revealed its limited capacity to protect the medium-term future of the region. A number of other initiatives and developments have contributed to trying to rein in overallocation (Box 12.2).

Although the MDB Ministerial Council became increasingly reluctant after the early years to respond energetically to the issues that needed to be managed, it did continue to commission major knowledge projects, which revealed the need for new policy and institutional reform. One such project was a study undertaken in 2006 by the Commonwealth Scientific and Industrial Research Organization (CSIRO) to assess future risks to inflows into streams and storages from climatic change, and the reductions that are being caused by the growth in farm dams, new plantation forestry projects, increased groundwater pumping and improved channel and irrigation management. In light of these factors, CSIRO predicted a decrease in streamflow of between 2.5 and 5 billion m^3 over the coming 20 years, equivalent to between 18 and 36% of long-term mean annual flow to date. The situation is not expected to stabilize, and predictions for the mid-21st century are for a reduction in inflows of the order of 4.5–9 billion m^3 (van Dijk et al., 2006). For purposes of comparison, current diversions for irrigation, which are officially considered to be too high, are of the order of 11–12 billion m^3.

Most of the management options needed to respond to these threats were not available in practice to the MDB Ministerial Council under the regime put in place in the mid-1980s. Most of the states individually had appropriate management options but chose not to use them (Arthington and Pusey, 2003).

At the national level, there has been an important set of reforms, which has committed Australia to steer towards a path of sustainable development, beginning with COAG National Strategy for Ecologically Sustainable Development (COAG, 1992). Its provisions drew heavily from the Bruntland (1987) report and adopted the precautionary principle as a guiding philosophy.

Reform in the water sector was tied to microeconomic reforms under the National Competition Policy (1994), which aimed to remove subsidies and ensure competition and economic efficiency. This eventually saw the restructuring of the irrigation sector, with Victoria already well on the way to mandating full recovery of operation and maintenance costs through corporatization[3] of the Rural Water Commission in 1994. In 1996, the state government backed away from plans to fully privatize the Rural Water Corporation; instead it broke it into a number of independent water supply corporations, with state and user oversight. Since 2001, the Victorian Department of Sustainability and Environment has gained an increasingly directive role in the activities and strategies of the corporatized rural water suppliers, partly through leverage associated with funding for capital improvements and

> **Box 12.2.** Responding to overallocation.
>
> There have been five main thrusts to reining in overallocation in the basin:
>
> - Imposition of the cap. In NSW, this has involved the development of detailed water-sharing plans in all water management districts, especially those with actual or potential overallocation, including groundwater (DLWC, 1998).
> - Enablement of water trading to reallocate water to higher-value uses. Owing to the drought, record prices in permanent trades have been achieved year on year since 2004, and water entitlement has become a means of collateral for farmers wishing to take investment loans (Rabo Bank, Sydney, 2004, personal communication).
> - Promotion of water-conserving technologies, including better measurement, reduced conveyance losses and more efficient on-farm practices that minimize recharge to shallow (and often saline) groundwater. In Victoria, the 'Foodbowl' project proposes better flow measurement to reduce overdelivery, conservation technology and practice, and system remodelling in exchange for entitlement transferred to Melbourne (interbasin transfer, 75 Mm^3), environmental flows in winter (75 Mm^3), and for reallocation in agriculture (75 Mm^3).
> - Administrative tinkering with the allocation process through the specification of revised river-flow rules, dam-operation rules, and inter-annual carry-over reserves for environmental allocation. In NSW, this has included restricting access to 'off allocation': pump diverters along the river have traditionally had opportunistic access to water in the river that is rejected by its intended users, usually because of changing weather or summer flood flows (Wijedasa *et al.*, 2001). Off-allocation amounts were not defined, but the authorization to harvest rejection flows was announced on an event basis.
> - Structural adjustment: since the early 1990s, marginal producers, often those with irrigated pasture for meat and wool production, and typically on saline soils (such as Pyramid Hill in north-west Victoria), have quit farming. Water trading has offered some improved compensation for some of these farmers, and may contribute to reducing negative environmental externalities. There was a proposal to fund some farmers to leave the land late in 2007, but the amount offered was considered too low and it has so far not been implemented.
>
> A main thrust of the recent $A12.9 billion Labour government plan (in 2008) is to reduce the allocated volume in return for system and on-farm investments in water conservation. This has also added impetus to the need to account for surface water and groundwater flows effectively, and to monitor and control the capture of runoff on farms. In all states, the ability to capture runoff in farm dams has been restricted through new licensing requirements.

water conservation efforts directed at reducing total allocations to irrigation and increasing those for the environment. In NSW, an initially reluctant irrigation community eventually opted for privatization of the three major systems (the Murrumbidgee, Colleambally and Murray irrigation systems). The main intention of privatization was to eliminate subsidies in service provision, and to a large degree this has happened, although continuing investment in water conservation (for environmental purposes) and drought relief could be regarded as subsidies in a different form. Interestingly, some irrigation systems, such as the Murray irrigation systems in NSW, managed to negotiate modest but favourable capital endowments to privatize.

In Victoria, the corporatized systems started well on full-cost recovery and did a lot of work on asset replacement and financing, and negotiated this with their users, resulting in significant increases in water fees. However, low allocations through the recent drought have severely impacted their investment schemes, since the money banked for future asset replacement comes from 'sales water' (i.e. the volume of water supplied above basic entitlement), which has, historically, been an additional 70% of volume, on average, in the Victorian system of allocation. There is evidence that the privatized systems in NSW are currently opting for aggressive modernization and improvement now, but pay less attention to long-term cost recovery and asset replacement.

The lessons of privatization do not yet emerge clearly because of the timing of the current drought. Interestingly, the states continue to be responsible for bulk water allocation, regulation and oversight, although the operation of the major storages (Lakes Hume and Dartmouth) is delegated to the Department of Water and Energy (NSW) and Goulburn–Murray Water (Victoria).

COAG (1994) also instructed the separation of water titles from land, to stimulate and encourage water trading, which was seen as an important means of reallocating water to higher-value uses within the irrigation sector. This alone was expected to accelerate the transfer of water from marginal agricultural enterprises to higher-value ones. Within the MDB, there is little competition for water from higher-value urban and industrial uses, as most of the demand is located on the coast, outside the basin. Thus, the target of these reforms was more economically efficient agriculture, with the expectation that some of the larger environmental externalities related to salinity would be mitigated through the exit of the more marginal producers. The economic reforms in water management set out by COAG were 'enforced' by an interesting combination of incentives and penalties. The most telling of these was to link progress in economic and service provision reforms with tranche payments of federal tax revenue given back to the states, with hundreds of millions of dollars at stake. There was an implicit assumption that these economic reforms would have a positive impact on environmental externalities.

Water trading has become increasingly active through the 1990s, due both to the separation of land and water rights under COAG and to very low water availability from 2000 to 2007. Water in Australia can be traded on the temporary market, where a farmer sells a portion of annual allocation to another user in one season or year. Water entitlements can also be traded permanently. Trades are registered and brokered by irrigation suppliers, estate agents and associations (such as the Murrumbidgee Horticultural Association). Although it has been argued that trade effectively reallocates water at the margin (Turral et al., 2005), with permanent trades accounting for less than 1% of total volume in any year, the cumulative effects of permanent trade are starting to be seen.

Temporary trading has been extremely active in different places at different times through the dry period from 2000 onward, with more than 30% of allocated volume traded in the temporary market in some valleys in NSW (Turral and Fullagar, 2006). If water allocations return to higher levels, the volume of temporary trading is expected to diminish. There are continued impediments to trade, due to infrastructural limitations and restrictions against trading water downstream if transmission losses are very high (for example, along the Lachlan River). Interstate trading is limited, due to continuing debate about exchange rates for volumes moving from one state or agroclimatic region to another (Etchells et al., 2004), and limited allocations over recent years. In the medium term, it is expected that trading will provide a useful mechanism for reallocating water flexibly in dry periods, and strategically within agriculture as the effects to climatic change are felt (NWI, 2005b).

In November 2000, the COAG agreed to a regional model for the delivery of the National Action Plan on Salinity and Drainage (NAP). Following this, the Natural Resources Management Ministerial Council (NRMMC) adopted a regional delivery model for the funding of environmental activities at a regional level, leading to the integrated implementation of both the NAP and environmental funding based on regional needs. In effect, this led to the institutionalization of catchment management authorities (CMAs) as the primary organizations responsible for natural resources management. The CMAs integrate public and private interests and, in Victoria and NSW, cover both irrigation and dryland areas, but it is fair to say that the focus of many CMAs is predominantly on land rather than on water management. Some parts of the basin (in Queensland) are not covered by CMAs. Formal catchment management has arisen out of the need to integrate the burgeoning number of Land Care groups with varied mandates and foci, and there remains a considerable challenge in coordinating the activities of different CMAs within one river basin, let alone throughout the MDB (M. Wood, DSE Melbourne, 2008, personal communication).

The principal driver underpinning the regional delivery model for NRM (natural resources management) was to 'harness the capacity of those closest to the problem on the ground, building on local knowledge, experience and expertise and enabling flexible and responsive solutions to local NRM challenges' (Senate of Australia, 2000). The key features of the regional delivery model include:

- The development of a framework that sets out the respective NRM roles for commonwealth, state/territory and local governments and the community.
- A shift from the funding of individual projects to funding outcomes determined through regional NRM strategic planning.
- Devolution of decision making to a regional level, i.e. a dispersed rather than a centralist approach, which allows for flexible decision making tailored to local conditions and needs.
- Introduction of national standards and targets to guide and provide direction for investment in NRM.
- A comprehensive accreditation, monitoring and evaluation framework to achieve consistent and acceptable standards of programme delivery.
- Encouragement of community capacity building through involvement in local NRM.

Altogether 56 NRM regions have been established across Australia. The boundaries for each region were agreed by the federal, state and territory governments. In the MDB region, they reflect state regional boundaries created under the legislation. In South Australia, the region has a board established by the state-enacted Natural Resources Management Act, SA 2004, which reports to the SA Minister for Water, whereas in other states the NRM boards report to the Minister for Environment.

The COAG-instigated reforms continued in 2004 with the National Water Initiative (NWI, 2005a) in light of continuing resource management, accounting and pricing issues. The NWI also recognized the importance of including interactions of groundwater and surface water into a more comprehensive accounting framework. This realization was prompted by increases in groundwater use, particularly in NSW, both in response to drought and, some feared, as a response to limitations on surface allocations imposed by the cap (MDBC, 1999; Sinclair Knight Merz, 2003). Interestingly, despite these concerns, recent estimates of cap compliance (MDBC, 2007) show that groundwater usage in NSW (the dominant groundwater user) peaked in 2001–2002 and declined by more than 35% in 2005–2006, with a similar scale of reduction in total use.

The Murray–Darling Basin Water Agreement (MDBWA) was signed at the COAG meeting held on 25 June 2004. The MDBWA set out the arrangements (The Living Murray) for investing \$A500 million over 5 years, commencing 2004–2005, to reduce the level of water overallocation and to achieve specific environmental outcomes in the Murray–Darling basin. The states were unable to agree on a funding formula with the commonwealth, and doubts emerged over the ability to buy-back this volume on the water markets with the funds available. This impasse may have been a key factor in the subsequent promulgation of the 2007 Water Act.

Attempt 3 – breaking with the past

Even before the release of the 2006 CSIRO study ('Sharing water resources of the Murray–Darling basin') dissatisfaction with the slow speed of reform had prompted the commonwealth and state governments to adopt the NWI at COAG's June 2004 meeting (COAG, 2004). A comparison of water management being conducted in the Murray–Darling basin and as projected by NWI reveals two very different philosophies. One is goal orientated in its approach to change (a so-called 'stretch strategy'), while the other is incremental. Stretch strategies accept that the needed capacity may not be available but use target setting to stimulate its development. The oft-quoted example is President Kennedy's decision to put a man on the moon. The incremental approach only sets goals that are already known to be achievable. Approval of the NWI means that a stretch strategy for Australian water management was endorsed at the highest political level. Whether that commitment can be made real has been a continuing question since 2004.

Concern about the state of the Murray–Darling basin was the primary motivation for the introduction of the NWI. In its philosophy and approach, the NWI is fundamentally different from the policy frameworks that have controlled water management in Australia since irrigation was first established in southern Australia in the late 19th and early 20th centuries. For nearly a century, Australian water management had been controlled by public officials and government ministers, who applied an administrative approach to the distribution of heavily subsidized water. During most of this period, it was governments and their officials who led the way in promoting increased water use. As part of the nation-building project that extends back to the mid-19th century, the aim was to use water to create new rural communities. Governments actively sought out people who would use the water made available by the publicly funded storage and distribution systems. Data about the volumes of water available, where it came from and where it was going were important for managers, but except in times of drought there was little concern about overextraction or the need to balance competing priorities.

In NSW, which continued to build irrigation works 20 years after 'full development' had been reached in Victoria, such schemes had considerable autonomy in the way they expanded the use of irrigation water. The result was the ad hoc development of many poorly documented entitlement and distribution systems, which reflected biophysical variations between regions and the idiosyncrasies of local communities and their water managers, who often stayed long term in the same place. Variation in entitlement systems continues today, although there has been a progressive standardization within each state. The reasons for different styles of water entitlement are embedded in the hydrology and storage available. The water allocation system in Victoria was more secure and better accounted for than in NSW: (i) due to a greater degree of regulation (both internally and on the main stem of the Murray); (ii) because most schemes were managed by one overarching body – the Rural Water Commission; and (iii) because a bulk allocation system had been put in place over the top of individual rights, beginning in the 1980s. For a number of decades, this was less of a problem than it might seem. The expansion was occurring during a time (1945–1985) that was significantly wetter than the first half of the 20th century; extractions were still at fairly moderate levels for the early decades in this period; and there was minimal water trading, which meant that inconsistencies between regions were not a significant issue. Nevertheless, as the audit shows, total diversions continued to increase slightly in Victoria in the 1990s, prior to significant reductions in the ensuing drought.

The context for managing water has altered dramatically. More than a decade ago, Australian governments, both Liberal and Labour, at federal and state levels, had undergone a philosophical transition from one of 'nation and community building' to a much tighter focus on the promotion of economic growth. At the same time, there has been an emerging consensus that many hydrological systems are now in serious uncontrolled environmental decline. This development has coincided with an expansion in the number of stakeholders determined to influence water policy, which has forced governments to step back from their previously close relationship with irrigation communities and increasingly adopt the role of arbiter between competing interests.

According to the NWI, the tensions between the many different demands that are placed on hydrological systems are to be managed through the development of comprehensive water plans. These water plans are to include secure water access entitlements, statutory-based planning, statutory provision for environmental and public-benefit outcomes, plans for the restoration of overallocated and stressed systems to 'environmentally sustainable levels of extraction', the removal of barriers to trade, clear assignment of risk for future changes in available water, comprehensive and public water accounting, policies focused on achieving water efficiency and innovation, capacity to address emerging issues, and many more elements (NWI, 2005b). The plans are to provide for the 'adaptive management of surface and groundwater systems' (NWI, 2005b), with their connectivity recognized where it is significant (NWI, 2005b). The states and regions within states have taken up this legislative mandate at different rates, in response to varying combinations of influences. Water-planning processes and the local committees are constructed in different ways, and different areas have individual funding arrangements. In South Australia, for example, all growers and urban users pay a water levy and, although this is small, it does engage the community.

Within each state, apart from South Australia, there are multiple jurisdictions dealing with water, with nine in NSW, incorporating 74 bodies, and seven in Queensland, incorporating 115 bodies. It is partly this sort of institutional mass that NWI architects hold responsible for what they see as the prevailing inertia in water management. However, there are many who have strong reservations about the wisdom and likely effectiveness of a federal administration, not least because of concerns about the skill base and the lack of understanding of fundamental issues and detail at federal level.

Many documents attending the NWI and the National Water Commission have a 'back to basics' enthusiasm that almost implies there has never been water planning in Australia, which is evidently far from the truth. Thus, water plans must take a comprehensive approach to managing hydrological systems in a sustainable way. The NWI explicitly states that the volume of flow needed to maintain environmental sustainability, whatever the level of modification negotiated in developing the plan, must be met before allocations for extraction are determined (NWI, 2005b). A key task is to define the requirements of environmental sustainability, and of the institutions able to ensure that sustainability is achieved and maintained. This gives the debate about the meaning of the concept of 'environmental sustainability' a new urgency. Extrapolating from the Brundtland definition of sustainability and the relevant NWI sections, there seem to be two minimal criteria that need to be met for a hydrological system to be defined as environmentally sustainable: the level of modification must be politically acceptable to society in general, and its environmental condition needs to be stable from a system-wide perspective.

The aim is to introduce a system of water management that will be durable for the long term, but its implementation has been difficult. According to the original policy in mid-2004, all hydrological systems subject to overallocation (a term whose meaning continues to be debated) were required to have water plans in place that would remedy that situation by the end of 2007. The first biennial assessment of the implementation of the NWI was released by the National Water Commission late in 2007. While concluding that progress had been achieved, the Commission made it clear that implementation was well behind schedule, with many difficult issues still outstanding (National Water Commission, 2007).

In addition, water plans must take account of indigenous issues by making arrangements for indigenous representation in water planning 'wherever possible', and provision for indigenous social, spiritual and customary objectives 'wherever they can be developed'. The plans should also include allowance for 'the possible existence of native title rights to water in the catchment or aquifer area' (NWI, 2005b). Water plans are also to provide a common currency that will allow entitlements to be traded from one region to another.

In January 2007, in response to failure to implement remedial programmes in the MDB, the then Prime Minister (John Howard) announced a $A10 billion national water reform package. It called for the states in the MDB to transfer their constitutional powers over water management to the commonwealth government so that a comprehensive basin-wide approach could be introduced. After some negotiation, NSW, Queensland and South Australia agreed, but Victoria refused. Eventually, the commonwealth government passed the Water Act 2007, which introduced the most substantial organizational changes since the reforms of the 1980s, and possibly since the first framework was put in place in 1914–1915. The medium- and long-term future of these proposals is by no means clear, as some of the political momentum behind them was part of a broader challenge to state authority across many other axes, including Aboriginal affairs, education, health and, especially, industrial relations. With a change in government, it is not clear how the agenda for a stronger federal bloc will transform, but it is unlikely to remain as now, especially as many of the 'promises' were not put into law. The succeeding Labour administration actually increased the $A10 billion 'plan' to

$A12.9 billion in April 2008 and, with some further assurances of a greater role and autonomy for the states, succeeded in getting the state of Victoria to agree to defer powers to a federal authority.

The new legislation is less comprehensive than originally intended but not dependent on referred powers. It will impose a sustainability-based cap on both surface water and groundwater, designed to be responsive to the expected impacts of climatic change, a basin environmental watering plan and a number of other reforms. For reasons of political expediency, existing water-sharing plans have been allowed to continue until their dates of expiry (in the case of Victoria this is as late as 2019), even if they are clearly not compliant with the NWI sustainability principles. The Water Act 2007 is the legislative cornerstone of the third attempt to introduce a comprehensive management framework for the region. It will be some years, however, before it can be assessed in terms of its success in introducing sustainable water management in the Murray–Darling basin. It is more likely that success will be achieved through a deft incorporation of state sensibilities, knowledge and capacity than through grand new initiatives. It remains to be seen if a less democratic process results in less innovation (as might be expected) and faster and more responsive progress towards sustainable water management.

The Water Act displaces the MDBC as the apex body with an independent Murray–Darling Basin Authority (MDBA), composed of a chair and four independent part-time members. There will still be a Ministerial Council to provide comment on proposals, but final decisions will be made by the Commonwealth Minister acting on the advice of the MDBA. Key functions of the Authority (see www.environment.gov.au/water/action/npws-act07.html) include:

- Preparing a basin plan for adoption by the minister, including setting sustainable limits on water that can be taken from surface water and groundwater systems across the basin.
- Advising the minister on the accreditation of state water resource plans.
- Developing a water rights information service which facilitates water trading across the Murray–Darling basin.
- Measuring and monitoring water resources in the basin.
- Gathering information and undertaking research.
- Engaging the community in the management of the basin's resources.

A key difference between the MDBA and the MDBC is that the former will be skills-based and not made up of jurisdictional representatives, and that unanimous agreement by all state representatives is no longer required before decisions can be made. The former Commission Office will continue to perform similar work as the technical and institutional coordinating and managing body for the basin, under the direction of the MDBA.

The basin plan will be developed in consultation with the communities, and will accredit state-based plans, but must have the following mandatory content:

- Limits on the amount of water that can be taken from basin water resources on a sustainable basis, known as long-term average sustainable diversion limits. These limits will be set for basin water resources as a whole and for individual sources, and are considerably lower than the limits set by the 1994 cap.
- Identification of risks to basin water resources, such as climatic change, and strategies to manage those risks.
- Requirements that a water resource plan will need to comply with if it is to be accredited under this Act.
- An environmental watering plan to optimize environmental outcomes for the basin by specifying environmental objectives, watering priorities and targets for basin water resources.
- A water quality and salinity management plan, which may include targets.
- Rules about trading of water rights in relation to basin water resources.

The Act establishes a Commonwealth Environmental Water Holder to protect and restore the environmental assets of the Murray–

Darling basin, and also outside the basin where the commonwealth owns water.

Challenges for the Future

The main challenges for the future concern the best way to reduce overall allocation in the basin and, more importantly, how to get all the states to comply with a significant reduction in water availability compared with recent history. It is no longer merely a question of complying with the 1994 cap, but of adjusting to significantly reduced allocations for the irrigation sector. The pressure to do this is mostly driven by current environmental allocation concerns, plus the expectation of reductions in mean annual flow of the order of 20–30% by 2100 under a range of climatic change scenarios (CSIRO, 2007), illustrated for Victorian rivers in Fig. 12.6. Much of the political rhetoric justifying the recent federal takeover of water management has been fuelled by a long and historically unprecedented drought, which has allowed the attribution of many woes to bad management rather than to a change of circumstances. However, the possibility that those changed circumstances represent the future is one that will make very significant changes to water use and management.

The institutional challenge is whether a more active and dominant role by the central government will deliver the expected results. Multilateral, negotiated and voluntary water sharing and custodianship of the basin have been noisy and slow to react, but innovative in many of their solutions. It can be argued that, in order to be acceptable to all parties, only innovative solutions, such as the Salt Credit Scheme and the cap (in its first incarnation), will emerge. Against that, it is clear that more painful adjustments, such as a revised cap, will prove difficult to negotiate and see through, especially in the time-frame now expected. The belief of the federal government is that it has the intellectual horsepower, political muscle and financial resources to succeed where it and others believe the MDBC/MDBMC has failed. This is probably a belief that is common to many central government elites, and their immediate technocracies, and often leads to impatience with detail and the preservation of considerable secrecy and minimal transparency.

At this stage, the process is too new and too

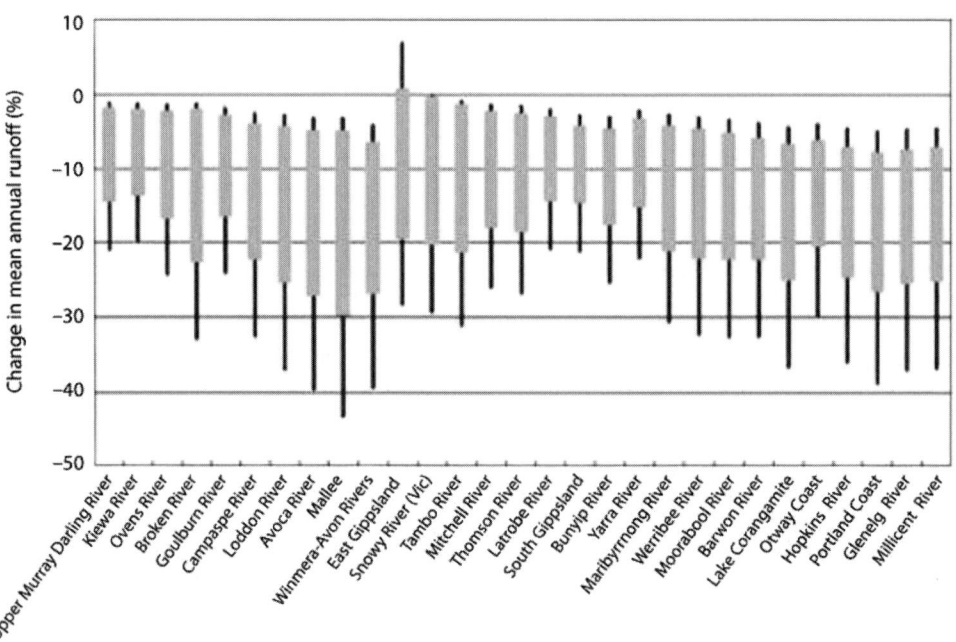

Fig. 12.6. Expected range of reductions in mean annual streamflows in 2030 in Victorian rivers, using an ensemble of ten global climate models (DSE, 2006a).

young for any comparisons to be made, but there are fears of declining skill and knowledge at large in the water sector and there is a need to hire many experienced and technically proficient staff at the central level. Logically, there are useful templates from recent history, through the commonwealth's brokerage of innovations such as the Salt Credit Scheme and its insistence on implementing the National Competition Policy, through a mix of 'carrot and stick' approaches associated with the tranche payments of federal tax revenue back to the states under COAG. It seems likely, given the past pragmatic history of Australian water management, that the states will ultimately retain more autonomy and responsibility in management of the MDB, with perhaps more forceful (and better informed) guidance and incentives from the centre.

Within agriculture itself, it is likely that existing adaptations to climatic variability will be needed more frequently and will also have to be developed further to adapt to the expected reductions in water availability. The prospect of higher evaporative demand and lower water availability is expected to lead to declining productivity (DSE 2006a; CSIRO, 2007).

Recent upturns in the real prices of commodities are already sending strong signals to the agriculture sector in Australia, which returns smaller farms to profitability. However, it remains to be seen whether this will stop the trend to larger and more intensive holdings, and to a more diversified and market-niche-oriented irrigation sector.

Australia, more than most countries, will face a daunting challenge to increase productivity in agriculture with smaller and more variable water supplies in the wake of climatic change. An ABARE study (Beare and Heaney, 2002) predicted significant reductions in the volume and value of irrigation-sector outputs in the MDB, which would not be mitigated effectively by increased water-use efficiency or by water trading, although water trading was modelled to be the better adaptation of the two.

A more fundamental structural challenge to the water sector is likely to emerge in the need to reformulate water entitlements as the impacts of climatic change unfold. Reduced water availability implies reduced allocations and further changes to inter-annual storage management in dams if the reliability of supplies to farmers is not to be seriously compromised. The current system is based on a sensible proportional allocation, which responds well to variations in water supply that follow inter-annual climatic variability. However, the hydrologic basis for the current entitlements is changing, and if emerging theories on a step change in climate over south-east Australia are correct (Kirby et al. (CSIRO) 2006; CSIRO, 2007) the need will be strengthened. The proposals for a lower cap, implemented by the MDBA, will reduce total available allocation, which may be further reduced once climatic-change-induced reductions in water availability become clearer. At the moment, the proposals to conserve water and reduce allocations in Victoria do not fully account for expected climatic-change reductions in water availability, and are likely to be further revised in the future, in the light of experience with conserving water through projects such as 'Foodbowl'. One of the requirements of the new Water Act is that the entitlement formulations in different states and regions should be unified. This proposal has obvious merit, but it is also clear that different formulations of entitlement have emerged to meet local conditions and needs, and that unification beyond a certain point is likely to be counter-productive.

While scientists, policy makers and farmers are struggling to understand and agree upon current environmental water allocations, climatic change will further reduce water availability for natural systems. Even harder decisions will have to be made to assess and reserve allocations for environmental flows if climatic change turns out as predicted.

The only emerging urban water transfer in the basin is for Melbourne, and planning seems to be well in hand to ensure reliable water supplies through a transfer from within the basin, over the 'Great Dividing Range' to Melbourne (DSE, 2006b). In economic and volumetric terms (75 Mm^3), this is almost a 'no-contest' from a within-basin perspective, and the capital costs of transfer ($A700 million for the pipeline and pumping stations) and compensation for entitlement ($A300, to be used for water conservation) are to be paid by Melbourne Water to Goulburn Murray Water as the bulk entitlement holder.

Conclusion

From the perspective of improving the arrangements for water resources management in developing countries, there are many useful lessons available from the past experiences in the MDB, despite, and because of, conflicts being experienced in response to overuse, inadequate environmental allocation and the emerging stress of climatic change. It is clear that there are strong benefits to having a well-accounted water allocation system that internalizes natural hydrological variability. But it is also clear that this sound basis has not prevented the MDB from getting into trouble, nor does it necessarily help in implementing solutions. It is certainly imperative to know enough about the dynamics of water availability and use for managers and policy makers to know the extent, likely impact and possible adaptive strategies to climatic change, and to balance productive use with sustainable ecosystems. Achieving agreement and consensus on how to do this is a different question altogether.

Innovative solutions to a succession of water resources management problems have emerged in Australia, sometimes with a lot of noise, debate and antagonism, but also with an underlying sense of pragmatism and recognition of wider issues. Public information has been at the heart of the debate, and science and research have helped guide and constrain political options within the realms of the practical and the achievable.

It makes great sense not to overallocate a water resource system. The emerging challenges for Australia, and the MDB in particular, with comparatively good information and relatively small numbers of well-endowed stakeholders, point to severe problems for over-allocated basins elsewhere. The task and cost of reducing allocations and usage are politically daunting, even more so where millions of poor are dependent on an increasingly scarce resource and where basins reach across national boundaries.

Australian water management is evolved and complex in comparison to those in most other countries, although that is often not realized or celebrated within the country. It is constantly evolving and adapting to changing needs, biophysical influence and public expectation. Recently, there has been heightened concern about the use of groundwater, and inadequate accounting of surface water and groundwater interactions, which have implications for major groundwater users, such as India and China, whose agriculture is much more dependent on groundwater than that of 'down under'.

Australia is entering a new phase of potentially top-down water administration, after nearly 50 years of increasingly bottom-up initiatives. The commonwealth has moved from strategic funding – through sponsoring water reforms over the top of those undertaken within individual states, and supporting bottom-up natural resource management initiatives, such as Land Care, CMAs and other community-based efforts – to a position of ultimate control and responsibility. It is not without irony that this should happen at the same time as the better aspects of Australian water management (devolution, voluntary consensus, technical soundness and open information) are promoted elsewhere. It remains to be seen what lessons this new policy direction will offer.

Notes

1 ICM (integrated catchment management) in this context is at a lower level of scale than river basin management and could be said to have a stronger focus on land than on water.
2 One of the perennial incentives for interstate cooperation has been federal funding for capital works. Interception of mostly natural saline in-flows to the river was one of the capital-funding levers applied in return for agreement to limit the salt discharge from each state – known as a salinity credit. Each state manages the salinity loads within the framework of its total credit. This measure was designed to leave the states free to prioritize salinity management strategies within their jurisdiction. The effluent from salt interception is pumped to large evaporation pans: there has been perennial interest and some commercial activity in salt production and marine fisheries at some of these sites.
3 Corporatization involves the development of private-sector business practice within a state-owned or partially state-owned enterprise. In Victoria, the irrigation systems are required to pay a dividend to the state government each year, as well as cover all operation, maintenance and future capital investment needs.

References

ABARE (Australian Bureau of Agriculture and Resource Economics) (2007) *Australian Commodity Statistics, 2007.* Canberra, Australia.
ABS (Australian Bureau of Statistics) (1998) *AgStats, ver. cat. no. 1353.0, ABS.* Australian Bureau of Statistics, Canberra, Australia.
Arthington, A.H. and Pusey, B.J. (2003) Flow restoration and protection in Australian rivers. *River Research and Applications* 19, 377–395.
Beare, S. and Heaney, A. (2002) *Climate Change and Water Resources in the Murray Darling Basin, Australia, Impacts and Possible Adaption.* World Congress of Environmental and Resource Economists, Monterrey, California, ABARE Conference Paper 02.11, Canberra, Australia.
Bellamy, J., Ross, H., Ewing, S. and Meppem, T. (2002) *Integrated Catchment Management.* CSIRO Sustainable Ecosystems, Canberra, Australia.
Bjornland, H. and McKay, J.M. (2002) Aspects of water markets for developing countries: experiences from Australia, Chile and the US. *Environment and Development Economics* 7(4), 769–795.
Blazey, P. (1972) *Bolte: a Political Biography.* Jacaranda Press, Milton, Queensland.
Bruntland, G. (ed.) (1987) *Our Common Future: the World Commission on Environment and Development.* Oxford University Press, Oxford.
Cape, J. (1997) Irrigation. In: Douglas, F. (ed.) *Australian Agriculture: the Complete Reference on Rural Industry.* Morescope Publishing, Melbourne, Australia, pp. 367–374.
Clark, S. (1983) Inter-governmental quangos: the River Murray Commission. In: Curnow, G.R. and Saunders, C.A. (eds) *Quangos, the Australian Experience.* Hale & Iremonger, Sydney, pp. 154–172.
COAG (Council of Australian Governments) (1992) *National Strategy for Ecologically Sustainable Development.* Australian Government Printing Service, Canberra, Australia.
COAG (1994) *Water Resources Policy.* Report of the Working Group on Water Resources Policy. Council of Australian Governments, Canberra, Australia. Full documentation available at www.coag.gov.au
COAG (2004) *National Water Initiative.* Government Press, Canberra, Australia.
Commonwealth Parliament (1917) *River Murray Waters Act, 1915 (No. 8). Government Gazette,* 17 January 1917.
Commonwealth Parliament (1993) *Murray-Darling Basin Agreement (1992) Act 38.*
Connell, D. (2002) Lake Victoria cultural landscape plan of management. CD published by the Murray–Darling Basin Commission, Canberra, Australia.
Connell, D. (2007) *Water Politics in the Murray–Darling Basin.* Federation Press, Sydney.
CSIRO (Commonwealth Scientific and Industrial Research Organisation) (2001) *Australian Dryland Salinity Assessment 2001.* Canberra, Australia.
CSIRO (2007) *Climate Change in Australia.* Commonwealth Scientific and Industrial Research Organisation and Bureau of Meteorology, Canberra, Australia.
Cullen, P., Whittington, J. and Fraser, G. (2000) *Likely Ecological Outcomes of the COAG Water Reforms.* Cooperative Centre for Freshwater Ecology, Canberra, Australia.
DLWC (Department of Land and Water Conservation) (New South Wales) (1998) *Water Sharing, the Way Forward.* Department of Land and Water Conservation, Sydney.
DSE (Department of Sustainability and the Environment), Victoria (2004) *Securing Our Water Future Together.* Victorian Government, Melbourne, Australia.
DSE, Victoria (2006a) Climate Change – What if the Next 10 Years are the Same? Power Point Presentation. Department of Sustainability and the Environment, Nicholson Street, Melbourne, Australia.
DSE, Victoria (2006b) *Sustainable Water Strategy.* Central Region Discussion Paper. Department of Sustainability and the Environment, Nicholson Street, Melbourne, Australia.
Eaton, J.H.O. (1946) A *Short History of the River Murray Works.* Stevenson, K.M., Government Printer, Adelaide, Australia, pp. 21–41.
Etchells, T., Malano, H., McMahon, T.A. and James, B. (2004) Calculating exchange rates for water trading in the Murray–Darling basin, Australia. *Water Resources Research* 40(12), W12505.
Haismann, B. (2004) *Murray Darling River Basin Case Study.* Background Paper. World Bank, Washington, DC.
Hallows, P.J. and Thompson, D.G. (1995) *The History of Irrigation in Australia.* Australian National Committee on Irrigation and Drainage, Mildura, Australia.
Hammerton, M. (1986) *Water South Australia: a History of the Engineering and Water Supply Department.* Wakefield Press, Adelaide, Australia, pp. 240–242.

Kemper, K., Dinar, A. and Blomquist, W. (2005) *Institutional and Policy Analysis of River Basin Management Decentralization. The Principle of Managing Water Resources at the Lowest Appropriate Level – When and Why does it (Not) Work in Practice?* World Bank, Washington, DC.

Kingsford, R.T. (2000) Ecological impacts of dams, water diversions and river management on floodplain wetlands in Australia. *Austral Ecology* 25(2), 109–127.

Kirby, M. et al. (CSIRO) (2006) *The Shared Water Resources of the Murray–Darling Basin.* MDBC Publication No. 21/06. Murray–Darling Basin Commission, Canberra, Australia.

Maheshwari, B.L., Walker, K.F. and McMahon, T.A. (1995) 'Effects of regulation on the flow regime of the River Murray, Australia'. *Regulated Rivers: Research & Management* 10(1), 15–38.

McMahon, T.A. and Mein, R.G. (1986) *River and Reservoir Yield.* Water Resources Publications, Colorado.

MDBC (Murray–Darling Basin Commission) (1987) *Murray–Darling Basin Environmental Resources Study.* Murray–Darling Basin Commission, Canberra, Australia.

MDBC (1995) *An Audit of Water Use in the Murray–Darling Basin.* Murray–Darling Basin Commission, Canberra, Australia.

MDBC (1996) *Cost Sharing for On-ground Works.* Murray–Darling Basin Commission, Canberra, Australia.

MDBC (1999) *Salinity and Drainage Strategy.* Murray–Darling Basin Commission, Canberra, Australia, pp. 2–3.

MDBC (2002) *Lake Victoria Cultural Landscape Plan of Management.* Murray–Darling Basin Commission, Canberra, Australia.

MDBC (2007) *Water Audit Monitoring Report 2005–6.* Murray–Darling Basin Commission, Canberra, Australia.

MDBMC (Murray–Darling Basin Ministerial Council) (1987) *Murray–Darling Basin Environmental Resources Study.* Murray–Darling Basin Commission, Canberra, Australia.

MDBMC (1988) *Salinity and Drainage Strategy.* Murray–Darling Basin Commission, Canberra, Australia.

MDBMC (1989) *Background Papers: Murray–Darling Basin Natural Resources Management Strategy.* Murray–Darling Basin Commission, Canberra, Australia.

MDBMC (1990) *Natural Resources Management Strategy, Murray–Darling Basin.* Murray–Darling Basin Commission, Canberra, Australia.

MDBMC (1995) *An Audit of Water Use in the Murray–Darling Basin: Water Use and Healthy Rivers – Working towards a Balance.* Murray–Darling Basin Commission, Canberra, Australia.

MDBMC (1996) *Setting of the Cap.* Murray–Darling Basin Commission, Canberra, Australia.

MDBMC (1999) *Water Audit Monitoring Report 1997/98.* Murray–Darling Basin Commission, Canberra, Australia.

MDBMC (2000) *Review of the Operation of the Cap.* Murray–Darling Basin Commission, Canberra, Australia.

National Competition Policy (1994) Library of the Australian Parliament. www.aph.gov.au/library/intguide/econ/ncp_ebrief.htm

National Land and Water Audit (NLWA) (2001) *Australian Water Resources Assessment: Surface Water and Groundwater – Availability and Quality, 2000.* Land & Water Australia, on Behalf of the Commonwealth of Australia, Canberra, Australia.

National Water Commission (2007) *First Biennial Assessment of the National Water Initiative.* National Water Commission, Canberra, Australia. www.nwc.gov.au.

NWI (National Water Initiative) (2005a) South Australia implementation plan. www.nwc.gov.au/nwi/docs/SA%20NWI%20Implementation%20Plan.pdf

NWI (2005b) An intergovernmental agreement on a national water initiative. Available at http://www.nwc.gov.au/www/html/117-national-water-initiative.asp

SA (South Australian) Government (2005) NWI – South Australian implementation plan www.nwc.gov.au

Senate of Australia (2000) www.aph.gov, au/senate/committee

Sinclair Knight Merz (2003) *Projections of Groundwater Extraction Rates and Implications for Future Demand and Competition for Surface Water.* Murray–Darling Basin Commission, Canberra, Australia.

Turral, H.N. (1998) *HYDRO-LOGIC? – Reform in Water Resources Management in Developed Countries with Major Agricultural Water Use: Lessons for Developing Countries.* ODI Research Study. Overseas Development Institute, London.

Turral, H. and Fullagar, I. (2006) Institutional directions in groundwater management in Australia. In: Giordano, M and Villholth, K. (eds) *The Agricultural Groundwater Revolution: Opportunities and Threats to Development.* CAB International, Wallingford, UK, pp. 320–361.

Turral, H.N., Etchells, T., Malano, H.M.M., Wijedasa, H.A., Taylor, P., McMahon, T.A.M. and Austin, N. (2005) Water trading at the margin: the evolution of water markets in the Murray–Darling basin. *Water Resources Research* 41(7), 88–93.

van Dijk, A., Evans, R., Hairsine, P., Khan, S., Nathan, R., Payday, Z., Viney, N. and Zhang, L. (2006) *Risks to the Shared Water Resources of the Murray–Darling Basin*. Murray–Darling Basin Report, Canberra, Australia.

Wijedasa, H.A., Turral, H. and McMahon, T.A. (2001) *Analysis of the Effects of Water Policy Changes to the River Diverters in the Murrumbidgee Valley*. Final Report. Centre for Environmental Applied Hydrology, University of Melbourne, Australia.

Williams, J. and Goss, K. (2002) Our difficult bequest: the collision of biophysical and economic reality, cultural values and public policy. In: Connell, D. (ed.) *Uncharted Waters, Murray–Darling Basin Commission*. Canberra, Australia, pp. 37–48.

Williams, J., Walker, G.R. and Hatton, T.J. (2002) Dryland salinisation: a challenge for land and water management in the Australian landscape. In: Haygarth, P.E. and Jarvis, F.E. (eds) *Agriculture, Hydrology and Water Quality*. CAB International, Wallingford, UK, pp. 457–475.

Index

Page numbers in **bold** refer to illustrations and tables

Abstractions
    control 192, 279–282, 286, 287
    drivers 186, 224, 225
    dry-season excess 182
    evolving 191
    excessive 188
    limits 26, 192
    methods 162
    rates 190
    results 257
    sources 266, 268
    unregulated 190
    see also Withdrawals
Access 28, 47, 60, 63–66, 153
    see also Rights
Actors
    Jordan River basin 38, 39, 41, 42, 43
    Lerma–Chapala basin 88, 90
    see also Stakeholders
Adaptation 115–116, 283
Africa, Sub-Saharan, non-equilibrium river basin systems 172
Agencies 226–227
Agreements 28, 43, 129–131, 221, 269–271
Agribusinesses 14, 224
Agriculture
    allocations within 13–14
    Australia 265
    Colorado River basin 137
    compensations 37, 40, 95, 140, 265
    contribution, Indian economy 218
    development, Jordan valley 23, 24
    double-squeeze 16–17
    El Haouareb Dam area 159–161
    evolving role 136–137
    fundamental transformation 141
    Great Ruaha River 176
    growth 5
    intensification 246–250, 257, 259–260
    large-scale 66–67
    low value perception 17, 37
    Merguellil basin 156–157
    modernization 5, 183, 202–203, 226
    Olifants River basin 49–50
    prestige 27, 38
    productivity 287
    profitability decline 27–28
    rainfed 157, 159, 219
    role, water management 140
    subdivision failure 67
    supply variation 211
    surplus expropriation 103
    transformations, India 218
    Tunisia 151–152
    types 14, 49
    veneration 103
    water demand 165
    water policies inclusion 55
    water use 231, 246–250, 257, 259–260
    see also Crops; Irrigation
Agro-climatic zones, Tunisia 147
Agro-pastoralism 51
Ain Ghazal, Jordan 23
Al Ghor 21
Al Zhor 21
Algal blooms 274, 278
Alliances 51, 52
Allocations
    administrative tinkering **279**
    bases 30
    changes, political decisions 255
    city versus agriculture 204–205

Allocations – *Continued*
  concession titles 89
  control 211
  dispute 81
  federalization 79–80
  increase 257
  installations' impact 161–163
  integrated approach 231
  irrigation expansion control 278
  KWDT award, technical limits 229
  local adjustments 229–230
  low-flow years management 229
  management 190
  mechanisms 89, 206–207, 231
  Murray–Darling basin **266**
  optimized 93
  perceptions 66, 257
  policies 89, 227–230
  priorities, non-agricultural 54, 55, 205
  procedures agreement, Krishna River basin 221
  proportional 287
  provisional 228
  reductions 10, 27, 92, 286
  rules 210, 260
  rural household 65–66
  scarcity response option 34
  strategies, India 232
  surface water 89, 90, 92–93
  systems, Murray–Darling basin 283
  water sufficiency management solution 173
  water use efficiency rationale 184–185
  Yellow River basin 114, 115, 118
  see also Reallocations
Alto Río Lerma Irrigation District (ARLID) 83–84, 90
Andhra Pradesh 220, 221, 226, 227, 228
Annals of History (SHiji) 102
Apartheid 66–67
  see also Differentiation; Segregation
Apportionment 129–133, 138, 278
  see also Allocations; Shares
Appropriation, prior rights 127–128, 129, 136–137, 206, 228
Aquifer, importance 206
Arabie dam (Flag Boshielo dam) 55, 56, 60
Arizona 132, 133–134, 139
Associations 27, 88, 92, 154
  see also Water Users Associations
Atequiza–Las Pintas aqueduct 87
Audit, water use 278
Augmentation
  approaches 224
  basin closure response option 10, 38, 95
  benefits 6
  incentives 139
  mechanisms 139
  options 38, 139–140

  projects 42, 230
  scarcity responses 34, 35–36, 38
  strategies 140, 232
  see also Reservoirs; Transfers; Waste-water, treated
Australia 263–288
Australian Bureau of Agricultural and Resource Economics (ABARE) study 287
Australian Capital Territory, Australia 266, 273

Babak, Ardeshir Ibn 198
Badia 22
Bajío 79
Balances
  categories 217, 219
  economic efficiency 224
  equity 60, 61
  expression methods 32–33
  finger diagrams 34
  flow imbalance 161
  supply and demand, (im)balance 31
  terms evolution 32–33
  Yellow River basin role 101
  see also Water accounting
Ballesteros, Luis P. 81, 85
Barrages 271
Basin, units concept 1, 85
Bedouins 24, 28, 29, 40, 41, 42
Behesh Abad tunnel 201
Benefits
  augmentation 6
  distribution 33–40
  equity 63
  sharing 60, 69
  shifts 210, 211
  water import/export 209
Betterment Schemes, Tomlinson Commission 55
Bhavani River basin Southern India 238–260
  see also Lower Bhavani Project
Bhima 217, 220, 226
Black Economic Empowerment 69
Blue water 32–33, 42, 79, 161
Boards 54, 59, 67
Boers 50–51, 52
Boulder Canyon Project Act (1928) 130
British influence 23–24, 51, 52, 53, 226
Brundtland report (1987) 280, 284
Bunds 162, 163
Bureau of Reclamation 136
Buying 10, 253, 260
  see also Trading

Cajas de agua 79
Canals
  Bhavani basin 242, 244, 247–249, 257
  central control rationale loss 105
  China 104, 109

diversions 240, 241, 253
  Jordan 24–25
  Lerma–Chapala Basin 84, 87
  releases 252, 253
Capital 164
Capitalism 13–14, 52
Caps 192, 279–282, 286, 287
Catchment
  assessment 162–163
  management
    agencies 15–16, 57–61, 69
    authorities 281
    forums 58–59, 67
    integrated 272–273, 274
Cauvery River basin 238, 240, 252, 254–255
  see also Bhavani River basin
Central Arizona Project (CAP) 132, 133, 134, 139
Centralization
  India 230
  Lerma–Chapala basin 79, 80, 82, 85
  local autonomy tensions 119
  North China Plain management 106
  Yellow River control 105
  see also Federalism
Chadegan reservoir 198, 201–202, 203, 204–205
Challenges, Yellow River basin 111–114
Change 4–6, 16, 110–118
Channel, shifts 101, 102, 105
Charges 50, 190
  see also Prices
China 99–120
Ciétenega de Chapala (Lake Chapala marsh) 81, 85
Civil Code, Iran 206
Clark, Sandford 270
Climate
  agricultural adaptation 287
  Australia 264
  change 124, 165, 287
  equilibrium 190
  hydrological variability confusion 165
  Jordan 25
  Kairouan 148–149
  Krishna River basin 215
  Merguellil River basin 148–149
  non-equilibrium 190
  Oliphants River Basin 49
  Tamil Nadu 239
  Tunisia 147
  vagaries, agriculture vulnerability 37
  variability 171–172, 287
  Zayandeh Rud River basin 197
Closure
  Bhavani basin 238–260
  competition induction 3

  defined 2–3, 196, 214, 232
  drivers 232
  externalities 16
  hydraulic mission relationship 75–96
  implications, multiple scales 258
  interconnectedness induction 20
  Krishna Basin **216**, 223–230, 232
  Lerma–Chapala basin 89–93, 95
  lower Jordan River basin 20, 31–34, 42
  overexploitation relationship 77–79
  prevention failure 229, 231
  process **3**
  response options 11–12, 37–38, 95, 115–117, 223–230
  results 3–4
  river basin management difficulties 13
  Wadi Merguellil basin 168
  Yellow River basin 118, 120
  Zayandeh Rud River 196, 206–210, 211
  see also Depletion

Coefficients 63–66, 100, 166, 232, 268
Coimbatore city 239, 243, 255
Collaboration 106, 107, 136
  see also Cooperation
Colonization 153, 168
  see also British influence; Europeans
Colorado River basin 123–143
Colorado River Basin Project Act 133–134
Colorado River Basin Salinity Control Act (1974) 133
Colorado River Compact (1922) 126, 129–130, 141
Colorado River Storage Project (CRSP) 133
Colorado–Big Thompson Project 133
Comisión Estatal de Agua de Guanajuato (CEAG) 93, 94
Comisión Nacional de Irrigación (CNI) 82–83, 84–85
  see also Secretaría de Recursos Hidráulicos (SRH)
Comisión Nacional del Agua (CNA), Mexico 88–90, 91, 92, 94
Comité de Defensa del Lago Chapala (Committee for the Defence of Lake Chapala) 87
Comités Técnicos de Aguas Subterráneas (COTAS), Mexico 93–94
Command area development (CAD) programmes, India 226–227
Commonwealth of Australia 264
Commonwealth Environmental Water Holder establishment 285
Commonwealth Scientific and Industrial Research Organization (CSIRO) 280
Communism, China 107
Communities 274, 285
  see also Indigenous people; Participation; Tribes

Community Advisory Committee to the Ministerial Council, Murray–Darling basin 273
Compacts, Colorado River 126, 129–131, 138, 141
Competition
    closure induced 255, 258
    farmers increased fears 257
    Great Ruaha River 171–192
    Jordan 42
    manifestation 41
    National Competition Policy (1994), Australia 280, 287
    Olifants River basin 50–51
    overcommitment result 228
    results 196
    water export induction 161
    see also Conflicts; Upstream versus downstream
Concessions 80, 82, 83, 89
Conflicts
    allocations 37, 92
    Bhavani basin 257
    causes 83–84
    farmer 257
    Great Ruaha Basin 175, 176
    hydropower/irrigation 175
    Krishna River basin 221, 228, 232
    Lerma–Chapala Basin 83–84
    lifting 253
    Madras–Mysore Dispute 254–255
    reallocations 184, 204, 253
    resolution mechanisms 89
    sharing 221, 228
    transfers 86, 90, 91–92
    upstream; downstream 175, 203–205, 222
    see also Competition; Legislation; Upstream versus downstream
Confucianism 5, 103, 104, 119
Conquests 51–52
Conservation
    Acts 53
    Commissions 16, 106, 108, 109, 115, 118, 119–120
    costs 40
    ecosystem 89
    environmental water 280
    limited prospects 42
    measures 27, 227
    mechanisms 139
    Office, China 104
    policies 12, 27
    programmes 12
    soil and water 151, 154–155, 161, 162–163, 164–166, 168
    species habitat plan 126
    strategies 140, 224–227
    technologies promotion **279**

water 173, 280
see also Savings
Constitutional Amendment, 73rd. (1993), India 230
Controversy, Lake Victoria 276
see also Conflicts
Conversion, volumetric **270**
Cooperation 106–107, 109–110, 119–120, 263
Coping, strategies, farmers 209, 229
Corporization 280
Corrales dam 83, 85, 86
Cost-benefit analysis 2, 389
Costs
    demand, unmet 62
    distribution, Jordan 33–40
    identification 2
    political 42
    recovery 27, 37, 69, 280
    scarcity response options selection 34–35
    sharing 269
    shifting 38, 40
    social and environmental 5
Cotton industry 268
Council of Australian Governments (COAG) 265, 280, 281, 282
Councils 16, 89, 90–91, 93
Credit 164, 225
Crises, Lake Chapala 10–11, 82–88, 93–94, 95
Croplands 152, 242, 243–246
Cropping
    double 201, 248, 255
    intensity 217, 243, 250, 260
    large-scale 52
    patterns 91, 159–160, 217, 250, 259
    regulations violation 247–248
    seasons 244
    systems 157
Crops
    Bhavani River basin 239, **244**, 247, 248, 249, 250
    choices alteration 250
    Colorado River basin 137
    exports 50
    Gobichettipalayam taluk **249**
    high-value 38, 50, 53–54
    increase, perennial 248
    India **218**
    irrigated, percentage, Houfia sector 159
    Jordan 22, 23, 25, 26, 27, **31**
    Krishna River basin 217, 227
    low-value 27, 37, 38, 42, 152, 162
    Murray–Darling basin 268
    Sathymangalam taluks **249**
    system of rice intensification 227
    tabias 162
    Tanzania 183

Tunisia 152
Usangu basin 177
Wadi Merguellil basin 157, **158**, 159–160
water-intensive 250, 259
yields increase 139–140, 162, 250
see also Croplands; Cropping
Cuesta-Gallardo, Manuel 81
Culture, China 99, 103, 104, 110, 119

Dams
   Bhavani River basin 243
   China 109
   Colorado River Basin 124, 125, 130, 133
   flood control role 155
   Great Ruaha River 182–183
   Krishna River basin 227, 230
   Lerma–Chapala basin
      Arcediano dam 92–93
      Corrales dam 83, 85, 86
      map, main dams **84**
      Solís dam 84, 87–88, 91, 95
   lower Jordan River basin 35
   Merguellil basin *see* El Haouareb dam
   Murray–Darling basin 265, 267
   Olifants River basin **48**, **53**, 54, **55**, 56, 60, 69
   Santiago River 86, 92–93
   Yellow River basin 110, 114, 117
   Zayandeh Rud River basin 202
   see also Storage
Darling River 265–266
   see also Murray–Darling basin
Dead Sea 25, 31, 32, 42
   see also Red-Dead project
Decentralization 90, 108, 110, 118, 230, 271
Decision making 2, 14–15, 389
Degradation
   awareness 5, 119
   causes 273–274
   impacts 222
   limiting 230
   livestock role 163
   progressive 182
Demand
   changing 256
   crash 189
   exceeding supply 201
   increase 138, 229, 248, 255
   management 88–89, 115, 173, 226, 227, 232
   projections 62–63
   tensions 283
   see also Supply and demand
Demand–supply equation 171
Democracy 56, 66–67, 68–70
Department for International Development (DFID), UK 179

Department of Water Affairs (DWA), South Africa 53
Department of Water Affairs and Forestry (DWAF), South Africa 48, 57, 58, 60, 61, 67–69
Depletion
   aquifer, Bhavani basin 256–257
   beneficial 32, 33, 178
   blue water 79
   causes 33–34, 42, 112, 219, 221, 229
   Colorado River basin 137
   curve, bending down 89–90
   increase 112–113, 218
   inflows, Krishna River basin 217
   non-beneficial 32, 178–179
   processes 32
   results 191
   see also Closure
Depoliticization, Jordan 29, 30, 41
Desalination 36, 42, 167
   see also Salinity
Development Trust and Land Act (1936) 52
Differentiation 51, 52, 53
   see also Segregation
Dirección de Aguas, Secretaría de Agricultura y Fomento (SAyF), regulations 84
Discharge
   alteration 181
   decrease **216**, 217, 218, 230
   determination 204
   flows 198
   monthly, Krishna Basin 219
   sewage 254
   unintended 126, 253
   wells impact 203
Discrimination 52
   see also Apartheid; Differentiation; Segregation
Diseqilibrium 171–172
Disi project 36, 42
Dispensation, democratic 68–70
Disputes 81, 254–255
Diversions
   cancellation result 249
   capacity 276, **277**
   exceeding use 276–**277**
   increase 278
   Lake Tiberius 22–23
   limits 285
   schemes, Godavari basin 230
   sustainability 285
   works programme 271
   see also Irrigation; Transfers
Drainage 80–81, 85, 86, 254, 275, 281
Drinking-water 86, 255, 258, 260
Drip-irrigation 160, 167, 227
   see also Irrigation

Droughts
    Australia 267
    Krishna River basin (2001-2004) 214, 223
    Lerma–Chapala basin 77
    Murray–Darling basin 278
    rights held in reserve for 278
    shortage sharing 138
    vulnerability 209
    see also Scarcity
Dry periods, solutions (irrigation) 103
Dry-season 187, 188
Dublin Principles 6, 14
Dushui, (Office of the Director of Water
        Conservancy, China) 104
Duty, water 190
Dynamics, rural 214–234

Ecology 38–40, 61, 125
Economy 110–111, **218**
Ecosystems 5–6, 89, 126, 214, 222
Efficiency
    balances, economic 224
    determinants 164
    development policies 14, 224
    end-use 27
    gains 207–208
    increase prospect, soil moisture 219–221
    irrigation 32, 42, 184–185, 191, 207
    paradox 140
    river basin 12
    water-use 32, 89, 258
Effluents 25, 254
Ejido (common property) 82, 83
El Haouareb dam
    agriculture water use 157, 159–161
    flood protection role 155
    flows assessment role 157, 159
    inflow decrease 168
    location 148
    releases 151, 167
    sedimentation 166
    water levels 149–150
    water losses 166
Electricity
    agriculture use **247**, 248, 259
    connection endorsement 249
    control, government 255
    costs 246, 254
    dependency 260
    free 238, 246–247, 253, 258–259, 260
    generation 182
    Operation Disconnection (1980-1982) 246
    power cuts 11, 181
    rationing 186
    tariff rates 259
    use limitation, water-pumps 250
    see also Hydroelectricity; Hydropower

Elevation, river beds 101–102
Employment 52, 54, 56, 65, 239, 256
    see also Labour
Endangered Species Act (1973) 134–135
Energy 54, **247**, 258–259
Engineering
    approach 42, 230
    hydraulics 106, 107, 108–109
    schemes 102, 117
    solutions, promotion 119
    training institutes, China 106
    water management tool 116–117
Entitlements
    executive authority lack 278
    formal quantification need 229
    overallocation 7–8
    reformulation 287
    sharing rules 269, **270**
    South Australia problems 271
    surplus water 221
    systems variation, Australia 278, 283
    trading 281
    transfers **279**
    see also Rights
Environment
    allocations 263, 264
    changes 181–182
    closure impacts 251
    concerns 273–274
    costs 5
    demand 173
    federal movement 134–136
    flows
        amounts, Cauvery Tribunal decision 255
        implementation 222–223
        management 276–277
        release requirements 256
        requirement 61–62, 132, 287
    human regulation consequences 254
    hydrograph modification consequences 125
    impacts mitigation 263
    irrigation, negative impacts 278
    plan 285
    preservation 223
    protection 61, 132
    reallocations 9–10, 263
    restoration 126, 137–138
    sustainability 62, 284
    trade-offs 61–62
    water use 114, 222–223
    see also Degradation
Environmental Impact Statement (EIS) 139
Environmentalism 133, 134–136, 184–185
Equilibrium behaviour **172**, 175, **189**, 190
    see also Non-equilibrium behaviour
Equity 14, 56–63, 65, 68, 224
Erode district, Bhavani basin 239, 243–244

Erosion 101, 148, **162**, 166, 182
Esfahan, Iran 16, 196–212
European Union, Water Framework Directive 6
Europeans, influences 126–127, 265
Evaporation
    depletion process 32
    exceeding rainfall 100, 215
    falling inflow factor 252
    losses 76, 157, 163
    reduction 275
    unproductive, limiting 207
Evapotranspiration 32, 49, 50, 178–179, 211, 239
Evictions 185
Exploitation 24, 25, 37, 75, 166, 205–206
    see also Overexploitation
Exports 50, 125, 209, 256
    see also Transfers
Expropriation 51–53

Farm
    types 49, 50, 67, 157, 159–160
    see also Agriculture
Farmers
    adaptations 229, 238–260
    Black 55–56
    community concept 28–29
    coping strategies 229
    increased competition fears 257
    political power 258
    water sources 217
Farming systems 13–14, 49, 152–153, 157, 177–178
Federalism
    allocations 79–80
    Commonwealth of Australia 264–265
    environment movement 134–136
    initiatives, irrigation impact 225–226
    interventions 15
    river regulations 80, 81–82
    water affairs process 80
Fertigation 32
Fesguias (irrigation channels) 153
Fish hatcheries, Oliphants River basin 49
Flag Boshielo dam 55, 56, 60
Floods
    cause claims 11, 91
    control methods 2, 103, 114, 150, 155
    infiltration 151
    management 114
    prevention 105
Flows
    amounts 49, 150, 198, 268, 278
    assessment 157–162
    calculations 178–179
        see also Water accounting
    ceasing 181

decrease causes 151
dilution control 271–272
distribution changes **163**
dry-season 182, 187, **188**
imbalance 161
into Ihefu wetland 187
lifting increase effect 259
measurement 124, 190, **270**
methods **219**
paths fluctuations 253
reductions 13, 25, 50, 126, **286**
regulation 115, 269
reporting 124
return, importance 253
seasonal patterns 278
sources 268
volumes 186–187
see also Environment, flows
Food 61, 242, 256, 287
Forestry 13, 50, 101, 103, 155, 269
French Protectorate (1881-1956) 153, 168
Friends of Ruaha 180

Gardens 200
Gavkhuni swamps 205, 210
Gazankulu 52
General Law on Communication Routes, Mexico 80
Geomancy (*fengshui*) 103
Geopolitics 29–30
Gini coefficient 63–66
Glen Canyon dam 124, 125, 133
Gold 51, 52, 127
Governance
    allocation mechanisms 206–207
    changes 136
    Chinese water history 102–103
    electricity control 255
    environmental, stakeholders 185–186
    irrigation narratives 183–185
    landscape, reshaping 135–136
    local 200
    non-equilibrium behaviour 188–190
    options 255–256
    river-basin-based approach 2, 85, 185–192
    structures 14–16, 230–231
    tensions 118
        see also Conflicts
    see also Entitlements; Licences; Permits; Rights
Grain transport 104, 105
Grand Canyon 124, 133
Grazing, rainfall patterns criticality 51
Great Leap Forward, China 109
Great Ruaha Power Project 182
Great Ruaha River basin, Tanzania 171–192
Greater Sekhukhune District Municipality 56
Green Revolution 14, 25, 224

Green water 157, **160**, 161, 162
Gross geographical product 49, 54, 63
Groundwater
    abstractions 33, 49, 112, 151, 221, 229
    access 224
    allocations **266**
    apportionment scheme omission 132
    complexity 256–257
    control 208
    depletion 204
    drawdown problem 112
    exploitation 24, 37, 205–206
    inflows 161
    irrigation 87, 216–217, 243
    levels 206, 247, 251–252, 253–254
    losses 178
    management 150, 166–167
    overdraft 42
    overexploitation 13, 75, 77, 93–94
    overuse 26
    quality 254
    recharge 13, 49, 175, 254
    surface water interactions 229
    use 238, 248, 257, 265, 266, 282
Groundwater Control Bylaw No. 85 (2002), Jordan 26
Grow More Food Campaign 242
Grupo de Ordenamiento y Distribucíon (GOD) 92
Grupo de Trabajo Especializido en Placecíon Agrícola Integral (GTEPAI) 11, 90–91
Guadalajara 10, 88, 95
Guanajuato State Water Commission (CEAG) 93, 94
Guarantees, constitutional, South Africa 56–57

Han Dynasty, water administration establishment 104
Harim 206
Hazelwood, A. 183, 184
Health 208
    see also Waste-water, treated
Heavy metals concentrations 56
    see also Pollution
Hehai Engineering Institute, Nanjing 106
Historically Disadvantaged Individuals 59
Homeland Constitution Act (1971) 52
Homelands, South Africa 55–56
Hoover dam 124, 125, 130
Horticulture 268–269
Howard Plan 264
Huai River 107, 108
Huanghe shuili weiyuanhui see Yellow River Conservancy Commission
Hume dam 265, 267
Hundred Schools, China 103

Hunting industry 50
Hyberabad 215
Hydraulic mission
    defined 4
    demise 88
    desirability 17
    first lake Chapala crisis 82–88
    heyday 85–88
    large-scale interests priority promotion 69
    Mexico 79–80
    overbuilding result 94
    results 4–6
    revolutionary irrigation promises 85
    state role, Olifants, River basin 53
Hydraulics 82, 85–88, 106, 108–109, 153–155
Hydro-politics 47–70
Hydrocracies 4, 15, 75–96
Hydroelectricity
    concessions 82, 83
    dam construction 82, 83, 84
    Lake Chapala crisis cause 81, 95
    large-scale water use 221–222
    Santiago River 81, 86–87
    schemes 153
    transfers 222, 265
    see also Electricity; Hydropower
Hydrology
    calendar 191
    change 181
    environmental implications 254
    interconnectedness 191, 192, 196, 200
    livelihoods implications 253–254
    Oliphants River Basin 49
    pathologies 12–13
    perspective, Bhavani River basin **240**
    Usunga plains study 186–187
    variability 165, 267
Hydropower
    capacity, Tanzania 177
    Mtera-Kidatu 177, 181, 182–183
    power cuts 11, 181
    provision, Arizona 133
    storage 242–243
    upstream water claims 179, 186–187
    water use competition claims 179
    see also Electricity; Hydroelectricity

Idealism, South Africa 56–57
Ihefu wetland 177, 187
Impact assessment 2
    see also Benefits; Costs
Imperial period, China 103, 104–105
Imports 140, 209
    see also Transfers
Incentives
    conservation 12, 139

measures 153–154
negative 40
reallocation from low-value to high-value crops 38
response to 14
supply augmentation 139
technical investment 154
private sector involvement 226–227
profitability 226
technologies adoption 116
water savings 154
see also Subsidies
Income 164
Incremental approach 282
India 214–234, 238–260
Indian Farmers' and Toilers' Party 246
Indian State Electrical Boards 259
Indigenous people 52, 127, 131–132, 276, 284
see also Apartheid; Bedouins; Segregation; Tribes
Individual actions 258, 260
Industrialization 222
Industry
    development, water management role 106
    reallocations 204–205
    water demand 205
    water priorities 54
    water use 32, 33–34, 113, 160, 161, 222, 243
    withdrawal figure 243
Inefficiencies 140
Inequity 58, 60, 63–66, 68
Inflows
    Dead Sea 25
    depletion, Krishna River basin 217
    falling **252**
    fluctuation 253
    Lake Chapala 78, **78**, 79
    net **33**, 217, 218, 223
    pump-irrigation effect 252
    reduction 90, 168, **252**, 280
    return, importance 253
    seasonality 198
    sources 31, 161, 197–198
    see also Flows; Rainfall
Infrastructure
    development 62, **155**–156, 216–217, 223
    hydraulic, construction 95–96
    implementation 69
    importance 192
    operation and maintenance costs 27
Initiatives 90–92, 225–226, 258, 282–284
Instability 101, 102, **172**–173
Installations, impacts/valorization 161–165
Institute for Crustal Studies (ICS) 140

Institutions
    centralization/decentralization 108
    changes 278
    conservation strategies 224–227
    creation 15
    engineering training 106
    history 106, 126–137, 269
    political stakeholders 179–180
    programmes 140
    reforms 69, 70
    responses 115–117, 271–274
    stagnation 67–68
    transformation 57–61
    tribal resource management replacement 28
Integrated approaches 76, 231, 272–273, 274
Interactions, Great Ruaha River 186–191
Interconnectedness
    across scales 210
    closure induced 20
    complexity 238
    ecosystems, water-users 214
    hydrological 191, 192, 196, 200
    increase 260
    socio-hydrological 203
    technology use effect 42
Interdependency 41
    see also Interconnectedness
International Map Trade Association (IMTA) 92
International Water Management institute (IWMI) 57, 67, 68, 179
Internationalization, China 106–107, 109, 110, 119–120
Interstate Royal Commission (1902), Australia 269, 270
Interventions
    closure responses 223–230
    externalities 12–13
    federal 15
    impacts 210, 223–224
    methods 153–154
    planning 23–25
    state 201–202
Investments 27–28, 153, 154, 231
Iran 196–212
Irrigated-areas
    Cauvery basin 255
    growth 31, 77–78, 79, 88, **174**, **181**, 255
    Jordan 31
    Krishna River basin 217
    Lerma–Chapala basin 76, 84
    Murray–Darling basin 268
    upper Great Ruaha River catchment **181**
    Usangu basin 181
Irrigation
    allocations reduction 92
    ancient networks superimposition **200**, 201

Irrigation – *Continued*
  China 104, 109
  development 4, 53–54, 65, 79, 221, 225
  establishment, Australia 265
  expansion 25, 183, 278, 283
  historical, Tunisia 152–153
  impacts 225–226, 278
  increase 111, 188
  intake monitoring 190
  losses 178–179
  management transfer 59–60, 75–76
  modernization 183
  outputs value **268**, 287
  perspective, Bhavani River basin **240**
  polluted water use 254
  potential 129
  private sector involvement 226–227
  profitability 226
  programmes, India 216–217, 225
  protective **225**
  quotas 38
  rainfed production replacement 157, 159
  revolutionary 82–85
  schemes 60, 65, 67, 152–153, 183
  shortfalls 62
  squeeze **9**
  state 30
  subsistence 55–56
  systems
    contracted out 207
    drip-irrigation 160, 167, 227
    gravity, historical development 239–242
    groundwater 87, 216–217, 243
    micro-irrigation 207
    small-scale 50, 59, 221
    supplemental 219
    surface-irrigation 23
    variety 24–25
  techniques 152–153, 155, 162
  technology adoption 14
  treated waste water use 25, 26–27, 34, 36, 37, 38
  types 24–25, 177, 249
  underutilization 232
  units, Mexico 76
  withdrawals percentage 49–50
  see also Canals; Lifting; Pump-irrigation; Tanks; Wells
Irrigation Department establishment (1903), *Zuid-Afrikaanse Republiek* (ZAR) 53
Irrigators, water markets viewpoint towards 137
Issues 10–16, 102
Ives, Joseph C. Lieutenant 123
Ivory 50

Jalalabad 199, 203
Jordan River 20–43
Jordan Valley Authority (JVA) 29–30
Jordanian Division of Irrigation 24
Juntas de Aguas (water boards) 84, 85

Kairouan 148–149, 150, **151**, 152, 156, 166–167
Kalingarayan canal 248–249, 252, 253, 254, 257, 258
Kalingarayan weir 239
Kanniyampalayam 240, 249–250
Kapunga irrigation scheme 183
Karnataka **220**, 221, 226, 227, 242, 255
Kerala 242, 255
Kilombero Sugar Company 177
King Abdullah canal 24–25
Knowledge 30, 203, 280, 286–287
Kodiveri 240, 248, 252, 253, 257, 258
Krishna Basin, South India 214–234
Krishna Water Disputes Tribunal (KWDT) 221, 223, 227, 228, 229, 231, 232
Kuhrang **197**, **198**, 201
Kukkukthorai River 242
KwaNdebele 52

Labour 29, 52, 108–109, 164
  see also Employment
Laja River 81
Lake Chapala
  agreement consultative committee 89
  crises 10–11, 82–88, 93–94, 95
  marsh draining 80–81, 85
  size and depth 76, 96
  storage 81, 91, 95
  volume fluctuations 77
  see also Lerma–Chapala River basin
Lake Mead 124, **139**
Lake Powell 124, 130, **139**
Lake Tiberius 20, **21**, 22–23
Lake Victoria 271, 276
Lakes, draining 86
Land
  Acts 52
  aggregate production factor 163
  allocation 69
  concentration, extreme forms 80
  cover, seasonal **245**
  early development, Jordan 23
  reclamation 80–82, 86
  redistribution 85
  reform 56–63
  tenure 82
  vivification 28
  white expropriation 51–53
Land Care programme 273, 281
Land Settlement Law (1933), Jordan 29
Land-use 31, 49, 111, 180, 243–250, 252
Landscape, changes 117, 217–221

Las Pintas canal 87
Law
　enforcement difficulties 166
　environment 134–136
　General Law on Communication Routes, Mexico 80
　groundwater 26, 132
　irrigation 30, 53
　labour 52
　land settlement 29
　Law of the River 126, 132, 138
　Public Law 480 (PL480), India 256
　reallocations mechanisms 138
　water management relevance, China 110–111
　see also Legislation; Water Laws
Lebowa 55, 56
Legislation 110–111, 153–154, 232, 257–258, 270–271, 285
　see also Law
Lenjan tunnel 201
Lenjanat 203
Lepelle Water Board 54
Lerma River 75, 83, 84–85, 90
Lerma–Chapala River basin 75–96
　see also Lake Chapala; Lerma River
Lesotho Highlands Project 54
Levees, raising 114
Ley General de Vias de Comunición (General Law on Communication Routes) 80
Liberalization, economic 40
Licences 26, 36, 57, **270**, 278
Lifting
　illegal 249
　impacts 253, 259
　increase 244, 248, 249, 259
　initiatives 258
　methods **246**, 248
　sources 249–250
Limits
　caps 192, 279–282, 286, 287
　cost transfers process 209
　diversions 285
　electricity use 250
　living with 137–140
　long-term average sustainable diversion 285
　new era of 123
　overcoming 139
　releases 255
　technical, KWDT award 229
Limpopo 47, 59–60
Linkages, hydrological: salinity; landscape 265
Litigation 130, 132, 138
　see also Law; Legislation; Water Law
Livelihoods 41, 152, 176, 253–254
　see also Employment
Livestock 50, 152, 162, 163, 180–181, 185
Livingstone, I. 183, 184

Loess Plateau, Yellow River basin 117
Log-rolling, political behaviour 6
Losers 63–66
Loskop dam 53–54, 55, 56
Lower Bhavani Project (LBP) 242, 247–248, 252, 253, 255–256, 257–258
　see also Bhavani River basin
Lower Colorado Multi-Species Habitat Conservation Plan 126
Lower Jordan River Basin (LJRB) 20–43

Maadi 198, 201
Madibira irrigation scheme 183
Madras–Mysore Dispute 254–255
Maharashtra 220, 221, 226, 227, 230
Malnutrition reduction 61
Management Details of Yellow River Water Regulating 115
Management transfer 59–60, 75–76, 88–89, 167
Mandate period, Transjordan 23–24
Markets 27, 136–137, 141
　see also Trading
Marsh draining, Lake Chapala 80–81, 85
Mbarali 180, 183
Merguellil basin 16, 147–168
Meters, water **270**
Mettur **240**, **241**, 242, 254, 255
Mexican Water Treaty (1944) 130
Mexico 75–96, 130, **131**, 132–133
Mfecane process 51
Micro-irrigation 207
Migration 26, 29, 41, 51
　see also Refugees
Minerals 51–53, 54, 127
　see also Mining
Ming dynasty, Yellow River regulation 104–105
Mining
　de Hoop dam benefit 69
　expansion 50
　gold 51, 52, 127
　growth 56
　water priorities 54, 55
Ministry of Agricultural development (SAyF), Mexico 82, 83, 84
Ministry of Tourism and Natural Resources, Tanzania 185
Ministry of Water and Irrigation (MWI), Jordan 25
Ministry of Water Resources, China 5
Minor Irrigation Programme 242
Mirabs 198, 211
Mismanagement 187
Mkoji catchment 190
Mobility, social groups 41
Modernization 5, 183, 202–203, 226
Monitoring and evaluation group, Lerma–Chapala River Basin Council 89–90
Monopoly 50–56

Mourhab valley 202, 207
Mozambique 48, 61–62
Mtera–Kidatu hydropower complex 11, 177, 181, 182–183, 186
Municipalities, water use 32, 33–34, 160, 161, 243
Murray River Basin Commission 15
Murray–Darling Basin Authority (MDBA) 285
Murray–Darling Basin Commission (MDBC) 273, 285
Murray–Darling Basin Environmental Resources Study 274
Murray–Darling basin (MDB), water use restraint 263–288
Murray–Darling Basin Ministerial Council (MDBMC) 273, 274
Murrumbidgee River 265–266
see also Murray–Darling basin (MDB)

NABARD (National Bank for Agriculture and Rural Development) 225
Nagarjuna Sagar 229
National Action Plan on Salinity and Drainage 281
National Bank for Agriculture and Rural Development (NABARD) 225
National Competition Policy (1994), Australia 280, 287
National Economic Commission (NEC), China 106
National Hydraulic Plan commission, Mexico 88
National Land and Water Audit (1998-2000), Australia 273
National Resource Management Ministerial Council, Murray–Darling basin 281–282
National River linking Project, India 224
National Strategy for Ecologically Sustainable Development, Australia 280
National Water Act (1988) South Africa 57, 60, 66, 68
National Water Carrier, Israel 23
National Water Commission, Australia 284
National Water Development Agency (NWDA), India 224
National Water Initiative (NWI), Australia 282–284
National Water Policy (2002), India 225
National Water Services Act (1997), South Africa 57
Nationalist Period, China 105–106, 107
Nationalization 201, 206
Native Administration Act (1927) 52
Native American (Indian) water needs 131–132
Native Land Act (1913) 52
Natural Resources Management Act, Southern Australia (2004) 282
Natural Resources Management Ministerial Council, Australia 265
Natural Resources Management Strategy 274–275
Naturalists 5, 103

Navajo Generating Station 133–134
Ndbele 52
Nekouabad schemes 203–204
Networks, hydro-social 95–96
New South Wales, Australia 266, 269, 283
Nilgiris district, Bhavani basin 239
Non-equilibrium behaviour 171, **172**, **173**, 175, 188–190, 191
Northern Sotho 52
Noyyal River 255

Oases 24, 196–212
Ocotlán pumping station 87
Office of the Director of Water Conservancy, China 104
Olifants River basin, South Africa 15–16, 47–70
Olifants River Scheme 56
Operations and maintenance (O&M) 27, 37, 154, 269, 280
Ordering and Distribution Group, Lerma-Chapala basin 92
Organizations 15, 16, 27, 28–29, 231
see also Associations; Institutions
Ottoman period, Jordan 29
Outcomes 63–66
Outflows 95, 268
Outline of Yellow River Harnessing and Development 115
Overabstraction 26, 278
Overallocations
causes 206
entitlements 7–8
Law of the River flaw 138
management 271–282
problems 288
responses **279**
water plans requirement 284
Overbuilding 6–7, 16, 17, 85, 94–95, 231
Overcommitment 16, **225**, **228**, 232
Overdrafts 8–9, 33, 42, 206
see also Overexploitation
Overestimation 90, 92, 184–185
Overexploitation
basin closure relationship 77–79
BouHafna aquifer 151
curve, bending down 89–90
flows reduction 13
hydraulic mission 75–96
impacts 41
management policies 167
results 3–4, 17, 232
water balance terms variation 32
wells expansion 203
Overextraction 79, 87

Panchayati Raj Act 230
Pangani River basin 179

Paradigms 4–6, 60, 66–67, 126–129, 134–137
Parageoplasia 171, **173**–175, 187–188, 191, 192
Participation 16, 76, 207, 226, 231, 232, 285
  see also Water Users Associations
Participatory Irrigation Reforms, India 226
Partners, foreign, Yellow River basin management 106, 107
Pathologies, hydrology 12–13
Patronage 30, 40, 42
Peace Treaty, Jordan (1994) 36
Pedi chiefdom 51, 52
People's Communes, China 109
Perceptions, changing 28–29
Permits 53, 56, 94, 206, 208
  see also Licences
Phlaborwa Water Board 54
Pipelines 67, 248, 271
Plans and planning 23–25, 108–109, 110, 189, 283–284
Plantations 239, 252
Policies
  fragmentation 231
  free basic water 57
  impacts 155–156
  implication, Bhavani basin 260
  protective irrigation, India **225**
  rural development 232
  settlement 4
  stakeholders 179–180
  water management, China 106
  white water economy creation 54–55
Politics
  allocations changes decisions 255
  blame 10–11
  China 116
  ecology 38–40
  geopolitics 29–30
  hydro-politics 47–70
  Jordan 26, 29, 30, 40, 41, 42
  local, India **225**
  Murray–Darling basin 269
  power 2, 258, 284–285
  stakeholders, institutions 179–180
  tribal organization 27, 28–29
Pollution 56, 75, 111, **112**, 208, 254
  see also Water quality
Polvaram-Vijayawada link 224
Population
  Bhavani basin 239
  Colorado River basin 124, **135**
  Great Ruaha River 176, 180
  Jordan River basin 23, 25, 26
  Krishna River basin 215
  Lake Chapala 79
  Lebowa 55
  Merguellil basin **156**

Murray-Darling basin 266
Olifants River basin 48, **53**, **55**, 56
Tunisia 165
Zayandeh Rud River basin 198, **199**
Porfiriato 80–82
Poverty 55, 56, 60, 61–62, 68, 117
Powell, John Wesley 128
Power see Electricity; Energy; Hydroelectricity; Hydropower
Precipitation
  Bhavani basin 239
  lower Jordan River Basin 22
  Murray–Darling basin 266
  Oliphants River basin 49
  Tanzania 175
  Yellow River basin 100
  Zayandeh Rud River basin 197
  see also Climate; Rainfall
Prestige, agriculture 27, 38
Prices
  agricultural commodities 287
  agricultural water 27, 37, 164, 226, 269
  desalination 167
  policies 154
  water use regulation mechanism 115, 116
  see also Costs
Privatization 27, 280–281
Productivity 62, 163–165, 287
Protests 40, 84, 246
Provincial Water Summits (2005:2006) 58
Public Works Department (PWD), Bhavani basin 242, 250
Pump-irrigation
  area statistics 243, 257
  electricity use 246–247, 248, 250
  impacts 151, 252–253, 257–258
  unauthorized 249, 250

Qanats 199, **200**, 201, 203, 204
Qing dynasty 104, 105
Queensland, Australia 266, 273
Quotas 27, 37–38
  see also Allocations

Rainfall
  Bhavani basin 239
  effective 33
  Esfahan 197, **197**, **198**
  evaporation losses 157
  Great Ruaha River catchment 176–177
  green water consumption, according to **160**
  Krishna River basin 215, **220**, 221
  Kuhrang **197**, **198**
  Lerma–Chapala basin 77, **78**, 79, 86, 87, 95
  lower Jordan River Basin **22**
  Merguellil basin 149
  New South Wales, Australia 266

Rainfall – *Continued*
   Olifants River basin 49, 51
   patterns criticality, agriculture 51
   ranges, Jordan valley 21
   runoff 161, 232
   tabias inflow 161
   Tanzania 175
   variability 13, 165
   variation sensitivity 79
   Varzaneh **198**
   Victoria, Australia 266
   worsening 165
   Zayandeh Rud River **197**
   see also Precipitation
Rainwater 61, 69, 157, 160, 226
Raising Irrigation Productivity and Releasing Water for Intersectoral Needs (RIPARWIN) 179
Ramsar Convention 269
Reallocations
   agriculture 38, 43
   benefit shifting 211
   caution 60
   control/monitoring lack 206
   drivers 232
   environment 263
   freshwater savings domestic use 27
   high economic value uses 13
   large-scale to other users 69–70
   market mechanisms role 10
   mechanisms, laws 138
   non-agricultural 12, 26, 141, 204–205
   option 34
   rights 199
   scarcity response option 37–38
   wells depleting aquifers 204
   see also Allocations; Transfers
Recharge 23, 151, 166–167, 168
Reclamation 36, 80–82, 86, 129, 136
Red–Dead project 11, 33, 34, 36, 40, 42
Redistribution 30, 85, 165, 210, 238, 260
Reforms
   allocation 60–61, 66, 67, 69
   Australia 5, 280, 282–284
   avoidance 61
   basin development 68–69
   complexities 75–76
   economic 281
   equitable 70
   institutional 69
   Jordan **39**
   Krishna River basin 224–225
   land 56–63
   Participatory Irrigation Reforms, India 226
   programmes, Tanzania 179
   socio-economic, entry point 66
   water allocation 60–61, 66, 67, 69
   see also Centralization; Decentralization

Refugees 24, 26, 29, 42
   see also Migration
Regional Commission for Agricultural Development (CRDA) Tunisia 154, 156
Regulation
   authority 227
   enforcement 167
   federalization 80, 81–82
   flows 115, 269
   hydraulics works 154
   interstate 15
   use 104–105, 115–116, 198–200, 267
   violation 247–248, 249
   withdrawals 84
Rehabilitation programmes 14
Releases
   careless 8
   control 255
   diversions percentage 253
   environment flows requirements 256
   flow distance 151
   frequency 255, 260
   quantity decision 252
   seasonal 167
Remote-sensing analysis, cropland areas 244–246
Rentier strategies 41
Rents, redistribution 30
Reopening 209–210
Report of the Cauvery Water Disputes Tribunal with the Decision 255
Requirements, estimation methods 50
Reserves, requirements 61–62, 63
Reservoirs
   capacity decline 124
   operation modification contention 139
   recreational use 49
   sediment trapping role 155
   spills 126
   storage capacity 35, **139**, 198, 216, 242, 255
   see also Augmentation; Storage
Resettlement programmes 24, 29–30, 53, 265
Revitalization programmes 60, 65, 67
Rice 183, 187, 188, 227
Rightholders 199
Rights
   absence 210
   calculation 190
   capture 229
   citizens, South Africa 57
   concessioning 80
   drought periods reserve 278
   fixed quanta system faults 190
   indigenous people 127, 137–138, 284
   instream flows 135
   low 257, 258, 260
   new, establishment 154

prior appropriation 127–128, 129, 136–137, 206, 228
proportional 192
qanat 199
renegotiation 255
riparian 207, 228
River Murray Waters Agreement 269–271
single volumetric 192
state, concern for 264
system design 189
trading rules 285
transfer 137
tribal 132, 153
see also Entitlements; Licences; Permits
Risks, identification 2, 285
River Basin Management and Smallholder Irrigation Improvement Project (RBMSIIP) 179, 184, 185, 189
River Murray 265–266, 278
see also Murray–Darling basin
River Murray Commission 270
River Murray Waters Agreement (1915) 269–271
Rudasht scheme 201
Rufiji River basin 179, 183, 189
Runoff
 alteration 13
 coefficients 100, 166, 192, 232, 268
 decrease **111**
 land-use changes effect 111
 mean annual
  Lerma-Chapala basin 76
  Tanzania 175
 rainfall relationships 157, 161
 storage 197–198
 surface, Merguellil basin 149–150
Rural Water Commission 280, 283
Rust de Winter scheme 53

Salinity
 causes 208
 control 133, 275, 281
 desalination 36, 42, 167
 dry 13
 international incident 132–133
 landscape hydrological linkage 265
 management 275, 285
 measurement, electrical conductivity units 275
 responses 271–274
 soil 38, 204, 205
 trends **272**, **276**
 wetlands 269
Salinity and Drainage Strategy 275
Salt Credit Scheme 286, 287
Santiago River 85–86, 92–93
Savings 27, 42, 90–92, 116, 154, 164
Scarcity
 causes 26

 claims, Great Ruaha River 186–187
 closure induced 3
 creation 50–56
 discourse 62–63
 generation 6, 16
 impacts 239, 260
 perception 66, 68
 response options 2, 34–40
 sharing 147–168
 Tanzania 175
 see also Droughts
Science, valorization, China 109
Secretaría de Agricultura y Fomento (SAyF) 82, 83, 84
Secretaría de Agricultura y Recursos Hidráulicos (SARH), Mexico 88
Secretaría de Fomento, regulations 81
Secretaría de Recursos Hidráulicos (SRH) Mexico 82, 85–88
 see also Comisión Nacional de Irrigación
Sedentarization programmes 24
Sediment
 flushing 113, 114
 impact control 114
 levels 101
 reduction 117
 removal 125
 reservoirs operation role 110
 transport 166
 trapping 155, 166
Segregation 52, 55
 see also Apartheid
Seguias (terraces) 153
Sekhukhune 56
Self-reliance, China 109–110
Settlement 4, 29, 152–153
Sewage 254
 see also Waste-water
Shares
 diversions, farmer share 252, 258
 establishment 255
 flow analysis basis 268
 interstate, Murray–Darling basin **266**
 modification 171, *173*, 175, 190–191
 see also Allocations
 rules 198, 199, 269
Sharing
 benefits 60, 69
 costs 269
 plans, Australia 285
 rules 269, **270**
 shortages 138, 139, 147–168
 time-share 30
 see also Allocations
Sharing water resources of the Murray Darling basin 282
Shortages 62, 138, 139, 143, 147–168, 175

Siltation 151
  see also Sediment
Sinks 13, 31
Skills, decline fear 286–287
Smallholders 14, 58, 184
Snowfall 22
Snowy Mountains Hydroelectricity Scheme 265
Social groups 152–153, 224
  see also Tribes
Société Nationale d'Exploitation et de Distribution des Eaux (SONEDE) 156
Socio-economics 28–30
Soil
  depth, *tabias* 161
  management 205
  moisture 219–221
  salinization 38, 204, 205
  storage capacity 157
  texture, Merguellil basin 148
  types, Zayandeh Rud River basin 197
Soil and water conservation (SWC), Merguellil River basin 151, 154–155, 161, 162–163, 164–166, 168
Solís dam 84, 87–88, 91, 95
South Africa 47–70
South African Development Trust, (South African Native Trust) 52, 54, 55
South African Native Affairs Commission (1905) 52
South Australia 266, 271
Special Forever programme 275–276
Specialized Working Group on Integral Agricultural Planning, Lerma–Chapala basin 11, 90–91
Species 125, 126, 134–135, 182, 188, 269
Squeeze, water 9, 16–17, 31
Stability 41, 172
Stakeholders 76, 179–180, 185–186
  see also Actors
State
  control, hydraulics works, Tunisia 154
  interventions 201–202
  irrigation 30
  law enactment, Colorado River basin 136–137
  regulation 15
  rights, concern for 264
  role, water development 53–55, 221
Storage
  accounting 277
  aquifers, Tunisia 147
  Colorado basin 124, 133
  Corrales dam 83, 85, 86
  hydropower 242–243
  Kairouan plain 150, 167
  Krishna River basin 216
  Lerma–Chapala basin 76, 77–79, 88, **91**
  loss, tanks 166

Massingir dam 48
minor irrigation 217
Mtera dam 182–183
Murray–Darling basin 267, 276–278
Olifant River basin dams 54
runoff 197–198
soil 157
Solís dam 84, 87–88, 91, 95
Wehdah dam 35, 39, 42
  see also Dams; Reservoirs
Stretch, strategy 282
Sturt, Charles 271
Subsidies
  agriculture 14, 38
  Colorado River Storage Project 133
  drip-irrigation 167
  drought relief 280
  electricity 253, 258–259
  groundwater use effect 238
  hydraulic mission 5
  low to high value agriculture conversion 38
  modern large-scale white farms 67
  see also Incentives
Subsistence 55–56
Suburbanization 26
Sufficiency, management 173
Supply
  constraints, runoff reduction contribution 111
  decline 173
  management 276–278
  projections 62–63
  rebalancing 258
Supply and demand 31, 111, 113, **174**, 276–278
  see also Water accounting
Surface water
  abstractions results 257
  accounting **178**
  agreement 92–93
  allocations 92–93, 257, **266**
  changed dynamics 252–253
  depletion 75
  development 221
  flows measurement **270**
  pumping impact 253
  resource development **241**
  supply evolution **202**
  use 199, 201, **202**, **266**
  withdrawals increase 259
Surface-irrigation 23
Surplus water 222, 228, 255
Suspension, river 101–102, 114
Sustainability, management 179, 187, 191
Sustainable management of the Usangu Wetlands and Catchment (SMUWC) 179, 187, 191

Tabias (terraces) 155, 161–162, 163, 165–166
Tamil Nadu 238, 239, 250, 255–256
Tamil Nadu Agriculturalist's Association 246
Tamil Nadu Electricity Board (TNEB) 242, 259
Tamil Nadu Water Supply and Drainage Board (TWAD) 254
Tanks
   assessment 162
   cascade 242
   intensive exploitation 166
   irrigation 215, 240, 242, 243
   rehabilitation programmes 14
   sediment trapping role 166
   small, construction 155
   storage loss 166
Tanzania 171–192
Tanzania Electric Supply Company Limited (TANESCO) 186
Taoism 5, 103, 104, 119
Technical and Economic Plan for Yellow River Comprehensive Utilization 108
Technocracy 69
Technology
   adoption incentives 14, 116
   changes 41
   emphasis 30
   hydraulic mission, driving force 94–95
   improvement costs 40
   introduction 24
   investments 27–28, 153, 154
   limits, KWDT award 229
   professionals 106
   valorization 109
   water-saving 116
   water-use aspects 28–30
Temperatures 21, 148–149
   see also Climate
Tennessee Valley Authority (TVA) 1, 5, 24, 85
Tepuxtepec dam 83, 84
Terraces 155
Terraces (*tabias*) 155, 161–162, 163, 165–166
Thengumarahada Co-operative Farming Society 242, 250
Time-share 30
Tiruppur city 255
Titles 281
   see also Rights
Tomlinson Commission, Betterment Schemes 55
Tourism, Oliphants River basin 49, 165, 185
Trade-offs 61–62, 67, 141
Trading 116, 269, **279**, 281, 285
Trajectories, processes 6–10
Transfers
   administrative fiat decisions 10
   agriculture, marginal to high value enterprises 281

   alternatives 40
   costs 36, 263
   engineering approach 42, 102, 117
   frequency 137
   historical implementation 147
   impacts 86, 252
   interbasin 86, 210, 224, 265
   irrigation use 223, 230
   management decentralization 88–89
   motivation, political 205
   non-agricultural
      causes 17
      cities 26, 204–205
      domestic and industrial 204–205
      hydroelectricity 49, 222, 242–243, 265
      tourism centres 165
      urban 25, 90, 165, 263, 287
   opposition 12
   scarcity response 36
   sector 90
   see also Diversions
Transformation 57–61, 118, 141, **218**
Transjordan 23–25, 40
   see also Jordan
Transport 104, 105, 155
Triangulation approach 259
Tribes
   -based society 152, 168
   access control 28, 153
   land and water control 28
   political 27, 28–29
   rights 132
   solidarity 29, 41
   see also Bedouins; Indigenous people
Tungabhadra sub-basin **220**
Tunisia 147–168
Tunnels 201

Unemployed 56, 65–66
Union Irrigation and Conservation of Waters Act (1912) 53
Union of South Africa 52
Upper Colorado River Basin Compact (1948) 130–131, 138
Upper Colorado River Commission 130–131
Upper Colorado River Endangered Fish Recovery Program 126
Upper Great Ruaha River catchment (UGRRC) 176, 177, **180**
Upstream versus downstream
   competition 175, 186–187, 203–205, 254, 256, 271
   efficiency gains 207
   rights-holder diversions control 135
   see also Conflicts; Interconnectedness; Redistribution; Transfers
Urban areas 54, 88, 160–161, 242–243, 255

Urbanization 222, 254
Usangu basin **176**, 183
  see also Upper Great Ruaha River catchment
Usangu Game Reserve 180

Variability, Great Ruaha River, Tanzania 171–192
Varzaneh, rainfall **197**, **198**
Vedas (prohibitions) 93
Victoria, Australia 266, 269, 283
Victorian Department of Sustainability and Environment 280
Village water use 199

Wadi Merguellil River basin, Central Tunisia 147–168
Warfare 50–51
Warlord period, Yellow River basin 105
Washing Away Poverty: Water, Democracy and Gendered Poverty Eradication in South Africa 66
Waste, of water, smallholders 184
Waste-water
  health risks 208
  productive use 229
  re-use 25
  reclamation 36
  treated, irrigation use 25, 26–27, 34, 36, 37, 38
  treatment programme, Lerma Chapala basin 89
  see also Effluents
Water
  availability **33**, 217, 288
  categories **35**
  consumption 164
  control shifts 217
  excess see Floods
  mobilization 154
  quantity, concerns, Lerma–Chapala basin 88
  security 268
  stress 111
  white expropriation 51–53
Water accounting
  green water *160*
  inadequacy, surface-groundwater interactions 288
  Krishna Basin **219**, **220**
  lower Jordan River basin 31–32, 33
  method **178**–179
  water balance categories 217, **219**
Water Acts
  Australia 263, **270**, 282, 284, 285, 287
  South Africa 47, 53, 54, 57–61
Water Allocation Scheme (1987), China 114
Water Audit 278, 279
Water Code, Merguellil River basin 153–154
Water for Food Movement 61

Water Framework Directive, European Union 6
Water for Growth and Development 58
Water Laws
  China (1988) 110–111, 115
  Iran (1968) 206
  Mexico (1992) 88, 89
  Olifants River basin (1998) 5
Water Policy (1987), India 225
Water poverty index (WPI) 63, **64**
Water and Power Resources Service 136
Water resources 23–28, 34, 150–151, 175–180, 215–221
Water Resources Regulatory Authority, Krishna River basin 227
Water Strategy Policy, Jordan 26
Water table levels 150, **151**
Water Table Watch 273
Water users 177–178, 221–223
  see also Actors; Stakeholders
Water Users Associations (WUAs)
  allocation rules adherence 89–90
  Alto Río Lerma irrigation district 90
  establishment 84, 115–116, 226
  management transfer 59–60, 167
  transfers objections 91–92
Water-harvesting, techniques 13, 161–163
Water-quality
  apportionment scheme omission 132–133
  decline 111–**112**
  deterioration likelihood 166
  health concerns 208
  improvements commitment 89
  issues 278
  management 285
  policies 275
  see also Waste-water, treated
Water–society relationships 1
Water-use
  assessment 157–161, 217
  audit 278
  categories 33
  complexity 256
  constraints 278
  domestic 222
  drivers 256
  intensification 243–250
  projections 33
  renegotiation 268
  restraint 263–288
  rural 54, **65**
  sectoral 33–34
Waterscapes 30, 217–221
Wealth, concentration 56
Wehdah dam 35, 39, 42
Weirs 239–240, 242, 248, 252, 253, 271
Wells
  depth 243, 247, 249

Erode district **246**
expansion, Zayandeh Rud River 203
failure impacts 254
fake 249
groundwater levels decline 206
increase 246
irrigation 244, 256–257
lifting methods **246**, 248
private, aquifer exploitation 166
tube 76
types 243
versus Qanata 204
water level changes **251**
Wet periods, solutions, (flood control) 103
Wetlands
cultivation encroachment 182
decline 182, 187–188
livestock keepers and fisherfolk removal 185
management 179, 187, 191
salinity 269
seasonal 177
see also Marsh draining
Whites 51–55
Winners 63–66
Winters decision (1908) 132
Withdrawals
agricultural 33–34
controlled renewable blue water 32–33
domestic 156
Ihefu wetlands 187
increase 33, 259
irrigation percentage, Limpopo Province 49–50

Lake Chapala crisis role 95
municipal and industrial 243
regulation 84
trends **35**
Yellow River basin **113**
see also Abstractions
Witwatersrand 51, 53, 54
Women 29, 50, 56, 58, 61, 254
World Wildlife Fund 179–180, 185, 191

Xiaolangdi dam 114, 117

Yarmouk River 20, **21**, 23, 25, 32
Yellow River Administration 105
Yellow River Available Water Annual Allocation and Main Course Regulating Scheme 115
Yellow River basin 99–120
Yellow River Conservancy Commission (YRRC)
allocations mandate 118
centralized water administration 16
engineering plans 109
establishment 106
institutional structure 108
international cooperation mitigation 106
internationalization commitments 119
management structures 115
Yü the Great (circa 145-90BC) 102–103, 104

Zacapu marsh 80–81
Zarqua River 20, 21, 23
Zayandeh Rud River, Iran 196–212
Zuid-Afrikaanse Republiek (ZAR) 51, 53